Second Edition

System Management

Second Edition

System Management

Planning, Enterprise Identity, and Deployment

Jeffrey O. Grady
JOG System Engineering, Inc.
San Diego, California, USA

CRC Press
Taylor & Francis Group
Boca Raton London New York

CRC Press is an imprint of the
Taylor & Francis Group, an **informa** business

FIRST INDIAN REPRINT, 2015

CRC Press
Taylor & Francis Group
6000 Broken Sound Parkway NW, Suite 300
Boca Raton, FL 33487-2742

International Standard Book Number: 978-1-4398-2013-1

Library of Congress Cataloging-in-Publication Data

Grady, Jeffrey O.
System management : planning, enterprise identity, and deployment / Jeffrey O. Grady. -- 2nd ed.
p. cm.
"Content of this book first materialized under two earlier titles: System Engineering Planning and Enterprise Identity and System Engineering Deployment."
Includes bibliographical references and index.
ISBN 978-1-4398-2013-1 (alk. paper)
1. Systems engineering--Management. I. Title.

TA168.G653 2010
658--dc22 2009050496

Visit the Taylor & Francis Web site at
http://www.taylorandfrancis.com

and the CRC Press Web site at
http://www.crcpress.com

Printed and bound in India by Nutech Print Services

FOR SALE IN SOUTH ASIA ONLY.

Contents

Part four: Generic planning

Part six: Program implementation management

List of figures

List of tables

Preface

Much of the content of this book first materialized under two earlier titles *System Engineering Planning and Enterprise Identity* (Grady, 1995) and *System Engineering Deployment* (Grady, 2000). These books offered that an enterprise should have an identity expressed in their vision statement, organizational structure, and written process description. Their content was more about program and functional management than system engineering in that they covered the definition of generic planning data in the functional structure and provided a transform into program-unique planning. Out of those writing experiences, the conversations that motivated them, and a growing concern that few enterprises were able to implement an outstanding systems capability, came a realization that it cannot be attained without simultaneous improvements in three areas: system engineering, program management, and functional management. So, this book combines the content of the two earlier books and fills in the gaps to supply a management focus for this series on system development.

The content of my five previous books *System Requirements Analysis* (Grady, 1993), *System Integration* (Grady, 1994), *System Engineering Planning and Enterprise Identity* (Grady, 1994), *System Validation and Verification* (Grady, 1997), and *System Engineering Deployment* (Grady, 1999) was written during a period of tremendous change in my life, my knowledge of system engineering, and in the world. As a result, the story line was in a state of flux throughout the five books. I was in the process of changing from employment in aerospace industry to running my own system engineering consulting company with the primary service being system engineering training. In the process of teaching hundreds of people in tens of companies and through several universities, I settled on a particular way of organizing all system engineering knowledge, that which I understood at least, into a few baskets, and I saw that there was near but not perfect alignment of this structure with my published work. Previously published work was reorganized to conform to the four fundamental elements of system development: (1) define the problem in good specifications, (2) solve the problem through the application of cross-functional teams, (3) prove that the solution complies with the specifications, and

(4) have all of this work accomplished in the context of a good technical management infrastructure.

Big changes were going on in the world during this period as well; chief among them was the defeat of the latest brand of totalitarianism called communism. This heroic action required a tremendous investment by the Western world, and particularly the United States, in a military capability that, over a period of 70 years (focused primarily in a purposeful attempt to do so in the 8 years of the Reagan presidency), stressed the Soviet economic and intellectual foundations to the breaking point. The author was an active part of this process as a U.S. Marine for 10 years and engineer in aerospace industry for 30 years with the clear focus of beating the Russians throughout.

While the military applications for the systems approach to problem solving did not end with this win, the new environment encouraged the application of these same ideas that had provided the heavy military stick to problems of a more benign nature in commercial industry. The author found himself working with companies that produced earthmovers, automobile braking systems, bodily fluid analysis equipment, general and commercial aviation electronics equipment, and locomotives. Very often, the message polished through 40 years of military and space-related work was not well received, and was often viewed as suspect or unclean being useful only for developing tanks, planes, bombs, and other ugly items.

I came to believe that the system approach to solving very complex problems, no matter the nature of the product line, can be summarized in four fundamental areas of interest: defining the problem in specifications, solving the problem through the coordinated application of specialized creative design skill and knowledge intertwined with generalist integration and optimization effort, proving that the resultant solution solved the right problem correctly through well-planned and implemented verification work, and sound technical management of the ongoing development work.

This simplified view of the world was in fairly good alignment with my published work, I found with some surprise, which happened not through conscious plan but apparently driven by the same forces that had shaped a particular integrated view of the whole knowledge domain.

Change will continue in all of our lives, of course, and probably at an increasing pace as time goes by. So, we have to try to form snapshots of a fleeting reality that will be useful over the longest possible time span. Hopefully, my books are a mutually consistent snapshot of the union of four human skill and knowledge sets, all of which must function well independently and together to solve very complex problems with useful products. These skill sets have generally been addressed independently in the literature though *Essentials of Project and Systems Engineering Management* (Eisner, 2002) and the work of Arnold M. Ruskin are exceptions with coverage of two of the four essential components. The other two are functional management and domain engineering management.

The concurrent engineering initiative went a long way toward saving system engineering from self-destruction, but there remains, at the time this series was brought together, a gulf between these four management perspectives that must be closed on a common understanding. There is a great opportunity here for the International Council on Systems Engineering (INCOSE), the Program Management Institute (PMI), and the several software societies to do some great bridge-building work.

Through conversations with engineering managers in industry (particularly Thomas Nagle, deputy GPS program manager for civil applications at the time this is being written), I have come to believe that there are three specific levels of skill that an enterprise must master in order to practice system development well. First, it must contain people who know what they are doing at the individual level. Second, they must have written down what their process is in some detail and ensure that their people understand it. Finally, they must have had some positive experience in applying this process on one or more programs—they must be capable of deploying their process to programs. An enterprise must satisfy all three levels of system development capability and to do so requires all four skill sets covered in this set with those doing the work interacting knowingly and cooperatively.

It is hoped that this book and the other three books listed in Table 1.1 will satisfy a need in industry for a desk reference useful to system engineers, program managers, and functional managers for information of immediate use in doing their very difficult work. Second, this book was intended to be used as text material in a system development certificate program offered by the author to industry. It may be useful for others involved in the same activities. This book can be used as the textbook for a first course in systems engineering named *Introduction to Systems Engineering* or *Foundation of Systems Engineering*, for example. In such a class, this book could be used as a basis for a class team project to write a system engineering manual (SEM) or system engineering management plan (SEMP) for a real or imaginary company or program and complete some program planning work using one of the work cataloging systems included. (Please visit http://www.crcpress.com/product/isbn/9781439820131 for more details.)

The central theme throughout this book is that an enterprise should have a well-defined identity in terms of its written internal procedures. It should have only one standard set of procedures (one identity) and these procedures should be repeated from program to program taking advantage of repetition as a means of improving personnel performance. The enterprise should actually use these procedures in performing its work, and work to incrementally improve them over time based on lessons learned on programs.

In the classes I teach using this book, I offer the students a SEM that they can tailor or re-write that attempts to blend the best of MIL-STD-499A, MIL-STD-499B, the U.S. Air Force integrated management system concept, the IEEE Standard for Systems Engineering 1220, Electronic Industries Association Systems Engineering Standard EIA 632, and ISO 9001 coupled with a generic planning and transform engine (a variation of the U.S. Air Force integrated management system) into a description of a powerful, integrated system development environment.

My experiences in industry as a system engineer and engineering manager in systems engineering acted as the selection and integration engine for the final result. I also benefited from conversations over the years with several members of the International Council on Systems Engineering and other practicing system engineers.

I hope you are able to use this prescription in your enterprise to great benefit. I confess that it may not be easy to understand the broad and deep sweep of the intended process. I have tried to cover a lot of territory within the confines of this cover. Even at that, there are many specialized areas in the book that are only briefly touched upon.

Jeffrey O. Grady
San Diego, California

Acknowledgments

There are other books available that describe essentially the same process covered in this book. Certainly, I derived a lot of benefit from these alternative descriptions of what many of us have experienced on the job. In this book and *System Requirements Analysis* (Grady, 2005), *System Synthesis: Product and Process Design* (Grady, in press), and *System Verification* (Grady, 2007), there are only a few unique elements I have added to a process that has been evolving for at least 70 years: (1) description of a program planning model in this book that coordinates with the enterprise documentation developed using system engineering principles, (2) emphasis on modeling as the source of all requirements and a call for better integration and optimization of the models used in requirements analysis especially across the hardware (HW)–software (SW) gap, and (3) insistence on a tighter linkage between the product requirements and verification planning in a book on verification.

Over the period 1993–2010 in classes presented at the University of California San Diego and Irvine, through public and on-site courses sponsored by UCCE, ATI, TTC, and iVille, and at companies and government agencies across the country contracted through JOG System Engineering, hundreds of students have contributed to my knowledge of the field of system engineering. It is true that if you want to learn something, you should try to teach it. This pattern may or may not be in the best interests of the student, of course, so I hope all of these students have benefited as much as I have from these experiences.

Just when you think you have mastered some concept and have concocted a foolproof way to explain it to others, a student will ask a question that shakes your infrastructure to the core. A new notion is built only to survive a short time. Out of this constant building and breaking process can come a better understanding fully tested. I hope that has been the case in the content expressed in this series of books. If so, a great deal of the credit must go to those hundreds of students who contributed to the flow. If not, then these ideas have simply been badly expressed by me.

As is so often the case, I have learned more from my failures than my successes. I appreciate the tolerance of a string of managers aware of those failures. They deserved better. Our employers deserve to have the job done right and well every time even the first time we attempt a particular task. Ideally, they would have provided some form of training or experience-gathering prerequisite to our first attempt, but in any case we have an obligation to prepare ourselves as well. Very likely my interest in training was motivated, subconsciously perhaps, by a feeling of guilt over my past failures in an attempt to encourage others to improve their skills and knowledge about our profession, thus reducing the probability of flawed performance on the job. While failure is often involved in our most effective training experiences, our employers deserve better.

Author

Jeffrey O. Grady has been the president of JOG System Engineering, Inc., San Diego, California, a system engineering consulting and training company since 1993, working with a wide array of companies with a broad cross section of product lines. Prior to this, he worked for 30 years in aerospace companies as a customer-training instructor at Librascope on antisubmarine rocket (ASROC) and submarine rocket (SUBROC) underwater fire control systems; as a field engineer, project engineer, and system engineer at Teledyne Ryan Aeronautical on unmanned day and night as well as strategic and tactical photo reconnaissance, SIGINT, strike, EW, and target aircraft launched from ground launchers, aircraft, and shipboard; a system engineer at General Dynamics Convair on advanced cruise missile; and a system development manager at General Dynamics Space Systems Divisions on space transport (Atlas) and energy systems.

Jeff has authored nine published books and numerous papers in the systems engineering field. He holds a bachelor's degree in mathematics from San Diego State University, a master of science in system management with a certificate in information systems from the University of Southern California, and a systems engineering certificate from the University of California San Diego. Jeff teaches system engineering courses around the country onsite at companies and universities including the University of California San Diego. He is an International Council on Systems Engineering (INCOSE) fellow, founder, and certified system engineering professional (CESP).

Acronyms

AAR	After action review
ABD	Architecture block diagram
ACWP	Actual cost of work performed
AFSC	Air Force Systems Command
AFSCM	Air Force Systems Command Manual
AIAA	American Institute of Aeronautics and Astronautics
ANSI	American National Standards Institute
ASIC	Application-specific integrated circuit
ASROC	Anti-submarine rocket
BAFO	Best and final offer
BCWP	Budgeted cost of work performed
BCWS	Budgeted cost of work scheduled
BIT	Business integration team
CAD	Computer-aided design
CAM	Computer-aided manufacturing
CAM	Cost account manager
CDR	Critical design review
CDRL	Contract data requirements list
CD ROM	Compact disk read-only memory
CEO	Chief Executive Officer
CIT	Corporate integration team
CM	Configuration management
CMM	Capability maturity model
CONOPS	Concept of operations
CPM	Critical path method
C/SCS	Cost/schedule control system
DAL	Data accession list
DBS	Drawing breakdown structure
DDP	Development data package
DFD	Data flow diagram
DIG	Development information grid
DNA	Deoxyribonucleic acid
DoD	Department of Defense

DoDAF	Department of Defense Architecture Framework
DoE	Department of Energy
DSMC	Defense System Management College
DT&E or DTE	Developmental test and evaluation
EAR	Enterprise architecture report
EDI	Electronic data interchange
EFFBD	Enhanced functional flow block diagram
EIA	Electronic Industries Alliance
EIT	Enterprise integration team
EMI	Electromagnetic interference
ERB	Engineering review board
EVS	Earned value system
FAA	Federal Aviation Administration
FCA	Functional configuration audit
FIPS	Federal information processing standards
FMECA	Failure mode, effects and criticality analysis
FOT&E	Follow-on operational test and evaluation
FRACAS	Failure reporting and corrective action system
GDSS	General Dynamics Space Systems
GIG	Global information grid
GSD	Grand systems development
IBM	International Business Machines
ICAM	Integrated computer-aided manufacturing
ICBM	Intercontinental ballistic missile
ICD	Interface control document
ICDB	Interim common database
ICN	Interface change notice
ICWG	Interface control working group
IDEF	Integrated computer-aided manufacturing definition
IEEE	Institute of Electrical and Electronics Engineers
IMP	Integrated master plan
IMS	Integrated master schedule
INCOSE	International Council on Systems Engineering (formerly NCOSE)
IPO	Input process output
IPPT	Integrated product and process team
IR	Infrared
IRAD	Independent research and development
ISO	International Standards Organization
IV&V	Independent verification and validation
IVP	Integrated verification plan
IVR	Integrated verification report
JOGSE	JOG System Engineering
KPA	Key process area

MBS	Manufacturing breakdown structure
MID	Modeling identification
NASA	National Aeronautics and Space Administration
NCOSE	National Council on Systems Engineering (earlier name for INCOSE)
NDIA	National Defense Industry Association
NFL	National Football League
OOA	Object-oriented analysis
ORD	Operational requirements document
OT&E or OTE	Operational test and evaluation
PBT	Program business team
PCA	Physical configuration audit
PDR	Preliminary design review
PERT	Program evaluation and review technique
PIT	Program integration team
PSARE	Process for system architecture and requirements engineering
PSL	Program specifications library
QA	Quality assurance
RAM	Reliability, availability, and maintainability
RAS	Requirements analysis sheet
RATO	Rocket assisted take off
RDD	Requirements-driven design
RFI	Request for information
RFP	Request for proposal
RPM	Revolutions per minute
SADT	Software analysis diagramming technique
SAIC	Scientific Applications International Corporation
SAR	System architecture report
SBD	Schematic block diagram
SDR	System design review
SDRL	Supplier data requirements list
SECAM	System engineering capability assessment model
SEI	Software Engineering Institute, within Carnegie Mellon University, developed the software CMM
SEIT	System engineering and integration team
SEM	System engineering manual
SEMP	System engineering management plan
SEMS	System engineering master schedule
SEPC	System engineering process criteria
SOW	Statement of work
SRA	System requirements analysis
SRR	System requirements review
SysML	System modeling language

TPM	Technical performance measurement
TQM	Total quality management
TRW	Thompson Ramo Wooldridge Inc.
TSA	Traditional structured analysis
UADF	Universal architecture description framework
UCI	University of California Irvine
UCSD	University of California San Diego
UML	Unified modeling language
USAF	United States Air Force
USC	University of Southern California
UV	Ultraviolet
VMR	Verification management report
VTN	Verification task number
WBS	Work breakdown structure

part one

The foundation of systems development

chapter one

Introduction to systems development

1.1 *The notion of system*

This book and the three books listed in Table 1.1 revolve around the development of a construct called a system.

A man-made system is a collection of objects that interact in accordance with a set of rules to achieve some purpose or function. The hierarchical organization of these objects into a whole-part pattern forms a product entity structure for the system. Everything in the system is in its physical product entity structure. Everything else in the universe that is not in the system product entity structure is in the system environment, as depicted in Figure 1.1. The entities in the system interact with each other and with the environment selectively through interfaces, causing the entities to be capable of accomplishing ends not possible using the entities disassociated from each other. A man-made system consists of some combination of physical elements, referred to as hardware; computer software that is an intellectual construct that operates within a hardware computer; facilities and real-estate improvements; and human beings accomplishing mental activity, decision making, and work in accordance with procedures to guide human performance.

Figure 1.1a shows the traditional system engineer's vision of a system, while Figure 1.1b illustrates a common software engineer's view of a system using the context diagram offered by the modern structured analysis (MSA) framework. Unified modeling language (UML) and system modeling language (SysML) practitioners would view the system in terms of the terminators replaced by actors gaining benefit from the system through use cases. All of these views of systems offer a sound starting point for the development of a system, even one that develops systems like the system development enterprise and its programs.

1.2 *Purpose and scope of the books*

This book and the three books listed in Table 1.1 capture the author's suggestions for how to develop product systems that solve very complex problems posed by customer needs. A customer need is a very simple statement that a person or an organization phrases from a mission-oriented

Table 1.1 Grand System Developments Books

Book title	Author/Year
System Management: Planning, Enterprise Identity, and Deployment	Grady (in press)
System Requirements Analysis	Grady (2005)
System Synthesis: Product and Process Design	Grady (in press)
System Verification	Grady (2007)

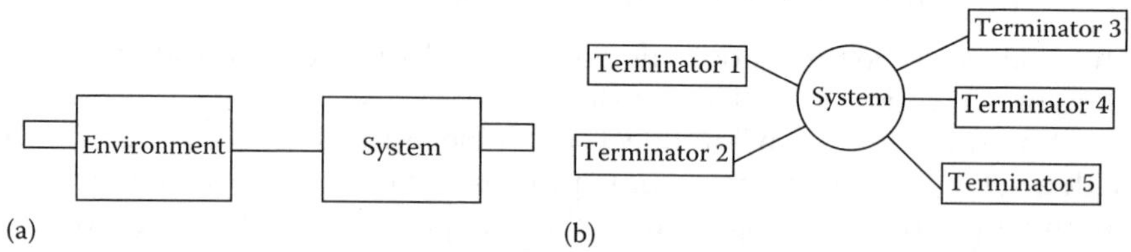

Figure 1.1 Ultimate system abstraction: (a) traditional system image and (b) MSA image of system.

perspective to communicate a capability they do not at the time have, but wish to possess. The work that intervenes between the statement of this need and the ability of a developer to deliver the resources and processes that will satisfy that need is called system development in this book and in *System Requirements Analysis* (Grady, 2005), *System Synthesis: Product and Process Design* (Grady, in press), and *System Verification* (Grady, 2007). This work is accomplished by a business enterprise through what is called a program. The process of using this developed system to achieve the original need is called system employment that includes the operational use of the system and the sustainment of the system. The latter includes the deployment/delivery of the system, maintenance and other logistic activities, modification from time to time, and the eventual disposal of the system at the end of its life.

The word "grand" is used to signify that there is something larger than simply the product system that must be created and delivered. Practitioners of this process will agree that it is quite a feat to deliver an optimized product on time that fully satisfies the need within available cost and schedule constraints. But, even this great accomplishment is insufficient. There is a grander goal, and that is to deliver the combination of a product and process that are jointly optimized. One step beyond this capability resides the enterprise that is optimized for developing optimized systems of products and processes. This is called the grand system, the union of the program process and program product multiplied several times within an enterprise across several ongoing programs, each optimized for a particular product line and customer base. The system development process includes four fundamental processes coordinated with this book and the three other books listed in Table 1.1:

1. Application of management skills across the other three activities to cause the program to produce the product on time with good quality, within cost, and with the features needed by the customer
2. Definition of the problem to be solved and the specific entities the system will consist of in a set of specifications
3. Synthesis of the content of the specifications in a trio of transforms, starting with transforming the specification content into a product design, then the design into a set of procurement and material acquisition and processing plans, and finally the materials into a collection of manufactured products
4. Verification that the product satisfies the driving requirements in the specifications

These books deal with man-made systems excluding natural systems that are not developed in the normal sense of that word. The best we can do with natural systems is to characterize them. A system engineer working on the Yucca Mountain nuclear storage site told the author that they considered the mountain as GFE, God-furnished equipment. They characterized the natural barrier to potential contamination getting into local groundwater supplies, decided what the barrier had to be, and introduced man-made augmentation to provide the needed aggregate isolation between the material stored and the local water table effective over a period of thousands of years. The development of any man-made systems or man-made augmentation of natural systems will surrender to the methods covered in these books, whether they are complex military weapons systems or relatively complex commercial systems like farming equipment and medical instrumentation.

System development is an organized problem-solving process for transforming needs into preferred solutions. It commonly requires many people to do this work where the problem is very complex, involving the application of many different technologies in search of the best solution. In such situations, the complexity of the product and the problem-solving task combine with customer expectations for minimizing the cost and the schedule to demand the deployment of an organized way to combine the different knowledge domains of many people into a commonly held understanding of the problem to be solved and a good solution to that problem. The heart of the system development process is bound up in the way that we humans organize and locate ourselves relative to one another, the nature of the problem to be solved, and the degree of success that we have in infusing a sense of common purpose into the work force thus formed. The development of systems is achieved through effective communication between many different people all working together on parts of the common problem with which they are presented. This requires a degree of organization, thus dooming ad hoc behavior on the part of the development team to failure, and offers a well-managed team of people implementing a well-planned effort some chance of success.

1.3 Jobs of interest in system development

It does require many different kinds of knowledge to develop solutions to complex problems for reasons that we will discuss in Chapter two. All of these people must contribute their special knowledge and skill for the greater good of the development effort. We can group most of these people into five critical jobs. People staffing these jobs have a key opportunity to contribute to program success or cause it to fail because of their contributions or lack of them. Where people staffing these jobs receive any formal training other than on the job, it is commonly tightly focused on their immediate discipline, mostly resulting in a knowledge gap between these disciplines. The enterprise that wishes to excel in the development of systems must find a way to cause these five classes of people to possess a common vision of the enterprise, programs in general, the program they are working on at the time, and the role that each plays in achieving program success.

As a result of mastering the content of this book and the three other books listed in Table 1.1, the reader would qualify as an enterprise technical architect, in the opinion of the author, a job not even recognized in most enterprises. This is the kind of person needed to act as the integration agent across the enterprise process and product being developed on one or more programs within the enterprise in accordance with a common enterprise process incrementally improved over time as a result of acting on lessons learned from programs previously completed.

As we will see, an enterprise must deal with two sets of needs that are not always free of conflict. Customers are interested in solving their problems stated in a written need statement or specification, vaguely described verbally, or with specificity somewhere in between. Often, customers cannot clearly describe what they need where there is nothing currently available that satisfies their need. The enterprise to which they appeal for a solution also is working toward satisfying their own vision statement (the brief, ultimate statement of enterprise functionality) that is commonly going to contain mention of some combination of making profit and delivering good products of value to customers in a particular product line while treating their employees and stock holders well.

It requires a tremendous effort on the part of the five casts of characters within an enterprise to encourage concurrent satisfaction of the customer need for each of several product systems being developed at any one time, their corresponding programs, and the enterprise vision statement. These books present an organized collection of ideas about how these five sets of people can optimize their contributions to the development of product systems. In the process of so doing, the author believes that we are also describing a new (to some) job classification referred to in these books as an enterprise technical architect.

It is very important that all of these people contribute to the whole in a cooperative way with clear knowledge of each other's responsibilities. All too often, people in these five professions are independently trained and remain isolated in practice. Their job functions are ill defined, sometimes overlapping and sometimes full of voids. The principal message in these books is that people in these five professions have very specific functions to perform in relation to development efforts called programs or projects, and they must knowingly cooperate toward satisfying enterprise and program goals in order for the enterprise and its programs to be successful.

1.3.1 *The general system engineer*

A system is a collection of things that interact to achieve a specific purpose. Systems may be simple or complex but will always involve two or more things joined through some form of interface, or relationship, working toward a common purpose. The organized process for creating systems covered in these books will result in the partitioning of the product system and the system development effort into coordinated subsets of product and related work. Each of these subsets will be assigned to a team of specialists charged with solving a particular problem that is part of the aggregate system solution. System engineers may be employed within these teams to integrate and optimize at the team level as well as at the system level to integrate and optimize across these teams and other boundary conditions.

System engineers are specialists in dealing with complexity. Many people think of system engineers as being a mile wide and an inch deep in knowledge, and this is true except for the fact that every system engineer must come to the job through some path seldom involving a university degree in system engineering followed immediately by work in that field in industry. So, most system engineers have some particular domain knowledge acquired during their years in college and early career work in the electrical, mechanical, or other engineering domains. The knowledge of all engineers is, as Robert Machol said in the first chapter of the *System Engineering Handbook* (1965) T-shaped in that they have common knowledge, or a breadth of knowledge, that permits them to communicate with others (language, mathematics, and science, for example) and deep specialized knowledge in some field. In all other knowledge domains, system engineers may be very thin but since they have mastered the art of talking to other people and listening to what they say and have a good mental attitude about the complexity and skill in using certain tools, they are able to place the ideas they receive into a pattern and observe conflicts and inconsistencies that call out for further study and resolution.

Successful system engineers have mastered a way of organizing products and processes into a simple partitioned pattern of input–process–output relationships. They recognize entities that can be thought of as activities or objects and correlate them with inputs and outputs. As a result, they are able to model complex situations even before they exist in reality and reach conclusions about their utility before a lot of resources have been expended to create them.

System engineers are generalists who enjoy dealing with wholes rather than parts. They can partition large problems and entities into related sets of smaller ones and synthesize many parts of a whole into an aggregate view. System engineers may not be qualified to accomplish design in any domain, yet they are able to understand the key facts in many different domains well enough to join together with the domain experts to cause them to achieve goals none of them could alone or without the help of the generalist.

1.3.2 Domain system engineers

As man expands his knowledge base, recognizing the limitations of his own individual fixed knowledge capacity, he will find it necessary to build more specialties with fincr granularity. Today, in system development enterprises, one finds not only domain engineers populating increasingly narrow but deeper knowledge spaces and their corresponding functional departments, but domain generalists who have mastered a broader band of knowledge embracing the work of several domain specialists. Software, telecommunications, optics, and avionics system engineers are examples of this reality. Of these, the software domain system engineers offer development programs the biggest challenge because so much of the hard functionality flows into that domain today, and so few hardware-dominated system engineers outside of the software fold understand their world well enough to communicate effectively with those who do that a wide chasm can develop between the hardware and software people, and between the system elements under development under their charge.

Of all the domain system engineers, the one that is commonly the best placed to understand the overall operation of the system in the most advanced and detailed fashion is the control system engineer. Thus, control systems theory is an excellent subject for study in a system engineering college curriculum and a good background for those who would do system engineering work at any level. This knowledge is seldom useful at the grandest level because the concepts, components, and relations are simply too complex to deal with in a rigorously detailed mathematical model. But control system knowledge is a good background. There are probably few control system engineers still working actively in this field in industry who only understand an analog implementation of a control system in hardware. The combination of a control system engineer who also understands the software implementation of control system theory is particularly well prepared for system engineering work in the twenty-first century.

1.3.3 The program manager

Enterprises achieve their work through programs that some enterprises may prefer to call projects where they are small in size. These organizational entities are focused on solving a specific problem most often for a particular customer but, in commercial enterprises, the customer may be a demographic group described in weakly understood specifics. The program succeeds by creating a product that the enterprise can sell, and receive adequate payment to more than support their existence. The enterprise will have to staff the program organization with people and identify someone to manage these people toward program goals: the program manager.

The program manager manages a program through other people. In Part three, we will consider several ways of organizing people and will pick a structure that requires the least complex relationships between work entities and the product entities that must be developed. This structure will involve a matrix of shared responsibilities, where functional departments supply resources to all of the programs and each program must manage these resources within the context of achieving program goals. So, the program manager has a temporary job that may last 6 months or 5 years, or more. He or she must bring together the available resources, organize and tune them to fit the situation, and manage these resources over time. At the conclusion of the program, all of the resources applied not consumed by the program must be returned to the source, the functional departments, hopefully better for the experience.

In reality, this is not a binary exchange occurring only twice (start-up and shutdown) in a program life cycle. As programs come into being, thrive through the engineering development process, move into a manufacturing emphasis, and then on to a customer support phase through the useful life of the product, different people are needed with different skills and knowledge. The tasks change and the people needed to accomplish these tasks change. The program manager, therefore, is faced not with a static situation to manage but one that shifts in time. In fact, the person selected to be the program manager in the initial program phases may be entirely the wrong person to manage the program in later phases. Many program managers who did very well in acquiring the business and managing a DoD demonstration validation phase program have failed badly when continuing on managing the program through the engineering and manufacturing development phase and production.

The author maintains that few enterprises have succeeded in perfecting the performance of system engineering and management work on their programs. The author also maintains that much of the reason for this inability rests at the feet of program managers in industry. Much of the system engineering work must be done early in the program but often yields little immediate gratification. The benefit of having spent the money early to accomplish work is often delayed, and a delayed gratification is as difficult for a system engineer to sell to a program manager as it is for a father to sell it to a daughter prevented from going to a dance. Despite the sure knowledge that program managers understand that money *spent well* early on programs reduces later risk, it is very hard for any program manager to act on that knowledge when he or she knows that they will encounter serious problems later and will need that money then. Money and time are both nonreversible commodities. Thus, a management reserve account and schedule slack are established and reserved until those serious problems arise. As valuable as a management reserve is in mitigating future risks, if its creation guarantees that funding will not be available to accomplish the front-end work that could preclude the need for a program reserve later, it may be somewhat counterproductive.

The program manager must demand front-end work be performed well and quickly, which requires that the enterprise possess the resources that will encourage this result and have in place the means to continuously improve these resources in response to program lessons learned. System engineering talent must exist, and it must be capable of producing good results rapidly and affordably.

1.3.4 The functional manager

In a popular form of enterprise organizational structure, called matrix, two management forces are at work: (1) program managers manage the day-to-day conduct of the enterprise programs and (2) functional managers run the departments that provide programs with the resources they need to be successful. In far too many enterprises, program work is managed directly by functional department managers or their surrogates assigned to programs. This happens because functional departments have an enduring continuity over time that programs do not have.

Over time, the functional department managers accumulate power and influence, causing the enterprise to operate in a fashion that is not in the best interest of the enterprise. In the author's view, functional managers should not be managing any program work directly.

Functional department managers should be providing all enterprise programs with the resources they need to succeed. This includes people who know what they are doing in the various functional department charter knowledge spaces. The right number of people is important, but all of these people must know what they are doing. In addition, the functional managers must provide programs with good tools and best practices with their personnel fully trained to exploit them both. To the extent that these functional managers can cooperate among themselves to encourage that an integrated capability is available for program managers to apply, so much the better.

The functional managers must also be held accountable for continuously improving the enterprise capability based on lessons learned from programs and independent work on their part to keep abreast of developments in their field of expertise. This is an application of control theory where programs apply functional department resources prepared specifically for the enterprise product line and customer base. As the programs apply these resources, data is collected, related to how well these resources are matched to program needs, and this data is used as an error signal to move the resource mix to a new position better suited to satisfy program needs. The enterprise activity is so complex and so many changes are constantly occurring that it is unlikely that this process will ever reach perfection in any enterprise, no matter how energetically they pursue improvement. But, if this effort is not pursued energetically, the enterprise capability will surely recede relative to its competition.

Program managers must manage the work that is going on today. New business people and functional managers must be constantly working to make sure there is a tomorrow.

1.3.5 The enterprise technical architect

In any case where specialization, partitioning, or decomposition is applied, there must be an integration and optimization force applied as well. By breaking large tasks into pieces, we adjust the depth and breadth of the complexity of such tasks to human proportions but, in the process, suffer the potential for the loss of insight into potentially valuable interrelationships between the pieces.

In every case where the valuable notion of partitioning or decomposition is applied, it is necessary that we also apply an integrating and higher-tier optimizing force. The system engineer is such an agent on programs coordinating across cross-organizational interfaces, linking up the results of related work products, and insisting on discipline in identifying and complying with approved requirements.

Whether the entities involved are products, programs, or the overall enterprise process, the enterprise technical architect is offered as the integration and optimization agent. In this book, we will introduce an organizational entity that will be new to many readers, called the enterprise integration team (EIT), that is the home of the enterprise technical architect.

1.4 Content

This book and the three other books listed in Table 1.1 collectively cover the whole system development life cycle. These books coordinate with the four system engineering courses that the author's company, JOG System Engineering, offers through universities, through short-course companies, and through direct marketing to companies. These four books address the four fundamental elements of the system development process from the system engineer's perspective that are applied to develop both products and corresponding processes based on a generic process capability that is clearly defined and constantly improved based on lessons learned from programs.

The related course materials for these courses are organized into three divisions. Division 1 is the textbook available from booksellers. Division 2 of the student manual is available from the author's company by enrolling in a course. Exhibits A, B, and C of Division 2 provide lecture, workshop, and course administrative materials, respectively. Division 2 of this courses also provides Exhibit D that offers a bibliography for all of the courses, Exhibit E that provides a combined glossary and acronym list for the whole series of courses, and Exhibit F that provides a map to the contents of all the books.

Division 3 offers sets of supporting documents. These documents are commonly supplied by a CD-ROM or available for download from a Web site. They are listed under each book described next. At the time this was written, this data was made available from the author's company through course enrollment. The plan is to make them available via download by anyone from one or more Web sites with no course enrollment precondition. The supporting data is provided in three variations of a set of documents commonly produced by system engineers on programs as follows: (A) templates for documents, (B) data item descriptions (DID) for conversion of these templates into real program documents, and (C) sample documents created in accordance with the DID for a specific fictional system used in class as a basis for workshop activity. Divisions 2 and 3 are not supplied with the purchase of this book.

1.4.1 System management

This book provides an introduction to system development and focuses on the management aspects of the functional departments and programs. Part one provides an introduction to this book and the other three books listed in Table 1.1 and discussed in this section, provides a rationale for the systematic methods that have come to be applied in system development based on available human knowledge, and concludes with a discussion of the fundamentals of system development.

Part two of this book presents a process handbook, the what and why of a process, a detailed example of a common process model, and a discussion of ways to document one's process model. A process model is a documented identification of all of the actions necessary to accomplish some goal. In the author's process model, a process flow diagram is used where each step in the process is identified by a block. Each of these blocks is named and carries a modeling identification (MID). The MIDs are composed of the letter F (for function), followed by a string of base 60 alphanumeric characters that uniquely identifies that process step for work-cataloging purposes.

All of the work in the enterprise, whether on programs or for the generic enterprise, is going to be accomplished by people in the organization. Part three of the book exposes several alternative organizational structures and builds the case for the matrix structure where the enterprise has several programs in the house at any one time, the problems that have to be solved by these programs are complex, and these problems require an appeal to many different knowledge domains. The enterprise knowledge base and generic process are linked to the functional department structure, and managing the program work is focused on a program organization using cross-functional teams staffed by people assigned to the program teams from the functional departments.

Program planning should proceed from a generic common process improved in the past based on lessons learned from past programs continuing on into the future as far as the eye can see. The generic process is placed under a central management authority and contributed to by all of the functional departments, as explained in Part four of this book. Planning components are crafted for the generic process based on the functional department generic process responsibilities. Program planning strings are crafted in Part five of this book based on the product entity structure. These two sets of strings are connected together in much the same way that one establishes traceability in specifications to form the program planning details.

Part six of this book organizes management techniques for the three fundamental system development efforts: requirements analysis, synthesis (design, integration, optimization, manufacturing, and material acquisition), and verification. Each of these matters is taken up again in detail in each of the books listed in Table 1.1. Part seven closes out this book with a discussion on assessment and improvement.

Division 3 for this book includes the following exhibits:

- G Enterprise vision statement
- H Enterprise architecture report
- I Enterprise system engineering manual
- J Program plan
- K Program work breakdown structure (WBS)
- L Program statement of work (SOW)
- M Program integrated master plan (IMP)
- N Program integrated master schedule (IMS)

1.4.2 System requirements analysis

Part one of the book provides an introduction to requirements analysis reviewing the generic process applied in the book but extending the detail for requirements analysis.

Part two establishes a foundation for four basic kinds of requirements analysis expanded upon in Parts three and four. This foundation includes some fundamentals, traceability relationships, and coverage of four strategies that can be used for requirements analysis, only one of which is recommended as a complete strategy.

One of these four strategies, structured analysis, is expanded upon in Part three, Traditional structured analysis, and in Part four, Computer software structured analysis. Following a discussion of a general theory of structured analysis, several different modeling structures are presented that collectively include most all of the models used in industry today for requirements identification. Some of these are effective where the product will be designed as a hardware entity, and some are useful where computer software is the intended product implementation.

Part five covers specifications where requirements are documented. The content and intent of standards for specifications and their utility as style guides are discussed. Applicable documents, which permit reference to other documents and the requirements they contain as applying to the development of the item covered by the specification, are also covered. Part six offers ways to implement a sound requirements management process, including performance risk mitigation, value management, requirements integration, and verification management (covered in detail in *System Requirements Analysis* [Grady, 2005]). Part seven covers the use of computer tools to support the requirements analysis activity. Part eight closes out the book.

System Requirements Analysis Division 3 data includes Exhibits G through O with the three branches noted.

- G User documents 1
- H User documents 2
- I System architecture report
- J System specification
- K Hardware item performance specification
- L Hardware item detail specification
- M Software item performance specification (software requirements specification)
- N Software detail specification
- O Technical performance measurement (TPM) plan and report

Subsequent to this book being published the author developed some new ideas related to universal architecture description frameworks (UADF). Three of these are discussed in the course formed by the use of: (1) functional analysis for both hardware and software, (2) MSA expanded to include process for system architecture and requirements engineering (PSARE), and (3) the combination of UML and SysML. The related text is supplied to students in a separate paper.

1.4.3 System synthesis

System Requirements Analysis (Grady, 2005) and *System Verification* (Grady, 2007) are dominated by system engineering activities but the work covered in *System Synthesis: Product and Process Design* (Grady, in press) is dominated by design. Each cross-functional team cooperates to develop a preliminary and detailed design but this process is propelled by the creative genius of design engineers who translate science, technology, engineering, and mathematical knowledge into design solutions for the problems defined in their specifications. Part one provides an introduction. Part two organizes the design process and its planning. Part three covers the product definition work more fully covered in the requirements book. The work covered in Parts two and three must be integrated and optimized at the grand system level, and this is discussed in Part four. Part five provides an overview of several specialty engineering disciplines to encourage system engineers to have a basic understanding of their contributions on programs. Part six discusses post-design activities, including material, manufacturing, and quality. Part seven discusses new initiatives that will influence future work in the synthesis portion of the development life cycle.

Student manual Division 2 coordinated with this book includes Exhibits A, B, and C that provide supporting data for the Systems Synthesis course. Exhibits D, E, and F are reserved for similar supporting data for a specialty engineering course. Division 3 provides Exhibits G through M offering template/DID/sample documents for the synthesis course including the following:

- G Interface control document (ICD)
- H Trade study report
- I Design concept record
- J Design review documents
- K Reliability, availability, and maintainability AAA report
- L Safety hazard analysis report
- M Mass properties report

1.4.4 System verification

Verification compares the results of the synthesis process with the requirements documented for the items designed, and we make decisions about the degree of compliance with these requirements. After an introduction in Part one, Part two covers item qualification while Part three covers acceptance verification. Part four covers system test and evaluation. Part five applies the verification activity to processes, and Part six closes out the four books.

Student manual Division 2 Exhibits A, B, and C provide course support documents while D, E, and F are voids. Division 3 Exhibits G, H, I, and J provide templates for integrated verification documentation:

- G Integrated verification plan (IVP)
- H Integrated verification management report (IVMR)
- I Integrated verification data report (IVDR)
- J Test and evaluation master plan (TEMP)

1.5 Series applications

The principal beneficiaries of these books will be practitioners in industry working as system engineers, program managers, and functional managers. These books are intended to contain practical information on what to do and how to do it in support of tasks common in the development of products and processes to satisfy customer needs. So, the principal application for the series is as a desk reference for system development practitioners of the kinds noted.

The author wrote these books specifically to act as textbooks for a four-course system development program of instruction that would satisfy the needs of the system engineering certificate program offered by his company. Lecturers in short-course companies and universities are, of course, free to employ these books in their own programs.

At one time, the author believed that an enterprise should develop its own system development process documentation, telling not only what must be done but how to do it as well. The author has concluded over a few years of experience in working with enterprises, however, that this is more than most can reasonably accomplish. This is the equivalent of writing a textbook, and while the author finds it easy to create millions of words, everyone does not. The people who understand the enterprise's preferred process and can write well are all too often needed to do everything else, including working on programs, providing little time for them to contribute to these documents.

In Chapter three of this part, we will discuss ways of documenting the enterprise process. The author's advice is to prepare a document that tells what must be done, essentially a process specification, and reference each step in the common process to the best how-to material available. These sources may include textbooks, in-house or external courses, videos, and experts within the company. Hopefully, this book and the three books listed in Table 1.1 will be one of these sources of how-to knowledge selected by enterprises.

The near-term application intended for these books at the time they were developed was to provide textbooks for a system engineering certificate program titled grand systems development (GSD) offered by the author's consulting company, JOG System Engineering, Inc.

Figure 1.2 shows the structure of the books used in the JOG System Engineering Grand Systems Development certificate program. This program consists of the four courses by the same titles as the four books discussed. In each case, Division I is the textbook component. Division II contains the presentation material, workshop guide, and course administrative information in Exhibits A through C. Division III includes one or more exhibits providing templates for documents related to the course. Several workshop sets are available, and the lecturer selects the set most closely related to the client's product line and customer base.

1.6 Closure

A man-made product system consists of two or more entities that interact in predetermined ways to achieve a predetermined function. System development is an organized process for creating man-made product systems based on a predetermined problem definition. The system development process applies selected specialized knowledge sets in an organized way to solve problems posed by the needed product system. Product systems are created by process systems. These process systems, also known as projects or programs, are managed by program managers who are

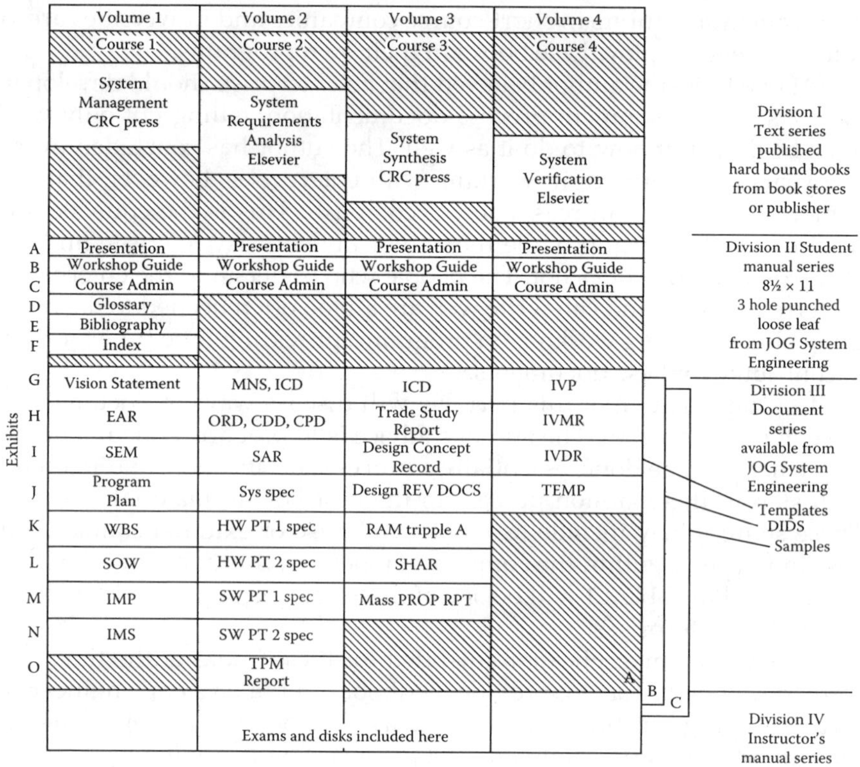

Figure 1.2 GSD student manual series.

responsible for designing, staffing, and operating these process systems. Programs employ cross-functional teams assigned to the major items in the product entity structure determined through structured analysis of the customer need. One program system level team acts to integrate and optimize across these several product teams. System engineers work in process systems on the system and item teams to develop the technical characteristics of product systems, focusing on the interfaces between the things in the product systems and the interfaces between the teams primarily populated by specialized people. This interface work in the process system is consummated through human communications, which unfortunately is perhaps the least effective interface on planet Earth.

An enterprise that develops systems is in itself a system as are its programs. These process systems can be architected using the same modeling methods that are effective on programs to define and develop product systems. Thus, this book extends the system architecting work covered in the author's three other books (*System Requirements Analysis* [2005],

System Synthesis: Product and Process Design [in press], and *System Verification* [2007]).

The entities of the enterprise and program process systems are organizations populated by humans who have mastered particular knowledge subsets. In the case of an enterprise that is organized in a matrix structure, these entities are functional departments organized around the knowledge subsets that provide all programs with people who know what they are doing, with good practices to guide the work performance on programs, and with good tools these people require to perform their work. In the case of programs, the entities are cross-functional teams staffed from the functional departments so as to understand the requirements appropriate to the entity to which they are assigned and organized around the product system entities.

This book extends the modeling capabilities discussed in the author's *System Requirements Analysis* (2005) book to the other two systems we must deal with in the development of product systems, the enterprise that applies programs to develop particular product systems and the programs themselves.

chapter two

*The human foundation of systems development**

2.1 *The expanding knowledge base and its effects*

Throughout all prior time up through much of the first half of the twentieth century, a single engineer could generally accomplish the complete design task for an entity, as illustrated in Figure 2.1a. This was true for two reasons. First, the problems posed for engineers to solve could be solved completely within the context of a single engineering discipline, such as electrical or mechanical engineering. Second, the expanse of knowledge required to solve these problems could be mastered by a single engineer with no help from specialists focused narrowly on particular facets of the engineering process. That is, the problems were relatively simple.

Beginning with World War II during the late 1930s and continuing immediately thereafter during the early cold war years of the late 1940s and the 1950s, the military problems posed for engineering solutions became increasingly complex and often would not yield to a solution solely within the context of a single engineering discipline. It was also found that it was not possible to depend on a single engineer in any one field mastering the complete range of knowledge, due to an expansion of the knowledge base for science, engineering, and weapon systems. As a result, it became necessary to apply the specialization paradigm narrowing the scope of the traditional engineering disciplines and adding others, like reliability. Engineers from several disciplines had to work together, as suggested in Figure 2.1b, each supported by a cast of specialists.

The intent was to cover the complete range of knowledge needed to solve the problems encountered in the development of increasingly complex systems. Enterprises dealing with these complex problems commonly formed functional departments specializing in each of the specialty disciplines needed for their product line and customer base. In too many organizations managed through a matrix, the mistake made was that the responsibility for program work was placed in the hands of the functional department management rather than people reporting directly to the program manager. Specialty engineering integration during development

* The material in this chapter is from Grady, J., *System Engineering Deployment*, CRC Press, Boca Raton, FL, Chapter 2, 2000. Used with permission.

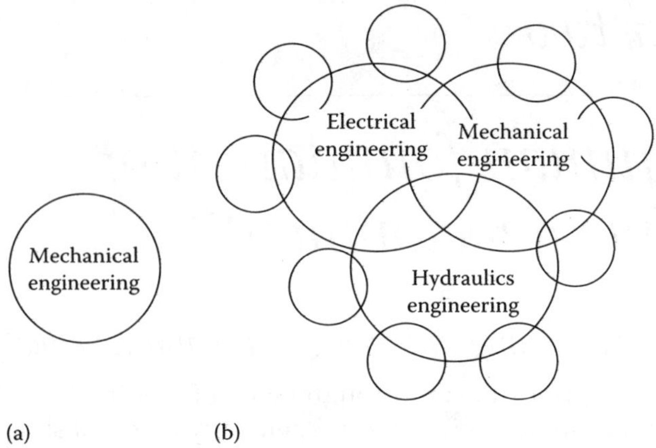

Figure 2.1 The world has changed: (a) yesterday and (b) today. (From Grady, J., *System Engineering Deployment*, CRC Press, Boca Raton, FL, pp. 17–25, 2000. With permission.)

became a very difficult problem with all the people physically collocated by functional departments and budget partitioned in a crazy quilt relationship between the items in the system (identified by a work breakdown structure) and the functional departments contributing to item development. This condition was attacked in the 1990s by movements called concurrent engineering and cross-functional teaming.

The difference between the two situations exhibited in Figure 2.1 is very significant in the emergence of the systems approach. It is at the very foundation of the system development process driven by the explosive expansion of man's knowledge base beyond the early specialization framework applied to the engineering field, consisting of civil, mechanical, electrical, and chemical engineering, forcing a further specialization of the people who do this work.

2.2 *Knowledge dimensions*

Specialization has more than a single dimension of depth. The knowledge of all engineers, indeed of all individual human beings, is, as Robert Machol said in the first chapter of the *System Engineering Handbook* (1965), T-shaped in that they have common knowledge, or a breadth of knowledge, that permits them to communicate with others (language, mathematics, and science, for example) and deep specialized knowledge in some field. There was a time on planet Earth when every soul on Earth knew everything that comprised man's collective knowledge base. This knowledge was very dull by today's standards, all related to survival, but all could master it because it was not very extensive. Over time, man has mastered a progressively

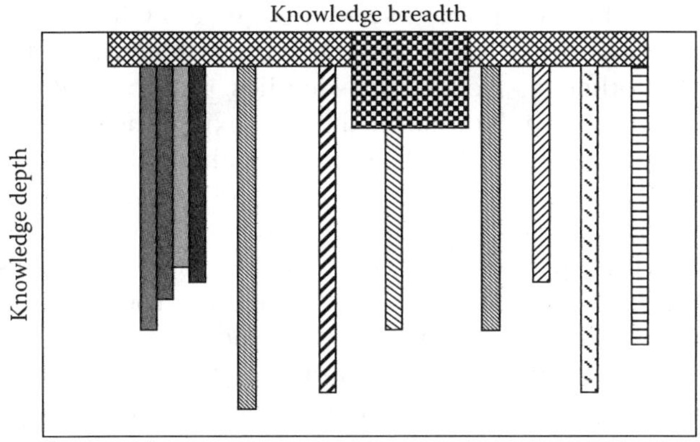

Figure 2.2 Breadth and depth of knowledge. (From Grady, J., *System Engineering Deployment*, CRC Press, Boca Raton, FL, pp. 17–25, 2000. With permission.)

greater amount of knowledge, passing a ceiling many thousands of years ago that matched the maximum capacity of the human brain. The universal solution to this problem that man had more knowledge than any one person could master was, of course, specialization.

We should recognize that a specialist cannot focus their limited knowledge capacity on depth in their field and also master a great breadth of knowledge. When we specialize in depth in a field, we are limited in other fields, as suggested in Figure 2.2. Any development program will require access to particular knowledge but not all knowledge. The nature of a program that the several specialists suggested in Figure 2.2 will be working on determines the kinds of specialists required. A project involving flight in the atmosphere will require an aerodynamicist, for example, whereas a computer system operating in a fixed facility probably will not. You can imagine that these several specialists might have some difficulty in communicating except for the shared knowledge at least including mathematics and a language (crosshatched pattern shared by all of the specialists) permitting them to discuss their common problem.

All specialists do not, however, specialize in depth. A system engineer (crosshatched knowledge space in Figure 2.2) specializes in the breadth of knowledge for the purpose of being able to understand and integrate the work products of the specialists. It is not possible for any person, even a system engineer, to master both the depth and the breadth; so there will be specialized fields about which the system engineer is not well informed. To the extent that the system engineer is thin from a knowledge perspective, problems not detectable by the specialists may become embedded in the evolving design. This can especially occur at the product interfaces.

Every system engineer has to come from a past experience that probably involves an engineering discipline providing a knowledge strength. The author has known system engineers with a background in the normal engineering disciplines (mechanical, electrical, chemical, and industrial), but others as well, such as textiles, physics, mathematics, economics, and journalism. No matter the past experience, the system engineer must accumulate sufficient knowledge about the many disciplines involved in a system development process to enable communication of ideas in these disciplines. In today's world involving an intense appeal to software, which promises to continue indefinitely, the system engineers, many of whom come from a hardware background, must especially extend their knowledge base into software modeling to the extent necessary to enable intelligent communication.

So, how does the system engineer master both the depth and the breadth sufficiently to provide a useful service on a program? Figure 2.3 demonstrates one route. You first take an inventory of your knowledge and compare it with the knowledge space required for common programs upon which you must work. Build a list of the areas not well understood, and identify someone in these fields whose knowledge you trust. You and the knowledge base formed by your personal support team create a whole system engineer with a broader knowledge mastery than any one real person can realize. It is not necessary to inform all of your personal team members of their selection, but it is helpful to write this list down on paper, and as people leave your team for retirement, other employment, or other reasons, you can consciously replace them with another person. This is an example of what many people call networking, of course, but networking for a very specific purpose, to make you a whole system engineer.

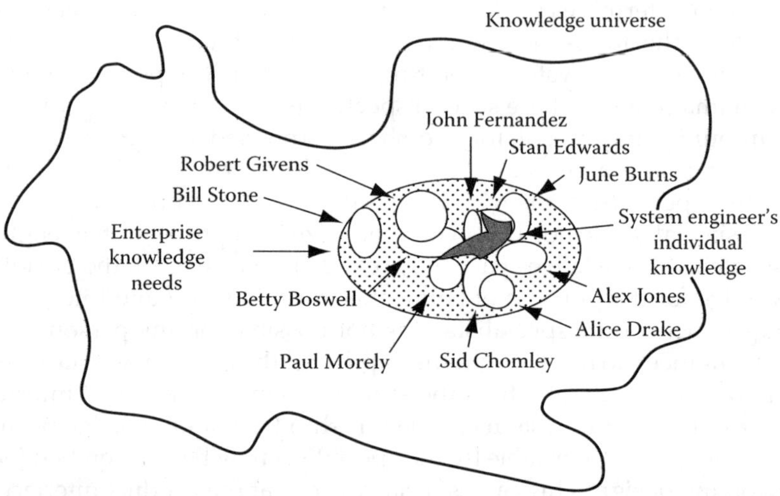

Figure 2.3 The whole person. (From Grady, J., *System Engineering Deployment*, CRC Press, Boca Raton, FL, pp. 17–25, 2000. With permission.)

The continuing expansion of man's knowledge base combined with an essentially static condition of human mental capacity for a very long time encourages an ever-expanding number of thinner and deeper knowledge domains, for example, electrical engineering being partitioned into many more specialized specialties. It is also encouraging the evolution of domain system engineers (software, telecommunications, and avionics, for example) with knowledge demands corresponding to the checkerboard portion of Figure 2.1.

Systems to solve complex problems cannot be created by individuals because of their demands for knowledge beyond what any one person can master even through the pooled knowledge technique discussed above that helps the system engineer in any product line. Also, it becomes necessary for an enterprise to assign more than one person of the many disciplines involved in these developments in order to get many things accomplished within the schedule required by the customer. The knowledge base for an enterprise, therefore, can best be represented by the model in Figure 2.2, with an added third dimension of density reflecting number of specialists in the several fields, as shown in Figure 2.4. A development enterprise may have one or more specialists in every field required by their product line and customer base, but may not be able to bring to bear

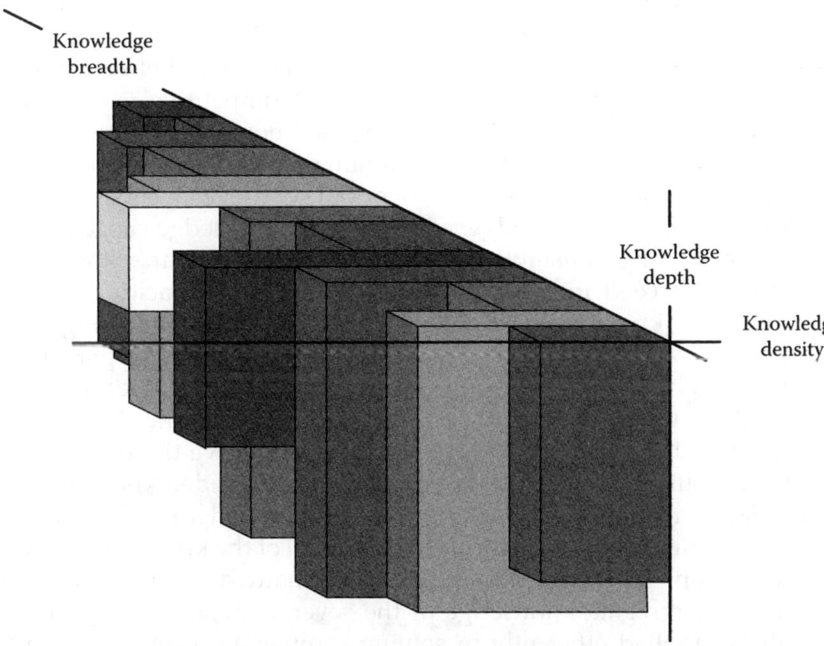

Figure 2.4 Organizational knowledge dimensions. (From Grady, J., *System Engineering Deployment*, CRC Press, Boca Raton, FL, pp. 17–25, 2000. With permission.)

some of these specialists on all programs because of a density problem. All of the specialists may be fully involved in other programs, and one program could suffer as a result.

The knowledge breadth dimension is the basis for the functional departments chosen for a particular enterprise based on the technologies needed for the chosen product line and customer base. The functional managers of each department must look ahead to program needs and ensure that sufficient talent will be available when needed. In order to support this need, the cost/schedule control system serving the enterprise must track planned work on all programs, so as to give program managers not only an understanding of program cost and schedule performance, but also the personnel demands in the different department domains over time in the future, based on the aggregate of all programs in the house and those that may be implemented. In the latter case, functional managers must attach a win probability factor for the purpose of determining personnel density issues.

2.3 Knowledge bond

A man-made system is a collection of things that interact to satisfy a predefined need, as noted earlier. The customer need for a system defines a problem that must be solved. Man-made systems are created by system development enterprises to satisfy these needs. It is generally accepted that possessing more knowledge relevant to a problem by such an enterprise is better than less in solving problems. Efficient and effective problem-solving organizations are rewarded in a competitive marketplace, encouraging system developers to master the knowledge related to particular kinds of systems on which the organization chooses to focus.

As noted earlier, there has been, over the past 10,000 years, an explosive growth in man's knowledge, while human knowledge capacity limit has remained fairly constant. A very long time ago, man's knowledge base came to exceed individual human knowledge capacity. Mankind has solved this knowledge deficiency problem when confronting problems requiring access to a larger knowledge base than any one person can master by applying the concept of specialization. Under specialization, individuals focus on subsets of knowledge such that no one among them has mastered the whole knowledge set needed to solve the problem but collectively some number of people do possess the needed knowledge.

While specialization solves the problem of knowledge coverage, it does so by partitioning or isolating the aggregate of all of the knowledge needed to solve a complex problem into the minds of many different people. The collective or aggregate knowledge of the several cooperating specialists can only be applied efficiently in solving complex problems if it can be shared. Humans have developed the ability to communicate by verbal and visual means in various media permitting the sharing of knowledge. It is intended that the specialized knowledge of individuals working together will be shared through some form of human communication. The generic specialized knowledge is less important in this communication than the system-peculiar consequences of that knowledge applied to solve the problem represented by the system need.

This knowledge-sharing process is enhanced where the people working together possess a shared vision and where the knowledge space corresponding to this vision is maximized. In the engineering field, some essential elements of this shared vision are a common spoken and written language, elementary logic, mathematics, and physics. To the extent that the people working together on a common problem can efficiently share problem-related knowledge in near real time, the fruits of their labor will be enhanced in terms of reduced risk of failure, less cost, shorter solution time frame, and greater effectiveness of the resultant solution.

The process of encouraging the intense cooperation of specialists toward the solution of a common problem, once referred to as specialty engineering integration, is now often called concurrent engineering or development. Clearly, this process depends on good communication between the specialized participants in order to expand the shared vision of the system under development. Unfortunately, this must be accomplished via the worst interface on planet Earth—human communication. Good communication can, however, be encouraged through several steps that recognize the fundamentals of human psychology.

1. It is known that the quality and magnitude of communication between people increases as an inverse function of the distance between them. So, we should locate people close together who must communicate a lot and, especially, the more complex the concepts that they must communicate.
2. Knowledge may be shared in a wide range of patterns, as suggested in Figure 2.5 showing communication paths between eight people, A through H. Figure 2.5b calls for everyone to communicate through a single person. Figure 2.5c encourages communication between anyone and everyone in each case. Figure 2.5a provides for communication in any pattern we may choose, assuming that any number of communication connections may be enabled for any one communication action. The problem with the configuration in Figure 2.5c is that many of the communication actions are of no interest to some people, so we are detracting from the focus of these people who need not participate in particular communication actions when we force them to do so. The problem with the configuration in Figure 2.5b is that the synergism of the group may be lost in that communication directly between individuals is discouraged. In order to make the configuration in Figure 2.5a work, we must adopt the goal of

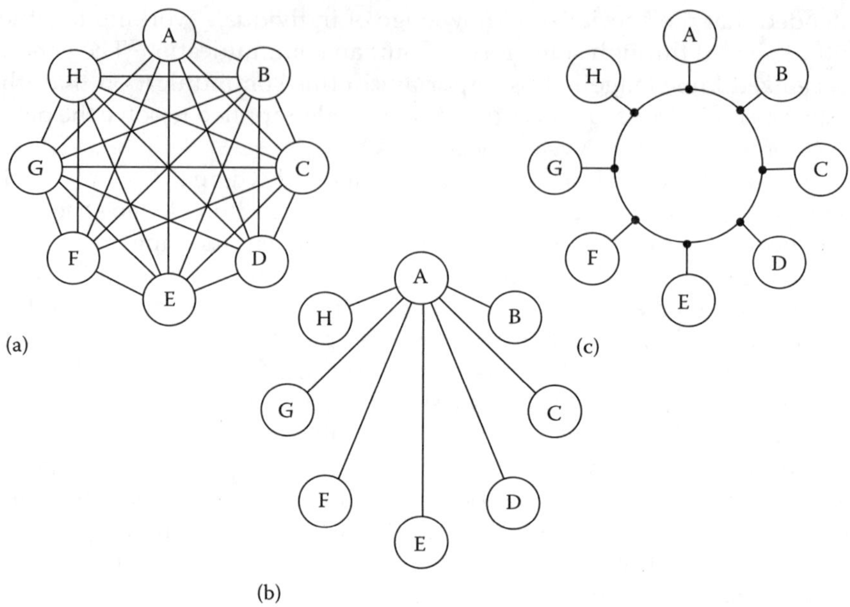

Figure 2.5 Some human organizational communication patterns: (a) universal selective, (b) integrator, and (c) simultaneous shared. (From Grady, J., *System Engineering Deployment*, CRC Press, Boca Raton, FL, pp. 17–25, 2000. With permission.)

maximizing the opportunity for communications while minimizing the need to do so. The magnitude of the shared vision determines the need. The greater the shared vision, the lesser the need to communicate; but the shared vision is achieved through communication.

3. We need to locate the people close together who have the most difficult communication task. All of the specialists of a particular kind, like reliability, working on the development of a system have a simpler communication problem between themselves than all of the specialists of different kinds working to develop a particular item in the system. So, we should organize cross-functional teams in accordance with the system product entity structure and collocate physically as a function of these teams. While computer networks can be helpful, despite their use of voice and video images, it is nearly impossible to improve upon the effectiveness of face-to-face discussion.
4. The knowledge retained in the minds of the people working on a development and the direct human communication network via voice and visual images by various media should be supplemented by machine-stored information connected via machine networking to encourage sharing of on-going work products in near real time. The

information on the machine media should be organized by the team and discipline coinciding with the organization of the product and the people working on the product development. The people must be encouraged to place their in-process work product in the machine storage media and update it in keeping with current conclusions.

2.4 Knowledge and its exchange

Programs represent strings of knowledge generation, accumulation, and sharing in time. Knowledge can exist in one of the four forms or nodes: (1) that which is stored in a human mind, (2) that which is stored in a digital computer, (3) physically stored information in the form of paper documents or computer storage media (tape, disk, etc.), and (4) the special case of knowledge resulting from the synergism of the group. Figure 2.6 illustrates

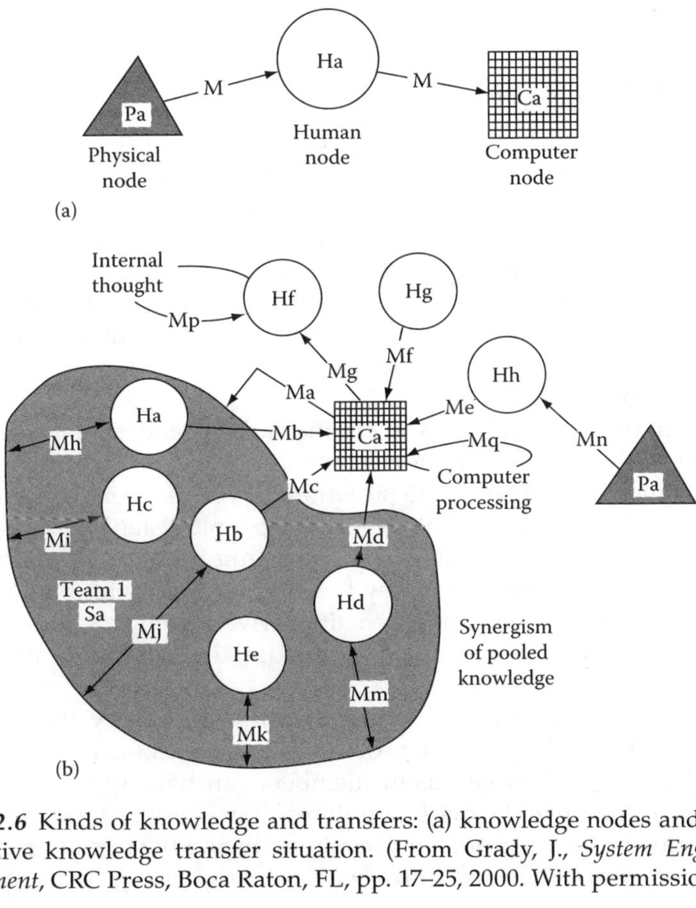

Figure 2.6 Kinds of knowledge and transfers: (a) knowledge nodes and (b) representative knowledge transfer situation. (From Grady, J., *System Engineering Deployment*, CRC Press, Boca Raton, FL, pp. 17–25, 2000. With permission.)

some sample situations in the transfer or exchange of knowledge between these nodes.

Figure 2.6a identifies three of these nodes coupled together through messages. The physical node Pa may be a paper document being read by human Ha who enters data into a computer Ca based on the new information derived from the paper document, possibly modified through an internal thought by Ha.

In Figure 2.6b, we have a more complicated situation where there are five humans who are inputting information into a computer (Hg via message Mf, Hh via Me, Ha via Mb, Hb via Mc, and Hd via Md). Human Mg is deriving information from the computer via message Mg and thinking about it via Mp to create new knowledge. Human Hh is deriving knowledge from a paper document Pa via message Mn. The data being entered is being processed to create new knowledge via message Mq. Several members of Team 1 are meeting in a group deriving information from the computer via message Ma and interacting as a group resulting in synergism Sa not possible by any other means.

2.5 *Individual versus group work*

The reader can think of many other combinations of these communication building blocks interconnected by messages in various ways. One might feel that maximizing communications might be the right action in all cases. However, if everyone assigned to a program maximizes their communications with everyone else on a program, the result could be to deprive everyone of any opportunity of doing any real work. For example, if we had a team of the six engineers working on a common problem as shown in Figure 2.7, we would find that there are hundreds of communication situations.

One of these situations is everyone talking and no one listening. Another six situations cover one person talking and the five others listening, as in meetings. Another 15 situations cover different combinations of 2 people simultaneously talking and 4 people listening. Yet another situation is all six people internally communicating, more often called thinking, signified in Figure 2.7 by the circular lines with both ends on the same human node. It is the benefits derived from this situation that we have to balance against the goal of maximizing communications on a program.

Individual specialists need time to think and apply their specialized skill and knowledge to the problem at hand, and forced, continuous, intense interaction between team members can have the effect of suppressing the specialized work that will result in useful information products of value to the program and thereby reducing the rate of progress of the program. On any one program, we have the challenge of maximizing

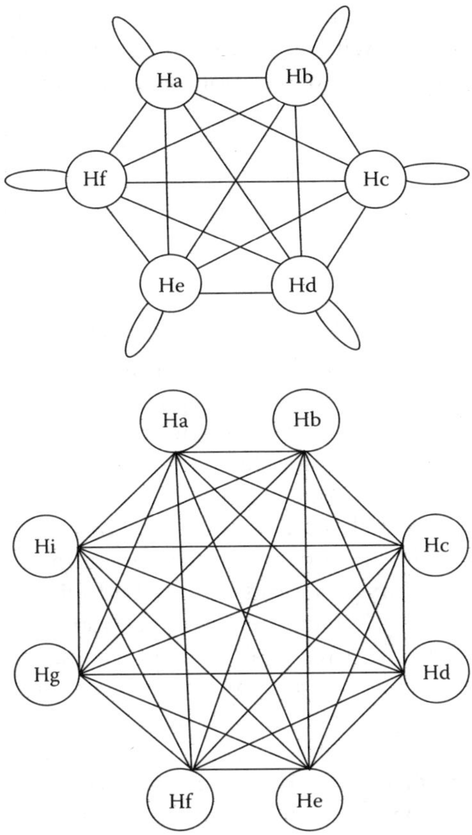

Figure 2.7 Communications possibilities. (From Grady, J., *System Engineering Deployment*, CRC Press, Boca Raton, FL, pp. 17–25, 2000. With permission.)

the opportunities for messages between all of the human nodes, while minimizing the need for them so as to provide time for the development of knowledge and related messages by the people representing the nodes reflected in Figure 2.7. This means, we wish to reduce the signal-to-noise level for all program workers.

All of these communication situations will not be accomplished with the same effectiveness in a real program environment. Some may be suppressed by the emotional makeup of the particular humans selected to staff the team, while others are overdone. An introverted engineer may not be assertive in communicating information he or she has produced or in gaining access to information needed from others, while an extrovert may force his or her correct or incorrect ideas on everyone else as well as demand the needed information from teammates no matter the schedule realities.

There are several paradigms that could be applied to regulate the communication process within a group of persons on a program. We might conclude that the communication process should not be interfered with in any way, relying on the participant to ask the right questions to the right people at the right times and provide the right answers to the right people at the right times. Alternatively, we may conclude that some form of facilitation or control is required to encourage efficiency in movement toward scheduled events.

2.6 Communication facilitation and facilitization

There is no more powerful communication mode than person-to-person conversation. The physical collocation of product-oriented team members from the several different disciplines encourages effective communication using this mode when it is needed either to ask and answer a question or to supply needed but unsolicited information. Formal team meetings should be called periodically to take advantage of synergism of the group, and informal discussion groups should be encouraged to handle problems as they arise. The space should be supplied with a computer projection capability; some wall space; and a projection screen, or a nonglare whiteboard surface permitting projected image markup or annotation, upon which to project the image. Wall space is also needed, upon which the team may place key drawings and data presenting the current big picture to the team.

If the team is relatively small and can be located in a single unobstructed, contiguous space, the team can effectively be in a meeting all the time with the people involved in the conversation changing over time as it moves from one area of their interest to another. A team in this situation can selectively be working fully in the individual mode, engaging in an all hands meeting, or something in between depending on the needs of the moment. The advantage of this arrangement is ease and speed of assembling the right conversation at any time. The disadvantage is that it can be a very distracting environment, within which the specialist must accomplish his or her detailed work. In the author's experience, he found that more experienced people could more often tolerate and even excel in this environment, whereas less experienced people needed a more isolated thinking environment.

The other extreme of team space is a fully cubicalized space such that everyone has a measure of privacy within which to work within his or her specialized field. Obviously, the disadvantage is the difficulty of interaction between members. The team using this arrangement will have to have a separate or adjoining area within which to meet, and this space should have a computer projection capability, as mentioned above.

In addition to direct voice communications, teams need to be able to extend their communications over a distance via telephone conversation, fax (or its equivalent), video exchange, and computer networks. Eventually, all these kinds of services will be integrated into a desktop instrument, by whatever name, the purpose of which is to seek out and share knowledge between the humans staffing a program development effort. This capability will not expand the individual's capacity for knowledge and, thus, solve the problems created by the specialization solution for human knowledge limitation. Rather, the continuing expansion of knowledge, despite the improvements in communication, will encourage a multilevel grouping of the application of system thinking. But the computer will vastly improve communication between people in the workplace. It will more closely integrate together the nodes of knowledge residing in the people staffing teams.

2.7 The enterprise knowledge base

The computer, once again by whatever name, will become the instrument of shared vision for teams and programs. The work that all of us individual specialists accomplish should lead to knowledge products composed of documents (drawings, text, database content, and so forth) that provide the results of that work to be shared. In the past, this sharing process has been imperfect, and the only aggregate repository has been in the minds of those working on the program. Each specialist has read and retained the information that was thought to be related to their work. This information influences subsequent communication between program personnel, each of which has been impressed with some subset of the aggregate program knowledge. In this situation, the shared program knowledge base, that which everyone knows, might reach 50%. A continued improvement of model-driven development computer support resources should result in coupling a much higher percentage of the aggregate human program knowledge into the aggregate model implemented in a computer network built by the team members, and from which all program team members can draw.

The work of all the specialists must be placed in a computer network so that it can be shared by all team and program members. The computer network thus becomes a member of the program and its teams. This member happens to be able to master a tremendous range and mass of information as far as remembering it faithfully is concerned. This member, however, can do very little with this information currently. The computer becomes a very valuable, though limited, team member with a great memory, and thus provides the program with the important shared vision function.

Books have, in the past, been the principal shared vision instrument for man's knowledge base, as illustrated in Figure 2.8. Through reading

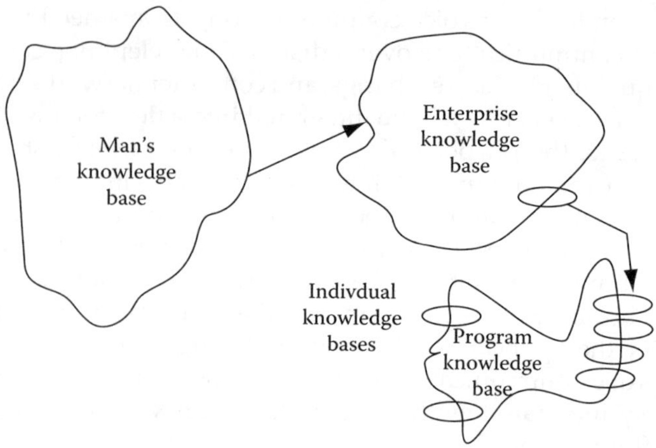

Figure 2.8 A progression of knowledge bases. (From Grady, J., *System Engineering Deployment*, CRC Press, Boca Raton, FL, pp. 17–25, 2000. With permission.)

books, we can share in the wisdom of others and thus develop with others a common perspective (not necessarily a shared belief). The computer is rapidly taking over this role from books for society, and it should move into this role for enterprises and programs.

The enterprise knowledge base is some subset of mankind's knowledge base focused on the enterprise's product line and customer base. This knowledge is retained in the minds of the persons working for the enterprise, but it should also be captured in a documented form available to all. The program knowledge base is initially generated through the assignment of personnel, those temporarily not employed on other programs as well as people from other programs. These people carry with them their knowledge about their home functional department's charter and how to do related work.

This enterprise knowledge base consists of two kinds of information, what-to-do knowledge and how-to-do knowledge. Both these kinds of knowledge at some level of indenture should be captured in written form and reside on the computer network. In the case of a specific discipline, this may amount to a checklist of things to do on a program and the sequence in which they should be done to satisfy the what-to-do knowledge. For each item, we also need to know how to do the related work. This may be in the form of an in-house manual; reference to a textbook or standard created by others; or reference to a course available in the company, from a local university, or otherwise accessible.

All of this information should be accessible from everyone's desktop. Specialists who have not yet mastered their specialty can move toward competency faster than they might without this information being easily available to them. Those who would manage the work of others can find out what all of these people are doing in terms of work products created at a time of their own choosing. System engineers will become better informed about how the various specialists contribute to the increasing store of knowledge on programs. The whole enterprise and its subsets, called programs, will achieve a shared vision that helps all of the participants to focus on achieving the enterprise vision in their work on each program.

In a military unit, the members are bound together by a sense of history of that service and unit, a tradition of selfless service, and the knowledge that failures will bring them down in the esteem of their fellows and may lead to punishment and failure as well. They have a shared vision that brings into focus what their purpose is and how to attain it. As a result, we find it possible to cause a group of otherwise sane persons to form up into a unit and march in cadence in response to their leader's commands. These same people will move into harm's way with energy, knowing fully well the personal dangers involved.

An enterprise should seek this same kind of dedicated service in its people, but cannot generally appeal to the same forces that shape the ideal Marine infantryman or an offensive tackle on a pro football team. A common vision shared between management and everyone else, a sense of common justice, a reward system responding to both individual and team efforts, enduring top management support for the current enterprise identity and methods to continuously improve it (rather than a temporary leap to each new fad), and a means to share the fruits of our specialized knowledge without fear of retribution will go a long way toward providing an enterprise with people who will repeatedly march into hell's furnace with a proposal or program manager. The fine tradition of a beer party from time to time might not hurt, though it may be unfashionable.

This bond can be very strong in both directions. As an example, a fine gentleman, Erich Oemcke was the engineering vice president at Teledyne Ryan Aeronautical for a period in the 1980s. He personally led three proposal efforts in a row, all of which the company lost. After the third loss, Erich gathered together all of the people in the department and told them that he could not expect them to follow him on another crusade as selflessly as they had on the previous three and, as a result, he was resigning from his position and the company. A lesser man would have blamed everyone else whether valid or not.

2.8 The system engineering art not yet fulfilled

The subject of this chapter has been people, how they may be organized, and how we can encourage them to communicate. System engineering is rooted in the reality that none of us can know everything because man has been so successful in exposing new knowledge and making it possible to pass it on from generation to generation. The available knowledge is more expansive than individual human knowledge capacity. System engineers practice their trade at the boundary location between different knowledge, product, and process domains. In the latter half of the twentieth century, man invented another form of knowledge retention, called a computer, and extended its reach through networks to form extensive tracts of knowledge.

At the time this portion of this book was being written in 2009, we had not perfected the ideal relationships between humans and machines in the retention and sharing of knowledge in the development of solutions to satisfy our needs. At some time in the opening period of the twenty-first century, a better relationship between the people and our machines will be realized, leading to significant improvements in our ability to solve very difficult problems dealing with unfulfilled needs and the development of systems to satisfy these needs.

chapter three

The fundamentals

3.1 Product system

A system is a collection of entities that interact to achieve a purpose, goal, or function. There exist natural and man-made systems. Some examples of natural systems are the solar system, man's circulatory system, and the Mississippi River system. The latter, of course, is a highly man-modified system, including dams, levees, and dredging. This book and the other three books listed in Table 1.1 discuss man-made systems. Man-made systems are created to solve problems observed by man. The problem may be relatively simple, generally leading to a simple system, or very complex, most often leading to a very complex system. The ideal is to create the simplest system possible to solve a problem—composed of the fewest entities with the most straightforward and simple set of relationships.

We may think of man-made systems as solutions to problems. At the first instance when we become aware of a problem that ends up being solved by a man-made system, we know relatively little about the problem and its solution. In order to solve the problem, we must accumulate information related to it, apply appropriate knowledge and thinking to this information, and make decisions about choices exposed. If the problem is simple, one person unaided may be able to apply their own knowledge to evolve a solution focused on a particular system synthesis. If the problem is complex, it will be impossible for any one person to solve it because of a knowledge deficit problem explained in the previous chapter. The solution of a complex problem will require two or more people, each with different specialized knowledge that collectively covers the complete knowledge space needed for the solution.

Man-made systems could be created by the person or organization that identifies the problem and must use the system that solves the problem. However, enterprises have evolved that specialize in creating systems for customers who identify a need and pay to have a solution developed and delivered. This system development book and the other three books listed in Table 1.1 are intended to apply to these enterprises that create systems as a business, providing methods for efficiently doing so. These systems are the products of these enterprises, so we call them product systems.

It is conceivable that a system could consist entirely of a process with no material content other than a human being performing the process steps so as to solve the problem originally perceived. This kind of system is not the principal target of these books, but will yield to methods included for developing process system elements keyed to their corresponding product systems. A product system once delivered will have to be supported by a process system involving logistics support, operation, maintenance, modification, and disposal processes.

Over the past 50 years, the evidence seems to support the idea that a better solution occurs when the product system and its corresponding process are jointly, simultaneously, or concurrently developed. There was a time when product systems were created prior to the development of the corresponding processes, leading to serious supportability problems. Today, an enterprise that develops product system solutions would do well to follow a concurrent approach, simply because it results in the lowest life cycle cost and the most effective solution.

3.2 Creator process system

Given that product systems are created by enterprises, these enterprises must do so in accordance with some kind of process. This implies that the product system is created in accordance with a carefully devised method following a planned sequence of actions. This need not be the case. It is possible for one or more individuals to join together for the purpose of solving complex problems with no plan in mind. This approach is often called the ad hoc approach. Disregarding the cost and time required to solve a problem, this approach can be very enjoyable for those employed in it. The problem is that it will almost always cost more, take longer, and result in a less satisfactory solution than an organized approach that has been carefully thought through and planned.

An enterprise that creates product systems has all of the characteristics of a system. It consists of things and people that interact to satisfy a goal. It has both process and material aspects. Its people and the organization they become a part of are the principal elements in the entity structure of this process system, but it also includes the facilities, tools, computers, software, and other resources these people bring to bear on the problems they confront.

If an enterprise is formed for no other purpose than to solve one problem and at the conclusion of that development activity the enterprise will be disbanded, it probably does not make a lot of sense for that enterprise to create an enduring process system other than the plan for the one specific product system that it is responsible for developing. If it is intended that an enterprise should survive indefinitely and develop more than one product system, it may make sense for that enterprise to develop a generic or common process system.

The alternative is for the enterprise to treat every product system development activity as an independent event failing to take advantage of lessons learned from prior development experiences. This kind of enterprise will probably not improve its product system development capability over time. It will probably have an improvement curve like the one in Figure 3.1a rather than that in Figure 3.1b. The enterprise would be better served to treat itself as a goal-striving system where the goal is captured in a vision statement. This vision statement might involve making profit, satisfying customer needs, or producing the best widgets in the world. It should be focused on a particular product line (because one cannot excel at doing everything) and customer base (for the same reason).

Any enduring enterprise should possess a single common development process that is improved based on lessons learned from all of its programs causing a continuously improving capability. This book is intended to show how this kind of capability can be put in place in an enterprise.

The vision statement is one of the two major drivers for the work that an enterprise performs, as shown in Figure 3.2. It is based on the needs of the enterprise. The other driver is a series of customer needs for the product systems that flow through the enterprise. The system development process implemented by the enterprise must satisfy both of these needs. Generally, these are not mutually exclusive goals. An efficient process that produces products that customers value will most often be consistent with the vision statement and produce products that satisfy customer needs.

The enterprise forms a program for each system it must develop. Each program must draw resources from the enterprise, and these resources

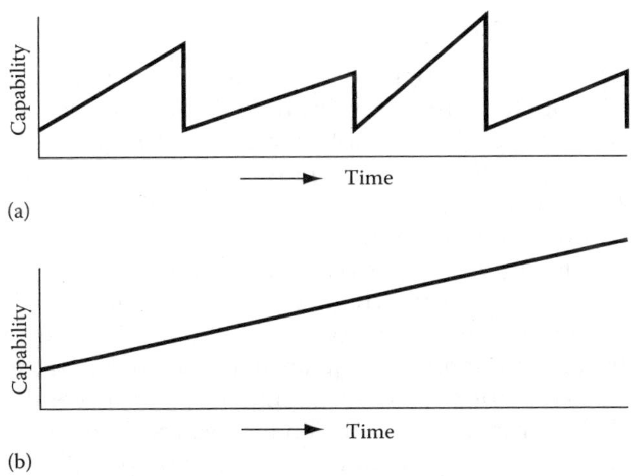

Figure 3.1 Enterprise improvement curve choices: (a) every program on their own and (b) common process under GSD.

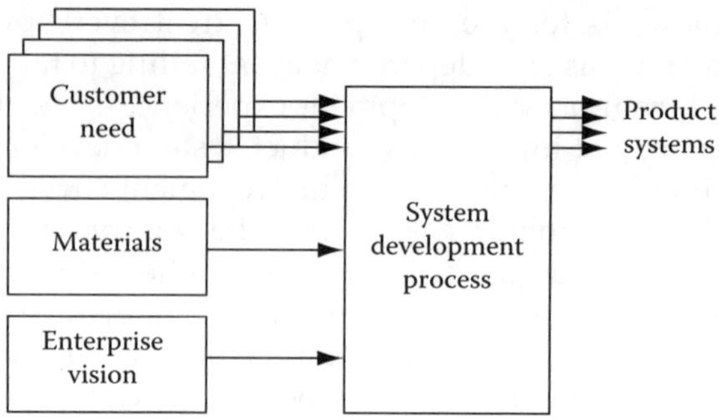

Figure 3.2 System development process drivers.

are finite. These resources can be most efficiently used on programs if they are standard and if all of the programs are applying essentially the same process. The problem is to craft a generic process that can be applied to all enterprise programs with little need to adjust or tailor the common process for each program. This generic process must include elements that are standard for all programs and avoid those that would have to be modified for each unique situation, leaving these for program unique definition. If the generic model must be very intensely modified for each program, it is not a standard. If it has such small substance that it must be extensively added to for each program, then it is of small value. There is a central space between these extremes where the generic model can be useful to provide an infrastructure for both functional department development and program planning development.

3.3 Grand systems

In many enterprises, the program product development process is crafted anew for each program with relatively little experience carryover from other programs or from a generic identity for enterprise programs. This book and *System Requirements Analysis* (Grady, 2005), *System Synthesis: Product and Process Design* (Grady, in press), and *System Verification* (Grady, 2007) are based on the premise that the most advantageous enterprise infrastructure is constructed by knitting together the relationships between several entities. In so doing, we create what the author refers to as a grand system, consisting of products and their process created concurrently in enterprise programs that apply a standard template in the creation of these programs. A program is an enterprise process for creating a product and process system. It should be formed out of a fusion between a generic enterprise process (which is constantly being improved based on program experiences) and a set of customer needs (one for each program) expanded through structured analysis to identify the major elements in the product system around which the generic planning information is fitted.

Product systems include material product and process elements related to the use of that product system once delivered. These process elements must be developed concurrently with the material elements in an integrated fashion with the whole optimized, rather than only the material elements optimized. But even this very noteworthy achievement is insufficient. Product and process systems are created by enterprises that are themselves product and process systems, which ideally should also be optimized for system creation in a particular product line and customer base. Finally, these pairs of systems should be compatible, and now we arrive at the grand system level formed by an enterprise system that creates multiple product and process systems efficiently and, generally, two or more simultaneously in different uncoordinated stages of development. In the process of doing so, they must integrate and optimize about the enterprise level and, in doing this well, are considered by the author to be performing GSD.

3.4 *Introduction to system development*

System development is the art and science of creating product and process systems. It includes the transformation of customer needs into a preferred system solution description in some tangible format and the conversion of that description into a physical reality suitable for its intended purpose. Enterprise systems form programs to accomplish this transform. These programs are staffed by enterprise personnel derived from, in the author's view, functional departments, each of which has been identified with a particular charter determined through structured analysis of the enterprise vision. As these department personnel are brought into the program, they bring with them a string of capabilities including good tools and best practices that are coordinated with the skill sets and knowledge of the personnel.

The process of creating product and process systems passes through four significant stages illustrated in Figure 3.3, starting with product and process definition (Define system, F41) and ending with Sustain system, F48, and Use system (F47). The key elements of the development activity relative to the content of this book and the other three books are accomplished in four major work elements:

a. Function F41. Define the problem through requirements analysis yielding a series of specifications.
b. Function F42. Solve the problem through synthesis of the contents of the specifications applying the latest and best technology, processes, knowledge, and skills. This includes integration and optimization of product and process entities throughout the product and process

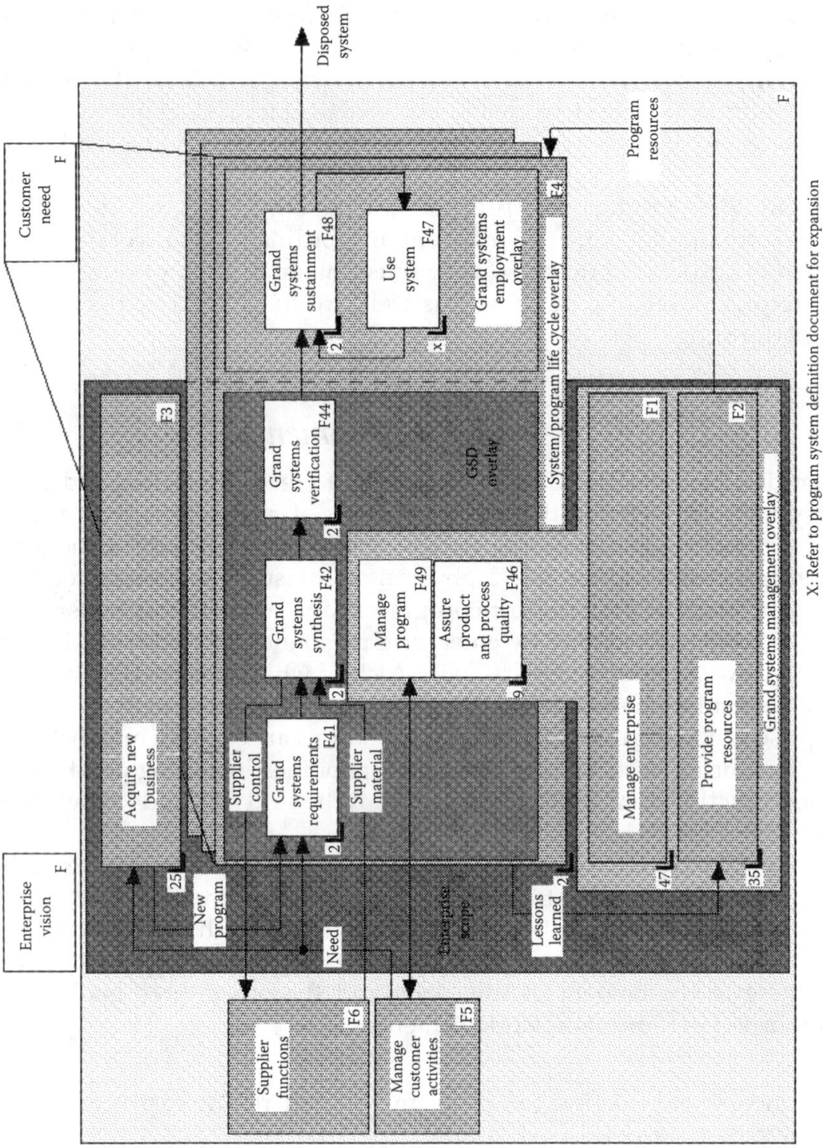

Figure 3.3 Life cycle master flow diagram.

hierarchy up to the system level. This includes design, material acquisition, and manufacturing.

c. Function F44. Prove that the resultant product satisfies the problem definition through verification involving testing, analysis, and other activities.

d. Functions F1, F2, F46, and F49. All of the work described in a through c must take place within the context of a sound technical and program management infrastructure.

There exists a final stage of the system life cycle where the system is employed, involving the use and sustainment of the system over its useful life. These books deal with development and does not extend into employment other than to understand it as a basis for the requirements appropriate to the entities to be developed.

3.5 Introduction to program management

Program management is a management science that defines program plans and controls their implementation to achieve predetermined goals. The person ultimately responsible for a program is called a program manager in this book. This person controls the resources the enterprise assigns to the program and is responsible for interaction with the customer as needed. The program manager is charged with delivering the material and processes as covered in a contract or other agreement at the specified time and within the planned cost figures. Also, the product and process delivered must have the performance features called for in the specifications approved for the program.

Program managers accomplish their assigned duties through the efforts of other people whom they manage. The overall program is partitioned into some number of elements oriented toward the product structure, determined early in the program through structured analysis. Teams are formed around these elements, and these teams are managed by team leaders in accordance with planning data oriented around these same elements in terms of a development task description, a cost estimate, and a schedule compatible with the technical element requirements for that item. Each top-level team leader reports to the program manager (lower-tier team leaders may report to higher-tier team leaders) and focuses on achieving their planning milestones. The program manager focuses attention on the aggregate plan and results, shifting priorities where needed to optimize the whole development effort.

3.6 Introduction to functional management

Programs need resources, and they should be able to depend on a functional department structure within their enterprise to provide these resources. Yes, a matrix management structure is encouraged for an enterprise with multiple customers who want complex problems solved for them. The functional department structure should be providing all enterprise programs with good people, good tools, and good practices, all coordinated across the enterprise. The functional management should not be given project management roles, as has been done in many companies with the program work done within the confines of the functional department space under the watchful eye of the functional manager. The functional manager is a resource expert who should be focused on providing the enterprise with the very best resources possible coordinated with other functional departments doing the same, each in their own area of responsibility.

3.7 Introduction to system engineering

When the problem that a program must solve is very complex, it is necessary to understand that complexity and to break that complexity into related parts such that the solutions of the small problems thus identified will lead to a solution to the whole, complex problem. Human beings have a limit to the problem density they can handle in terms of depth and breadth. One can understand a very broad problem so long as it is not necessary to master it in much depth. Similarly, one can master a lot of details so long as it is not necessary to master many other matters peripheral to that one detail.

We are all specialists, as we will discuss in Part two of this book. This is so for some very understandable reasons, and there seems no alternative to this fact of life. Programs will be staffed by specialists, and there is no alternative. Most of the specialists assigned will be deeply qualified in narrow fields, like structural analysis, reliability, and mechanical design, as a function of the applied technology an enterprise must bring to bear on the contracts it commonly engages in. Programs need these people to develop the detailed knowledge required to solve the complex problems involved in system development.

There is one other kind of specialist that programs need, and that is a generalist or system engineer. System engineers are skilled in interacting across the boundary conditions that the other specialists respect whether they be in the product system or the program system involving processes and responsible teams. System engineers are specialized in understanding complexity and in communicating with other more narrowly specialized human beings about complexity in ways that they are capable of understanding things they need to know to be successful in their own field.

The system engineering process respects three fundamental steps in solving problems: define the problem, solve the problem, and prove that your solution is a good solution to the original problem. The first is consummated in specifications. The second is accomplished through integration and optimization during the synthesis process (product design, material acquisition, and manufacturing). The third is accomplished through detailed plans for verification of the product against the specifications. Management support techniques are included within each of these three spaces to help present accurate information to the program manager and staff to support good decisions that encourage a successful program.

3.8 Standards

The system engineering discipline description transitioned through several military and industry standards over the 1990s. At the beginning of this decade, MIL-STD-499A had been in force for some time and had become a normal definition of the systems approach in industry serving the U.S. Department of Defense (DoD) and its suppliers. There was no commercial standard nor was there a great deal of interest in the process as a formal approach to product development in many commercial firms. Efforts were underway to replace 499A with 499B, but these efforts happened to coincide with a conclusion within DoD that it should work toward applying commercial standards wherever possible. About the time 499B was ready for approval, DoD had canceled many management and manufacturing standards and was rapidly converting to performance specifications where suppliers were told what DoD needed and not how to do the related work. In this environment, 499B was not approved within DoD and the U.S. DoD now has no system engineering standard of its own.

The Electronic Industries Alliance (EIA) took over and edited MIL-STD-499B slightly and released it as an interim standard EIA 632. Subsequently, EIA 632 was completely redone recognizing 33 requirements for a sound system engineering process. In parallel with this activity, IEEE released IEEE 1220 that was intended as a commercial standard. It was assumed that EIA 632 would evolve as an aerospace standard and IEEE 1220 would cover the commercial marketplace. Surprisingly, many aerospace firms elected to apply IEEE 1220. In the late 1990s, it appeared that U.S. military/aerospace and commercial firms were moving toward acceptance of a common development standard and the world toward an International Standards Organization (ISO) standard for system engineering or development. It should be noted that as standards are extended to cover a broader range of activity, they tend to move away from capturing best practices and toward reflecting a least common denominator.

ANSI/EIA 632 and another EIA standard 731, in the author's view, offered industry an ideal system engineering standards set. The final version of ANSI/EIA 632 was carefully crafted to define the essential characteristics of a sound systems approach with no prescriptive content. It is a process specification with 33 requirements. The enterprise could use this standard as one basis for designing a process for developing a product. It could then apply the rest of the systems approach to verify that its programs implement its process faithfully. EIA 731 provides a capability maturity model for system engineering, and thus can be used as the verification medium. These standards will continue to evolve, but the EIA model, intended or not, providing a process specification and the verification machinery to determine if a particular program process design complies with a process specification, is a good model. It happened that 731 was not keyed to 632; therefore, one could apply ANSI/EIA 632, IEEE 1220, or any other standard as the basis for their process and assess its program implementation using 731.

As the work on this book was completed in 2007, the work in several societies had produced an integrated capability maturity model called CMMI and ISO/IEC 15288, Life Cycle Management—System Life Cycle Processes. Standards tend toward increasingly broader coverage and inevitably narrower coverage in specific areas of interest.

3.9 Something more than management competence

In the author's opinion, there are some fundamentals that a development organization must master to have a chance of being a grand system developer. The development organization should

1. Employ a common process derived from the enterprise vision statement using a modeling approach
2. Apply a three-step process within one's common process of defining the problem, synthesizing a solution, and verifying that the solution satisfies the problem statement
3. Be organized into a functional department structure oriented around the kinds of knowledge needed derived from common process functionality and product line needs and explained in a set of functional department manuals mapped to tailored customer-preferred standards
4. Possess common process planning documentation derived from common process functions concatenated with the performing functional department identification
5. Employ an organized means to transform its common process into a program-specific plan using the IMP-IMS format ordered with planning strings composed of the common process strings concatenated with program planning strings formed by program, phase, program event, and product entity identifier (WBS)

6. Employ cross-functional teams staffed from functional departments in accordance with related planning strings in the IMP, oriented about the product entity structure, and led by team leaders reporting to the program manager
7. Recognize that one must apply integration and optimization wherever partitioning has been applied to size a problem to human dimensions and implement a program integration team (PIT) to accomplish that work at the top program/system level and an enterprise integration team (EIT) to accomplish that work at the enterprise level

These fundamentals are fully covered in this book. But there is another characteristic that the enterprise top management must be absolutely possessed with, which is not covered in this book. This is an intensity of focus toward a goal of continuous improvement and an ability to motivate employees to excel often only seen in repeating NFL Super Bowl winning teams or successful Japanese automobile firms. Despite the evidence provided by the relative success of Japanese auto firms compared to American auto firms in the 1980s through the 1990s at least, American management practices appear to be focused on the near-term growth rather than on a long-term growth. Managers at all levels, in any country for that matter, have an obligation to make what they are responsible for better over time and to leave a job with the enterprise possessing a better capability than when they started their tour. Management jobs should not be looked upon only as opportunities for a manager to climb another rung on the ladder, but as an opportunity to improve the organization managed. There are many books that focus on this problem, but few companies, executives, and management schools seem to have implemented the steps necessary to encourage American industry to excel over the long haul. The resources covered in this book should be thought of as providing a platform on which great managers can accomplish their work. Great managers may have a degree of success without this platform, and mediocre managers may get by with it, but put the two together and one's competitors will be in very great difficulty.

3.10 Changing paradigms

This book and the other three books all listed in Table 1.1 were all constructed based on the experiences of one system engineer collected over a period of 10 years as a U.S. Marine, 30 years as an employee of four system development firms, and 15 years as a consultant and trainer in system engineering over the span of 1954 through 2010. The author believes that the content describes an effective way for an enterprise to develop good systems, but it should be recognized that the content is a snapshot in time assembled by the author. Never over this 56 year span of time did the author ever experience in a company the union of all these ideas effecting work performance and product consequences.

This period began with the common use of the slide rule by engineers and is concluded with the application of networks of computers and supporting applications providing a tremendously useful supplement to the humans forming teams on programs. Over this time span, the machinery changed, and the methods of organizing people to accomplish development activity, and the methods applied by these people all evolved in an attempt by companies to reduce the cost and time of development, produce a better product, and encourage an environment where many people, each possessing some subset of man's accumulated knowledge, could cooperate toward a goal much larger than any one of them could master.

Many initiatives were cast upon the fabric of the system development process in these enterprises. In some cases, they stuck and were integrated into the preferred process, and in other cases, they were cast aside. At any one point in time, it might have been possible to collect together what industry in the aggregate might have been able to agree described the preferred process, but it would likely have been necessary to state the degree of agreement about this process in statistical terms. So it is today and will continue to be.

This book and the other three books describe a process to which the author would apply the term traditional system development process that is thought to be effective in the development of any system in the early days of the twenty-first century on planet Earth. Will man improve upon this picture over time? Certainly. Some of these improvements will be evolutionary, and a few may be of a revolutionary nature. But it is hoped that those who would improve this process would do so from a perspective of knowledge of the past. The author of these books has attempted to describe a particular system development paradigm applied within system development enterprises. The effect on the targets of these many development efforts has had the appearance of a canister rather than a solid shot from a well-aimed cannon. Those that have done it well through effective management of teams of people well schooled in their domains and experienced in the methods employed by the enterprise have, hopefully, profited well from their customers and have survived to continue their efforts while working toward a new plateau.

part two

System development process

chapter four

Process fundamentals*

4.1 The ultimate process step—The enterprise vision

When we develop unprecedented product systems, we should begin work from the grandest vision of the system description, the customer need. The reason for this is that it is the way users of systems become aware that they are not currently capable of satisfying some unfulfilled need. We commonly do not think of the details and then synthesize them into a whole. No, we start with a grand vision and expand the details working from the known toward the unknown, from the simple to the complex. The customer need is a simple statement telling what must be accomplished by some system yet to be determined. Through a structured analysis process, we can expose increasingly greater detail about the system to be, reaching an understanding of the problem to be solved captured in specifications.

This process can also be applied to systems that already exist but must be modified to satisfy a significantly changed need. In this case, the physical system already exists, so the product entity structure is known. It is still helpful to understand what the need is and to expand it into a functional view of the details. Just as in the case of an unprecedented system, we can allocate the exposed functionality to things in the existing product entity structure, gaining insight into things that the existing system cannot accomplish. As a result, we can make decisions about which elements in the existing product entity structure are no longer useful, which are okay as is, and which need to be modified in particular ways. In reengineering a system, we can still work from the needed functionality in accordance with Louis Sullivan's ideas, but we need to recognize the realities of the existing product entity structure and look for the least disruptive way of achieving the functionality differential between the as-is and the new needs.

This product system development logic also works for other kinds of systems, process systems, that are used to create product systems for several different customers at the same time. The ultimate functionality is a little different, however. Product systems are created by enterprises that function for various reasons. The management of an enterprise may

* The material in this chapter is from Grady, J., *System Engineering Deployment*, CRC Press, Boca Raton, FL, Chapter 4, 2000. Used with permission.

perceive that its purpose is to achieve a profit, provide products of value to customers, increase stock price, or capture market share, for example. We will refer to this ultimate enterprise process system functionality as the enterprise vision statement, akin to the product system customer need statement. This statement must be generic in nature in that the single enterprise may, at any one time, be involved in the development of seven different product systems.

Commonly, the management of multiple program enterprises permits each program manager to invent their own new process with little interference, leading to little chance of progressive aggregate enterprise improvement over time based on lessons learned. It is important that a multiple-program enterprise have a generic or common process that can be adjusted to satisfy program-peculiar needs. This generic process should be derived from the vision statement, just as we derive product system functionality from the customer need. Figure 4.1 illustrates how the customer need and the enterprise vision statements apply to a generic process captured in a life cycle flow diagram for the enterprise. This process system applies to every product system that the enterprise will ever create. The enterprise vision is the ultimate process functionality expanded to identify four top-level process system functions, each of which must be further decomposed and defined, as suggested in the figure. These four fundamental functions correlate with this book and the other three books listed in Table 1.1.

These two ultimate functionalities must coexist in the enterprise. While the enterprise attempts to satisfy multiple customer needs through several programs in different stages, it must also seek to satisfy its own vision statement, and these may not all be mutually consistent. This is the reason why this book encourages an enterprise integration team—to adjudicate between all these different expectations and encourage the best approach in the aggregate referred to as GSD.

This book employs functional flow diagramming as the means to express the enterprise process. We could have alternatively chosen to apply the same model within the context of UML/SysML in that these models now permit the activity diagram to be drawn horizontally or vertically. We could also alternatively have used a process for system architecture and requirements engineering (PSARE), formally referred to as the Hatley Pirbhai model, that uses bubbles to represent functionality rather than blocks. Some would encourage the use of an IDEF-0 diagram or an enhanced functional flow block diagram (EFFBD). As we will see, we will actually be using a combination of EFFBD and IDEF-0, illustrated by a simple functional flow diagram and backed up with a set of matrices that provides the complete story without the profusion of lines that would result without the supplementary matrix data.

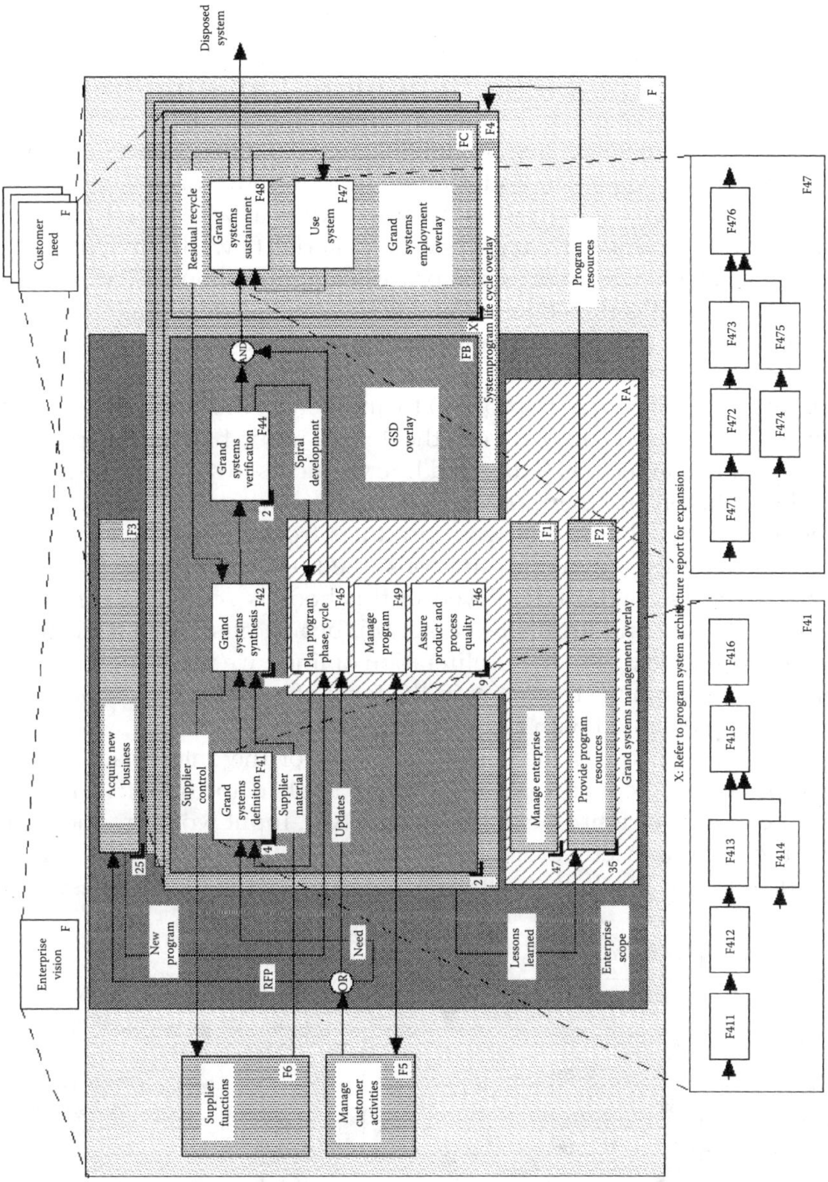

Figure 4.1 Vision and need statement relationships.

4.2 What exactly is flowing?

We describe a process in terms of a series of function statements, illustrated using blocks connected by directed line segments. In reality, we use a process diagram with the understanding that the process steps are in one-to-one correspondence with the functions, so as to be able to employ a single functional flow diagram for the whole life cycle involving development and employment. In a functional flow diagram, we commonly do not think of any material objects flowing from one function to another, except that time has passed and that we need to move on to accomplish some other function. In a process diagram, we deal with a real physical situation, where information, energy, or material flows from activity to activity. The blocks on our common process diagram should be viewed in this light. So, what is then flowing in our common process diagram?

The common process diagram consists of blocks that represent program work that produces work products, and it is these work products that flow from one function (process) to another. In this book, we refer to the inter-task relationships between the flow diagram blocks in a split format. These are connected by directed line segments to show the sequence of implementation and they are imagined to be connected by directed line segments to show the flow of work products from a source activity to a destination activity, where they are consumed or used in some way. The common process diagram employed in this book is actually a combination of an EFFBD and an IDEF-0 diagram, but because such a diagram would be unusable due to the proliferation of lines, we use a subset of the complete diagrammatic treatment in the form of a simple functional flow diagram supplemented with a set of five matrices.

Figure 4.2 illustrates the complete nature of the enterprise process model employed. Sequence-related flow is applied to the left side of a block and exits the right side of the block (FX-1). Input work products are applied to the lower left corner of the task block, and output work products exit the task block at the top right corner of the block. These two

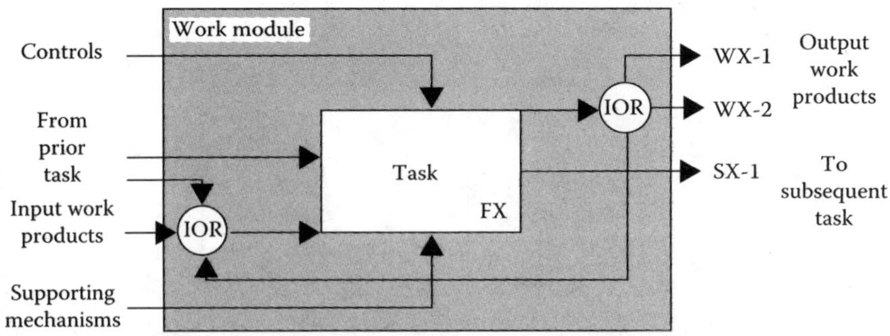

Figure 4.2 Complete process flow block.

line sets cover the EFFBD aspect of the diagram. A directed line segment applied to the top of the block is used to identify controls applied to the task, as in the example of a task procedure or regulation. Inputs applied to the lower edge of the block correspond to supporting mechanisms such as tools, facilities, or utilities. These two inputs comprise the added IDEF-0 aspect of the flow diagram.

While it would be possible to apply the diagrammatic features shown in Figure 4.2 in actually building our process diagram, the result would be too visually complex to be useful. It is suggested that we partition all of the features of the diagram into several modeling artifacts. A simple functional flow diagram revealing only the blocks and sequence lines is a subset of the kind of diagram shown in Figure 4.1 and is recommended as the model for our common process diagram.

The work product flow aspects can be effectively shown in a pair of matrices. Figure 4.3 illustrates a work product dictionary structure defining the work products. The suggested method of identifying the work products uniquely is to replace the source task modeling identification (MID) letter F with W and attach a specific base 60 character to the end following a dash assigned in an ascending order for as many as 60 work products for a given task.

Figure 4.4 shows a work product transfer matrix that reveals a source and a destination for the flow of all work products. In this matrix, a given product entity will come from a single source, but may be applied to many tasks as an input, as is obvious in the case of a specification. Every work product should have a single source and it must have at least

MID	SOURCE	TYPE	NAME	DESCRIPTION
W4142-1	F4142	Document	Specification	Document that defines requirements for an entity.
W4211-1	F4211	Verbal	Engineering ideas	Conversation takes place between electrical engineer and mechanical engineer about electrical connector type.
W4334-1	F4334	Product	End item	Physical product entity delivered to customer.

Figure 4.3 Sample work product dictionary.

MID	SOURCE	DESTINATION	APPLICATION
W4142-1	F4141	F4231	
W4211-1	F4211	F4212	
W541-2	F541	F4621	

Figure 4.4 Sample work product transfer matrix.

one destination. This is a fundamental characteristic of a lean process, of course. If a work product has no user, it should be deleted. The application field could include the way the work product will be employed in the destination task. The reader can, no doubt, think of other fields that may be useful in such a matrix.

Figure 4.5 illustrates a complete common process diagram if it were drawn with all of the lines discussed in this chapter for the four top-level functions shown in Figure 4.1. It should be pointed out that in order to be fully consistent, the lines associated with the control and supporting mechanism lines should reflect the same potential linkages shown for the sequence and work product flows. In most cases, these control and supporting mechanism lines will come from other tasks in the process but, in this figure, these are all shown originating from outside the enterprise. So, if these lines were drawn with the same assumptions applied to the sequence and work products (eight more OR symbols and the related lines), the diagram would be even more complex and unusable.

In order to simplify the diagrams, control and supporting mechanisms matrices are used to also supplement the simple flow diagram, leaving us with a simple functional flow diagram connected by the most significant sequence oriented directed line segments. These two added matrices list the control and supporting mechanisms inputs linked to the name, description, and task source of the input. Thus, the overall expansion of the enterprise vision becomes the relatively simple construct shown in Figure 4.1 that, combined with the matrices, is actually a powerful combination of an IDEF-0 diagram married to an EFFBD. Combining this data with a detailed description of each task in a function (or task) dictionary, one has a very comprehensive definition of the system development process the enterprise intends to apply to its programs.

The key piece of information to be linked to the tasks in the function dictionary is the responsible functional department that must supply all programs with the resources needed to be successful in performing the task in the form of people who know what they are doing relative to task goals, good practices, and good tools. The task descriptions should, therefore, include information relative to all of these elements. Where a task requires people from multiple departments working on a team, the matrices will show multiple listings. The function dictionary or enterprise RAS will tell which of the departments should be asked to lead the task.

An example of how all of the process defining data can be brought together and published, is provided by the JOG System Engineering Grand System Management course student manual Division 3, Exhibit H (enterprise architecture report).

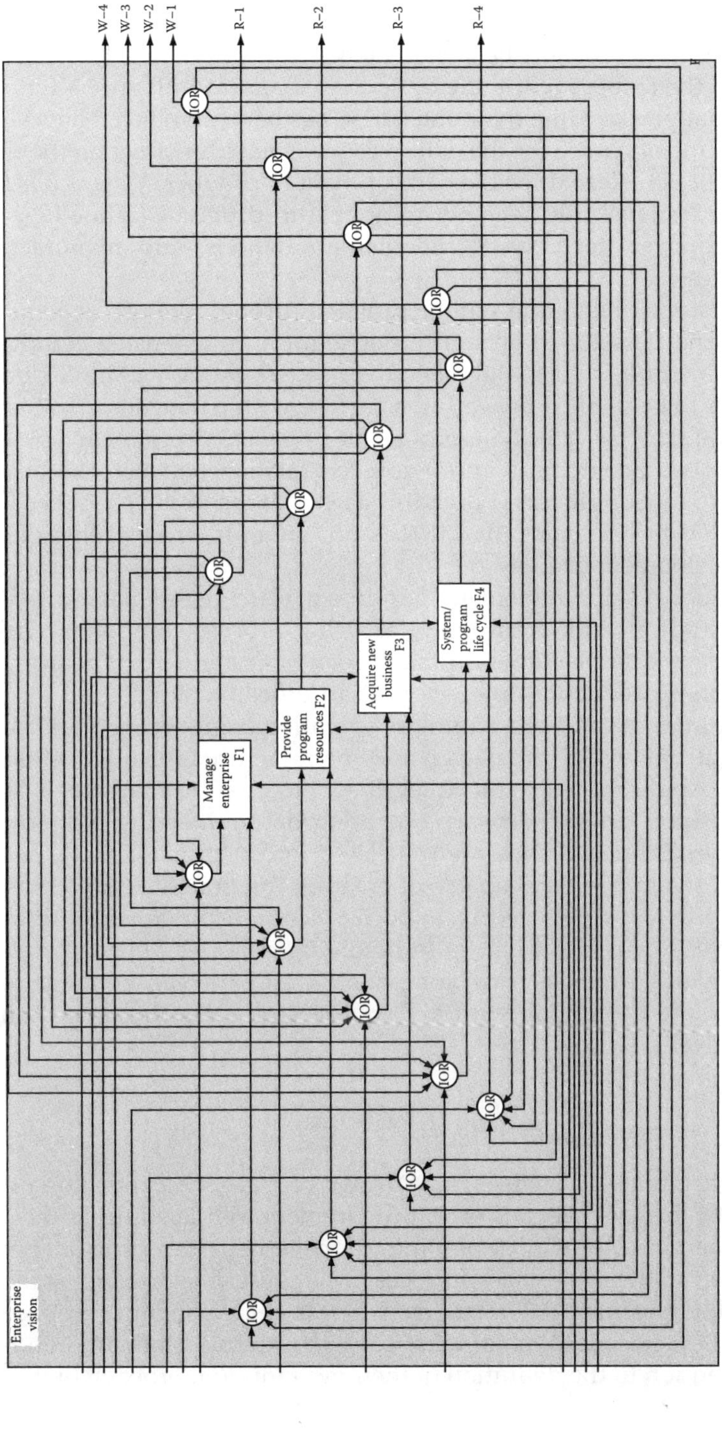

Figure 4.5 Enterprise common process diagram fully visually illustrated.

4.3 Product line effects

The reader will probably be able to identify his or her development experience in the context of the life cycle flow diagram in Figure 4.1, no matter in what product line their enterprise has been involved. However, the product line of choice for the enterprise will have an effect on the details of the process selected for the development of product systems. Different kinds of systems will demand an appeal to different technologies and related analysis techniques, models, simulations, and manufacturing techniques.

A customer need that will be satisfied through the use of some kind of seaborne vessel will generally not require an extensive aerodynamics analysis unless very high speed is needed, though a similar process will have to be applied beneath the surface with a noncompressible fluid medium. Likewise, aerodynamics is not a needed discipline in spacecraft development. So, there is an obvious correlation between the need for many different engineering domains as a function of the product concept and related technologies. But, that is not the only product line effect in one's development process.

Product system implementation in software versus hardware determines a radically different manufacturing process component. Instead of metal-working machines on the factory floor and circuit board manufacture on a production line, we find an office facility staffed by many software analysts and programmers operating engineering workstations. Even electronic hardware design and manufacture falls under this same pattern. Application-specific integrated circuits (ASICs) are designed using desktop design software operating on engineering workstations. The resulting lines of code are translated into a format directly compatible with the machinery capable of creating integrated circuit chips. The engineering steps involved in building electronic circuitry in this fashion are very different from those involving old-style "black boxes" full of wiring and large electrical components, and you can no longer easily carry out the design job with a T square and a drafting board and be competitive.

4.4 Customer base effects

The Department of Defense (DoD) evolved a rigorous acquisition process over many years of trial and error procurement. While doing so, they built up a tremendous storehouse of standards covering everything imaginable. Every step in the multiphased process was prescribed in these standards referenced in contractual documents. In this environment, aerospace contractors became adept at compliance with external standards and paid scant attention to the definition of their own internal process. In the early

1990s, the DoD became aware that they were not being well served by some military standards and began a course of action to adjust to commercial standards wherever possible.

The logic was inescapable. The upkeep of 30,000 standards was a drain on funding for development. The government was having trouble creating and maintaining standards in some areas where commercial practices were rapidly changing, like software and electronics, increasingly the key to successful military systems. Industry driven by a commercial marketplace, full of demand for innovative electronics products, was evolving parts that were as good as those called for in military specifications but not specifically identified as acceptable for military program procurement. At one time, the DoD was responsible for purchase of most of the integrated circuit production and could steer the production base at will. In a world where children buy more electronics than Uncle Sam, the DoD lost a lot of their leverage. Thus the move to commercial off-the-shelf designs integrated into military systems and the reliance on commercial standards for manufacturing and management as well as other fields.

In this new world, aerospace enterprises were expected to have a written process description and to actually follow it. This was a significant change from the 1980s when a contractor would have to write plans for implementation on a contract on which they were bidding, but might not be held strictly accountable for following them while under contract. This was especially true in the system engineering domain. Every room in the Pentagon and all of its annexes could likely be wallpapered with the pages of very good system engineering management plans never implemented on programs that won the competition.

A customer may contractually interfere with an enterprise's definition of its internal processes through a contractual imposition of external standards, or an enterprise may be held accountable for following its own internal process definition determined through customer audits of their process. Alternatively, a customer may not express any interest in what happens internally as long as the delivered product satisfies their requirements.

This book will cover a process definition approach that permits an enterprise to follow its own internal process no matter which of these kinds of customers they encounter. The problem is that the enterprise must have a written process description. Here too, this book will help in the creation of one that any customer will find acceptable. Customers that concern themselves about an enterprise's internal process reflect past concerns from programs that yielded bad products late at an overrun cost. It does not take many of these experiences to develop a great interest in your contractor's process in an attempt to foresee risks and avoid them. If you have a sound process and have a way of applying it to programs, you will survive any audit your customer wishes to apply.

Also, over many years, the DoD evolved a multiphased contracting approach that continues to have a great deal of merit. This is a risk-mitigating structure that recognizes that early in a program we do not know very much, certainly not enough upon which to base production decisions. In a phased approach, some work is accomplished to define the product and process followed by a major customer review. Based on the results of this review, decisions are made about the next phase that might include canceling the complete project. It is well known that the biggest life cycle cost in any system occurs in operation and maintenance (O&M), termed "use system" and "sustain system" in this book. The next-biggest cost occurs in commonly recurring production followed by a detailed design in terms of program cost.

Development cost, while it can be large by our individual standards, is generally small relative to the later cost elements noted above. So, where a customer lets a contract for $5 million to study a new program potential, these $5 million spent early may very well save $2 billion in system life cycle cost. Figure 4.6 illustrates this phenomenon showing that if system engineering skill is applied well early on a program, the result should be reduced production and O&M cost, thus significantly reducing total system or life cycle cost.

Many large system acquisition organizations employ a stylized acquisition process. Figure 4.7 illustrates a multiphased structure with DoD and National Aeronautics and Space Administration (NASA) phasing

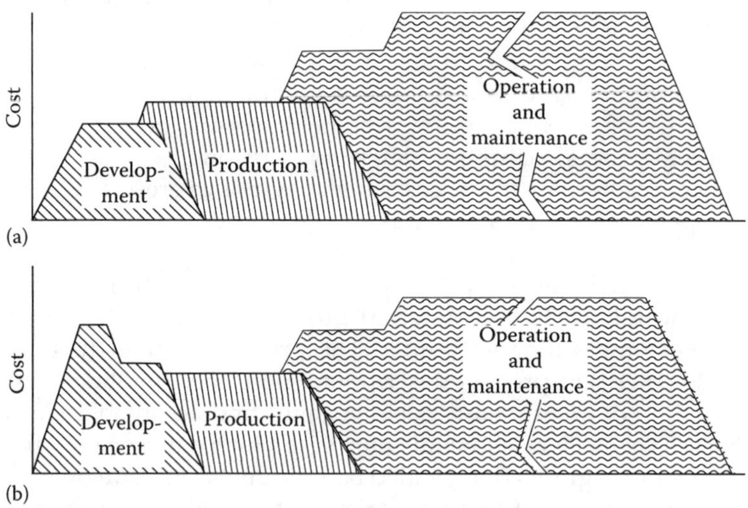

Figure 4.6 Relationship between system engineering and life cycle cost: (a) distribution of cost over the system life cycle and (b) reduced life cycle cost effect from system engineering work.

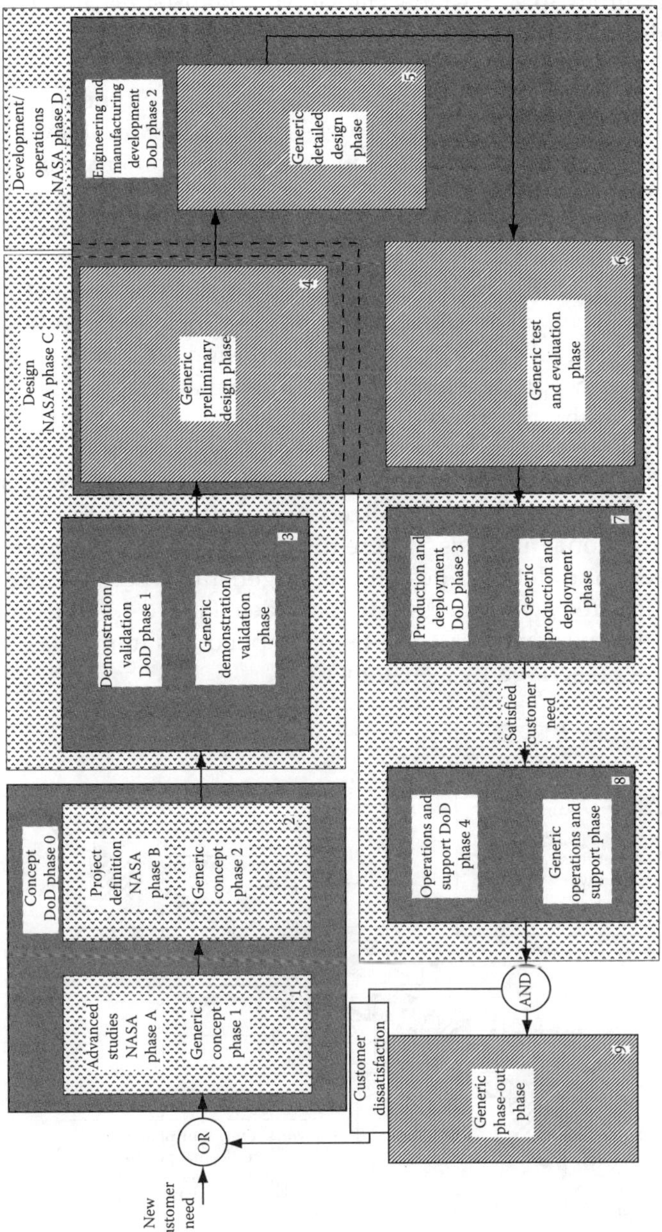

Figure 4.7 Multiphased program structures.

structures common to the decade of the 1980s overlaid. The author built this diagram when he was the manager of systems development at General Dynamics Space Systems Division in the 1980s, because he wanted to assure himself that it would be possible to possess a common process and apply it to multiple customers. The reader can see in Figure 4.7 that if the developer organized a common process focused on the micro phases corresponding to the lowest-level blocks in the diagram, they would be able to tinkertoy them together to satisfy the development process needs of either kind of customer.

When customers apply a multiphased acquisition approach, the enterprise's process description must be able to handle it. The process description encouraged in this book is configured so as to be compatible with any phasing concept. The steps in the process included can be applied cyclically, but generally reflect a three-step design, or synthesis, phasing arrangement: concept development, preliminary design, and detailed design. Commonly, the concept development activity is accomplished concurrently with the early requirements analysis work in a top-down cyclical fashion, where product entities identified through some form of functional analysis are conceptualized and act to refine the continuing lower-tier functional analysis. Where the risks permit, one could shorten this to concept and detailed design but, in general, where the problem is complex, this three-step process is a good idea. Figure 4.8 illustrates how the author sees the generic process being applied to a multiphased acquisition process prescribed by a customer.

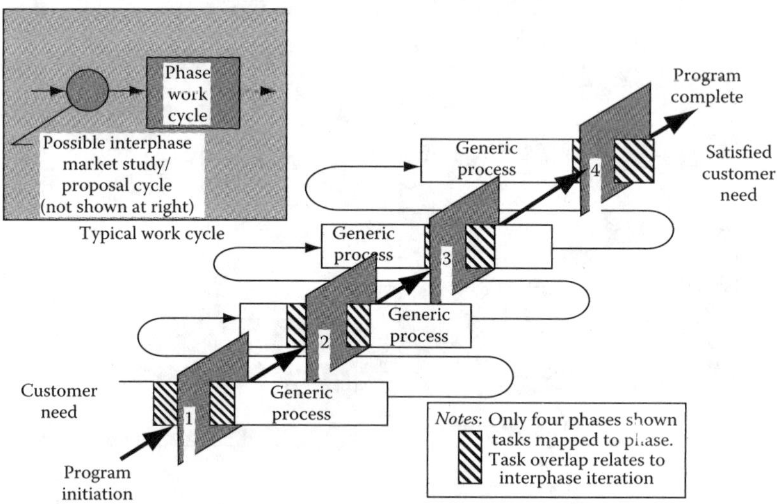

Figure 4.8 Program phasing and generic process relationships. (From Grady, J., *System Engineering Deployment*, CRC Press, Boca Raton, FL, p. 162, 2000. With permission.)

In early program phases, some of the early work in the generic process is accomplished. In the next phase, some of this work must be redone, because in the process of doing the work in the first phase, the program team learns enough to know that the solution could be improved. This is called iteration of the evolving solution. In addition, the program would have applied requirements analysis at the system level in the first phase and will have to apply this same process in the next phase for the next lower tier of the system. This same process continues until every item in the evolving product entity structure has been defined in a specification. The generic requirements analysis process may have to be applied 150 times throughout the whole program, in each case, taking advantage of the previous analytical work products while developing 150 specifications.

This process continues from phase to phase until the product system is delivered at the time the complete generic process has been consumed by the program. What is achieved through program phasing is the progressive reduction in program risk while avoiding the obligation of customer funding until program risks are thoroughly understood and minimized.

4.5 *Specific customer program staging/phasing patterns*

We will look at three current large acquisition agent views of their acquisition process. The reader should note that all of these processes reflect the life cycle that is a significant advance over the earlier views that these three organizations more tightly focused on in the development contract period.

4.5.1 *Department of Defense*

The DoD has long defined very rigorously its system acquisition process. The phasing process in effect in mid-2007 is illustrated in Figure 4.9 with some preferential editing on the part of the author, including denoting the phases with stage numbers, such as S21. These stage numbers will be used later to clearly define all work that must be accomplished on a DoD program. In a later chapter, we will map the generic or common process work to customer program phases, permitting both contractor and acquisition agents to clearly identify their view of work independently.

4.5.2 *Federal Aviation Administration*

The Federal Aviation Administration (FAA) applies a different program staging approach, as illustrated in Figure 4.10. Disposal was added by the author for consistency with other acquisition life cycle models and the JOGSE common process.

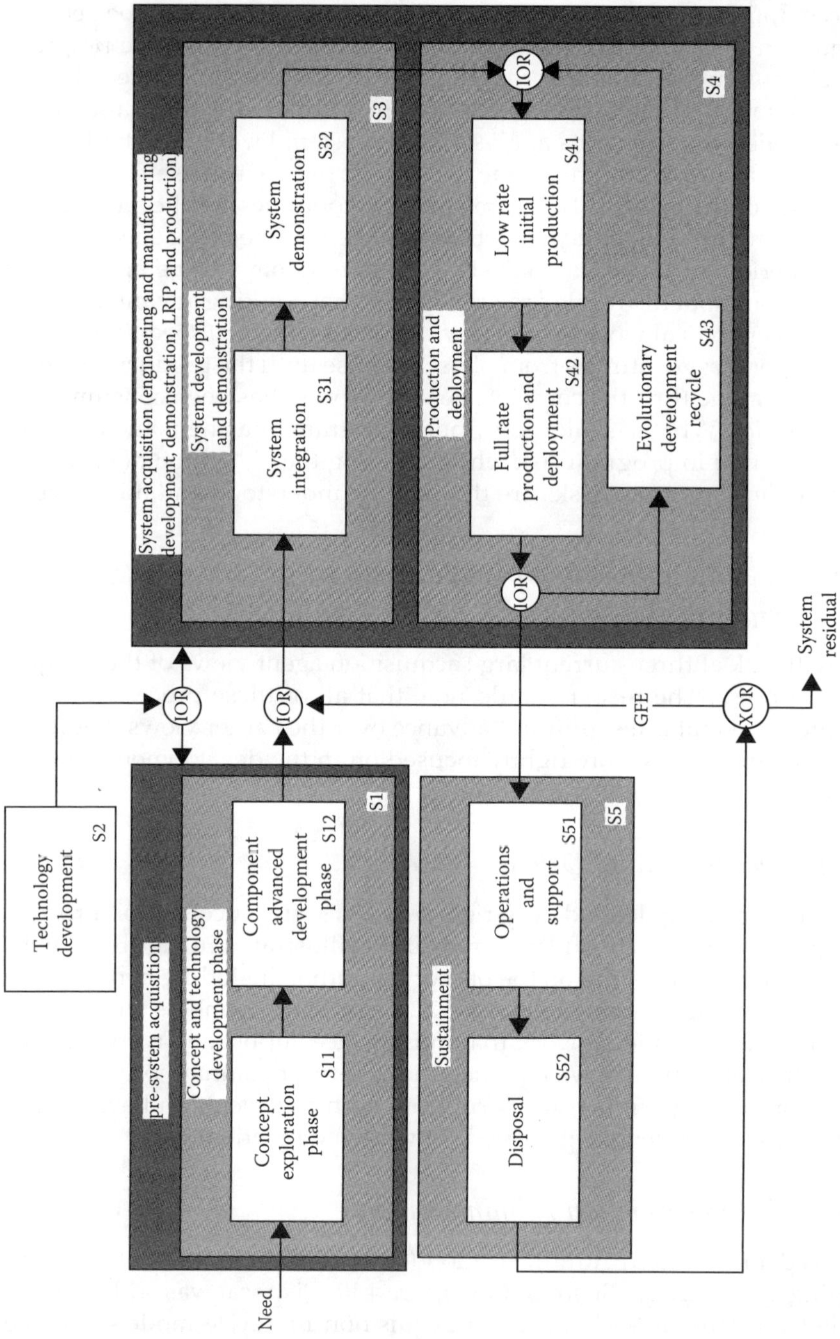

Figure 4.9 DoD acquisition life cycle model.

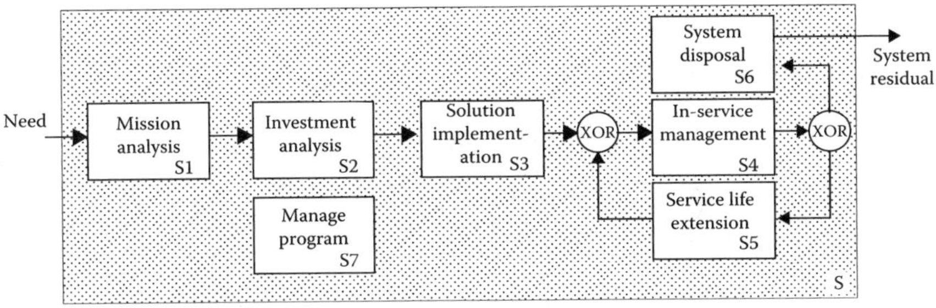

Figure 4.10 FAA acquisition life cycle model.

4.5.3 National Aeronautics and Space Administration

The NASA phasing model that came in effect in the year 2001 is illustrated in Figure 4.11. In each of these three phasing models, a single enterprise with a clear knowledge of its common process would be able to map that work into each of the phases as the program matures.

4.6 Structured process analysis and process expansion

The vision statement, composed in one or a very few sentences, obviously is not expressed in sufficient detail to fully express the generic enterprise process. We need more detail. A structured analysis, or decomposition, process can be applied to provide this detail. Several different structured analysis models are offered in the author's requirements book, but we will select one of them, process or functional flow diagramming, for this application. An enterprise might very well select a different structured analysis model from those in the requirements book with no loss of validity. In Section 4.2, process and functional flow analysis will be differentiated, but for our purposes in this book, they will be treated synonymously.

This decomposition begins with the vision statement that should give us some ideas for a list of n things that must be done in order for us to be able to claim that we had satisfied the vision statement. Then we should try to connect these process steps in some kind of flow network, showing which of them must be accomplished as precursors of other process steps. Some of these activities may have to be accomplished concurrently, together, or in parallel.

There are a lot of different ways in which this top-level enterprise functionality could be listed, but for purposes of further discussion we must agree on one. Any enterprise must accomplish a few things or pass from the scene. It must have an existence even when there are no programs in house and must retain knowledge determined by the product line and customer

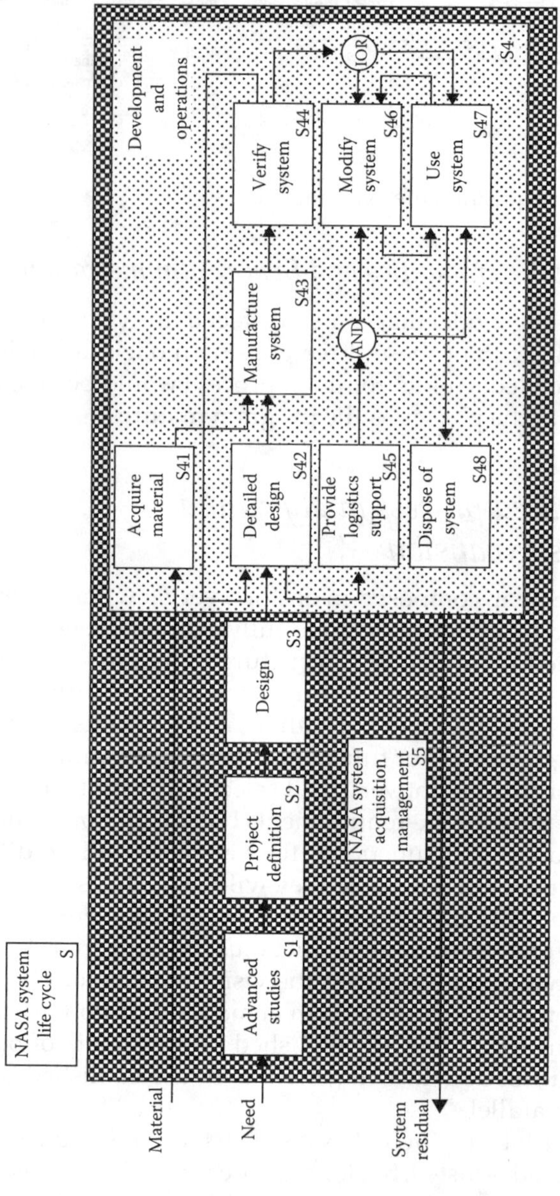

Figure 4.11 NASA acquisition life cycle model.

base selected in keeping with the vision statement. Also, the enterprise must be led and managed. All of this functionality can be allocated to the enterprise top management and a functional organization structure.

The enterprise must constantly be seeking new business, for the business it has at any one time is fleeting. If it is successful in completing a program, the income from that customer will cease and the lifeblood of commerce will have to flow from another program. This function can be satisfied by an organizational entity that seeks new business.

Finally, the enterprise must run programs to develop, manufacture, and deliver products of value to customers. At any one time, there may be several programs in the house, all at different points in their life cycle. Each of these programs must be managed by a program manager to whom the functional organizations have provided appropriate resources based on the needs established by the program for which they can pay. Figure 4.5 illustrates a generic process at a very high level that would satisfy this very-high-level implementation of the vision statement.

The new business acquisition process may be augmented by customer participation in the management of the acquisition where the customer is highly organized, as in the case of the DoD. Where the customer base is not organized, as in commercial sales, the enterprise may have to internally satisfy this function.

The enterprise vision shown in Figure 4.1 is identified as the process system function, F, and sub-function IDs have been assigned to illustrated subordinate functions. As shown in Figure 4.1, we can now continue to decompose F4 to fully characterize the enterprise's generic process for development of new product and process systems. Functions F1, F2, and F3 can similarly be decomposed. It does not make a lot of sense to try to generically decompose functions F47 and F48 because they are always going to be very system peculiar. This is precisely where the generic and program-peculiar process definitions must come together.

Functions F1, F2, and F3 can and should be captured in the generic process description. This can be done in the context of repeatable modules that can be applied to each specific item in the product system. Some parts of function F4 can be generically decomposed, but not all. As noted earlier, the requirements analysis process must initially be applied to the system. Thereafter, this same process must be applied to all of the things immediately subordinate to the system. This process flows down to lower levels, resulting in many applications of the requirements analysis process module that can be generically characterized. There are different ways we could decompose function F4, but the author has chosen the following:

F41	Define system
F42	Synthesize system
F421	Design system

F422 Material operations
F423 Manufacture components and assemble end items
F44 Verify system
F46 Assure product and process quality
F47 Use system
F48 Sustain system
F49 Manage program

These generic process modules can be linked with specific things in a particular product system by linking them to product entities in program-peculiar planning. The result then becomes part of the program process for developing the program product. Planning strings will be created in the form P2-S2-E3-A12-F422 that, in this example, covers material operations work (F422) for program 2 in stage (phase) 2 ending at program event 3 and dealing with product entity A12. This composite planning string illustrates how we can merge together customer-oriented program staging (phasing) with generic enterprise work definition. In Part three of this book, we will return to the enterprise functionality exposed here and, in the next chapter, to determine what functional organizations should exist and for which parts of this exposed functionality they should be responsible.

Figure 4.12 expands on the definition of function F4 sub-functions. Functions F47 and F48 must be expanded in the context of the customer need statement for the particular program and system. Function F5 is the business of the customer on programs developed for large customers, like the DoD, but may be part of the top management function for a commercial firm. *System Requirements Analysis* (Grady, 2005) focuses on the expansion of function F47 during the system definition process (F41) to give insight into the necessary characteristics of the product system, including the product system entities that the generic process modules will be linked to so as to form program process planning. It looks like there is a bootstrap process at work, but it can function effectively in the context of early program work in function F41 being accomplished at the system level that identifies the major things in the product entity structure, to which later a program process function F41 work can be accomplished based on the earlier planning oriented around the product entity structure.

Function F41 work must also develop the program-peculiar process aspects of functions F47 and F48. This includes logistics support (training, technical data, spares processing, maintenance, and maintenance levels and locations) and operational techniques.

Figure 4.13 illustrates this simple yet confusing collection of ideas. The principal confusing element is that the program process must create both the program product and the program process and part of the work in developing the program product definition is used to organize the program process. The high-level product definition is fed back to the

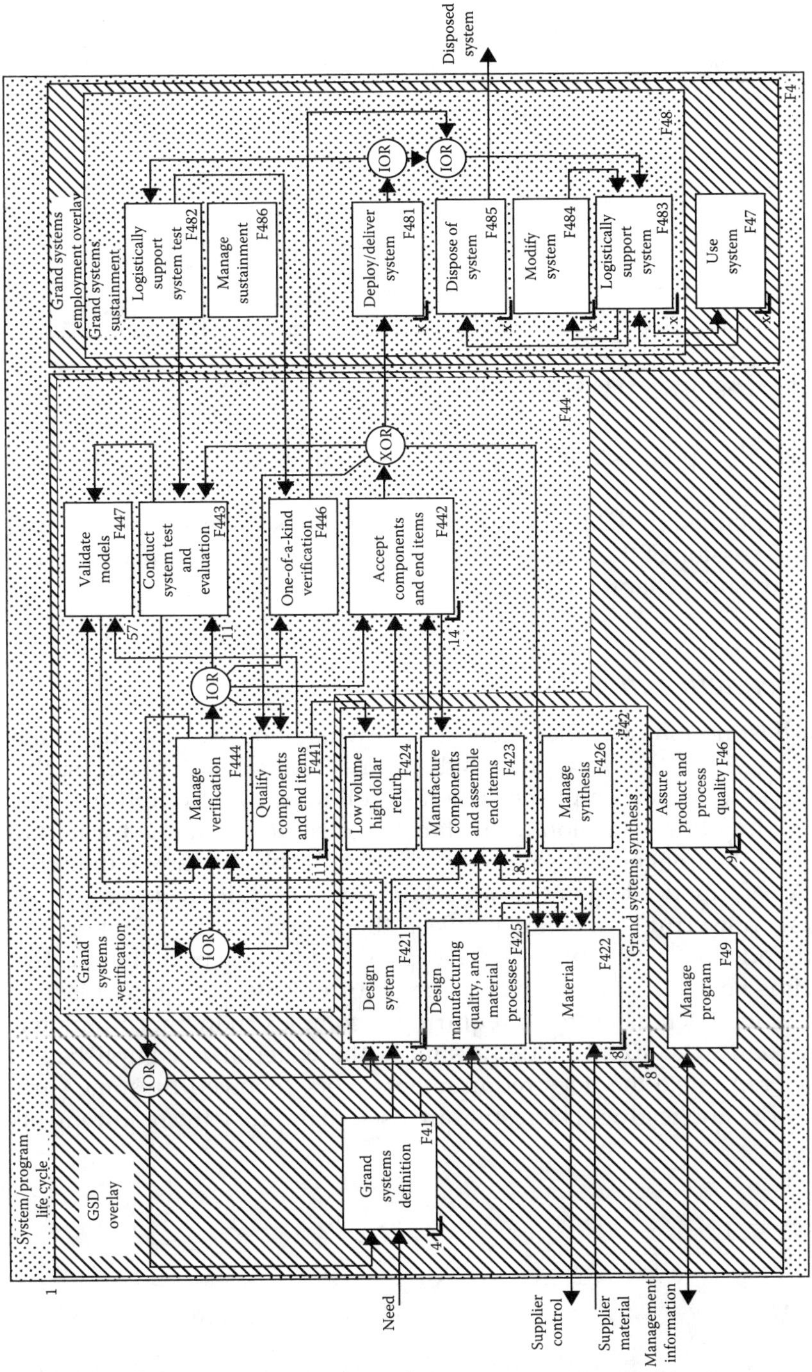

X: Refer to program system architecture report for expansion

Figure 4.12 Enterprise common process.

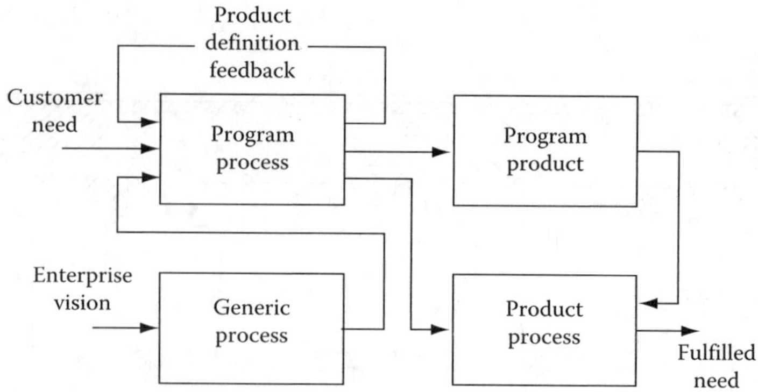

Figure 4.13 Aggregate view of product, process, and program.

early program process definition work to combine with the generic process definition to form a low-level program process definition. This fact will be explained in detail in Part four of this book on program planning after a discussion of the organizations that will accomplish the work identified in the next chapter.

4.7 Sequence models

The earliest notion of work sequence was what is now called the waterfall model. All of the prior process diagrams in this chapter suggest this model, where one accomplishes one task followed by accomplishing the next task in a linear fashion from start to finish. It was named waterfall because of the imagery of water flowing over the blocks shown in Figure 4.14a. A "V" model bends the waterfall in the middle at design, but it permits a powerful connection between the requirements definition and the tail-end verification process. The spiral model is especially effective where it is very difficult to develop the requirements as a prerequisite to design work, such as systems requiring a complex man–machine interface. Actually, the spiral model is simply the waterfall accomplished at a higher rep rate. One does requirements work followed by some early design work and testing. The results are used to improve the requirements

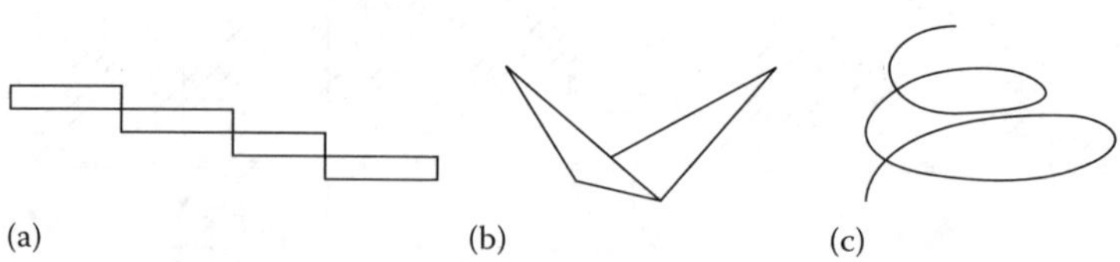

Figure 4.14 Sequence models. (a) Water model, (b) V model, and (c) spiral model.

model. We repeat this process as necessary to evolve a good design. The spiral process recognizes the low cost of criticism.

4.8 Documentation media

Chapter five continues the process expansion beyond that shown in Figure 4.12, producing a complete generic process diagram contained in the author's Grand System Management course student manual Division 3, Exhibit H, Appendix A, offering a generic process specification. This document is a definition of the process system that the enterprise represents. Exhibit H is referred to as an enterprise architecture report (EAR) and is similar to a product system architecture report (SAR) that is encountered in *System Requirements Analysis* (Grady, 2005). It contains the basis for the enterprise organization and clearly defines all of the work the enterprise will have to perform on programs and in support of its own continued existence.

This process specification should be a fundamental input to the program planning for any program in that it supplies the generic process definition as well as the map of the functional organizations to this process, yielding the resource strings to pull into the program along with the processes.

4.9 Common process versus agility

An enterprise is encouraged to be capable of great agility by many enterprise consultants today, and it is a good capability to possess. The author does not subscribe to this capability at the total loss of interest in common process, however. The author's view of common process may be way too much on the detailed side, even inhibiting agility if followed to this depth. It is clear to the author that the basic sequence of requirements, synthesis, and verification accomplished within a well-managed process occupies the ground floor of common process. Somewhere between this extreme and the extensive model offered by the author in the Grand Systems Management course student manual Division 3, Exhibit H, there is a happy hunting ground compatible with organizational agility.

It is not intended that an enterprise implement every common process task on all phases of all programs, of course. During early phases of programs, it is front-end tasks that will be appropriate and, on some programs, some tasks can be tailored out without undo risk. The common process definition is intended to offer the program planning team a shopping list of available tasks, and the team must apply sound management skills to design the program to achieve the best customer value, affordability, and the needed degree of assurance of success.

chapter five

Systems development description*

5.1 The ultimate enterprise functionality

Normally, we do not develop an enterprise in accordance with system principles. An enterprise is most often born small and grows, possibly even despite the owner's faults, into something that satisfies customer needs at a price and with a dependability that customers like. Ideally, we would, when starting an enterprise, identify an ultimate function that the enterprise is intended to satisfy, which the author refers to as its vision statement. Then, applying Sullivan's principle of form follows function, we could expand the definition of this ultimate functionality through the structured decomposition and allocation of the more detailed exposed functionality to organizational entities that must satisfy the allocated functionality. The functional organizational structure of an enterprise is akin to the product entity structure of a system being developed. This organization that creates product systems is composed of organizations of people who possess knowledge and apply it toward specific goals. The vision statement, expressing its ultimate functionality, is allocated to the complete enterprise just as the customer need—expressing the ultimate functionality of the system the customer requires—is allocated to the top element of the system product entity structure, the system. Now, as in system development, we must describe the enterprise functionality in some detail.

If we were about to establish an enterprise, this might be a fine way to proceed but here you are stuck in the middle of an existing enterprise that is not working well. Is there any hope for you and your enterprise short of starting all over? Well, this process can be applied to reengineer an existing enterprise just as the system approach can be applied to develop a new product or reengineer an old one. In so doing, in either case, we have to recognize that there is a current reality on the physical plane that is not going to go away with a snap of the fingers. There is a transition process that is required to move from the existing enterprise organizational architecture and functional responsibilities (the as-is) to some new combination (the to-be). Just as in reengineering an existing system, we have to determine what entities should remain as they are, which ones must be modified, which ones must be deleted, and what new elements must be

* The material in this chapter is from Grady, J., *System Engineering Deployment*, CRC Press, Boca Raton, FL, Chapter 4, 2000. Used with permission.

added. It may be that the move from the present reality to a new identity cannot be made in one step leading to a growth path composed of several steps permitting some adjustments in the process. At the end of Chapter four, we discussed how this transition could be made while continuing to serve the needs of several programs in various stages of completion.

Whatever we conclude the detailed functionality of the enterprise to be, it should have been expanded from an ultimate functionality. The author will use the phrase enterprise vision to denote this ultimate functionality. It represents a brief description of what the enterprise must achieve just as a customer need statement is the ultimate system function telling what the system to be produced must achieve. Figure 5.1 illustrates these relationships in their most fundamental form. The enterprise vision is the ultimate function that drives the enterprise process for creating product systems. This enterprise must function within system development and business environments while transforming material and an understanding of customer needs into a product system that satisfies that need over some period of time. Someone or some organization will operate the product system within a predetermined product system environment satisfying their need possibly drawing logistics support resources from the creator enterprise.

In the end, and the end must come for every man-made system, the residual elements of the system will either be disposed of or recycled into a new system, perhaps one that will continue to satisfy the old system's functionality but more commonly some increased capability.

At least some part of the enterprise vision statement should involve an interest in satisfying customer needs through programs focused on those

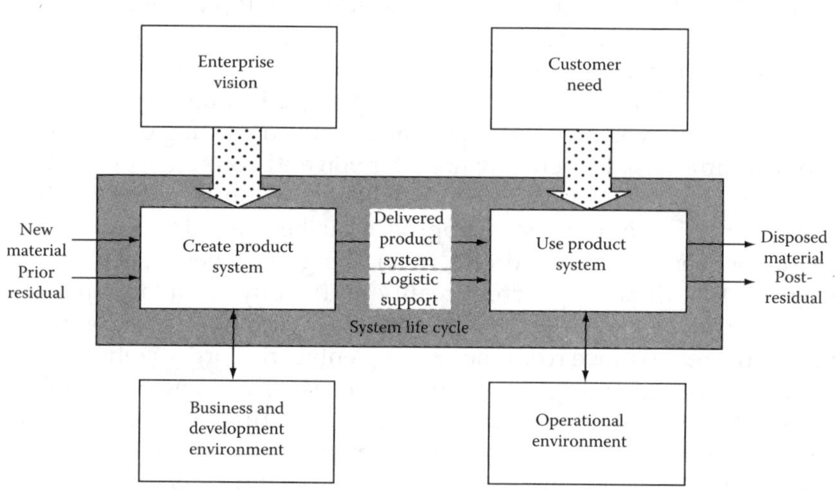

Figure 5.1 The ultimate system functional relations. (From Grady, J., *System Engineering Deployment*, CRC Press, Boca Raton, FL, pp. 43–63, 2000. With permission.)

needs. An enterprise may, at any one time, be in the process of satisfying more than one customer need, and this is the kind of enterprise this book envisions, a multiple program enterprise. By definition, in this book, the enterprise entity that responds to a customer need is called a program.

From the creator's perspective, we can look upon the creation and operation (or use) of the system as functions within the life cycle of the system. Some systems are operated by their creator, and others are operated by others. Space launch vehicles, for example, are often operated by their creator whereas fighter or commercial transport aircrafts are commonly operated by the customer. A single program may serve multiple customers as in the case of the Boeing 777 program building commercial aircraft for American Airlines as well as many other airlines. A single customer may also have interests in more than one program within a particular enterprise as well. In this book, we will focus on the generic relationship between an enterprise and a customer through the window offered by a single program. The elements of this single view can be multiplied in any number of ways to appreciate more complex situations in which people find themselves in industry.

5.2 First-tier functionality expansion

The create product systems function is a variation on or subset of the enterprise vision statement and will be allocated to an enterprise capable of responding to customer needs as suggested in Figure 5.2. This is a mindless allocation. The enterprise is, by definition, the entity that will

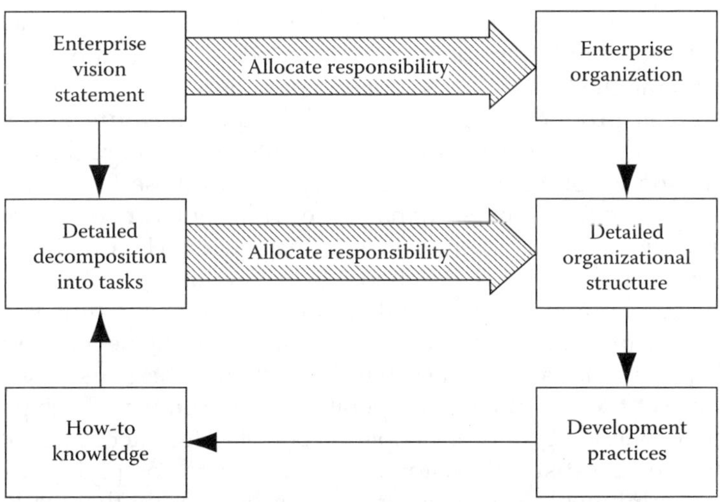

Figure 5.2 Process development and allocation. (From Grady, J., *System Engineering Deployment*, CRC Press, Boca Raton, FL, pp. 43–63, 2000. With permission.)

satisfy the vision statement, after all. But, the conscious recognition of this trivial relationship is not a senseless action. It is important to recognize a condition of logical continuity throughout the enterprise structure, and this step connects the creative spark of genius giving birth to the enterprise through its vision statement with the complete detailed description of the enterprise.

Subsequent steps require us to expand on the functionality needed by the enterprise in order to achieve the enterprise vision. These details will take the form of a multilevel functional flow diagram in this chapter. We will then, in later chapters, allocate the identified functionality to organizational entities thus expanding the detailed definition of the organization. This process is exactly similar to the activity one would apply on a program to understand the needed system characteristics and product entity structure. In the case of a product system, the product entity structure is a hierarchical arrangement of physical product entities arranged in a parent–child structure. In the application of this approach to the development of the enterprise, the organizational entity structure is a hierarchical arrangement of organizational entities that will have responsibilities for accomplishing work associated with the tasks expressed on lower-tier flow diagrams.

We might inquire that since the organizational diagram is going to be expressed in a hierarchical diagram, why do not we use a hierarchical list of functionalities rather than a flow or sequence orientation? This approach can be used, but the danger is that it will lead to a bad organizational design. The flow orientation, being different from the hierarchical organizational structure, encourages a degree of isolation between the efforts to understand the needed functionality and define the organizations responsible. The use of hierarchical structures on both sides tends to encourage a bidirectional coupling rather than the intended unidirectional relationship following the form follows function paradigm.

The create product system function illustrated in Figure 5.1 is but one of at least four fundamental functions our enterprise will require in order to satisfy the vision statement. Figure 5.3 shows these four fundamental functions. Someone must manage the whole enterprise toward satisfying its vision statement, enterprise management. The enterprise must execute its current business. We choose to group this work into something called programs, and we assume that more than one at a time will exist. Each of these programs is responsible for transforming new material and possible prior residuals into a system that satisfies their customer's need within the context of a particular operational environment. Each program requires a cohesive group of people responsible for executing program work directed at satisfying customer needs through complying with contractual commitments as well as plans and schedules for that work. There may be many programs at work at a time, and programs may contain subprograms sometimes called projects.

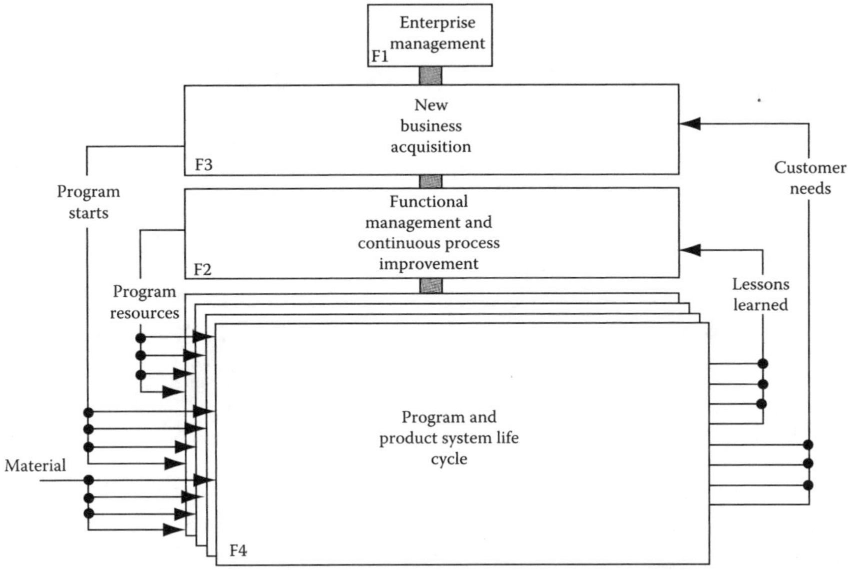

Figure 5.3 The four basic enterprise functions. (From Grady, J., *System Engineering Deployment*, CRC Press, Boca Raton, FL, pp. 43–63, 2000. With permission.)

The enterprise requires a means for new business acquisition, that is, gaining new programs. This function keeps abreast of potential customer needs and coordinates them with enterprise capabilities focusing energy on achievable business relationships through customer contact, proposal development, and other marketing activities. As a beginning of this analysis and allocation process, let us allocate responsibility for the four functions exposed in Figure 5.3. Figure 5.4 identifies organizations in our enterprise that will be responsible for the functions exposed in Figure 5.3. Only three programs are identified, and there could be many more.

The map should be obvious to the reader. The only allocation that is a little abstract is the allocation of the resource maintenance and continuous improvement function to the functional organization of Figure 5.4. This organization is responsible for the acquisition, maintenance, and extension of the enterprise's knowledge base to prepare it for work related to its customer base and product line identified in its vision statement. The functional organization will provide programs with the resources they need to succeed. The marketing organization is, of course, responsible for acquiring new business.

The programs are focused totally on their customers and cannot be depended upon to maintain and improve the enterprise's knowledge base. This is the responsibility of the functional organization in a matrix organization. We need both of these focuses in our enterprise: the focus on

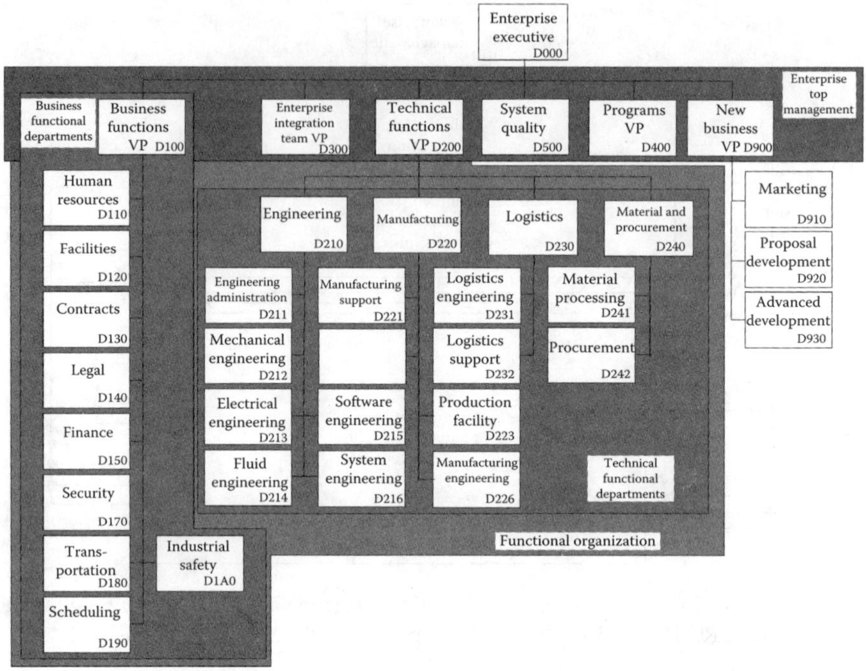

Figure 5.4 Top-level organizational diagram. (From Grady, J., *System Engineering Deployment*, CRC Press, Boca Raton, FL, pp. 43–63, 2000. With permission.)

satisfying customer needs through program organizations and the focus on maintaining and improving the enterprise and its knowledge base through the functional organization. The functional department structure should provide every program with people who have the knowledge appropriate to the program technology and skills related to that department's area of responsibility within the enterprise, best practices improved over time, and good tools coordinated with company practices and personnel knowledge.

Results from programs must be consciously reviewed in the context of a continuing search for excellence in program execution. An element of functional management must review the results and draw conclusions about the best improvement actions to take to improve program performance. This may entail the improvement of the knowledge base of its people, better practices, or better tools.

Figure 5.5 expands in Figure 5.3 for an enterprise that happens to have four programs in house at the time. Post-residual material may flow in from the development environment as customer furnished from prior systems past their prime and new material from suppliers based on purchase orders or contractual relationships. Post-residual material may also

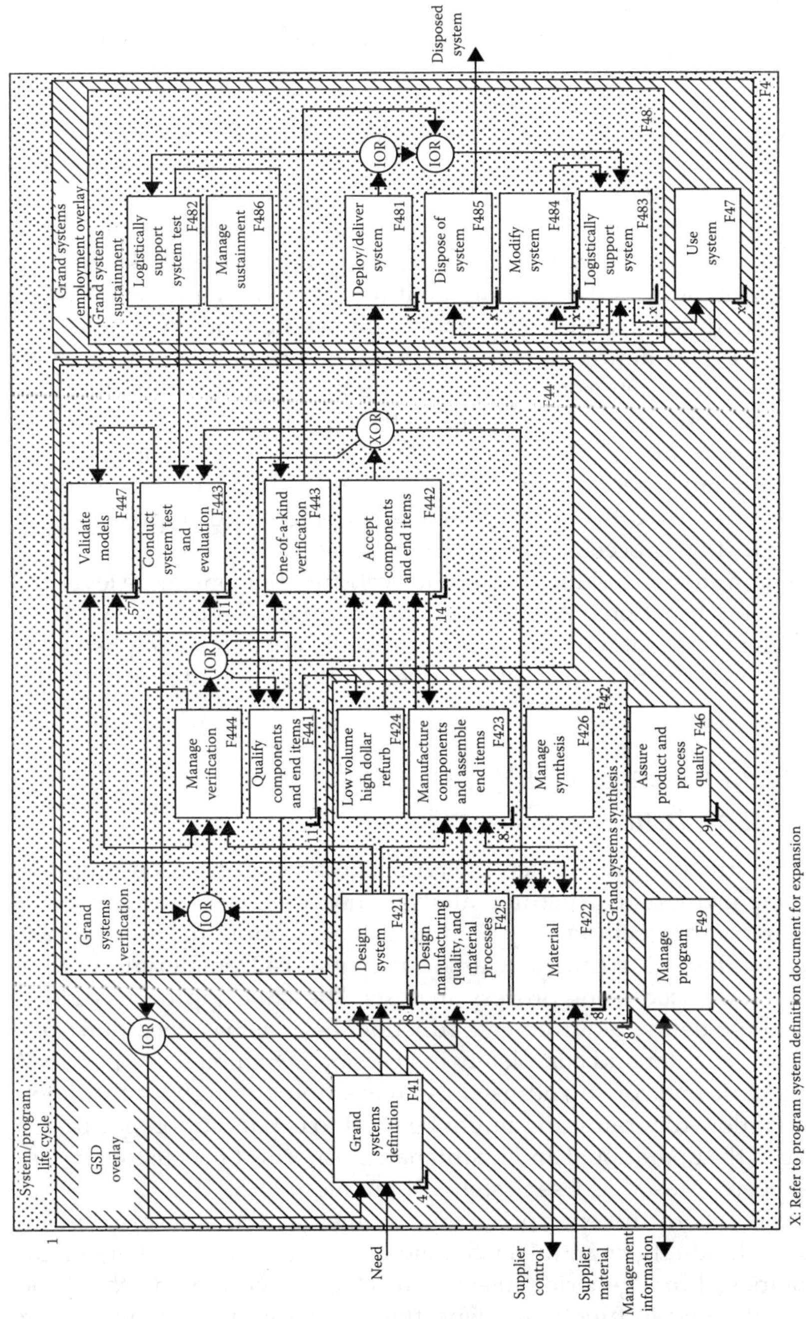

Figure 5.5 Expansion of first-level functional detail. (From Grady, J., *System Engineering Deployment*, CRC Press, Boca Raton, FL, pp. 43–63, 2000. With permission.)

flow in from systems previously created by the enterprise. Other important interfaces also exist between the development environment and the program and between the business environment and the other enterprise functions.

In Figure 5.5, we have identified the lower-tier functionality specifically, and these functions will be addressed in the next section. We have differentiated the normal development cycle from the modification development cycle identified as function F484. These modification processes are actually small programs that should follow the same pattern we define for a full program. Generally, these modification programs will take in product previously delivered and combine it in some way with new product to yield a modified product that can accomplish new or expanded functionality or performance figures. The previous product may actually be brought back to the enterprise physically for modification or modification kits and instructions may be shipped to the customer locations for installation. The normally complex nature of system development programs becomes evident when more than one of these modifications are underway at one time in different stages of development and implementation while production proceeds.

It remains to expand the system life cycle function one more level and describe these functions. We can then allocate these functions to organizational elements that will require certain knowledge to accomplish work associated with the responsibilities driven by the allocated functionality.

5.3 The marriage of enterprise and system functionality

Each of the functions illustrated in Figure 5.5 can be further decomposed using some method of structured analysis, not necessarily the functional flow diagramming technique applied so far. For example, the procurement and material handling functions associated with F422, material, could be further decomposed using process analysis. The same approach could be applied to the define requirements function, F411, as has been done partially in this chapter and completely in Division 3/Exhibit H/Appendix A for the generic enterprise and used as an example throughout this book. Functional flow diagramming might be used to expand upon our understanding of the operational and maintenance aspects of the customer need that has been phrased as Use System, F47, in Figure 5.5.

Here is where the enterprise process merges the enterprise vision and the individual program need statement. The vision statement can be decomposed in a generic sense as this chapter will do and that functionality allocated to functional departments that will be called upon to supply program resources to accomplish those functions. The use system function is program specific and must be decomposed as a program task with allocations flowing to product entities. Program process analysis may also develop insights into unique aspects of the process generically represented as part of the enterprise process diagram.

There are any number of structured analysis models that could be used to decompose function F47 for specific programs. If the product system were substantially a software system, one might apply a software-oriented modeling technique such as Hatley-Pirbhai or object-oriented analysis. The supporters of these two modeling techniques would, on the other hand, encourage the use of their model regardless of the nature of the system, whether hardware or software in nature. An enterprising analyst might even apply these techniques to enterprise modeling rather than the flow diagramming approach chosen by the author.

The author differentiates between functional flow and physical flow analysis as follows: the former is a sequence of things that must happen generally crafted prior to the time you know what resources will be available to accomplish those things whereas the latter is a physical analog of the process with known resources flowing in space and time through the process diagram. In a functional flow diagram, the directed line segments only indicate a sequence in which things must happen. In a process diagram, the directed line segments can be interpreted both in terms of sequence and a flow of some form of physical material through the process.

The system life cycle function will be expanded, identified in terms of functions in this book, but it is understood that there is a one-to-one correspondence between the corresponding functional and physical models. This commonly is not usually the case between the use system function (F47) and the physical process defined after one understands the physical system product entity structure. Thus, we could reidentify all of the define system functions Fn as processes Pn, where n is a string of base 60 characters. The author may use the words process and function interchangeably in this regard but does not intend that they are synonymous in a more general analytical situation. At the lower tier, these blocks will be referred to as tasks. All of these words are describing the same kinds of things, the work that must be done by the enterprise on programs.

5.4 *Functionality allocation*

The functionality exposed in Figure 5.5 is allocated to the organizations illustrated in Figure 5.4 as shown in Table 5.1. The functions map to organizational entities responsible for providing related resources to programs. While depending on the functional organizations to provide the programs with the resources needed to accomplish the functionality, it is the programs that are generally responsible for implementing work in accordance with functional department practices and the

Table 5.1 Allocation of Enterprise Functionality

Function ID	Name	Resp. Dept.	Description
F1	Enterprise management	D0000	Top-level management of the enterprise.
F2	Functional management	D1000 D2000	Organized focused on the knowledge base the enterprise needs to create products in its chosen product line and customer base.
F3	New business acquisition	D9000	New program starts are acquired from potential customers.
F4	System life cycle	D3000	The transform from need to delivered product proven to satisfy customer defined requirements.
F41	Define system	D2100	Transform the customer need into a preferred system definition.
F42	Synthesize system	D2000	
F421	Design system	D2100	Requirements defined in function F41 are transformed into designs suitable for manufacture and procurement.
F422	Material	D2400	Transform engineering definition of the product into a stream of material and services needed to manufacture the system. Temporary storage of material is included.
F423	Manufacture system	D2200	Transform engineering, manufacturing, material, and quality product definitions into a physical product.
F44	Verify grand system	D2500	Evaluate product items and system for compliance with requirements and produce evidence of a condition of compliance. Three classes are included: item qualification, item acceptance, and system test and evaluation.
F46	Assure product and process quality	D5000	Assess the performance of enterprise products and processes and assure that they comply with requirements.

Table 5.1 (continued) Allocation of Enterprise Functionality

Function ID	Name	Resp. Dept.	Description
F47	Use system	D2300	The system is operated and maintained throughout its life by the user. Logistics may be called upon to support field use with field engineers.
F48	Sustain system	D2300	Deliver products and services that ensure that operational elements of the system remain fully operational.
F481	Deploy/ deliver system	D2300	Move the product system from the creating organization to the operational organization. This may involve commercial distribution or military deployment.
F482	Logistically support system test and evaluation	D2300	Logistic support is supplied during system test and evaluation with factory deport support.
F483	Logistically support operational system	D2300	Logistics support to user organizations in the field.
F484	Modify system	D2300	Change an existing system in planned ways to accomplish improved functionality.
F485	Dispose of system	D2300	Transform any hazardous elements into a nonhazardous state and convert residual material into other applications or properly processed sanitary fill.
F49	Manage program	D4000	Management skill is brought to bear on all of the other functions encouraging the people responsible for those functions toward satisfying the goals established for them in planning.

Source: Grady, J., *System Engineering Deployment*, CRC Press, Boca Raton, FL, pp. 43–63, 2000. With permission.

program contract. The exact manufacturing and test responsibilities may be allocated in many different combinations but in this example we are assuming that the facilities involve manufacturing facilities and test laboratories that can best be preserved and managed through the functional structure responding to program needs.

The author evaluated many alternatives to the allocation of the verification function to a verification engineering organizational entity including placing it in a quality organization. This alternative was eventually shelved because it violated a sound principle of quality assurance that they should assure quality rather than accomplish work. If they were accomplishing actual work, then who would assure the quality of that work ending in a circular argument. Some readers would be more comfortable with verification being referred to as test and evaluation but that term suggests that the organization is only interested if the verification work entails testing. Interestingly, what MIL-STD-490A referred to as inspection DoD renamed as examination in MIL-STD-961E so they could use the word inspection in the quality assurance sense meaning a general verification task that, in a specific case, might be implemented as a test, analysis, examination, or demonstration. It is intended that the verification engineering organization provide programs with people who can plan and implement the complete verification process as well as provide the personnel and laboratories for verification work that happens to employ the test method. This matter is discussed in considerable detail in *System Verification* (Grady, 2007).

We now must identify more finely the needed functionality and associate that functionality with functional organizational entities responsible for the related work in the context of programs responding to customer needs. The remainder of this chapter provides an overview of the enterprise life cycle model developed by the author from the observation of and participation in many development programs.

5.5 *Define system, F41*

To further expose the needed functionality, we have but to study each of the functions listed in Table 5.1 and consider in more detail what things have to happen to encourage success in the function we are expanding. This can continue down to some level of granularity best determined in the organizational plane. When we are confident that we understand the full functionality of all of the organizational entities we have created and that we have created all of them that are necessary to accomplish the vision statement, we may conclude that we have exhaustively examined the needed functionality. Figure A-1 in Appendix A of Division 2 Exhibit H expands on the functionality expected from our enterprise in some detail and a map is included in Table A-1 between the exposed tasks and responsibilities coordinated with the functional organizational diagram shown in Figure C-1 (Division 2/Exhibit H/Appendix C). Some of the lower-tier functionalities is examined in some detail in subordinate paragraphs. The reader may conclude differently than the author about the best way to define one's process, but the example included in this book will, hopefully, offer encouragement for crafting a detailed definition that does satisfy the enterprise's process needs as a basis for its functional organizational needs and common program process.

The Grand Systems Requirements function, F41, consists of six subfunctions as shown in Figure 5.5. They are F411 customer needs analysis, F412 universal architecture description framework (UADF), F413 requirements analysis, F414 concept development, F416 manage requirements, and F417 clone specifications. In the first function, the enterprise seeks to understand the real or potential customer needs. In F412, the program models the problem and solution spaces for the system define the system in terms of a set of entities of which it shall consist and a set of modeling artifacts from which all requirements may be derived. In F413, the requirements are defined with precision. In function F414, we develop design concepts for selected entities to gain confidence that it is going to be possible to satisfy their requirements and as a basis for the continuation of lower-tier problem space modeling. In F416, the requirements are captured in the planned template and, subsequent to a review, are published in paper or electronic media. Validation takes place in this function as well as several other management-related actions.

5.5.1 Customer needs analysis, F411

Working with the users of the system to the extent possible the developer comes to an understanding of the user needs and acquisition agent resources relative to satisfying those needs. This work involves an initial modeling activity, discussions with the user, and a translation of what may be imprecisely stated needs into contractually valid system-level requirements.

The beginning of a new development starts with a need recognized most often by a user who concludes that there is something they need to do that their current resources do not support. A new program may entail solving an unprecedented or a precedented problem. In the former case, the problem has no current solution, and the solution will involve new development and design. In the latter case, there is an existing reality that has been providing service for some period of time. The program will likely involve some form of modification of that existing system. In either case, it will often be necessary for the user or its acquisition agent to perform an initial analysis. This study work may also be contracted with a firm resulting in the development of a preliminary system specification, an operational requirements document (ORD), the concept of operations document (CONOPS), or initial capabilities document (ICD).

Mission analysis is an important subprocess of an unprecedented process for translating a customer need into a critical mass of information about a preferred system solution that will satisfy the intended mission. The principal responsible functional department is system engineering for this activity. This process may begin with only a customer need conveyed by a potential customer through a request for proposal or deduced by the enterprise based on marketing analysis. In this case, the need acts as the ultimate function for the intended life cycle of that particular system and program but in particular for the use system function (F47) and is allocated instantly to the top product entity named system and expanded to identify lower-tier functionality which, in turn, is allocated to lower-tier elements in the product system thus forming the physical view of the system.

Also, in mission analysis, we must seek to understand the one or more ways the customer will use the system to be developed, and the results of this analysis feeds the development of the functional view of the system helping to visualize design concepts for the elements of the system that must satisfy higher-tier functionality. Part of the customer interaction process is a question and answer activity that attempts to bring forth from customer representatives answers about their intended application. A key part of the mission analysis process is the development of alternative solutions and trade studies focused on identifying the preferred solution among those studied. Figure 5.6 illustrates this process.

The need is expressed by one or more life cycle master flow diagrams, and the difference between them is expanded as necessary to differentiate between them. We then allocate the top-level functionality to things that would accomplish that functionality forming one or more product entity structures (ABD) for each functional view. For each of these views, we should identify the corresponding top-level interfaces (SBD), environmental factors (E), a logistics concept (L, possible basing differences, for example), a brief scenario (S) communicating pictorially or in the form of an event list what happens in the employment of the system, and a set of descriptive data (D) upon which the selection decision will be made.

5.5.2 *Universal architecture description framework, F412*

Figure 5.7 expands on the grand system definition function (F41) of Figure 5.5 for generic program X, and it also illustrates subordinate functions. This same set of processes may very well be going on in parallel at different points in the life cycle on several development programs at the same time in the enterprise. The term grand system refers to a need to develop both program product and program process concurrently and optimize

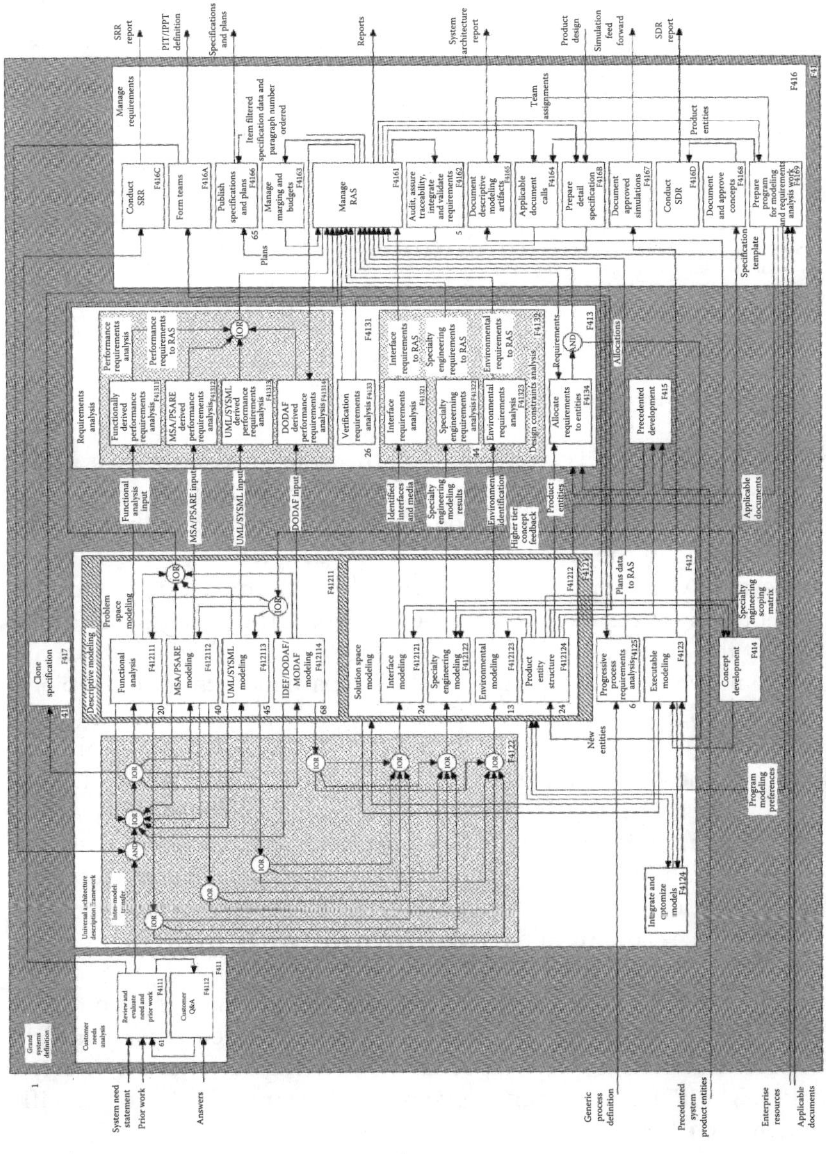

Figure 5.6 Structured selection of the preferred concept.

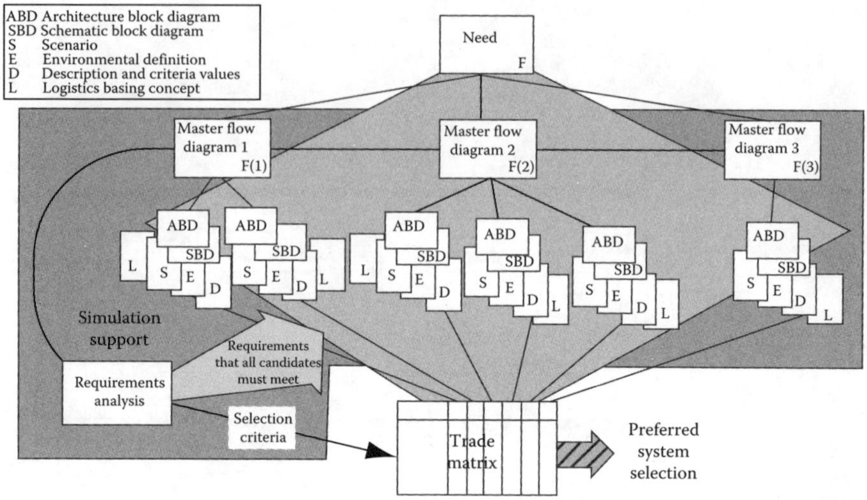

Figure 5.7 Grand systems definition.

at this grand system level composed of both while the enterprise is doing so for all programs. In this chapter, we will give some examples of this allocation process leading to functional department charters that enables the reengineering of the enterprise in the next chapter.

The term UADF is applied to a process that demands that all requirements appearing in specifications be derived from models and the selection by the developer of one or more modeling methods that are collectively comprehensive. One set of problem space models, for example, is UML and SysML.

Function F41 includes two fundamental tracks not entirely clear from the diagram. These two tracks focus on the two kinds of teams needed to create product systems. The grand system is the responsibility of the program integration team (PIT) led by a senior system engineer assigned to the program and selected by the program manager. This person may essentially be the chief system engineer for the program. Some of the tasks in our process diagram must be accomplished by the PIT at the system level and subsequently applied by the several integrated product and process teams (IPPT) for lower-tier elements. The PIT should accomplish the F412 function for several top-level specifications such that each IPPT can be formed with its top-level specification having been already created by the PIT. The PIT also assigns development responsibility for specific system items to each IPPT.

Through requirements analysis we uncover a set of requirements that all candidate solutions must satisfy and a set of selection criteria used as the basis for selection. A trade matrix is used to succinctly capture the intersections between alternative characteristics, derived from the candidate descriptive data, and selection criteria. A decision maker selects the preferred system from the candidate systems studied based on their relative ranking in the trade matrix. The trade study process is thoroughly covered in *System Requirements Analysis* (Grady, 2005) and *System Synthesis: Product and Process Design* (Grady, in press).

Each IPPT continues to expand the problem space model begun by the PIT to gain insight into item performance requirements and the physical product entities at the lower tiers. Solution space modeling work completes the content of lower-tier specifications in terms of interface, environmental, and specialty engineering requirements. This process will have to be performed repeatedly for each of the several items for which a team is responsible. The allocation of functionality results in the identification of things that have to be fitted into the physical model of the system, the product entity structure. Requirements analysis sheets (RAS) in paper or computer screen format, are the principal inputs and the outputs include

1. System architecture report (SAR)—Captures the system diagrams defining system composition. It includes (1) functional flow diagram illustrating needed system functionality, (2) the aggregate RAS capturing the allocations of functionality to product entities, (3) product entity structure diagram, a hierarchical block diagram defining the things in the system, (4) schematic block diagram defining all of the relationships between all of the things shown on the product entity diagram, and (5) any other documentation products of the system definition process.
2. Drawing breakdown structure—The engineering drawing overlay of the product entity structure telling what engineering drawings will be produced.
3. Specification tree—The specification overlay of the product entity structure telling what specifications will be prepared and what format they will follow.
4. Manufacturing breakdown structure—The manufacturing overlay of the product entity structure defining the groups of things moving from one major production area to another.
5. Configuration/end item list—Identification of the things through which the program will be managed.
6. IPPT responsibilities—Team responsibility boundaries relative to the architecture.
7. Work breakdown structure—The product component of the WBS is a finance overlay of the product entity structure upon which is based the whole program plan.

Throughout this work, PIT must integrate and optimize the evolving requirements pool and act to cause the modeling work accomplished by several teams to appear as one analysis. This can be accomplished by PIT being responsible for an SAR contributed to by all of the teams. As specifications are reviewed for release, their structured analysis basis must also be reviewed and, if approved, the latter is accepted into the SAR as the specification is released.

5.5.3 Requirements analysis, F413

All of the modeling work identifies modeling artifacts from which requirements must be derived. These requirements should follow a simple pattern of identifying the attribute to be controlled, a numerical value and unit, and a relationship between the attribute and the value such as equals or less than or equal to.

Every requirement identified in every specification is screened for ways to prove that the design is compliant. This process starts with each requirement that is extended into a verification requirement linked to verification tasks for which plans are developed, implemented, and results of accomplishing those plans documented in reports. The reports provide the evidence of compliance. This process begins with PIT development of system verification requirements and planning data followed by IPPT implementation of the same process for their areas of responsibility under the coordination, integration, and review by the PIT.

During preliminary design work to follow the requirements work, this is focused on qualification planning and during detailed design the verification planning work is focused on acceptance planning work. All of the design and analysis documentation developed for an item is completed as a coordinated package and moved through a formal release process that ensures that all enterprise documentation requirements have been satisfied in the package and that the content is mutually consistent within the context of the item and the system at large. Concurrent with program product development, we must also develop the program processes such that they are mutually consistent. The teams work to understand the process problem in the form of process requirements expressed in process specifications and plans.

In the case of a precedented development, diagrammed in Figure 5.8, there is an existing architecture that may have been providing satisfactory service to the customer for some time. For some reason, the customer has found it wanting in some fashion, generally because some factor in the environment of the system has changed. An analysis of the physical processes the system applies is generally helpful followed by a comparison with those that will be necessary in order to satisfy the new functionality expected for the new system. One can then study the existing architecture and determine which of several actions are appropriate: remove the item from the architecture, modify the item to satisfy new functionality, and

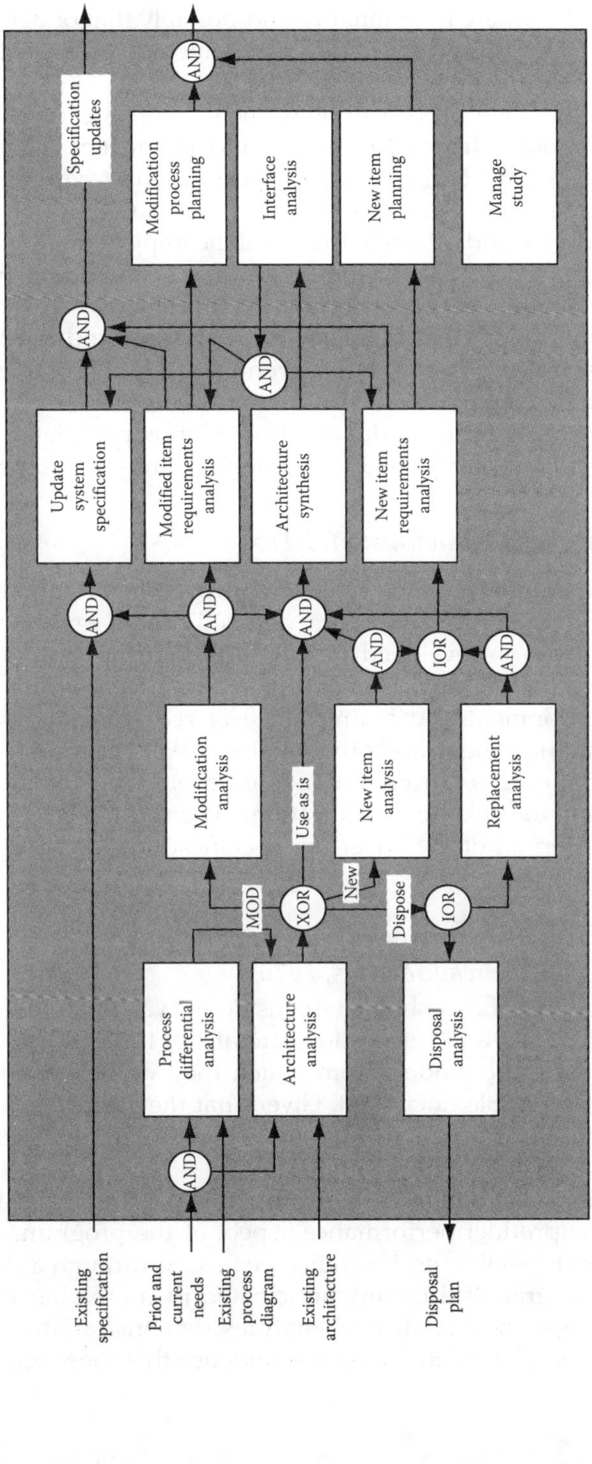

Figure 5.8 Precedented development work.

retain the item as it is currently designed. In addition, we may have to add new items to satisfy new functionality and possibly that of deleted items. The precedented and unprecedented development processes are covered in *System Requirements Analysis* (Grady, 2005).

The unprecedented process applies one or more structured analysis problem space models, from which functions that must be satisfied by the system may be defined and from which performance requirements may be derived. Some models are effective where the solution will be implemented in hardware and others where it will be implemented in computer software. The enterprise employing UADF will fashion their framework from available models that are collectively comprehensive. In general, this will include one model that is effective for systems and hardware and another that is effective for software. Given that all fairly complex systems will involve some combination of both, it will be necessary to begin the analysis using the systems model and transition appropriately to a software-effective model, so Figure 5.7 provides for inter-model transfers.

5.5.4 Manage system definition, F416

Each specification must be reviewed and approved by PIT or the program manager prior to the beginning of any detailed design work.

Specifications are developed for all things in the systems product entity structure that is also determined in this process. This task also covers the development and maintenance of requirements traceability, requirements management, and other related tasks. The understanding is that the PIT applies this process initially to develop the system specification and possibly the next tier specifications. Then all the IPPTs apply this process to develop all of the lower-tier specifications for which they are responsible.

5.5.4.1 Audit, assure traceability, integrate, and validate requirements, F4162

Several kinds of traceability should be considered: vertical or parent–child, longitudinal to design and verification, lateral that links the requirements in specifications to the models from which they were derived, sources, rationale, and applicable document. Given that the requirements for any one item are derived by many people, it is necessary to integrate across them to achieve an optimum result.

Requirements validation is a subset of the program risk activity focusing on the product performance aspect of the program. All of the requirements in top-level specifications are passed through a risk filter to determine the degree of difficulty anticipated in complying with them. Where a concern is uncovered, tests, analyses, demonstrations, and examinations are accomplished to improve confidence that those requirements can be complied with. If it is going to require some period of time and a fairly involved processes to validate a requirement and the requirement is a very significant driver, it should be considered for selection as a technical performance measurement (TPM) parameter that can be tracked over time and coordinated with specific action plans to cause closure between the requirements value and current demonstrated capability. All of this work should be complete prior to holding the preliminary design review.

In this book and the other three books listed in Table 1.1, the author has allocated the verification process to a Verification Engineering department. This covers all item qualification, item acceptance, and system test and evaluation. It does not include the development evaluation testing that engineering must accomplish to gain confidence in their ability to satisfy challenging requirements in program specifications. Requirements validation is an engineering activity that should be accomplished by the same people who write the requirements.

5.5.4.2 Publish specifications, F4166

All of the work illustrated in Figure 5.7 is initially accomplished by the PIT for the system and top-tier items. F4114 is repeated by the assigned teams for the items for which they are responsible; however, many times it is necessary to completely define the corresponding problems. All of the IPPT requirements work is coordinated and integrated at higher system levels by the PIT. This may include team reviews chaired by the PIT or audits conducted on team results in process. PIT should maintain traceability data initially established by the team responsible for the specification. PIT must also manage the margin and budget accounts explained in the author's requirements book.

The PIT reviews all product specifications, process specifications/plans, and product concepts. During the development of these documents and concepts, PIT must coordinate between and across responsible IPPTs to ensure that needed information is exchanged between the teams. This is encouraged by making it clear to the IPPTs what interfaces exist in the product that touch the product items for which they are responsible and in pairing up the teams to attack those interfaces that join product entities for which they are jointly responsible. This work is accomplished through cross attendance in IPPT meetings.

5.6 Synthesize system, F42

Synthesis consists of three fundamental activities: (1) design the system, during which time the needed parts for the system are determined, (2) manufacturing synthesis, and (3) the procurement of needed material. So, we have design, material, and manufacturing synthesis as noted in Figure 5.9.

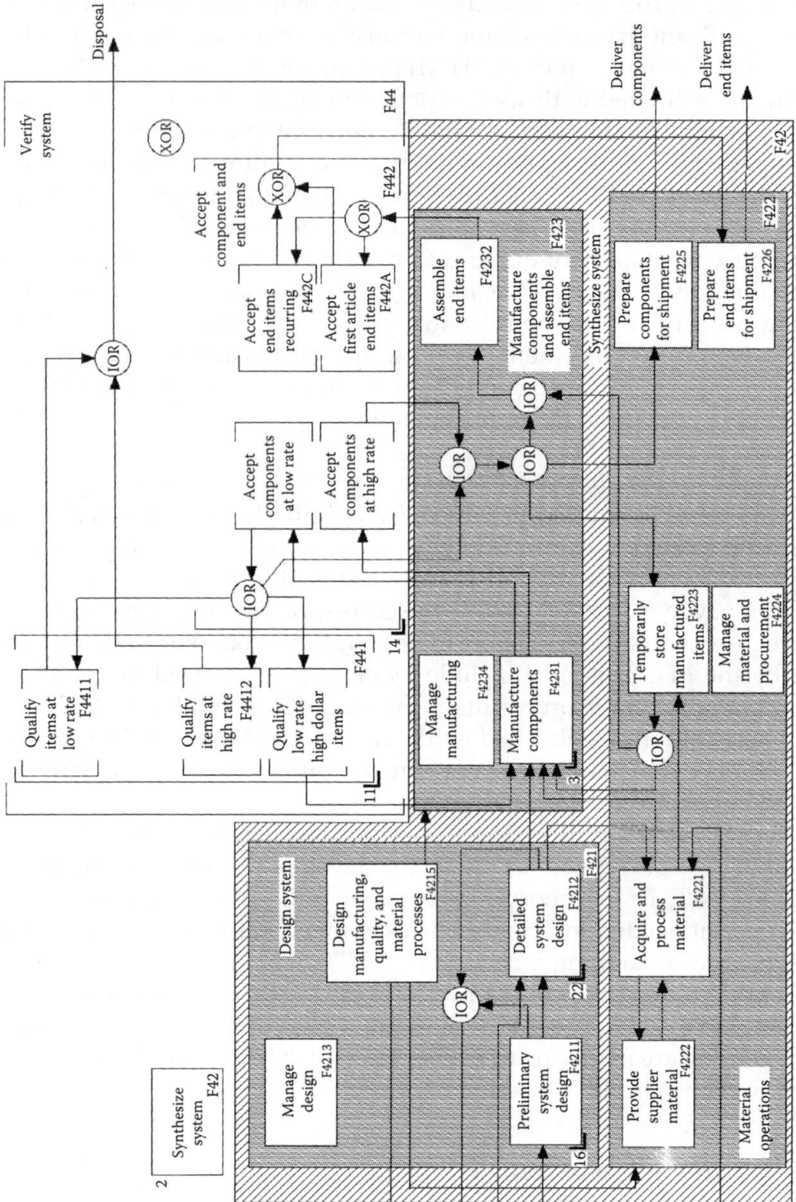

Figure 5.9 Synthesize system flow.

5.6.1 *Design system, F421*

Design work is accomplished in concept development (a part of the requirements analysis process covered in *System Requirements Analysis* [Grady, 2005]), preliminary design (F4211), and detailed design (F4212). The design work covers the product and related processes as well. The design work of all of the teams is integrated and optimized by the PIT. Item design and related process designs are reviewed and approved as a prerequisite to detailed design go-ahead.

5.6.1.1 Preliminary design, F4211

Figure 5.10 offers a detailed view of this process. A trade study process is encouraged where the preliminary design selection will not yield easily to sound engineering judgment. At the end of function F42111L5, the actual preliminary design work is accomplished in the task identified as document preliminary design, F42111L7. All of the work in this function (F421) is contributed to by people from many specialized functional departments and must be managed as a cross-functional IPPT.

The PIT does not do design work in the conventional sense. In Figure 5.10, the IPPTs are contributing to the process design work but the PIT must be in overall control of that evolving design. Also the PIT is accomplishing all of the integration and optimization work across all of the teams for product and process design.

The preliminary design system integration function, F42114, is allocated to an entity called a PIT in a program setting. However, every program activity must have a home in the functional organization that is responsible for providing all programs with personnel, tools, and practices corresponding to that responsibility. Therefore, these functions are allocated to an enterprise integration team (EIT) functional department yet to be fully characterized. The PIT is initially responsible for system definition (mission analysis, technical program plan development, and system specification preparation), applying one cycle of function F411, and subsequently for overall system development and integration once it has assigned integrated product team responsibilities to system elements and initially staffed the teams. Many functional departments will have to cooperate to accomplish all of the PIT work as is the case for all of the teams established on a program. But, to see this multiplicity of functional department responsibility clearly, one must look one or two levels deeper corresponding to the level that work is actually accomplished.

A large part of preliminary design integration and optimization revolves around reducing risk, as shown in Figure 5.11. Also, the work of the IPPTs is studied from the perspective of several disciplines and inconsistencies identified for resolution by PIT.

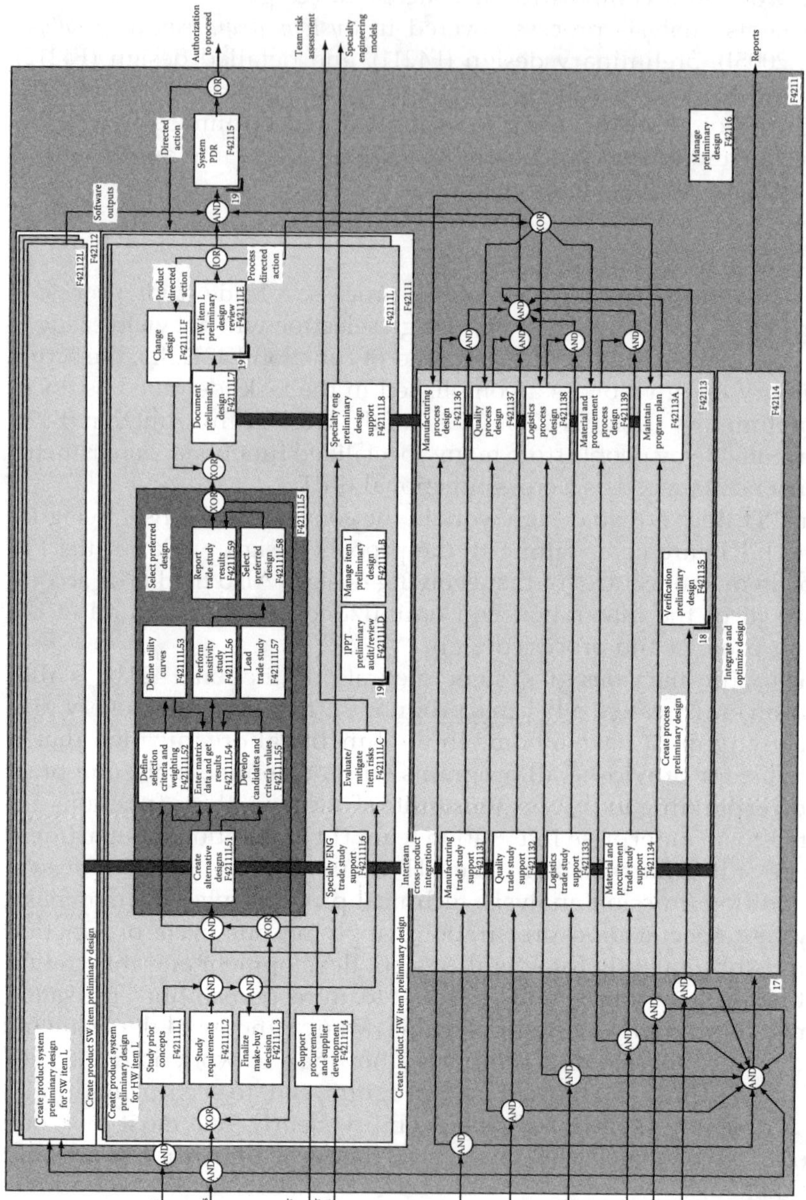

Figure 5.10 Preliminary design. (From Grady, J., *System Engineering Deployment*, CRC Press, Boca Raton, FL, pp. 43–63, 2000. With permission.)

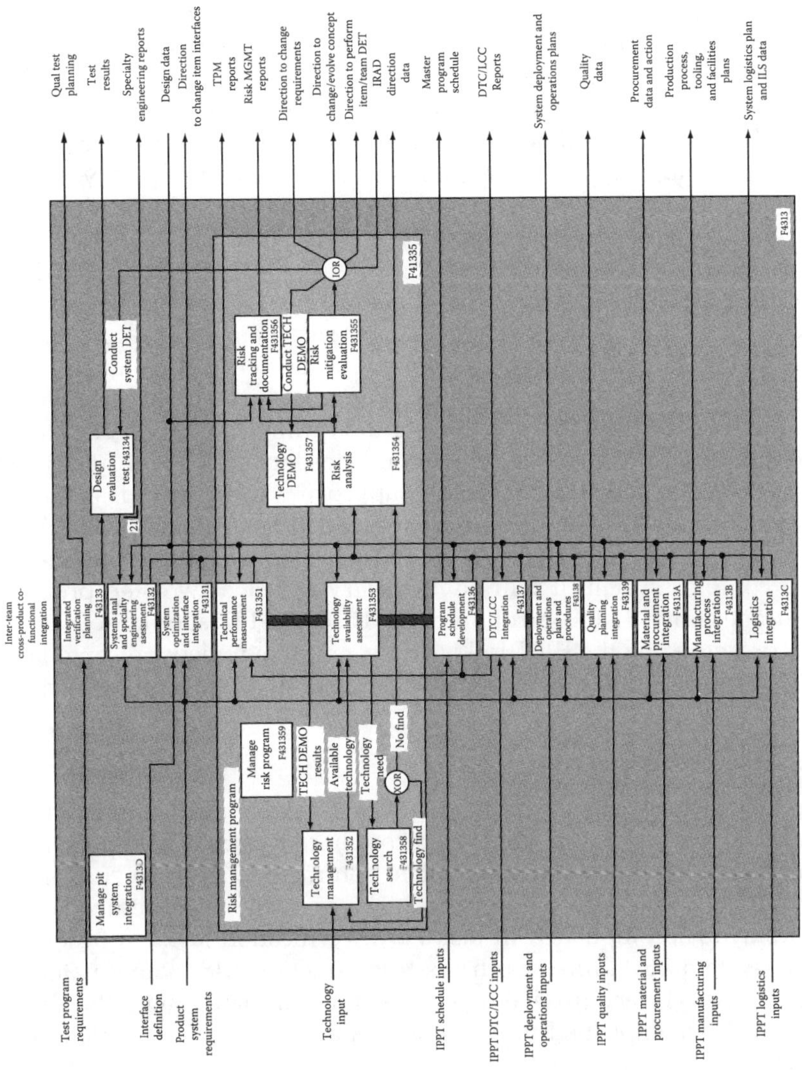

Figure 5.11 System integration and optimization during preliminary design. (From Grady, J., *System Engineering Deployment*, CRC Press, Boca Raton, FL, pp. 43–63, 2000. With permission.)

5.6.1.2 Item team detailed design, F4212

Approved preliminary design work is expanded to completely describe the preferred solution and submitted to a detailed design review as a prerequisite to manufacture and related procurements. Figure 5.12 offers a view of this process. As in preliminary design, these tasks are accomplished by specialists from many functional departments coordinated by IPPT management within the context of the team and across the teams by PIT accomplishing integration and optimization work.

The PIT accomplishes similar activities as in preliminary design phase with emphasis on detailed design feature compliance with requirements, evolving physical interface compatibility, manufacturing, quality, logistics, and verification (qualification and acceptance) detailed planning.

Generally, two levels of design reviews are held in a development effort: a system design review and item reviews. The former is held to expose the system design concept and determine if it will satisfy system requirements within a context of controllable risk. Item reviews are generally held in two sets, preliminary design review and critical design review. The former demonstrates that the performance requirements can be satisfied within cost and schedule constraints and commits to the development of the detail design. These two reviews are closeout actions for functions F4211 and F4212, respectively. The latter approves the detailed design and commits to manufacturing the resources needed for test and evaluation. In a simple system, one design stage and one review may be sufficient, but in a complex system it will be necessary to use a two-step process to encourage sound risk control. In either case, the PIT should hold item reviews presented by the IPPT followed by a system review presented by the PIT for the program manager that closes out all of the item reviews.

5.6.2 Material operations, F422

All product material identified in program engineering documentation is acquired from approved sources through contractual and purchase relationships with other firms, associate contractors, and the customer. This material is made available in accordance with schedule needs for manufacturing accomplished in task F423, verification tasks in function F44, and system sustainment purposes in function F48. Some material may also have to be acquired and expended in association with the engineering work accomplished to gain confidence in satisfying challenging requirements.

5.6.3 Manufacture system, F423

The product is produced in a physical sense. Material acquired in task F422 is processed in accordance with engineering and manufacturing planning

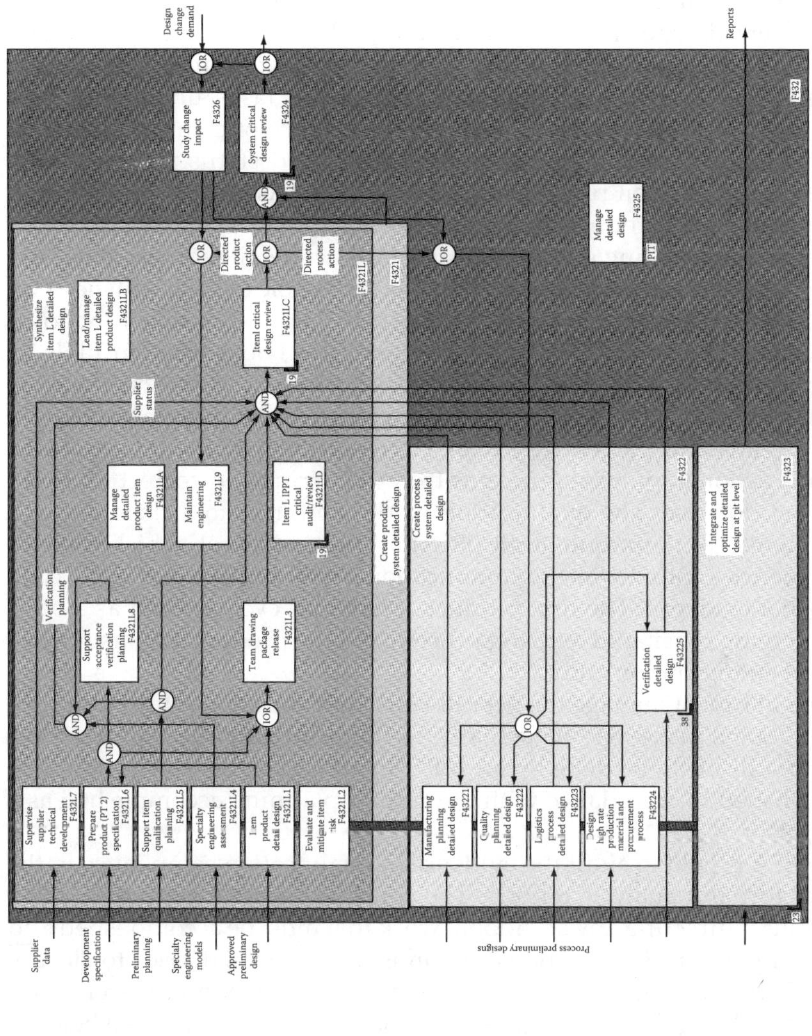

Figure 5.12 Detailed design. (From Grady, J., *System Engineering Deployment*, CRC Press, Boca Raton, FL, pp. 43–63, 2000. With permission.)

crafted in tasks F41 and F421. The manufacturing functional department is responsible for this activity for physical product manufacture. Software manufacture (coding) is an engineering responsibility.

5.7 Verify system, F44

The system engineering process involves three fundamental steps: define the problem (in specifications), solve the problem (through design), and prove that the design solved the problem (verification through test and analysis). Verification planning accomplished in tasks under functions F41 and F42 are implemented to develop evidence of compliance that is accumulated, correlated with planning data, and status reports developed. The verification process consists of three cycles, item qualification to prove that the product design is adequate, item acceptance to prove that the manufacturing process satisfies product detail requirements, and system test and evaluation to prove that the whole system satisfies system goals. Engineering is commonly the principal responsible functional organization, but in this book and the other three books listed in Table 1.1 quality assurance is made responsible for all formal verification work.

Management skill is brought to bear to implement the content of verification plans and procedures, track status of the activity, report status, and take action to prevent problems from occurring and correct for problems that do arise. The qualification work is brought to a close through a functional configuration audit (FCA) at the team level, which reviews the evidence captured in the qualification work and reaches a decision about that evidence. The first product acceptance cycle is used as a basis for reviewing the manufacturing process and the product that results in a physical configuration audit (PCA).

The PIT must manage the overall work flow during qualification verification. Some tasks may be actually accomplished by the PIT, but most of them will likely be done by an IPPT. In all cases, this work should be accomplished in accordance with the careful planning accomplished as a prerequisite.

A FCA is held to audit the evidence of qualification compliance in the form of test and analysis reports. The conclusions are captured in meeting minutes directing any additional work that must be accomplished and identifying any residual actions that must be completed prior to closure of the FCA. Generally, FCA closure is necessary to permit system-level testing involving the item in question. One or more FCAs may be held for each IPPT concluding with a system-level FCA to close out any remaining open items. Alternatively, in a smaller system, a single FCA at the system level may be adequate.

The first article through the manufacturing process should be subjected to a PCA to verify that the manufactured article complies with all manufacturing, engineering, and quality requirements. Subsequent production articles are simply subjected to acceptance and shipped.

5.8 System sustainment, F48

Throughout the development period, logistics support resources as well as product resources are defined. These resources from the manufacturing (F423) and material acquisition (F422) functions are applied to support field use of the product system through all or some portion of the product life cycle. At some point, the logistics support responsibility may shift from the producing contractor to the customer, which will require the resources to continue the support through product life. Sustainment is the principal responsible functional organization.

5.8.1 Logistically support system, F482 and F483

When the system enters system test and evaluation verification, it will require logistics support since things fail in test as well as normal use. Therefore, during logistics planning work accomplished earlier, provisions must be made for spares, special support equipment, technical data, and training relative to the system test crew and process. In the interest of cost and lack of availability of an abundance of resources so early in the program, compromises may have to be made such as providing factory support as the depot, the use of some factory test equipment rather than final field support equipment planned, the use of results of logistics analysis directly rather than polished technical data, and the cannibalization of parts to satisfy immediate spares needs. It is generally unwise to provide lavishly for resources in this situation because the results of system test and evaluation will commonly require changes in them.

When the system completes system test and evaluation verification and any needed design changes have been completed, the logistics resources must be ready to flow to outfit user initial operating capability and support any user testing that must be accomplished. Thereafter, the system must be supported over its lifetime. Commonly, at some point, logistics support will become normalized with the user taking over responsibility. Prior to that point, contractor personnel may be employed in the maintenance of the system or in only providing consultation through field engineering services.

5.8.2 Deliver/deploy product system, F481

The manufactured system resources must be moved from their point of manufacture to their point of operation and placed in initial operation.

This will commonly involve a transportation function. Deployment is accomplished in accordance with plans made during functions F411 and F421. Sustainment is the principal responsible functional organization.

5.8.3 Modify product system, F484

Based on customer needs, the system is modified. This will commonly involve coordinated hardware, software, and procedural changes. This function includes the whole process of defining the change, implementing it in design and manufacturing, and verifying compliance. This is essentially a mini program focused on a specific change that may involve a minor change or a substantial update to the system.

5.8.4 Dispose of system, F485

During the development of the system, thoughts of disposal may not be very popular with the program manager who is spending all of his or her energy to bring the program into being. It is necessary to develop the disposal process as part of task F41, however, since it may very well imply requirements for the product and supporting resource base. The system is analyzed for hazardous materials, and such materials once identified are disposed of in a safe manner. Ideally, these materials should have been identified and minimized in function F42. Some system resources may have residual capabilities that can be applied in new systems either as-is or after modification or update. Other system resources may be scraped. Logistics is the principal responsible functional organization.

5.9 Use system, F47

The system is placed in operation as planned in function F41 and verified in function F44. Over the life of the system, a need will materialize to modify the system for improved performance or to correct problems observed in operation and maintenance. Logistics is the principal responsible functional organization.

5.10 Manage program, F49

The overall program is managed toward program goals defined in program planning. The principal functional department is programs. The program manager works with an initial staff to build the program definition and staffing commonly during the proposal effort or early program studies that are either customer or enterprise funded. This book tells how the program manager may clearly define his or her program based on cataloged enterprise resources mapped to specific program planning identifiers. In so doing, the enterprise functional resources are clearly identified and to the extent that the program manager can pay for them through program man-hours, the program acquires the resources needed to implement the program in accordance with their plan and schedule.

The program plan, developed during preprogram work (proposal development or commercial program development), must be maintained accounting for changes driven by improved knowledge about program needs. PIT and program office must cooperate to accomplish this work to ensure that technical work demand is coordinated with resources available to accomplish that work.

All program documentation other than memoranda should be formally released, controlled, and correlated to particular design baselines defined by specific document sets or computer database versions. The configuration of design representations (models, simulations, mock-ups, etc.) must also be controlled. During the work accomplished in function F41, there is normally no physical product, so this work is documentation oriented.

5.11 Ensure product and process quality, F46

The traditional function of a quality assurance activity is to inspect produced product relative to some standard and determine if the product complies with that standard. In order to be effective, this quality function must be independent of the manufacturing agent. The practices effective in performing this function are well described in Juran and Godfrey (1999). If we should apply these practices relative to the product, does it not make sense to apply them to the process?

Individual and team efforts to accomplish program work cannot always be depended upon to follow program plans and enterprise practices. Commonly, there is no force other than program management to monitor and react to this problem, and it is not all that uncommon that it is program management that is encouraging the deviation from standard enterprise practices defined in functional department manuals. The author believes that the effective performance of system engineering work on programs and the continuous improvement of performance in doing that work demands program adherence to written functional department practices in order to encourage common process, employee work repetition, and all the benefits derived from common process.

Therefore, the enterprise quality assurance function should be applied across both product and process entities. Process implementation should be checked in accordance with some plan that is both effective and affordable. In this book and *System Requirements Analysis* (Grady, 2005), *System Synthesis: Product and Process Design* (Grady, in press), and

System Verification (Grady, 2007), quality assurance is made responsible for capability maturity model (CMM) application or other assessments of enterprise processes and their implementation. In this regard, it is important that it be verified that enterprise processes compare favorably to world-class practices defined in recognized standards for particular practices and that the programs implemented within those enterprises follow enterprise practices faithfully.

chapter six

Process documentation*

6.1 Enterprise generic documentation tree

The enterprise process could be documented in several different ways. The author suggests the development of an enterprise process specification, such as that included in Exhibit H, containing the flow diagram (Exhibit H/Appendix A), the functional organization diagram (Exhibit H/Appendix C), the function allocation matrix showing the correlation between process steps and the functional organization departments (Exhibit H/Table A-1), and a sketch showing how programs shall be organized using integrated product teams and program integration teams (PITs) (Exhibit H/Figure C-4). This document becomes the driver for functional department action to document their components of the enterprise identity.

The union of all functionality mapped to a particular department should be organized into a department charter defining the responsibilities of the department. At some level of indenture, all of these charter statements should be collected into one or more documents enlisting the work that must be accomplished by department personnel on programs that require that activity. All of these charters could also be included in the process specification. Figure 6.1 shows a top-level functional department documentation tree beginning with the enterprise vision statement.

The numbers in the lower right-hand corner of the document blocks indicate the responsible functional organization drawn from Division 3/ Exhibit H/Figure C-1. All of these responsibilities are obvious from the organization diagram except perhaps for the enterprise process specification. As mentioned earlier, the enterprise process owner is the enterprise integration team (EIT) also responsible for the enterprise development manual. Other lower-tier manuals not shown, in the interest of illustrating all of the engineering-related manuals in one simple figure, may also be necessary.

Some critics will complain that ISO 9001 encourages that all enterprise documentation be subordinate to a quality manual but the author believes this to be an unrealistic and flawed interpretation. However important the quality function is, it is not everything, and all of the overall enterprise responsibility should be recognized at the higher levels of

* The material in this chapter is from Grady, J., *System Engineering Deployment*, CRC Press, Boca Raton, FL, Chapter 6, 2000. Used with permission.

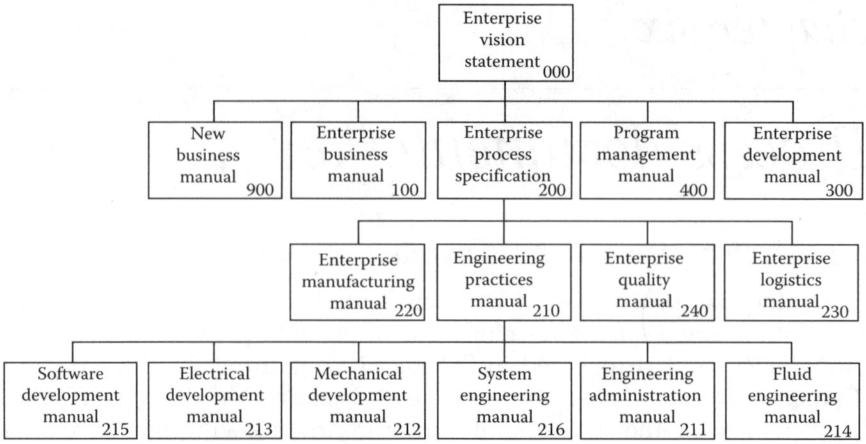

Figure 6.1 Enterprise process documentation tree. (From Grady, J., *System Engineering Deployment*, CRC Press, Boca Raton, FL, pp. 63–85, 2000. With permission.)

documentation. Another person may well elevate the quality manual one level to the first tier, and that is simply an individual choice as is the whole configuration illustrated in Figure 6.1. The precise configuration of this diagram is not important in isolation. It is important how this documentation is structured relative to the process responsibilities and functional organizational structure for a given enterprise.

6.2 Practices documentation content and responsibility

The manuals shown in Figure 6.1 must contain information telling what must be done on programs. They are internal standards for the performance of program work. They are an extension of the department charters contained in the enterprise process specification. The arrangement of content in the overall manual structure is determined by the way enterprise functionality has been allocated to the functional organization. Every function should be allocated to one or more functional departments and, where it is allocated to more than one department, one of the departments must be selected as the principally responsible department as illustrated in Figure 6.2. Function F has been allocated to departments W, X, Y, and Z. Department X has been identified as the principal department.

Department X has the responsibility of leading the effort to document function F's work requirements with the support of departments W, Y, and Z and the extension of that work definition to their individual departments. Department X is also responsible for selecting or developing the tools to be used by all four departments in the accomplishment of this work on

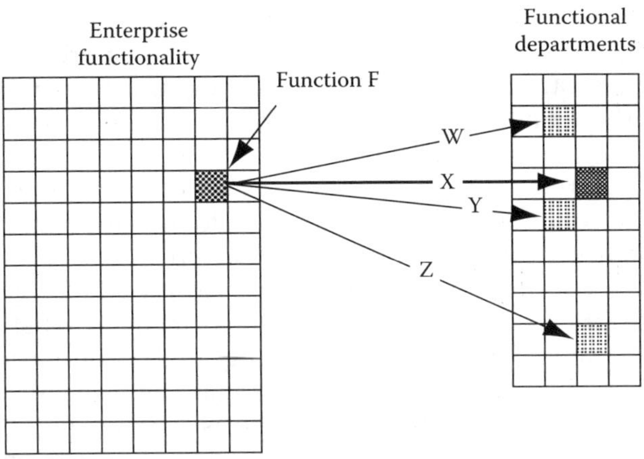

Figure 6.2 Principal functional department. (From Grady, J., *System Engineering Deployment*, CRC Press, Boca Raton, FL, pp. 63–85, 2000. With permission.)

programs and referenced in the practices manuals. Finally, department X must also develop or select the means by which persons in all four departments will acquire the skills necessary for performing the work corresponding to function F. This may be the selection of a textbook covering the activity and purchased for enterprise department libraries or the development of a training course. Department X becomes the project engineer for function F, leading the efforts of several departments for that task.

6.3 The process specification

The process specification is referred to in other parts of this book as an enterprise architecture report (EAR) that defines the enterprise process, organizational structure, and the map between them. This is in keeping with the intent to treat the enterprise and its programs as systems and to reengineer them in essentially the same fashion that the programs are encouraged to develop product systems using models to satisfy customer needs. In this and other books by the author, programs are encouraged to publish system architecture reports (SAR) with essentially the same content as called for in this book for the EAR.

6.3.1 Our past is not always a happy one

Companies often have prepared process specifications on manufacturing processes like heat-treating and printed wiring board treatment. But all too often they have not captured their engineering development process in a similar specification. The result is that there is no clear basis for program peculiar planning. Lacking this basis, each new program is planned as if it were the second program ever planned. Commonly, the company does not want to repeat the horrors of a previous program, so they redesign themselves for the next program. They never take the time to clearly define their process, so this pattern is repeated over and over again.

As a result, the company has no basis for continuous improvement, for there is no common process upon which to base those improvements. It may remain unclear precisely how each functional department in a matrix organization should contribute to each program; thus, the functional organization may not be constituted correctly to satisfy enterprise needs on programs. The resources that programs need in terms of qualified people, good tools, and best practices may not be available from the functional organization. In short, the enterprise will not be optimized relative to the product system development work with which it must deal.

In desperation, the company may well surrender to the siren song of process-based organization structures. Some companies have concluded that their functional organization is not serving their needs and replaced the functional organization with a basis in something called processes. The engineering department might be renamed as product development department and include a somewhat broader range of talent. The theory is that the notion of department builds walls around fiefdoms that are defended and fought for by their managers in ways not in the best interest of the enterprise, for personal ego motives for example. The departments are thought to be inwardly looking rather than cooperative in nature. This criticism is all too often valid.

The functional departments are replaced by process groups, and for a time, the enterprise may actually function quite well driven by the youthful energy of recreation. Over time, the same forces that resulted in the ills that encouraged this change will, lacking energetic management attention, revert to the same "department attitudes" that were previously observed. The only difference will be that the fiefdom boundaries will have changed.

The author maintains that in the conversion process, some damage will have been done to the enterprise. The original functional organization was likely based on the way the enterprise had decided over a long period of years to partition man's knowledge base in support of its programs. The new process-oriented organization will likely not overlay man's knowledge base in a pattern as clearly as the earlier one did. The knowledge needed to accomplish work in the new organizational structure may not match the way the rest of the world has partitioned knowledge (as in universities, for example), making it difficult to find new people who are ready for employment in the new structure. We will have to put more resources into worker training.

The functional organization, by whatever name, should be oriented toward man's knowledge base and how we traditionally partition it to specialize. Universities and societies commonly set the mold for this, and to the extent that we deviate from this mold, we cannot help but take on some process hysteresis losses relative to our competitors who have organized traditionally.

6.3.2 *The current reality*

Lest the reader misunderstands from the previous discussion, having a clear process definition is desirable. The author only suggests that it not be selected as a model around which to organize the functional department structure in an enterprise managed through a matrix. In the next part, we will take up an optimum organizational structure with clear lines of responsibility for everything the enterprise must accomplish for its own well-being and to the benefit of its customers. It will be based on two fundamental realities.

First, we have to recognize the way man commonly partitions knowledge to take advantage of the largest possible labor market while minimizing our retraining burden. Second, it should be clearly coordinated with the work that the enterprise must accomplish on programs. In the next part, we will rejoin this story after having exposed in this part a process definition that can be allocated to the functional department structure formed initially by knowledge domain needs based on customer base and product line.

The criticism of those who would encourage process-based organizations that functional departments grow into fiefdoms is valid, but it is not because of the use of the words "functional" or "department." It is because they are managed by mere mortals, and that is the best we will ever be able to do when it comes to selecting managers. As a result, the same end is forecasted for their process organizations as well. There is, however, another way.

The problem is not with words but with a failure to recognize and hold this truth in front of us through time, that functional organizations survive over time while programs are born and die. Functional managers who remain in position over a long period of time acquire power and distort the organization for their own benefit. To combat this, the top management must make it clear to functional managers that this danger exists and prevent organizational changes that might result in a lapse of watchfulness. The purpose of functional management must be very clearly identified and enforced. Functional managers owe the organization and its programs three things for the honor of serving in that position:

a. People who know what they are doing in the kind of work for which the department is responsible. They should have the right knowledge, understand their part in the generic process, and be able to deliver the results of their work on programs in a timely way.
b. The best possible tools available coordinated with personnel training to use them.
c. Best practices that clearly identify what has to be done generically, in what sequence, and links to source knowledge for how to do those tasks.

These resources must be coordinated, of course, within the domain of the individual department (such as electrical design, for example) but also across the departments as will be covered in Part 7.

6.3.3 *Structured analysis in your future?*

Enterprises are seldom born large and confused. They get that way over time driven by shifting demands and lagging responses compounded. Ideally, an entrepreneur would apply a structured analysis to his or her vision before moving into the marketplace. This would be expecting a great deal from a highly creative person capable of imagining and implementing something that has never before existed. People who are capable of doing this are commonly not well equipped to handle the details suggested in this book to form an enterprise based on the intended functionality.

Many of these entrepreneurs succeed through their genius and a lot of help from others, and their organization grows in size and value. Over time this organization will exhibit symptoms of inefficiency because the responsibilities are not clearly defined. At that point, the enterprise is a good candidate for reengineering. While the process described in this book will work in forming a new enterprise, the bad news is that it likely will never be used for that purpose because of the nature of the people who are capable of forming new enterprises and the way they are wired. The good news is that this process can be applied just as well to enterprise reengineering for all surviving firms. Division 3/Exhibit H offers an enterprise engineering process prescription following a structured analysis process.

6.3.4 *A representative enterprise generic process specification*

Division 3/Exhibit H offers an example of a generic process specification complete with representative process and organizational data under the title of EAR. This document is paginated within the Exhibit H cover sheet as a separate document that may be removed in paper or electronic media and edited for internal use. This process specification contains the requirements for the enterprise process and organizational structure. The design of the process falls into the several manuals appearing in the enterprise documentation tree.

6.4 *Engineering practices manual*

The engineering functional department should have a general manual providing policy and guidance on the implementation of engineering work on all programs. This book does not offer a copy of such a document.

6.5 *System engineering manual*

6.5.1 *Preparation*

This book is focused on system engineering and the deployment of it into programs, so we will discuss only the development of the system engineering department manual. The development of the other documentation can follow a similar pattern. There are several sources of manual content. One could begin with *System Engineering Planning and Enterprise Identity* (Grady, 1995) that includes a complete system engineering manual (SEM) in paper and computer disk formats. The disk copy can be edited as seen fit by the company, or it can simply be used as is. Alternatively, one could turn to an available national system engineering standard and use it as a model. IEEE 1220 was intended as a commercial system engineering standard and EIA 632 was initially intended as a replacement for MIL-STD-499B that was prepared but not approved by DoD. MIL-STD-499A or 499B could be used as well since the basic process discussed in all of these standards is essentially the same. The Defense System Management College (DSMC) has crafted several versions of its system management guide that could also act as a guide for one's SEM development.

The supplementary data supplied in the JOG System Engineering student manual Division 3/Exhibit I related to this book includes a complete SEM in paper media and an available set of files provides this document in Microsoft Word for Macintosh and Windows machines. This document is different from the SEM included in the author's earlier book in that it is actually a process specification rather than a textbook providing how-to information. The author reluctantly concluded that the normal enterprise would be hard-pressed to create what is essentially a textbook in this field and should focus on something along the lines of a process standard like EIA 632 and complete the linkage to the how-to knowledge through reference to other documents and sources. The SEM in Division 3/Exhibit I is based on the structure of EIA 632.

If you choose to use one of these existing documents as a standard for preparing your SEM, you could simply scan it and begin editing it to your needs assuming it is not a copyrighted document. Some of the documents identified above are copyrighted. When starting with an external standard also you should make sure that you have addressed the results of your enterprise functional analysis and allocation. It may be necessary to rewrite portions of the source material, add material, and delete other material in order to craft the final document. An organized way to do this is to develop a map between the functionality uncovered in your analysis and manual content thus establishing traceability between the two.

Rather than begin with an external model, you could begin with the list of allocations made to the system engineering organization. A brain storming session, attended by several respected system engineers, can be very helpful in focusing these allocations into an outline for the SEM. This exercise can be implemented by inscribing each function on a yellow sticky back note and initially placing them in some kind of outline arrangement rather than beginning from scratch. The team should try to move to consensus on the way the functions have been hierarchically organized, which ones are major points and which ones are subordinate.

Another way to do this is to apply computer projection where the functions are objects on a desktop that can be moved around relative to each other. Periodically, you should save numbered baselines so that you can go back to earlier versions if necessary.

Once the outline is agreed upon and it has been verified that it covers the allocated functionality, you can start writing the document. Make assignments against the outline and schedule the document development.

A third approach for developing the SEM is to draw upon some of the past program SEMP that you have prepared. Even if you never followed them, they may have good content. You may be able to cut and paste a quality SEM from all of this material with a few remaining original writing or editing jobs remaining. All of this work is of no value, of course, if the discipline and management force is not present in the enterprise to enforce the use of these manuals on programs.

6.5.2 External standards mapping

As you build your SEM, you should find out what standards of system engineering your customers prefer and map the content of those standards to your evolving SEM. You should end up with one or more matrices that show for every element of your customer's preferred standards where the customer can find the corresponding coverage in your SEM. It is possible that you will choose not to do some work covered in the customer preferred standard, so when you quote a job to the customer you should make it clear what parts of their preferred standard you will apply and what parts you will not. You should have a reasonable explanation why you have chosen not to perform deleted parts of their preferred standard.

Complete this job by tailoring the customer-preferred standards for agreement with your manual. This can be done using the legalistic or in-context tailoring methods. In the legalistic approach, simply make a list of the changes you would make in the document, 1 through N. Each of these statements covers a change such as "13. Delete paragraph 3.4.5.3." or "Change 13 pounds to 15 pounds in paragraph 4.5.3.2.1." Your traceability data should be to the tailored standard. It should be noted that tailoring should not be applied to any laws or regulations that your customers require to be respected in a contract. This is called breaking the law.

6.5.3 The discipline for process compliance

If your organization has no history of a written process definition or of following its own process documentation, something has to change, preferably before you commit resources to preparing a new generic SEM. Given that you prepare an SEM, what will ensure that its content will be followed? Top management must be prepared to expend some energy to cause the content to guide behavior. If management is not prepared to expend the necessary energy, then the attempt to improve process control will fail. One of the secrets of the effective management of process improvement actions is to limit the scope of the change at any one time to that which can be easily communicated and accomplished. Management must communicate a simple message repeated over and over with conviction transmitting some evidence of displeasure where the direction does not have the correct effect and offering some form of positive encouragement where people perform as desired.

People at the working level in engineering organizations often endure periodic intense management devotion to new initiatives sweeping through industry, contribute their interest and energy to comply, only to find after a brief period of time that the management is no longer interested in that initiative. If this process is repeated often enough, those people become reluctant to participate energetically. As you prepare the SEM, also prepare the management for support of the process defined therein.

6.6 The program management manual

The enterprise should have a generic program management manual providing policy for program management applicable to all programs. A principal of value to system engineering would include a demand that effective system engineering techniques be applied early in a program. In general, program managers should implement the common process approach because the alternative will fail to provide benefits of continuous process improvement, improved implementation through repetition of process, and a pervasive understanding among employees that they know what is going on.

Program managers should be clearly identified as the principal enterprise person on a program. At the same time, program managers should be cautioned that they must apply the common approach defined by functional management while continually seeking out processes that are not providing them with full value commensurate with their program budgets.

6.7 Practices deployment to programs

Later in the book we will introduce a document called an integrated master plan (IMP) that identifies all program work. Some customers encouraging the use of this kind of planning document suggest that it contain two sets of information called the work definition and narratives. The latter includes text and graphics that explains how the program work is going to be accomplished. The author maintains that the enterprise should build a generic set of practices and support how-to information, which is essentially the same on every program. The program planning will tell what has to be done, by whom it shall be done, when it must be completed, and maybe why. Each program task is an implementation of a generic task identified in the enterprise process specification and about which generic practices exist extending to tools to be employed, and how-to information references. Where a customer requires the development of an IMP or other form of program planning data, the contractor should reference generic data for access to these details rather than creating a new process description for every program. If you must craft narrative information as an integral part of the IMP, this data should be created by simply cutting and pasting from the source generic documents.

For each program task, we should identify a generic process description. One can picture a matrix making this transform of generic planning data into program planning data. In that the program data is going to employ tasks from your generic process diagram applied to items that must be developed for a specific system, much of this matrix can be canned and simply edited for the program. You should not have to create any new how-to data for a new program, rather apply the generic data. If a contract requires work never before accomplished on prior contracts and not captured in your generic data, you will have to build new descriptive coverage and may wish to embrace this as part of your generic process description.

6.8 Process quality assurance on programs

All of the maturity models encourage an enterprise to have a written process description and to follow it. If the reader follows the prescription in this book, the written process will be made available. It is even harder to follow the second step of actually doing the work in accordance with the written instructions. An enterprise requires a process quality agent to ensure this happens not just for the manufacturing process but for all of the other parts of the whole development process as well. The author has mapped that responsibility to the quality functional organization. The quality member of the PIT should periodically assess program compliance with generic and program-specific planning data and report the results to program and functional management for possible corrective action.

6.9 Functional metrics

Concurrent with the development of the practices, we should consider what measurements can be made when implementing the practices on programs to indicate the quality of performance of the process. When writing requirements for inclusion in specifications, it is a good idea to also craft the verification requirements at the same time. The reason for this is that it forces you to tell how you will determine whether or not the design that evolves satisfies the requirements. In the process of trying to phrase the verification requirement, you find out a lot about the quality of the product requirement. If the requirement is not quantified, for example, you will find it very difficult to write requirements for proving that the product satisfied the requirement to say nothing of the difficulty the designer will have trying to craft a design responsive to it. The ultimate bad requirement might be, "The product must work well and last a long time." This would be a very difficult requirement to verify compliance with.

The author recommends that you apply this same approach in creating your practices documentation. As you create a paragraph for the SEM, write another paragraph about how it can be verified that the program complies with that content. If possible, craft a metric defining with mathematical precision how to make the measurement and what it means. The author has never seen a practices manual that included a verification section along the lines of a specification, but it might not be too bad an idea.

6.10 Process improvement

Programs present the enterprise with tremendous opportunities to observe the quality of their performance, and all too many enterprises disregard the information available to them about their quality of process performance. Sure, you need metrics that act as process test points feeding you information about process performance details that can be tracked and compared over time to detect changes in performance. But you also need to apply human assessment of the process on programs using quality engineers who can compare generic and program-peculiar planning with

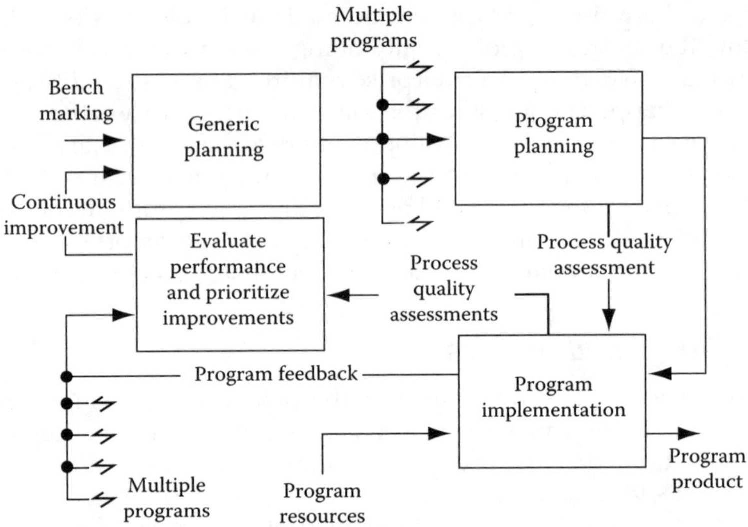

Figure 6.3 Continuous process improvement. (From Grady, J., *System Engineering Deployment*, CRC Press, Boca Raton, FL, pp. 63–85, 2000. With permission.)

actual performance on programs and identify any differences between them. This is the error signal of continuous process improvement. The program feedback lines in Figure 6.3 provide this error signal characterized by the metrics supplying data for review.

Ideally, we would have perfectly crafted practices and plans, and we would implement them absolutely faithfully. The real world, however, is not built to match this perfect model. There will be two sources of errors in the observation of program performance. First, the generic and program-specific planning data will include flaws or imperfect content. Second, we may or may not be following our planning data. Clearly, there are four possibilities here: (1) good planning data faithfully followed, (2) good planning data not followed, (3) bad planning data faithfully followed, and (4) bad planning data not followed. Of these choices, we would hopefully prefer the first one listed.

We have to decipher the error signal to determine which of these cases apply in order to decide what to do about program performance that we conclude can be improved. Our action in each case should be as follows:

1. Kudos to the program and the functional departments providing the generic planning data covered by the audit.
2. Critical feedback to the program to follow your practices with the possible replacement of one or more team leaders and even the program manager in flagrant cases.

3. Improve the planning data as quickly as possible and caution the program to be more watchful for practices in need of improvement.
4. Improve the planning data, either generic or programmatic, as quickly as possible and caution the program that their performance would be incorrect if their planning data had been satisfactory and to come forward in similar situations to report imperfect data rather than simply not following it.

By following this prescription, the enterprise will progressively improve its performance. The scenario painted above does not apply only to written practices, of course. It also applies to improving the staff competence, encouraged through process repetition, but this also requires seeking out the talent needed in the marketplace, weeding out, and separating or reassigning those who cannot perform in their current assignment and improving the tools applied on programs.

part three

Organizational structures

chapter seven

Introduction to organizational structures*

7.1 Updating matrix management

In order to discuss system engineering planning, we must make some assumptions about the organization through which we will deal. After all, the object of our planning work will have to be executed by this organization. There are three principal organizational structures common in industry: (1) projectized, (2) functional, and (3) matrix. In a projectized organization, the personnel are assigned to each program that has hiring and termination decision-making authority. In a functional organization, all personnel are assigned to departments that specialize in particular kinds of work, such as engineering, production, and quality assurance. The matrix organization approach attempts to marry the positive aspects of each of these approaches and avoid the negative aspects. This marriage is often less than perfect in practice, but this is not a slam against matrix management, rather a case of a bad implementation of it.

The author supports the matrix management arrangement for large companies with multiple programs. The matrix is characterized by (1) functional departments, led by a supervisory hierarchy, possibly including chiefs, managers, directors, and vice presidents, who provide qualified personnel, tools, and standard procedures to (2) project organizations responsible for blending these resources into a set of effective product-oriented cross-functional teams called integrated product and process teams (IPPTs), and managing these teams to achieve program success measured in customer terms.

The functional departments provide a pool of qualified specialists trained to apply department-approved standard best practices and tools in the development and production of products appropriate to the company's product line and customer base. The functional departments are charged with the responsibility to continuously improve the company's capability through small incremental improvements in personnel skills, tools, and procedures based on lessons learned from program experiences and continuing study of available technology, tools, methods, company

* The material in this chapter is from Grady, J., *System Engineering Deployment*, CRC Press, Boca Raton, FL, Chapter 6, 2000. Used with permission.

needs, and the capabilities of competitors. Company programs are internal customers of the functional departments. Note that this division of labor should result in a very lean functional management structure that is in tune with the times.

Program IPPTs are organized about the product entity structure reflected in the product entity diagram overlaid by the product-oriented program work breakdown structure (WBS). The teams are selected and formed by program management. These are led by persons selected by program management or by the teams from the personnel assigned with the approval of program management. Once identified, the team leader is responsible for (1) molding assigned personnel into an effective team, (2) concurrently developing the assigned product requirements, followed by (3) concurrently developing a responsive product design, procurement, manufacturing (tooling, material, facilitization, and production), test, operations and logistics, and quality concepts relative to the entity for which these are responsible. Some of these functions may have to be collected at the system level under a program integration team (PIT) that is also responsible for integration and optimization across the teams.

IPPT personnel must first focus on team building matters in order to form an effective cross-functional group. When the IPPT approach is first begun, there is a tendency to place an exaggerated emphasis on this step. You have to get through this phase as fast as possible to get your focus back on the work that these teams must do.

Next, the team must focus on product and process requirements analysis. The product of this work should be documented and approved by program management. Ideally, the teams would apply a structured, top-down requirements analysis process such as that described in *System Requirements Analysis* (Grady, 2005). The team must, concurrently with requirements work, develop alternative concepts responsive to the requirements as a way of validating and demonstrating understanding of the maturing requirements. Where there is no clear single solution, the team should trade-off the relative merits of alternative concepts. The team must be very careful not to influence the requirements identification too strongly by concept work (i.e., it must avoid leaping to familiar past point designs), but should take advantage of valid concept-driven requirements identified. This requires a delicate balance that should be assertively monitored by a system IPPT, called the program integration team (PIT) in this book and the other books listed in Table 1.1, and by program management at internal reviews. The requirements must also drive manufacturing, quality, logistics, material, test, and operational process design work as well as product design.

Only after approval of the requirements and complying preferred concepts at an internal design review, and at appropriate customer and/or independent verification and validation (IV&V) reviews, if required, teams should be authorized to proceed with concurrent product detailed design work and test, manufacturing, tooling, material, operations, logistics, and quality process design work. The team must implement all of the assigned tasks within budget and schedule constraints producing documentation that clearly defines what must be produced and how it shall be produced, tested, verified, and used to comply with the predefined requirements.

The functional management staffs programs with qualified personnel appropriate to the tasks identified on the project and the budgets for these tasks relative to the several functional departments. Personnel must be assigned to programs with a reasonable degree of longevity, because IPPTs require personnel assignment stability to be effective. The reason for this is that the principal work of teams is communications, and communications networks are fairly easily fractured by personnel turnovers and these fractures require time to heal.

The day-to-day leadership of personnel assigned to programs should be through the IPPTs and their leaders rather than through the functional departments. Personnel assigned to one team throughout a complete quarterly or semiannual personnel evaluation cycle should receive an anonymous evaluation from their fellow team members coordinated by the team leader. Personnel assigned to more than one team in one evaluation period can receive evaluations from all teams within which they served. Each functional chief should review all of the program team evaluations for persons from his or her department and integrate this data into department ranking and rating lists used as a basis for all administrative actions (training needs, compensation adjustments, promotion/status quo/setback decisions, and program assignment considerations).

Functional department supervisory personnel and senior working personnel under their guidance should monitor the performance of the personnel assigned to programs from their department and provide coaching and on-the-job training where warranted. The functional management also should provide programs with a source of project red team personnel used by the program manager to review the quality of program performance in the development of key products. Functional department chiefs can then use this experience as a source of feedback on improvement needs appropriate to current standard procedures, tools, and training programs.

Functional chiefs, managers, and directors should be encouraged to follow the performance of personnel from their department on programs, but should be forbidden to provide program work directions to these personnel, that is, the functional management may provide help and advice to their program personnel in how to do their program tasks, but should not direct them in terms of what tasks to do nor when to do them. Only the program-oriented IPPT leaders and lower-tier supervisors, if any, should be allowed to direct the work of team members on programs.

Functional management personnel, in this environment, are rewarded based on the aggregate performance of their personnel on programs (in terms of CDRL submittals, major review results, budget and schedule performance, and noteworthy personal efforts recognized by project management) and the condition and status of their department metrics, which may include depth, breadth, and the quality of standard procedure coverage for the department charter, toolbox excellence, and personnel training program effectiveness.

Program budgets are assigned to IPPTs by program finance. The IPPT leaders are not only responsible for the technical product development tasks defined in our program planning documentation, but the team budget and schedule as well. IPPT leaders report to the program manager. Teams are generally, but not necessarily always, led by engineering personnel during the early project phases (when the principal problem is product concept development and design) and later by production personnel (when the principal problem is factory oriented). Some companies have solved this leadership problem with product-oriented engineering-led development teams through the article inspection team first and factory- or facility-oriented production teams thereafter.

7.2 A model program organization structure

In our model of the perfect world, each project includes two or more IPPTs and one PIT. Some people prefer the name system engineering and integration team (SEIT) in place of PIT, and these readers can use this term or any other in place of PIT. The author avoided the term SEI because it has a special application in large contracts where the customer contracts with a firm to perform system engineering work for them.

The PIT should be led by someone qualified to serve as the deputy program manager drawn from available senior personnel. During the early phases, this person should be someone with extensive engineering experience. In later program phases, this position should be reassigned to someone with an extensive production experience. The senior engineering person assigned to the PIT in some companies would be called the program chief engineer, and that is the case in this book. This person is responsible for monitoring all engineering work on the program and coordinating changes in engineering personnel team assignments through the PIT and IPPT leaders. The chief engineer is the principal product technical decision maker for a program. Other major functional departments will have persons assigned to the PIT with similar responsibilities. The chief engineer is the engineering leader of the PIT and could be the PIT leader as well in early program phases.

The PIT is responsible for technical direction and integration of the work products of the IPPTs toward development of a complete product.

This includes (1) performance of initial system analysis and development of program-level documentation (such as the system specification and other high-level specifications, program plans, system architecture and interface block diagrams and dictionaries, and system-level analyses); (2) mapping of the evolving system architecture to IPPTs and formation and staffing of the teams; (3) review and approval of IPPT requirements documentation; (4) granting authorization for IPPTs to begin design work based on an approved set of requirements, concepts, schedules, and budget; (5) monitoring the development of interfaces between team items; and (6) development of interfaces between the complete product and external elements (system environment and associate contractor items).

The PIT, like the IPPTs, reports to the program manager, as illustrated in Figure 7.1. A project business integration team (BIT) includes all of the project-level administration functions, such as scheduling, finance, configuration management, data management, personnel, program procedures, project-level meeting management and facilitation, action item management, program reference document libraries, information systems services, and the program calendar of events. The BIT could, of course, be merged with the PIT.

IPPTs organized at the system level may have to further decompose items for which they are responsible in order to reduce the problem space to workable proportions. In this event, the system-level IPPTs (those illustrated on the first tier of Figure 7.1, which report to the program manager) will create their own sub-teams, like IPPT 2 in Figure 7.1, and acquire any additional personnel needed to accomplish sub-team tasks. Sub-teams may be fully staffed for independent work or rely heavily on the parent-level team for specialists. Sub-team leaders are referred to as either IPPT leaders or item principal engineers as a function of the magnitude of the task.

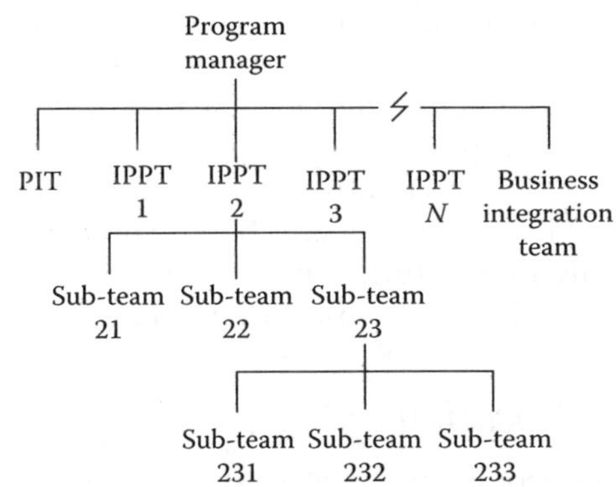

Figure 7.1 Program team structure. (From Grady, J., *System Engineering Planning and Enterprise Identity (SEPEI)*, CRC Press, Boca Raton, FL, p. 27, 1995. With permission.)

At the lower system levels, such as an on-board computer or a valve assembly, the person assigned the development responsibility is referred to as a principal engineer (as under sub-team 23 in Figure 7.1). The principal engineers draw upon the specialty engineers assigned to the parent IPPT (or sub-team) for related specialty work. IPPTs are responsible for the integration of the work of all sub-teams created under their authority. The PIT has this same responsibility over all IPPTs at the system level.

During the early development phases up through the FCA, the chief engineer should normally lead the PIT and be the principal product system-level technical decision maker subject to management by exception by the program manager. The PIT is the internal reviewing and approving authority for the system IPPTs. The PIT must review each system team's requirements and concept sets before the team is allowed to proceed to concurrent design.

Each IPPT leader is the principal technical decision maker for the IPPT and should review and approve subordinate team requirements, concepts, and designs. Each IPPT is responsible for the development of all its internal interfaces and jointly responsible, with the team responsible for the opposing interface terminal, for all their external interfaces. This responsibility extends from the PIT at the system level down through IPPTs, sub-teams, and principal engineers as a function of the scope of their responsibility.

Phantom IPPTs can be formed for items to be developed by associate contractors and work accomplished with this team through what is commonly called an interface control working group (ICWG). The senior member of this team will be selected for their interface development and interpersonal skills. On small programs, the ICWG may be a subordinate task of the PIT.

Teams or sub-teams may also be formed around major supplier items, such as a rocket engine, a guidance set, or other high-dollar or schedule-critical items. Leaders of these major supplier items may report to the program manager like other system-level teams or be organized as sub-teams to a propulsion or guidance system-level team. The team leader of a supplier team might be selected from procurement later in the program, but early on it should be an engineer. These teams might be treated like principal engineer items as a function of the dollar amount, development complexity, schedule criticality, or program policy.

7.3 Physical collocation options

The preceding text encourages a matrix structure with a very thin functional organization and a strong program organization through which the enterprise gets its contract work accomplished using the process, personnel, and tools provided by the functional organization. The understanding is that the personnel will be physically collocated in product teams in program work spaces. Many companies that have tried to implement this structure from a starting point as a matrix organization with a strong functional management structure, have found it very difficult because it de-emphasizes the functional organization where the power resided at the time the change was attempted.

7.3.1 The virtual organization

William H. Davidow and Michael S. Malone in *The Virtual Corporation* (1992), sketched an interesting alternative to the physical collocation of the people that may find adherents in the management of strong functional matrix organizations. A virtual team could be formed of members not necessarily physically collocated. The members would be connected via networked multimedia computers capable of simultaneous real-time video, audio, and computer data. With this equipment, a team could virtually function as if it were physically collocated.

The author's first thought after reading Davidow and Malone was that people could cooperate as a team from their homes and between different business units using the virtual concept. But a company could conceivably function under the IPPT concept while remaining collocated by function and gain some benefit from economies of scale in personnel assignment across several programs. This is an example of how you can become out of step with the advancing waves of technology and knowledge. Let us say that you conclude that you must de-emphasize your functional structure and team on programs. You succeed at this after a year of in-fighting with your functional managers. Then you find that you have the computer resources to enable your workers to cooperate via virtual teams from their original functional organizations.

Please read on before you conclude that this is the best alternative in organizing your personnel. This argument is only made to emphasize the point that you should be very careful, without being paralyzed, before committing to the latest management technology. The waves of change are striking the beach at a much faster pace than in times past.

7.3.2 Virtual functional organization

We might argue that if we can form virtual teams for programs, we could form virtual functional organizations as well using the same resources, leaving the personnel physically collocated by program. This is true. The thin functional management structure could periodically gather with its personnel collocated by program on a network version of a functional department meeting. In these brief meetings, the program people could discuss their problems, new approaches to their common process, and ways to use their tools. These meetings would be an important part of the continuous process improvement plan. So, we have our choice between the virtual product team (people physically collocated by function working on product teams) and virtual functional departments (people physically collocated by program team cooperating about the functional axis via networked computers).

7.3.3 Flexible optimum collocation policy

Between these two choices, it would seem sensible to place the emphasis on virtual functional organizations and collocated product teams. The relationships are more complex between the product team personnel than they are between the members of the same functional department. Given that face-to-face communication is more effective than even the most effective virtual capability, we should choose to align the physical collocation with the most complex relationship. So the author's recommendation is to physically collocate wherever possible by product team on programs.

There will be particular specialties in every company and on every program that cannot be supported by a single team or even a single program. For example, it may be difficult to justify a full-time reliability engineer, a mass properties engineer, or a nuclear survivability and vulnerability engineer on every team on a program. In these cases, the isolated specialists can effectively contribute to many teams via the virtual team concept and the company can gain some cost benefit from the resultant economy of scale.

7.4 Resistance to IPPT

There are many reasons why the organizational structure described here is very difficult to put in place in a company currently organized to perform autonomously within their functional organizations. These reasons include the following: (1) It disturbs existing organizational power relationships, (2) a conclusion is drawn that it is in conflict with the Department of Defense (DoD) cost/schedule control system (C/SCS) criteria defined in the DoD Instruction 5000.2, Section B, Attachment 1, and the WBS concepts defined in MIL-STD-881B, (3) the IPPT creates staffing problems for the functional organization, and (4) the IPPT interferes with an efficient personnel evaluation process. Let us look at each of these reasons.

7.4.1 Human resistance

Many of those currently in functional management roles will perceive this prescription for distributed power as a personal, professional, and career threat. If they remain in the functional management, they will see their relative power in the organization decline, as it must, to implement concurrent development. This will commonly result in resistance from those now in functional management roles. This resistance is inevitable. There is no best way to make the transition other than to announce it as far in advance as possible and openly discuss opportunities for those currently in power under the new system. It is, of course, entirely possible that at least some of those in the functional management should not be allowed to survive the change.

The long-range view of this shift, which many will not be open to see, is that a shift to the kind of organizational structure described above will attract a different collection of people to functional management positions. These positions should focus on providing programs with qualified people trained in the standard techniques defined for a function and skilled with the standard tools supportive of these methods. Personnel selection and training, continuous process improvement, and tool building are the principal roles of this new kind of functional manager.

Many of those now in functional management positions will not wish to continue working in the new environment. Some will prefer to migrate to program centers of power (program office, IPPT leadership, or program functional leadership roles). This is not a disaster, though some may perceive it to be; it is simply a rebalancing of the power centers. Those for whom power is a motivator will migrate to it wherever it resides.

Those who have worked in a matrix management structure for many years have observed many shifts in the balance of power between the functional organization and program organizations. These are not uncommonly driven by the personalities of those in power rather than by some sound rationale for improvement in the company's capability. The change we are discussing here requires an enlightened management whose first priority is to satisfy the company's customers and needs for company profitability. They must have a willingness to make changes toward these ends despite a possible temporary setback in the attainment of personal goals. This is a lot to ask of perfectly normal human beings populating management positions, and it will not often be satisfied in the real world. On the other hand, the change may be a necessary prerequisite to the company's survival in a very competitive world, and this can be a very good motivator.

Obviously, the movement into an IPPT requires a high level of support and leadership in an organization. Anything short of total commitment at the top will prolong the transition or destroy best efforts from any courageous and selfless people at the lower level. If your employees have endured a few aborted crusades in the past, it will require an even stronger attachment to your goal and some early success to build the perception of the inevitability of this goal.

7.4.2 C/SCS criteria conflict

7.4.2.1 What are the criteria?

The DoD has prepared a criteria, called the C/SCS criteria, for accounting and reporting systems that contractors must support in order to be considered for an award of a government contract. The C/SCS criteria listed in the DoD Instruction 5000.2, Part 11, Section B, Attachment 1 (formally in DoD 7000.2), includes five criteria: (1) organization, (2) planning and budgeting, (3) accounting, (4) analysis, and (5) revisions and access to data. This term is more commonly referred to as earned value currently.

Some companies following the matrix management model have concluded that there is a fundamental conflict between the C/SCS criteria and the integrated product development. The conclusion is that they must relate all program cost in a matrix between their functional organization and the contract WBS, which is a hierarchical organization of all products and services covered by the contract. Given that they wish to manage programs using integrated product development teams focused on WBS items, they cannot give the leaders of these teams the control of the budget allocated to these items, because they feel the criteria requires them to allocate it to cost account managers (CAMs) oriented toward their functional organization. Without budget control, the IPPT structure is doomed to failure.

If there is a conflict between these criteria and the IPPT, it resides in criteria 1 or criteria 3 as a function of how we interpret the phrases "functional organizational elements," "organizational structure," and "organizational elements." None of these terms are specifically defined in the DoD I 5000.2, Part 11, Section B, Attachment 2. Definitions for the terms "performing organization" and "responsible organization" do not refer to either the functional structure or the program structure of a matrix organization, but only to a "...defined unit within the contractor's organization structure..." The DoD does not require a contractor to organize under any particular model (matrix, projectized, or functionally). So, the criteria must be implementable by companies following any of the three kinds of organizational structures.

The DoD I 5000.2 criteria elements that relate to direct costs and organizational structures are listed below. Note that the numbering and lettering of these steps is different between DoD I 5000.2 and DoD I 7000.2, but the content is the same except that the contract WBS acronym used in DoD I 7000.2 is spelled out in the newer version.

1. Organization
    b. Identify the internal organizational elements and the major subcontractors responsible for accomplishing the authorized work.

c. Provide for the integration of the contractor's planning, scheduling, budgeting, work authorization and cost accumulation systems with each other, the contract work breakdown structure, and the organizational structure.
e. Provide for the integration of the contract work breakdown structure with the contractor's functional organization structure in a manner that permits cost and schedule performance measurement for contract work breakdown structure and organizational elements.

3. Accounting
   c. Summarize direct costs from the cost accounts into the contractor's functional organizational elements without allocation of a single cost account to two or more organizational elements.

7.4.2.2 *Alternative approaches*

The premise is that there is a conflict between this criteria and the use of an IPPT, or what the DoD finance community calls a work team concept. Given that this is a true premise, there are several alternative arrangements we should consider in reaching a condition of compatibility between this criteria and the IPPT.

7.4.2.2.1 Status quo Traditionally, in a matrix organization, we allow the CAM, assigned from a functional perspective, to manage a functional department's budget, and we hold him or her accountable for variances to a program manager responsible for one or more product-oriented budget segments, a role referred to in some companies (and in this book) as a WBS manager. The total program cost is decomposed into intersections between functional departments and the WBS. Program Office WBS managers manage the WBS columns of the matrix, and CAMs manage the organizational rows of the matrix. The IPPT structure seldom overlays this matrix in total alignment with either matrix axis. IPPT leaders are cut out of the budget management process and cannot possibly manage their team without budget control.

7.4.2.2.2 Change the criteria The DoD finance community and many other customer organizations are wedded to the criteria till the end, and maybe properly so. This was born out of contractor abuse and it will require decades of highly ethical behavior on the part of contractors to erase the stain of past excesses. Let us assume that we cannot change the criteria in any substantial way nor can we change it in any way in the near term.

7.4.2.2.3 Functional organization suppression The criteria do not attempt to define what a functional organization is, so we can suppress our functional organization in a matrix organization and establish a matrix between the WBS and IPPTs on programs. Probably, a suspicious government finance analyst would consider this a perversion of the criteria motivated by an attempt to avoid scrutiny in some way. The certification of our cost control system would be hard won. It would also make it difficult for the functional management to get insight into their personnel needs, since there would be no way for them to accumulate or project the future demand for their department's services.

7.4.2.2.4 Three-axis matrix It appears that we have three axes of interest in our data, IPPT, WBS, and the functional organizations, and that we are trying to manage with a two-dimensional matrix. So why do we not deploy an accounting system that can handle three axes or a mapping of budgets in three ways instead of the present two? While it does not appear that this would be out of step with the criteria (we would have an organizational structure to which we could map the budget), it would likely be hard to certify the system because of its unusual nature. This would also require some new development for the accounting system.

A variation on this theme could assign pseudo department numbers to IPPTs on programs, but other systems interfacing via department numbers may be disturbed. This alternative has a lot going for it actually. This approach can be implemented by simply assigning a unique department number to each of the IPPTs on a program. As the program comes alive, we transfer personnel from their home functional departments to the IPPT departments. Each IPPT department manager can function as a CAM, giving them the control we are seeking. The program office WBS managers provide control from the WBS perspective.

7.4.2.2.5 Projectized organization There is nothing in the criteria that requires us to organize in a matrix. Our company can be projectized, which would allow us to assign the organizational matrix axis to our program IPPT structure. Such a company suffers from a difficulty in making continuous improvements because it has no one methodology—no central focus for its specialized functions. Some companies avoid this problem by splitting each program into a separate business unit and allowing them all to develop their own identity and process.

7.4.2.2.6 Power to the WBS manager It is not uncommon for the contract WBS to be under the authority of the finance community in government and contractor organizations. This results in customer finance people coloring technical system composition decisions by virtue of forcing a particular product organization before the engineering community has evolved the most cost-effective product organizational structure based on the need. As a result, the WBS may be at cross-purposes to the product organizational structure about which the contractor would like to establish IPPTs.

Customer and contractor finance computing systems are not sufficiently flexible to be able to tolerate the adjustment of the WBS to align with a technically superior structure; so, the program moves forward with little interest in the WBS by the technical community. The WBS manager, therefore, can become a figurehead position in the program office, divorced from the direction of the technical effort, focused only on cost and schedule concerns. Meanwhile, the IPPT leader, focused on the technical aspects, cannot control his or her team because he or she has no budget authority. No one person has the authority to manage the integrated product development.

The right answer to this problem gives the budget and schedule responsibility as well as the technical responsibility to IPPT leaders. The first step in assuring this capability is to place the responsibility for at least the product component of the WBS in the hands of the PIT. This is not to say that the finance community should be stripped of all voice in this matter. Rather, the decisions on the WBS must be cross-functionally derived and not influenced totally by a finance position. Finance should be represented on the IPPT and, thus, have a channel through which to communicate their valid concerns. This joint responsibility between contractor finance and the development team cannot be successful if the customer finance people stonewall and insist on a predetermined WBS.

In the process of making WBS decisions between alternatives, the IPPT should weigh both financial and technical aspects of any particular choice, but almost always choose the alternative that best enables the IPPT. This alternative will generally be characterized by minimizing or simplifying the cross-organizational interface intensity between elements under the development management by different IPPTs.

With this arrangement in place, we can then proceed to merge cost, schedule, and technical control of the IPPT into one person. Let us accept that we need to retain the functional organization axis of our matrix organization in our cost-accounting matrix. This means that we must use the WBS axis to provide for IPPT leadership and control. Instead of making functionally oriented CAMs responsible for budget management, we associate with this job only a responsibility to collect costs for functional personnel planning and staffing, decision making, and financial reporting. Since the product WBS is structured in accordance with the product functional and physical structure, we now have an alignment between the WBS and the basis of assigning IPPT responsibilities to the organization of the product. The WBS manager is the obvious person to take on the role of an IPPT leader, with cost, schedule, and technical leadership.

Under this arrangement, functional organizations cannot be held accountable for cost variances. Since the WBS manager is in complete control, the WBS manager must be accountable for variances. Functional managers and directors should not be unbraided for failure to satisfy program schedules and cost targets. This is a WBS manager's responsibility, a program responsibility. Functional managers should be held accountable for providing qualified people, good tools, good procedures, all three of these being mutually consistent, and the continuous improvement of them based on program lessons learned. This division of labor results in a very lean functional management staff.

7.4.3 IPPT-stimulated personnel staffing problems

Some people hold that the IPPT creates staffing problems for the functional organization, and this is true if in moving to the IPPT the functional organization is stripped of any knowledge of future budget availability. The functional organization needs this information in order to be able to ensure that the right number of people with the right kind of training and experience are available when the time arrives to use them on a program.

Some of the alternatives we considered above, frankly deprive the functional organization of the information it needs to satisfy future program needs. The last one considered and recommended, does preserve functional management access to this knowledge.

The IPPT can also result in IPPT leaders contracting with the wrong functional organization for a particular kind of work. Some small companies may be able to handle the resultant volatility, and it can even be a source of good restructuring ideas in a rough and tumble, but potentially effective fashion. Large companies will generally have difficulty with this and should have a clear definition of functional department charters with an energetic enforcement of these charters by the functional management when program management attempts to deviate.

The reason that these boundaries have to be jealously guarded is that you wish to hold the functional organization responsible for continuous process improvement concurrently with high standards of performance. A functional manager cannot be held accountable for maintaining a company's proficiency in a particular specialty if a different department is being contracted to do this work on programs. The suppression of deviations from the functional charter responsibilities is not necessarily, however, the best response in all cases where this problem arises.

It is not uncommon that these occasions will be driven by a fundamental flaw in the way the company has assigned charters for specialties. So, when an incident occurs, the functional and program management should first consider if there is some value in considering an alternative organization charter responsibility map. If there is not, then the functional perspective should almost always win. At the same time, during periods when one or more disciplines are understaffed with respect to the demand for work, they may find it useful to establish temporary agreements with other departments to provide personnel to do their work. This same arrangement can be useful at other times as a means to create effective system engineers, or simply to increase the sensitivity to the needs and concerns of other disciplines, while creating personnel with broader qualifications to improve flexibility in satisfying changing program needs.

7.4.4 Personnel evaluation problems

Now, we know that some people, the late Dr. Deming, chief among them, are convinced that the devil himself or herself designed the personnel evaluation system used in much of industry. This system results in the functional management ranking all department personnel based on the functional manager's perspective of relative worth. This ranking is then used as a basis for salary increases, promotion lists, and, to some extent, work assignments that can influence future evaluations by virtue of experience gained through assignments. It is not uncommon that the evaluation criteria is flawed and applied in an irrational or uninformed fashion besides. Commonly, functional managers simply wish to get through this exercise as quickly and as effortlessly as possible.

If you remove people from their functional organizations and physically collocate them with IPPTs, you make it difficult for functional managers to observe their performance. If we believe the current IPPT/TQM literature, we accept that performance evaluation should be done by team members and not by the functional management anyway. So, how do we provide for personnel evaluation?

If we preserve the functional axis, we must accept that the functional management must be responsible for evaluation because they have the responsibility to provide programs with qualified people. However, the functional management need not be the only source of input for the evaluation data. If an employee were assigned to one and only one IPPT for a whole evaluation period (typically 6 months or 1 year), then the evaluation could conceivably be done by the IPPT members in some fashion, as Dr. Deming would suggest.

What happens when a person works fractionally on two or more teams over the evaluation period? It will not be uncommon for some specialists in traditionally low-budget specialties to work like this, but most will fit the pattern. We simply need some mechanism for the accumulation, merging, and distribution to functional department chiefs of team-derived evaluation data. Also, it would likely be useful to increase the frequency of evaluation events to perhaps quarterly, with this data merged in some way into annual figures for pay and promotion determination.

This is a valid concern for entry into an IPPT work environment, but we may have a little difficulty working up a computerized approach to acquire the data from people assigned to teams, collect and assess this data, and provide it to functional managers in a fashion that encourages rational decision making in rewarding the personnel for performance. It will require a change from present methods that generally are arbitrary, counterproductive, and otherwise, just terrible. Good riddance!

7.5 Model matrix for this book

Figure 7.2 illustrates the overall organizational structure that we will assume. One of the several program panes is expanded with program reporting in the vertical dimension and functional reporting in the horizontal dimension. The program managers report to the company or division executive as do the top people in the functional organization, on the left margin.

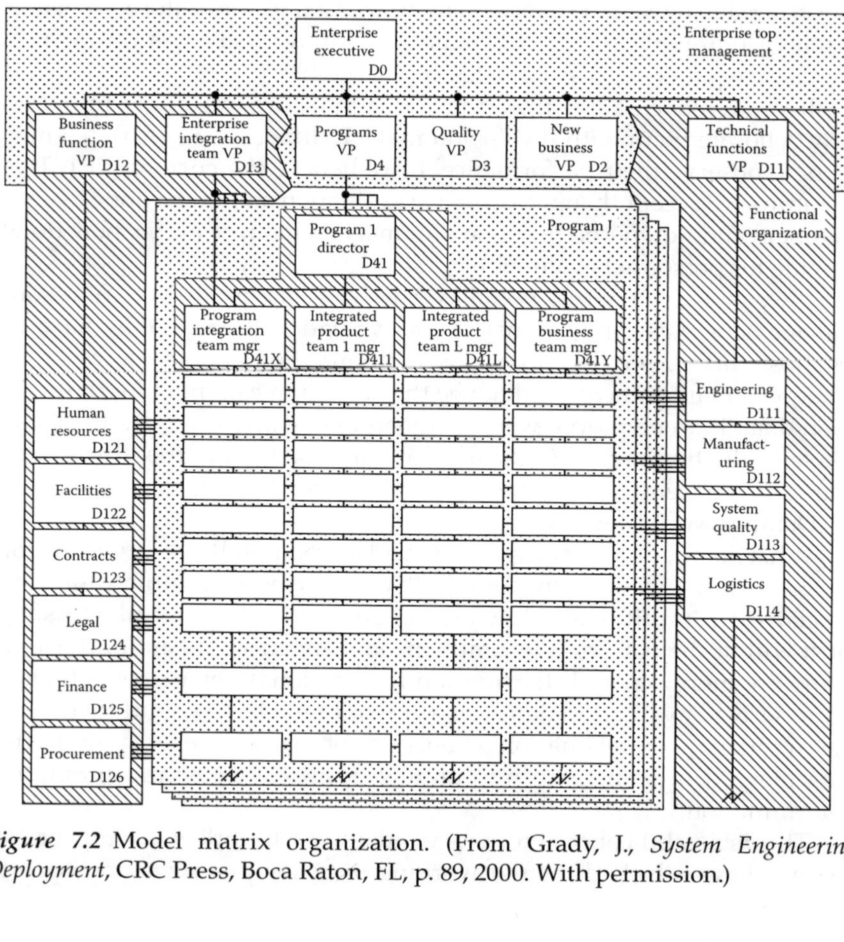

Figure 7.2 Model matrix organization. (From Grady, J., *System Engineering Deployment*, CRC Press, Boca Raton, FL, p. 89, 2000. With permission.)

Each program has two or more integrated product development teams aligned with the product structure that is coordinated with the WBS. Each team is staffed by the personnel derived by the program from the functional departments. In many companies, the functional chiefs/managers would control the tasking of the people staffing the intersections of the matrix. In our model, we will insist that the functional department chiefs specifically not provide program task assignments or work direction. This must come through the program channel traceable to an IPPT leader, also serving as a WBS manager.

Each program pane also includes a PIT responsible for system-level integration between the IPPTs and a business team responsible for program administration, finance, and contracts.

7.6 Enterprise integration team

If our company commonly has multiple programs running and a common set of resources, such as its manufacturing plant, the company management should consider themselves an integration team, which the author calls an enterprise integration team (EIT). This team must balance the available resources to serve the greatest good for all programs in the aggregate. This team can be formed as a working-level group that handles day-to-day work managed by an executive function composed of those who report to the company president or chief executive officer (CEO), or the division general manager, or the top-level staff can be thought of as an executive for whom this team implements the collective will. Another alternative that the author prefers is that this is a functional department reporting to the top management acting as the system agent for the enterprise, working through functional managers and the PIT on each program to integrate and optimize the enterprise.

The EIT acts to integrate the work of the entire PIT. It proactively seeks out conflicts in program plans with respect to company resources and stimulates interaction between conflicting programs to resolve the source of conflict in a timely way, avoiding cost and schedule impacts. This should be the principal work of the top management, rather than bureaucratic and administrative organizational matters.

This book encourages a very thin functional department structure, the managers of which are collectively responsible for providing programs with a sound and common product development, manufacturing, and verification process, and personnel trained in its implementation using available tools matched to the process and personnel capabilities. They are also the agents of change responsible for operating an effective continuous improvement process to upgrade tools, personnel education, and practices. However, the many independent actions of the functional managers may not always be in the best interest of the whole enterprise. The EIT should be the ultimate process owner, recognizing that the parts of it are owned by the functional departments. The EIT must function to cause the whole enterprise process to be as effective as possible and be involved in the selection and implementation of improvement actions.

In the case of a multidivision corporation, a corporate integration team (CIT) could be formed to accomplish the same purpose at the corporate level as the EIT is intended to serve at the division level. The utility of this integrated-team notion begins to run thin at some point, however. At higher levels of teaming, you are dealing with higher levels of abstraction and generality. There is a danger that in seeking a common practice at a high level (say corporate), you may force lower-level organizations (divisions, for example) into inefficient operation. While this book encourages a standard process, it does so for a single business entity based on that enterprise's situation in the marketplace.

So, the challenge for large corporations is to determine the optimum granularity of their division structure with respect to the product and customer base they seek to serve. As this optimum granularity is achieved, each business unit should be encouraged to optimize its identity for their plant and the market segment for which they are responsible. An efficient way to apply the CIT notion would be to make the CIT principally responsible for adjusting the granularity of the division responsibilities, determining when is the right time to consolidate or further decompose business units based on the current and anticipated business situation. This team should be very wary of trying to apply common practices on its divisions, however. Divisions are formed because of differences, and these differences are important.

This whole notion of integration teams is a recognition that we humans tend to look inwardly at our responsibilities rather than look across the gaps between our work and that of other organizations with whom we have to deal. This is true between an IPPT on programs, individual engineers within an IPPT, and at every level of an organizational structure. It is one of the key reasons that interface development is so difficult on programs. We should recognize that system engineers are specifically put on this Earth to work on these gaps between our knowledge and organizational entities, and make them responsible for stretching those at the boundaries to work across the gaps. Organizationally, this can be done by insisting that every IPPT be the integration agent for any lower-tier teams, assigning a PIT to do the cross IPPT integration and integration with the outside environment, and assigning an EIT to pull the parts of an enterprise together, its functional departments and programs.

chapter eight

Enterprise knowledge requirements and the functional organization*

8.1 The knowledge transform

Knowledge is what makes it possible to make progress on programs. Generic knowledge in the many specialized disciplines required on programs involved in solving very complex problems is applied to the problem to be solved and translated into program-specific knowledge. In order to accomplish this transform, an enterprise must begin the program with the correct generic knowledge mastered by the people assigned to the program, and these people must transform the aggregate of their knowledge into the information needed to solve the problem characterized by their customer to produce a product of value to that customer as suggested in Figure 8.1.

This transform process is accomplished in performing the planned work on the program, and several later chapters will focus on the related planning problems. In this chapter, we attempt to describe what knowledge the enterprise needs as the input to this process. This enterprise knowledge base is what drives the hiring done by the enterprise and the training it should accomplish for the staff. This population of employees should have been selected based on ideas that are traceable from the enterprise vision crafted in response to decisions concerning the intended enterprise customer base and product base. The people in an enterprise are the instruments through which everything happens. We are the physical architecture of the creator system banded together into functional organizations by a common knowledge space.

8.2 Generic knowledge needs

An enterprise that develops systems to solve complex problems requires the services of people specialized in some fields regardless of the product line and customer base. This includes all of the business-related disciplines: legal, contracts, administration, accounting and finance,

* The material in this chapter is from Grady, J., *System Engineering Deployment*, CRC Press, Boca Raton, FL, Chapter 6, 2000. Used with permission.

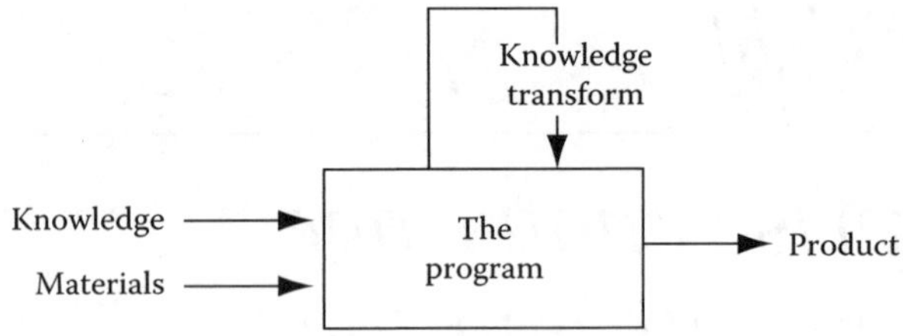

Figure 8.1 Knowledge transform. (From Grady, J., *System Engineering Deployment*, CRC Press, Boca Raton, FL, pp. 31–41, 2000. With permission.)

human resources or personnel, training and development, security, and transportation.

Given that we assume that the enterprise must produce products based on designs crafted by itself and others, we clearly need an engineering and manufacturing function and the related knowledge spaces. The acquisition of the products produced by others will be achieved through the contracts function previously identified.

We will want to ensure that our product is characterized by quality, so we will need that focus in our enterprise. This interest in quality should extend to process as well as product and embrace the system test and evaluation work. All system development activities produce documentation, so we will also need a means to maintain track of that documentation and clearly define its meaning relative to our product solution. We need configuration and data management knowledge to control, communicate, retain, and protect this documentation.

The enterprise, like an organism, must provide for its survival through a continuing acquisition of new business and a continuing improvement of its internal capability to successfully accomplish the work acquired through marketing efforts.

8.3 Product line base

The product line and the related technologies determine the kinds of engineering and technical knowledge needed in the enterprise. Generally, the product line shapes the need rather than prescribing it. Today, most enterprises involved in system development will generically need electrical engineers, mechanical engineers, and software engineers. The enterprise will need very different electrical engineers, however, if it deals with electrical power transmission rather than microwave communication instruments. A somewhat different mechanical engineering focus is needed for bridge design than for aircraft structural design. This is not always obvious, however. The gifted designer of the wings of the long-surviving BQM-34 target drone built by Teledyne Ryan Aeronautical from the early 1950s through the present was respectfully known as Bowman the bridge-builder.

In addition to the degree of product specialization needed in the basic engineering disciplines, there is also a need for analytical specialists as a function of product. If the efficiency of motion of the product relative to the Earth's atmosphere is important, then we need aerodynamics. If the fluid is water rather than air, we need people with the same general knowledge but with a different medium, and every aerodynamicist may not be equally knowledgeable for both mediums. On the other hand, an aerodynamicist skilled in Earth's atmosphere work would most often be able to adapt to a program dealing with Martian atmosphere penetration.

8.4 The customer base

The technical disciplines needed in the performance of contract work may be slightly influenced by the customer and his or her preferences. A particular customer may require that certain kinds of work be done that the enterprise might not otherwise perform. For example, the structural integrity required of a naval aircraft are much more demanding than that of a land-based aircraft due to the nature of the air field. Corrosion control may also be a much more important discipline in the case of the naval aircraft case. These influences may not change the knowledge mix but can change the relative importance of those disciplines.

8.5 Organizing possibilities

Knowledge is necessary in the program development organization, and this knowledge can exist in one of four configurations (individual human, physical media, computer, and shared) as we have seen in Chapter seven. There are three fundamental ways to organize an enterprise that is focused on a particular product and customer base, and in the structure we choose there must be an agent for enterprise knowledge acquisition and maintenance.

8.5.1 Projectized enterprises

We could elect to make each project a separate entity, each responsible for their own people, practices, and tools. The advantage in this arrangement is that the project can focus itself totally on the customer, ideally leading to a superior product. Clearly, this enterprise will have great difficulty developing a single enterprise-level process about which it can focus continuous improvement. One program or another may excel and bring in considerable profit for the enterprise while others do not do so well. It will be relatively difficult to bring all of the programs up to a similar level of excellence because of the differences between the programs. It may also be inefficient for a program to take advantage of personnel available from another program because of organizational, practice, and tool differences between programs. This structure does not provide for a knowledge agent. It simply focuses its knowledge needs on immediate program needs. It acquires and sheds knowledge as a function of the work demand.

8.5.2 Functionally organized enterprises

A functional organization is structured around the knowledge that is needed and does have a knowledge agent, functional management. The problem with this kind of organization is that the work is accomplished in the context of functional management. Program work is decomposed in this context and parceled out to functional managers. It is very difficult to coordinate this work in a cross-functional sense because all managers jealously guard their own space. The solution to this problem on DoD programs was intended to be the work breakdown structure that partitions all funds in two orientations by product entities and by functional responsibilities. A package of work may require contributions by 15 different departments, and if that work were managed with the physical collocation of those persons contributing, it would overcome the criticism implied.

8.5.3 Matrix-organized enterprises

In the matrix structure, there is a pair of axes reaching into a program product organization and the functional departments providing the people to accomplish the program work. Obviously, there are two orientations here, and we could pick either one. The matrix strength can be along the functional axis where people are physically collocated by functional department with all of the problems of work coordination of the functionally organized enterprise. Alternatively, the project could be the strong axis, and this is the most effective approach. The people are physically collocated on the product team they serve and are managed primarily from the program axis for however long they are charging time to the program. This supports the cross-functional integrated product development concept supported by the author and many other observers.

At the same time, the matrix provides an agent for knowledge in the functional structure, which should provide all programs with coordinated sets of good people, good tools, and good practices. The functional axis has the responsibility to maintain and continually improve the resource base consisting of people, tools, and practices.

We can begin to see the functional organization evolve out of the mist from the generic, product line, and customer base needs discussed earlier. We should collect the knowledge we need into organized chunks because it must be mastered by people who will do work on our programs, and we know that these people will have to be specialists due to the human knowledge capacity deficit problem relative to man's knowledge base. True, we could allow our employees to develop individual knowledge skills that appeal to them but we would realize considerable difficulty in managing the whole enterprise. The assignment of these specialists to program work would require matching individual knowledge to program problems rather than known collections of knowledge. Depending on the specific persons available for work on a particular program, we may have to choose a different set of persons for other work. This would be a tremendously complex problem made simple by standardizing on the knowledge spaces our enterprise purchases on the labor market.

It is the functional organization structure that offers the enterprise a means to chunk knowledge into standardized pieces as suggested in Figure 8.2 such that members of these departments can be treated more or less as interchangeable modules of knowledge encouraging great flexibility in staffing programs. The way we organize must respect the knowledge needs of our programs coordinated with the availability of knowledge workers. The latter is determined largely by universities and the way they decide to break up all knowledge into departments but also influenced by the industry as a whole and the way they choose to organize knowledge. While we might conclude that we need one department full of designers with a mix of fluid, mechanical, and electrical design skills, this kind of person may not be available on the labor market. Over time, our employees may fit into the desired pattern but we probably would not be able to reliably recruit new workers from other companies or from universities in the desired combination of disciplines.

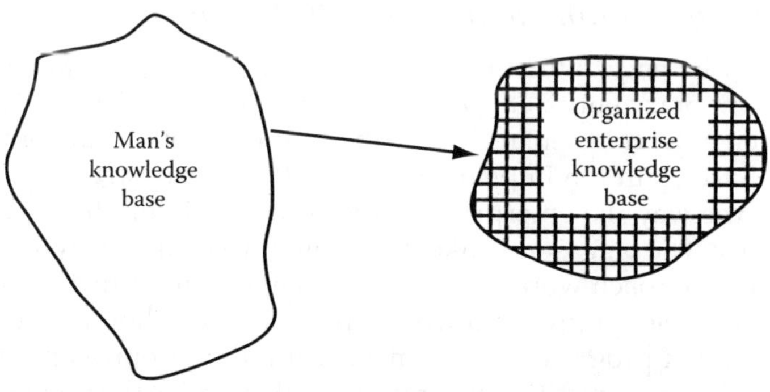

Figure 8.2 Organized knowledge. (From Grady, J., *System Engineering Deployment*, CRC Press, Boca Raton, FL, pp. 31–41, 2000. With permission.)

So, there are two forces shaping our selection of a functional organizational structure. First, there are the knowledge domains we require and our vision of the most efficient way to combine those knowledge domains for management purposes and program application of that knowledge. Second, there exist marketplace constraints driven by university organizational structures that are not always responsive to industry needs.

Table 8.1 begins the process of identifying top-level knowledge spaces within which we will collect the detailed knowledge domains determined by generic needs and the chosen product line. These organizations have been numbered for eventual development into department numbers for use in coordinating tool, practices, and personnel development and work responsibilities. Table 8.1 also correlates these organizations with respect to responsibility for the functionality expressed in Figure 4.12. The organizations marked with an X in the responsibility column cooperate in satisfying or supporting the functional management and continuous process improvement activities managed by the enterprise integration team (EIT) functional organization. These organizations are responsible for providing programs managed through the programs department with resources needed to accomplish program goals in terms of qualified personnel, good tools, and sound practices.

In order to associate the more detailed organizational entities with process responsibilities, we will have to coordinate functional responsibilities with the process that was developed in Part two of this book. The F column of Table 8.1 tells principal functions for which the department in responsible. For students with access to Division 3/Exhibit H/Appendix C it includes a detailed functional organization diagram and tabular list providing brief charter statements. Exhibit H/Appendix A defines in Table A-1 the mapping between functionality and functional organization responsibility.

8.6 Program implementation through teams

In an enterprise employing the matrix structure, every program organization is a temporary entity. It is true that some programs last for years, but they are generally transient relative to the endurance of the functional organization. Success in the implementation of the matrix structure is determined by the way the enterprise allocates responsibility to the two axes of the matrix. The biggest mistake that companies make in trying to make the systems approach work on programs is in making functional departments responsible for program work. Program work should be under the responsibility of program management using the resources provided by functional departments. The program should organize the people derived from the functional organizations into program teams determined by the architecture of the product. These teams are commonly called integrated

Table 8.1 Primitive Functional Organization

Dept.	Department name	Responsibility	F
0000	EXECUTIVE	Enterprise management	F1
1000	BUSINESS FUNCTIONAL	X	F2
	Human resources		
	Facilities		
	Contracts		
	Legal		
	Finance		
	Security		
	Transportation		
	Scheduling		
	Industrial safety		
	Administration		
2000	TECHNICAL FUNCTIONAL	X	F2
2100	ENGINEERING	X	F421, F41
2110	Engineering administration		
2120	Mechanical engineering		
2130	Electrical engineering		
2140	Fluid engineering		
2150	Software engineering		
2160	System engineering		
2200	Manufacturing	X	F423
2210	Manufacturing support		
2220	Manufacturing operations		
2230	Manufacturing administration and records		
2300	SUSTAINMENT	X	F47, F48
	Logistics development		
	Logistics support		
	Operations		
2400	Material and procurement	X	F422
2500	Verification engineering	X	F44
3000	ENTERPRISE INTEGRATION	Process owner	
3100	EIT		
3200	Information systems		
4000	PROGRAMS	X	F49
5000	QUALITY	X	F46
9000	NEW BUSINESS	X	F3

Source: Grady, J., *System Engineering Deployment*, CRC Press, Boca Raton, FL, pp. 31–41, 2000. With permission.

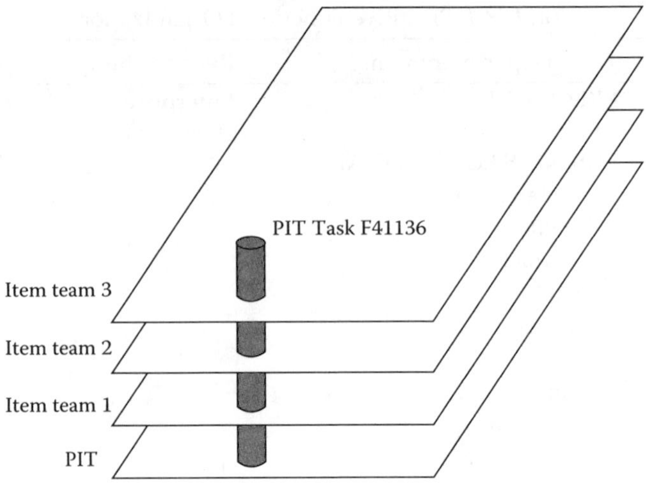

Figure 8.3 Cross-team integration. (From Grady, J., *System Engineering Deployment*, CRC Press, Boca Raton, FL, pp. 31–41, 2000. With permission.)

product and process teams (IPPT). These teams are responsible for all of the development work associated with the items for which they are responsible.

The IPPTs should be established by a system team referred to in this book as a program integration team (PIT) responsible for engineering the complete system. This primarily involves integration work across the team boundaries. The PIT may establish four IPPTs based on the top-level architecture. As the work unfolds on these teams, it may become necessary to establish sub-teams within one or more IPPT. These sub-teams function identically with the first-tier IPPT except that the first-tier IPPTs act as the integration agent for all of their sub-teams. This process can continue to expand beneath the first-tier IPPTs until a procurement decision terminates the expansion string or the items below the IPPT in question are not sufficiently complex to warrant a complete team.

The PIT interacts with all item team tasks in the third dimension (in and out of the page). Imagine the PIT process diagram overlaid in different planes with the item team diagrams with cross-organizational integration occurring in the vertical dimension as suggested in Figure 8.3. This task involves budget and margin analysis in the author's common process diagram. The IPPTs must support the PIT in the performance of this work by providing information and responding to direction.

8.7 Building modes

If the reader is going to form a new functional organization for a new enterprise starting tomorrow, the ideal way to proceed would be to first define the functionality of the organization to be formed in that the organization is a system, and the best way to develop a system is to follow the pattern of form follows function. The functionality of this system can be expressed in terms of a process flow diagram to be developed in Part two of this book. This is a string of blocks connected by directed line segments showing sequence. The blocks represent tasks that must be accomplished. The knowledge needed to accomplish each of the tasks on the process diagram should then be mapped to the task blocks in a fashion identical to the allocation of system functionality to product architecture in the development of a product system.

Concurrent with this allocation process, we need to define the map of knowledge to functional departments. Since the process blocks have been identified with respect to the knowledge needed to accomplish related work, when we allocate (assign) these process steps to functional organizations, they carry with them the knowledge demand previously attached as shown in Figure 8.4.

So, we can assemble a three-way relationship between process tasks, functional departments, and kinds of knowledge in the enterprise knowledge base piecewise by pairs. In this chapter, we are building a beginning view of the relationship between top-level functionality and a top-level view of a functional organization (Table 8.1) that would be useful in any development organization. The organizational names suggest particular knowledge spaces completing the top-level trio of ideas. In the next chapter,

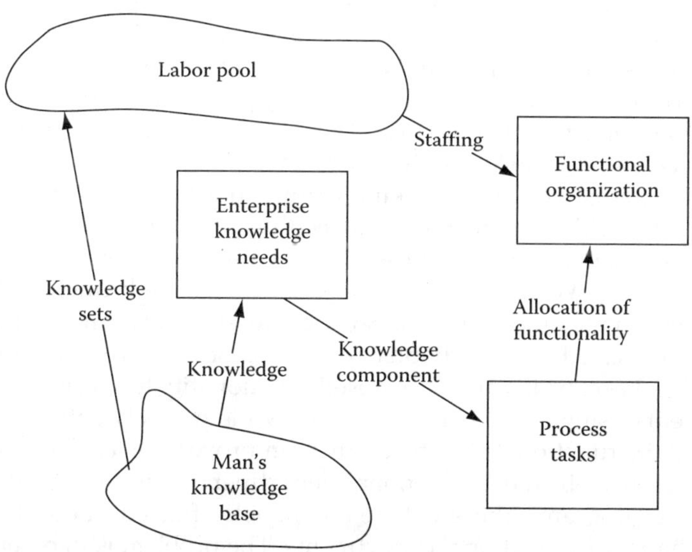

Figure 8.4 Knowledge mapping. (From Grady, J., *System Engineering Deployment*, CRC Press, Boca Raton, FL, pp. 31–41, 2000. With permission.)

we will complete the process definition by coupling enterprise functionality to the functional organization. This will still be a generic organization with the general skills needed in enterprises that seek to solve very complex problems, where we find a pronounced need for the application of the systems approach. A particular enterprise will probably have to tailor that structure based on their specific product line and customer base.

Given that our new functional organization has been created as indicated, we now need to staff it. That can be done by simply hiring people from the labor pool with the combinations of skills and knowledge associated with the functional department charters we have established. In order for this to be as easy as possible, the charters we elect in the functional organizations should be passed through a reality screen to ensure that they align with currently available knowledge domains. It may be necessary to refine the knowledge map to bring about coordination between our needs and the real world. The functional organization is now ready to supply all enterprise programs with good people, but must bring about a condition of synergism between its tools, practices, and people through training assuming that the other two components have been acquired as well.

The tools are a matter of selection by someone, and the practices have to be written describing what has to be done and how to do it coordinated with the process blocks at some level on indenture. So, we must have a rule of responsibility for acquiring these two components. The recommended rule is to identify all functional departments that must contribute to each process task through the allocation process (there will often be more than one allocation for a given task). Where there is more than one functional department responsible for achieving a task, we must identify a principal department. That principal department should be given the responsibility for tool selection for the task and initial preparation of the task practices. Other departments then must learn how to use the preferred tool in accordance with the task description.

The reader is, in reality, working within an existing reality defined by an enterprise with ongoing work proceeding in accordance with contractual commitments. How would we improve such an organization using these principles? We first must understand the needed functionality that your programs should perform. So, we can still perform the front-end analysis suggested in this chapter regardless of how we may apply that knowledge. We can allocate the exposed functionality to specific functional departments using the existing structure while possibly gaining insights into ways the functional structure can be improved. Given that almost all of the personnel in these functional departments will almost always be assigned to programs, there will be few people affected in the short term by any changes in functional departments. The problem is in determining to what extent any changes should be applied to programs. The following sequence is suggested to make the initial adjustment to the new structure:

1. Define needed enterprise functionality based on customer base and product line defined in the vision statement.
2. Allocate exposed functionality to functional organizations that will be responsible.
3. Identify the transform between existing functional structure and the new one. What responsibilities flow from what functional department to what functional department? If necessary, rename any departments as a function of the nature of their final responsibilities.
4. Reassign personnel as necessary to balance the available employee skills and knowledge with the responsibilities of the departments. Apply this new functional structure to all new proposals and programs and bring it on board all existing programs as early as possible, not necessarily right away.
5. Apply the new process description on all new proposals and programs and review existing program contracts, budgets, and schedules to determine to what extent it should be applied, possibly not at all on some programs.

Over time, all surviving programs will come under the new process. This revolutionary step should be viewed as a one-time activity. Thereafter, the enterprise must apply continuous process improvement based on lessons learned from programs to make incremental changes in process, personnel, and tools.

This situation is similar to the reengineering of a product system that remains in service while it is being improved. The product is redesigned and enters a major modification or a series of modifications to achieve new functionality. In that situation, we seek to understand how the initial functionality has changed and allocate the new functionality to the architecture, beginning with the initial architecture. Where the new functionality can be allocated to an existing item, we do it. Where there is nothing in the initial architecture that will accomplish a particular function, we have to consider modifying something that now exists or creating a new item in the architecture. We should consider eliminating anything that has not had any functionality allocated to it.

Once the resources become available to actually implement the changes in operational use, they have to be phased in so as not to disrupt the ongoing flow of work. For a period of time, the organization will have a dual charter, and we must focus our interest on those areas where process tasks have been allocated differently in the initial and new situations. Each of these conditions should have a specific decision made regarding department responsibility on each program applying that task. Figure 8.5 illustrates the pattern of decisions that must be made. The dark gray intersections reflect the principal department while the light gray intersections correspond to contributing departments. The white intersections

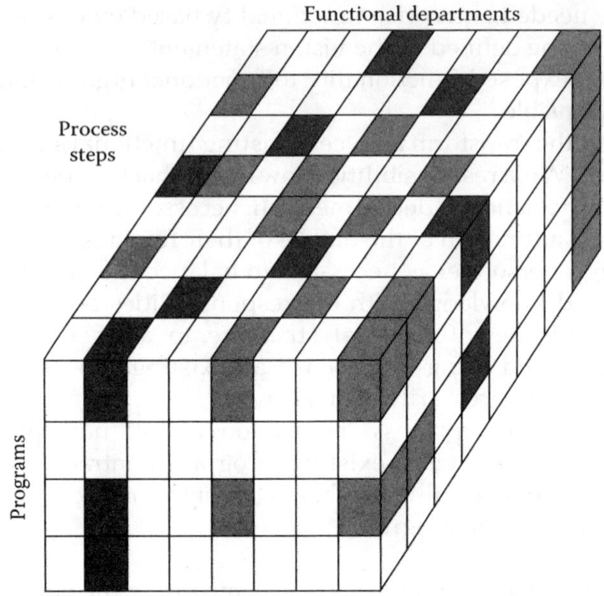

Figure 8.5 Program responsibility map. (From Grady, J., *System Engineering Deployment*, CRC Press, Boca Raton, FL, pp. 31–41, 2000. With permission.)

correspond to voids meaning that the indicated functional department has no responsibility for that task.

Where a program has not yet applied a task, it can be shifted to the new alignment provided there is no cost impact in conflict with the original program plan. If there is, shame on us for improving a task so as to increase its cost of implementation. If the task has already been completed on a program, it can be shifted to the new alignment with respect to that program as well. The problem arises where the task is underway on programs. In each such case, we should ask the impact of applying the new alignment. If it is slight, we should go ahead. If it is more significant, then we should delay the new alignment until later when the impact has lessened, perhaps until the task is complete on the program in the extreme.

The process of changing the alignment on a program may involve a transfer of personnel between departments that is of no consequence to a program organization, a change of practices, or a change of tools. The latter two actions can have a significant effect on program work and should be the basis for the decision on timing of introduction of the new alignment on programs.

In a real enterprise applying continuous improvement, there will be a condition of process instability in effect at all times, but this instability should be minimized. The rate of change cannot be allowed to be so great that this instability becomes too great. Where great revolutions are to be made, they must be phased in as new programs come into the enterprise. The old processes in this case will fade out as the programs in house at the time the change is made fade out or the tasks in question have been completed.

All of these problems in transition suggest that an enterprise should have a process agent, a process owner who can regulate the overall improvement process. The author maintains that the functional organization should be responsible for providing programs with good people, good tools, and good practices. However, as in all distributed responsibility situations, we need an integrator and overall owner. That agent, in the author's view, is an entity called an enterprise integration team, the ultimate process owner for the enterprise.

chapter nine

Functional organization charters and relations*

9.1 Functional department formation

The functional departments of the enterprise should be formed focusing on the aggregate collection of generic tasks mapped from the generic process steps covered in Part two to the functional organization structure. The functional department managers are made responsible for documenting their department charters based on the top management direction to them to provide programs with well-qualified persons, good tools, and proven best practices. They must determine the knowledge base needed to satisfy this requirement and, thus, the qualifications of the people appropriate for membership in their department. Where multiple departments share responsibility for a given task, a lead department is identified that should provide charter leadership for that task in terms of tool selection, training, and practices.

As discussed earlier, in a more realistic situation where the enterprise is taking the first step to implement an improvement process moving from a prior ad hoc department charter structure, there must be an orderly movement from the current methods to a continuous ribbon of improvement toward perfection. There will be some initial resistance observed from functional managers, because the change will represent instability, a fear-inducing force in any organization. The enterprise executive must prepare current functional management for the transformation by causing them to recognize that there will be some instability, but that this offers new opportunities rather than the seeds of career disaster.

The current functional managers must come to understand individually what drives them professionally and personally. If the organization has had a strong functional axis in the matrix where the functional managers actually managed their department's portion of program work either directly or through department surrogates, they have to realize that the new role of functional management is going to carry with it much less organizational power and prestige. If these are important to the individual, one should probably consider a change of assignment and

* The material in this chapter is from Grady, J., *System Engineering Deployment*, CRC Press, Boca Raton, FL, Chapter 5, 2000. Used with permission.

the top management must be tuned in to the need for some assignment exchanges. Power and prestige will probably, in this situation, migrate to some extent from the functional organization to the program organizations in the form of program managers and team leaders.

9.2 First-tier functional departments

In Chapter eight, we identified the first-tier functional departments in Table 8.1, and these are repeated in Table 9.1 for convenience. These represent our top-level view of the partitioning of man's knowledge into common sets for mastery by people. Each of these departments will presumably include subdepartments reflecting a finer degree of specialization necessary to master the whole field of system development skills and knowledge.

9.3 Expansion of the functional organizational hierarchy and charter development

You will need to perform an expansion of your enterprise functionality based on your product line, customer base, and preferences, all summarized in the enterprise vision statement, as shown in Figure 9.1. The other beginning point for this structured process for defining an enterprise infrastructure begins with the knowledge base necessary to solve the kinds of problems anticipated in light of the vision statement. One way to organize this knowledge in a functional organization structure for an enterprise involved in solving complex problems requiring the application of engineering skill is shown in Table 9.1. We now need to fill in the details covering what these functional organizations should bring to program work.

The author has created an example of these details based on having worked in several large engineering organizations, observing many more.

Table 9.1 Top-Tier Functional Structure

Org. ID	Department name
0000	Enterprise executive
1000	Business functional
2000	Technical functional
2100	Engineering
2200	Manufacturing
2300	Logistics
2400	Material and procurement
2500	Verification
3000	Enterprise integration
4000	Programs management
5000	System quality
9000	New business

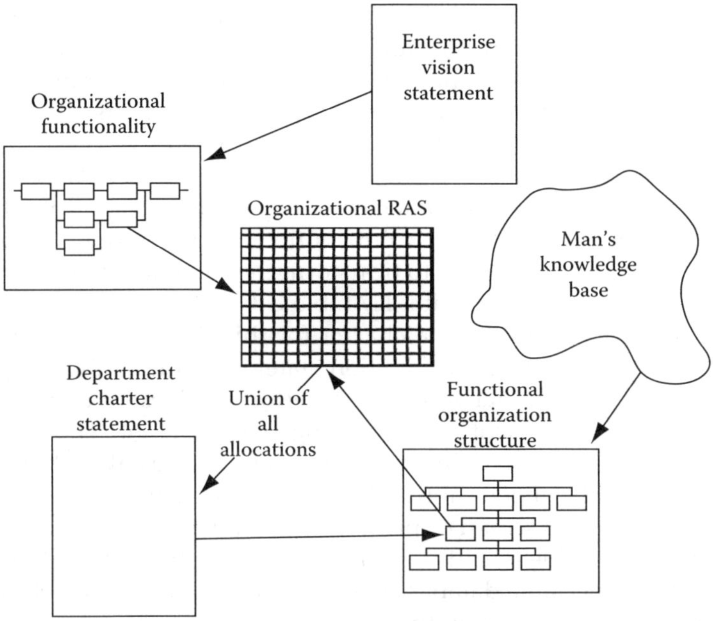

Figure 9.1 Organizational structured analysis flow. (From Grady, J., *System Engineering Deployment*, CRC Press, Boca Raton, FL, pp. 69–78, 2000. With permission.)

The results are displayed in Exhibit H/Appendix A. Figure A-1 sheets 1 through 70 model a functional process flow diagram. Table A-1 allocates enterprise functionality to functional organizational entities. Figure C-1 of Exhibit H/Appendix C shows the resultant functional organizational structure.

The department charters were derived by aggregating all of the functionality allocated to them in Exhibit H/Table A-1 and working back and forth between the needed functionality and physical organizational forms a few times. This is an iterative process that you will find very difficult to complete in one analytical cycle. While doing this work, you must keep several documents in a state of agreement: the enterprise functional flow diagram (Exhibit H/Figure A-1) identifying needed functionality, the function allocation matrix (Exhibit H/Table A-1) marrying the needed enterprise functionality to the organizational structure, the functional department diagram (Exhibit H/Figure C-1), and the related department charter matrix (Exhibit H/Table C-1). You will find that the discipline of the process suggested will evolve a consistent set of data that comprehensively addresses all of the work that will have to be accomplished on any program in your company's normal product and customer mix.

The charter statement is a high-level summary description of what the department is responsible for, the department version of the enterprise vision statement. The details are provided in Exhibit H/Table C-1, and this should be the basis for the development of department practices supportive of the department charter. As an example of the transform from allocated functionality to department charter, Table 9.2 lists all of the functionality allocated to department D216, Requirements Analysis. This is a subset of Exhibit H/Table A-1 that extracts department D216 line items. The RESP column tells the character of department responsibility from the set L = Leadership and C = Contribution or cooperation. Department 216 now has the responsibility to transform the content of Exhibit H/Table A-1 into its component of Exhibit H/Table C-1.

One approach to crafting a charter for this department would be to simply aggregate all of these function statements into a paragraph. It is recommended though that you make an effort to digest or distill this raw input into a very few sentences covering the whole lot. The details are always available to anyone interested in them in your allocation matrix. Ideally, this data has been captured in a relational database so that it can be brought up on the screen or printed out in any way desired. For example, we should be able to call the Table 9.2 view of the data in Exhibit H/Table A-1 from the same database system, just different views of the same table filtered in accordance with some criteria.

Refer to Division 3/Exhibit H/Table C-1 for the department 216 charter driven by the Table 9.2 content. Yes, there appears to be a lot of functionality missing in the charter. You have a choice of setting up the charter statement on any level of allocated functionality you choose. You can try to combine dozens of lower-tier allocations into a synthesis or evolve a charter statement from a relatively few high-level functions.

9.4 The popular knowledge map

It is important to keep in mind while building the picture just sketched, how the rest of the world has partitioned the knowledge base to which your enterprise will need access to staff the organization. True, the enterprise could organize themselves however they wish, but there is some advantage in reflecting the norm in terms of being able to hire people off the streets, from other companies, and from college campuses who do not have to be extensively retrained to fit your mold. The point is not that you should simply always respect the job marketplace, but it should be one of the factors considered in building the enterprise charter statements.

9.5 Program knowledge interfaces

It is true in developing a product using the structured analysis process that you predetermine the interfaces that will have to exist in the physical product by the way you allocate functionality to the things in the product

Table 9.2 Department 216 Functionality

Function ID	Function name	RESP
F41	Grand systems definition	D216
F411	Customer needs analysis	D216
F4111	Review and evaluation need and prior work	D216
F4112	Customer Q&A	D216
F412	Universal architecture description framework	D216
F4121	Descriptive modeling	D216
F41211	Problem space modeling	D216
F412111	Functional analysis	D216
F4121111	Adopt a functional model	D216
F4121112	Study higher-tier concepts	D216
F4121113	Hierarchical functional diagramming	D216
F41211131	Identify tier M functions	D216
F41211132	Expand tier M + 1 for branch N	D216
F4121114	Functional flow diagramming	D216
F41211141	Identify needed sub-functions	D216
F41211142	String functions into sequence	D216
F4121115	Alternative functional flow diagramming	D216
F41211151	Enhanced functional flow block diagramming	D216
F41211152	Behavioral diagramming	D216
F41211153	IDEF-0 diagramming	D216
F412112	MSA/PSARE modeling	D216
F4121121	Create environmental model	D216
F41211211	Prepare purpose statement	D216
F41211212	Develop functional context diagram	D216
F41211213	Develop scenario	D216
F4121122	Create behavioral model	D216
F41211221	Create level M DFD	D216
F41211222	Enter bubble P-spec data in RAS	D216
F41211223	Define lines and stores	D216
F41211224	Create companion CFD	D216
F41211225	Superimpose super bubbles on DFD in RAS	D216
F41211226	Enter bubble C-spec data in RAS	D216
F41211227	Define entity interfaces	D216
F412113	SysML/UML modeling	D216
F4121131	Build context diagram	D216
F4121132	Use case analysis	D216
F4121133	Build scenarios	D216
F4121134	Accomplish opening dynamic analysis	D216

(*continued*)

Table 9.2 (continued) Department 216 Functionality

Function ID	Function name	RESP
F4121135	Accomplish secondary dynamic analysis	D216
F412114	IDEF/DoDAF modeling	D216
F4121141	Joint modeling of all views	D216
F4121142	Joint modeling of technical views	D216
F4121143	User modeling of problem space using operations views	D216
F4121144	Contractor modeling of problem space using system views	D216
F41212	Solution space modeling	D216
F412121	Interface modeling	D216
F4121211	Evaluate interface needs from an architectural perspective	D216
F4121212	Evaluate allocated functionality strings	D216
F4121213	Identify interfaces	D216
F4121214	Define interface terminal media	D216
F4121215	Document interface identification	D216
F4121216	Evaluate cross-organizational interface impact	D216
F412122	Specialty engineering modeling	D216
F412123	Environmental modeling	D216
F412124	Product entity synthesis	D216
F4121241	Product entity diagramming	D216
F4121242	Engineering drawing breakdown	D216
F4121243	Manufacturing breakdown structure	D216
F4121244	Work breakdown structure	D216
F4121246	Responsibility overlay	D216
F4121247	Configuration item selection	D216
F4121248	Drawing release schedule	D216
F4121249	Specification tree development	D216
F412124A	Prepare SAR content	D216
F4122	Inter-model transfer	D216
F4123	Executable modeling	D216
F4124	Integrate and optimize models	D216
F4125	Progressive process requirements analysis	D216
F41251	Deployment planning analysis	D216
F41252	Schedule planning analysis	D216
F41253	Disposal planning analysis	D216
F41254	Material and procurement planning analysis	D216
F41255	Bottom-up verification planning	D216
F41256	Quality planning analysis	D216

Table 9.2 (continued) Department 216 Functionality

Function ID	Function name	RESP
F41257	Manufacturing planning analysis	D216
F41258	Logistics planning analysis	D216
F41259	Top-down verification planning	D216
F412591	IPPT specification verification work	D216
F4125911	Make verification entries in RAS	D216
F4125912	Define verification strings	D216
F4125913	Define item verification tasks	D216
F4125914	Select verification level	D216
F4125915	Map strings to tasks	D216
F4125916	Maintain item verification traceability	D216
F4125917	Integrate verification planning data	D216
F412592	PIT level work	D216
F4125921	Prepare and maintain task matrix	D216
F4125922	Prepare and maintain compliance matrix	D216
F4125923	Prepare and maintain item matrix	D216
F4125924	Select verification level	D216
F4125925	Prepare integrated verification management report	D216
F4131	Performance requirements analysis	D216
F41311	Functionally derived performance requirements analysis	D216
F41312	MSA/PSARE-derived performance requirements analysis	D216
F41313	UML/SysML-derived performance requirements analysis	D216
F41314	DoDAF-derived performance requirements analysis	D216
F41315	Verification requirements analysis	D216
F413151	Select verification method	D216
F4131512	Write verification requirement(s)	D216
F4131513	Perform analysis	D216
F4131514	Rewrite section 3 paragraph	D216
F4131515	Assign section 4 paragraph number	D216
F4131516	Make entries in traceability matrix	D216
F4131517	Make entries in compliance matrix	D216
F4132	Design constraints analysis	D216
F41321	Interface requirements analysis	D216
F41322	Specialty engineering requirements analysis	D216
F413221	Logistics support analysis	D216
F4132211	Availability requirements analysis	D216

(*continued*)

Table 9.2 (continued) Department 216 Functionality

Function ID	Function name	RESP
F4132212	Reliability requirements analysis	D216
F41322121	Reliability modeling	D216
F41322122	Reliability allocation	D216
F41322123	Failure mode, effects and criticality analysis (FMECA)	D216
F4132213	Maintainability requirements analysis	D216
F41322131	Maintainability modeling	D216
F41322132	Maintainability allocation	D216
F4132214	Human factors requirements analysis	D216
F4132215	Logistics factors analysis	D216
F41322151	Logistics concept development	D216
F41322152	Personnel and training requirements analysis	D216
F41322153	Support equipment requirements analysis	D216
F41322154	Technical data requirements analysis	D216
F41322155	Spares requirements analysis	D216
F413222	Safety requirements analysis	D216
F413223	Security requirements analysis	D216
F413224	PMP requirements analysis	D216
F4132241	Parts engineering analysis	D216
F4132242	Materials and processes analysis	D216
F413225	DTC/LCC requirements analysis	D216
F41322A	Aerodynamics requirements analysis	D216
F41322B	Thermodynamics requirements analysis	D216
F41322C	Mass properties requirements analysis	D216
F41322D	Structures requirements analysis	D216
F41322E	Guidance accuracy requirements analysis	D216
F41323	Environmental requirements analysis	D216
F4133	Allocate requirements to entities	D216
F414	Concept development	D216
F416	Manage requirements	D216
F4161	Manage RAS	D216
F4162	Audit, assure traceability, integrate, and validate requirements	D216
F41621	Assess specification for completeness	D216
F41622	Check for technical risk	D216
F41623	Check for cost and schedule compatibility	D216
F41624	Screen requirements for structural correctness	D216
F41625	Formal validation	D216
F416251	Change requirement	D216
F416252	Design evaluation testing	D216

Table 9.2 (continued) Department 216 Functionality

Function ID	Function name	RESP
F416253	Demo or inspection	D216
F416254	Analysis	D216
F416255	Technical performance measurement	D216
F4162551	Terminate TPM track	D216
F4162552	Track and manage TPM	D216
F4162553	Design evaluation test	D216
F4162554	Demo or inspection	D216
F4162555	Analysis	D216
F41626	Screen requirements for necessity	D216
F41627	Manage requirements validation	D216
F41628	Repair sentence structure	D216
F41629	Discard requirement	D216
F4162A	Integrate requirements	D216
F4162B	Assure traceability	D216
F4163	Manage margins and budgets	D216
F4164	Applicable document calls	D216
F4165	Document-descriptive modeling artifacts	D216
F4166	Publish specifications and plans	D216
F4167	Document-approved simulations	D216
F4168	Document-approved concepts	D216
F4169	Prepare program for modeling and requirements analysis work	D216
F416A	Form teams	D216
F416C	Conduct SRR	D216
F416D	Conduct SDR	D216
F417	Clone specification	D216

entity structure. Well, it is likewise true that you will predetermine the communication interfaces that must exist on programs between the people from the different departments by the way we allocate enterprise functionality. The organizational requirements analysis sheets (RAS) in Exhibit H/Table A-1 identifies all of the allocations of functionality to functional departments. This does not entirely tell how people from these departments will have to cooperate in performing these tasks on programs, but these interactions are predetermined by Exhibit H/Table A-1 content.

9.5.1 *Intra-task relationships*

So far, we have used a process flow diagram to define what the development organization must do, and we have allocated this functionality to a development organization architecture called a functional department organizational structure. It remains to give some idea about how the elements of the resulting organization will work together to accomplish the process goals, and this requires some understanding of the interfaces between the people from the functional departments when they ply their trade on programs. Part of this story is told by the allocation pattern on Exhibit H/Table A-1 that identifies organizational sets jointly responsible for process steps. Where this matrix shows a single functional organization as the only entry for a particular process, there is no difficulty in understanding the organizational behavior—they simply have to do it all, and the process dictionary in Exhibit H/Table A-1 tells this story. The problem appears when multiple functional departments are called upon to contribute to a process step. Where this is the case, one of these should be identified as the lead functional department, and this is the beginning of the behavior definition activity.

Before continuing with this work relationship discussion, the idea should be reinforced that the program work should be accomplished by cross-functional teams, referred to in this book as integrated product and process teams (IPPT). These teams are staffed by people assigned from functional organizations, and they contribute the skills and knowledge related to the functionality component allocated to their functional department, but the team is physically collocated and led toward their program goals by a program team leader. The inter-functional department relationship that must be identified is intended to be accomplished in the context of this team structure on programs.

The intended behavior could be communicated through the organizational RAS only, functional organization n-square diagrams, or functional organization schematic block diagrams, as illustrated in Figure 9.2. In Figure 9.2a, the fragment of the RAS provides an indicator of needed cooperation, but does not communicate what must be communicated. It does define who is the agent responsible for process development and implementation on programs in that D2161 is identified as responsible for leadership (L) and others are to cooperate (C). For a given process involving several functional departments, we could craft a schematic block diagram or an n-square diagram for department contributions for task F41114, as illustrated in Figure 9.2b. The contributing departments that must cooperate are listed on the vertical axis, and this same listing is assumed on the horizontal axis. Normally, these elements would be identified down the diagonal, but the diagram used here is too small for that. The diagonal intersections are marked to indicate the internal actions necessary for them to do their work. In this example, people from all of the other departments listed must accomplish work to achieve their goals for process F41114, and all of these results are used by department D2162 in producing the final product from this task, in this case a specification.

Function	Functional departments	Role
F41114	D2123	C
F41114	D2124	C
F41114	D2134	C
F41114	D2135	C
F41114	D2143	C
F41114	D2134	C
F41114	D2151	C
F41114	D2161	L
F41114	D2162	C
F41114	D2164	C
F41114	D2165	C
F41114	D2166	C
F41114	D2167	C
F41114	D2168	C
F41114	D2211	C
F41114	D2311	C
F41114	D2411	C
F41114	D2421	C

(a)

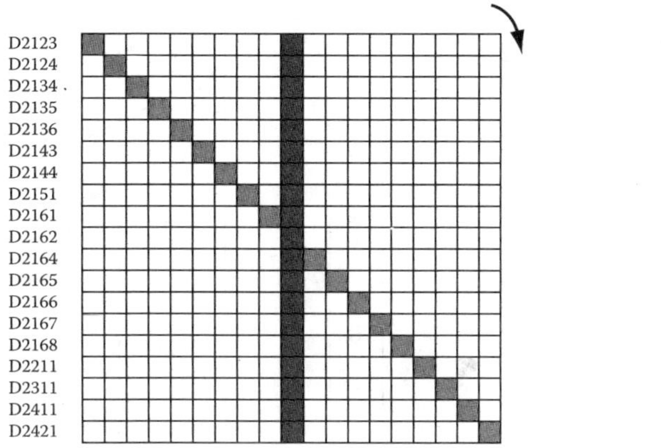

(b)

Figure 9.2 Enterprise organizational interrelationship definition models: (a) organizational RAS and (b) organizational n-square diagram. (From Grady, J., *System Engineering Deployment*, CRC Press, Boca Raton, FL, pp. 69–78, 2000. With permission.)

If we looked more deeply into the subtasks for F41114, we would find progressively fewer different departments involved and a need for less interaction between them. We would find one task, F4111447122, Reliability Allocation, for example, that will be accomplished entirely by the RAM Engineering Department, D2164. Its output is a reliability figure for each item in the product system, and these figures flow into the specifications.

Figure 9.3 offers a fragment of another intra-task exposition model. It is an operational sequence diagram illustrating that department U begins a process passing on some results to department V. The reader can follow the sequence from that point. Each entity (department in this case) has a separate column. The analyst places activity symbols in the columns corresponding to work responsibility and connects the activities in sequence. This kind of diagram identifies two or more channels on one axis, and time flows on the other axis (down in this case, though the diagram could be laid out horizontally). This diagram is called a sequence diagram in the unified modeling language (UML). The columns are called swim lanes and represent what are called physical classifiers that represent the entities that software will be written for, such as classes, components, and nodes. These kinds of diagrams were in use long before being adopted in the activity diagrams of the UML.

All work may be in a simple serial pattern or require concurrent activity by people from two or more departments. The analyst may use

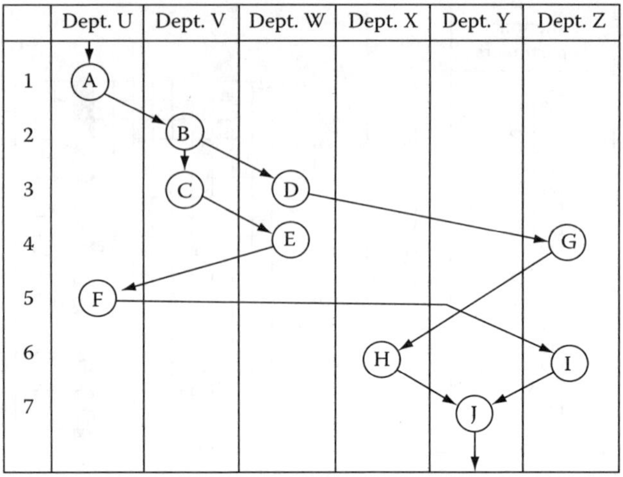

Figure 9.3 Operational sequence diagram. (From Grady, J., *System Engineering Deployment*, CRC Press, Boca Raton, FL, pp. 69–78, 2000. With permission.)

different symbols to indicate different kinds of actions, such as perform physical work, make a decision, or carry out analysis. This diagram could be supplemented with a task dictionary to explain what it means for department W to perform task D. The numbers down the left column suggest a clock running in some units of measure, like hours, days, or weeks. Vertical bars could be used to indicate the period of performance of the tasks where they last longer or shorter than one time increment.

9.5.2 *Inter-task relationships*

While the process definition and allocation activity described in this book should produce a thoughtfully complete accounting of the needed work linked to clear responsibility, there is one more question that we should ask of each process identified as a cross-check. Each valid task in an enterprise process must produce some value added, or it should not be included in the process. The outputs of each of these tasks should be identified and linked to the responsible functional department (the people on the program from that department). At higher levels of task identification, there may be many task products that must be contributed to by people from many different departments, but as we flow further down into the process hierarchy, we will find a point where most task products appearing as task outputs are the work of people from a single department or at least the clear overall responsibility of a single specialty.

In that the process diagram recommended in this book is constructed as an expansion, it stands to reason that higher-tier tasks will require a great deal more intra-task, interdepartment coordination and activity than lower-tier tasks that tend to be single-department oriented. There are several ways one can study the needed inter-task relationships. The process diagram itself gives us some information by virtue of the flow indicated. But, this format is not helpful in showing all of the data flow because it adds so many lines that the sequence intended is hard to spot visually.

A process n-square diagram could be used, but where the tasks in question are contributed to by people from different departments, it may not be clear who is responsible for the interfaces indicated. This could be listed in a separate dictionary.

Exhibit H/Table A-1, the organizational RAS, identifies for each line item, which is a program process step hooked to a functional organizational responsibility, one or more task products that must result from the related work. These are the significant accomplishments of the tasks. These are the things a manager should expect for the cost of supporting that task with money and schedule time. These products flow to other tasks, and a thorough task plan would also identify the terminal tasks for these task products. This step has not been included in the tabular data supplied with this book, but could be linked to the tasks as input requirements, illustrated as data flow in a requirements-driven design (RDD) diagram, or illustrated in an IDEF-0 portrayal of process either as resource inputs on the lower edge or controlling influences on the upper edge of terminal task blocks.

The two other diagrammatic treatments referred to in the previous paragraph lend themselves to illustrating not only the process flow but the interfaces between those responsible for accomplishing the indicated work. They are input process output (IPO) or RDD and SADT or IDEF-0. IPO and SADT were originally created as software analysis models, but were respectively adapted by Mack Alford as the basis for the Ascent Logic system engineering tool RDD-100 and the U.S. Air Force as the basis for an integrated computer-aided manufacturing (ICAM) definition (IDEF) language.

In the IPO/RDD application, the process diagram runs down the page like an old computer programming flowchart, as shown in Figure 9.4. In addition, we include a lateral connection for data flow into and out of these processes that have been derived from external to the process illustrated or from a prior process step within it. Note that the flow diagram is run vertically on the page rather than horizontally as in Appendix A. When the work is performed, there is a work product that flows out from the right side of the block. A work product, which often will be a document in whole or in part, flows from source tasks to tasks where that work product is consumed or applied.

It should be noted that RDD-100, which used this behavioral model, could be used to capture the generic and program process flow definition

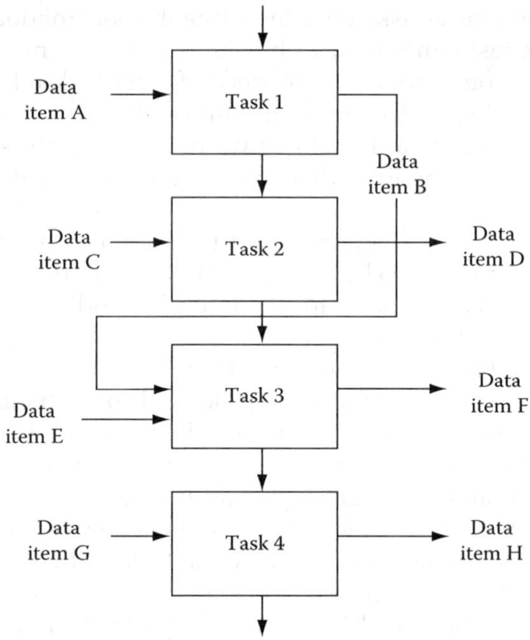

Figure 9.4 IPO/RDD application. (From Grady, J., *System Engineering Deployment*, CRC Press, Boca Raton, FL, pp. 69–78, 2000. With permission.)

with associations to the contributing functional organizations just as it could be used to capture product functionality and allocation to product architecture. When the program process is thus defined, RDD-100 would permit simulation of the program before you attempted to implement the program (as in the proposal period) yielding the information useful in verifying the plan. Unfortunately, Ascent Logic that provided RDD-100 went bankrupt. However, Vitech's CORE continues to provide an effective computer application for this purpose. Intellectually, CORE rotates the diagram 90° from RDD-100, putting the functional axis in the horizontal. These models are explained in more detail in *System Requirements Analysis* (Grady, 2005).

The SADT or IDEF-0 approach is illustrated in Figure 9.5. It consists of a flow diagram that steps across and down the page. Directed line segments can also enter the process blocks at the top to control the activity or into the bottom to provide resources needed in that process step. IPO/RDD and SADT/IDEF-0 offer ways to link functionality with data or work products, but one might still ask, "How do the people from those five different departments required for this task accomplish the work within the box?" It may be that an expansion of that block to a lower level using one of these two techniques will make the answer

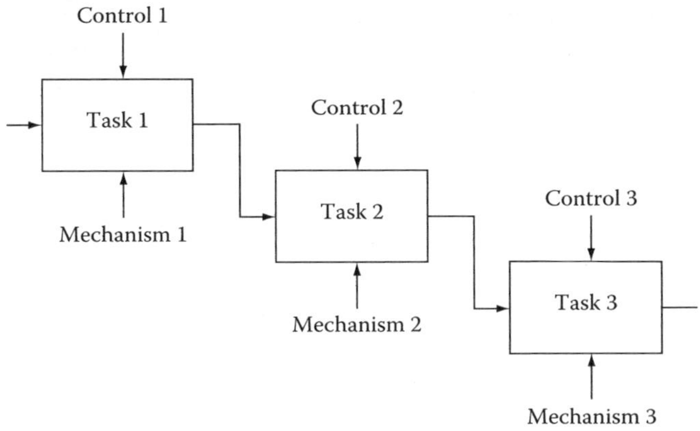

Figure 9.5 SADT/IDEF-0 application. (From Grady, J., *System Engineering Deployment*, CRC Press, Boca Raton, FL, pp. 69–78, 2000. With permission.)

clear. Otherwise, an application of n-square or operational sequence diagramming within the lowest-tier task may achieve the desired degree of process specificity.

There may be a balance point between the detailed process specificity and an intuitive common vision that would make it unnecessary to expand the process responsibility definition beyond some point determined by good judgment or common sense, and the practitioner should be alert to that acceptable level of detail.

Readers may say that they have experienced many of the tasks contained in the process diagram included in this book and can think of many cases where there simply is not any tangible task product that can be identified. The author rejects this conclusion. A tangible product can and should be identified. It may be as simple as a memo of conclusions that should be captured in the program library, and this documents what the person who did the work communicated verbally to the person who needed that information. If there is no tangible product, then there is no task and should be no budget. This message will cause even the most reluctant engineer or manager to invent a value-added reason for their existence in the process. This is a fundamental application of the Lean initiative, of course. Every task must produce a real and useful work product, or it should not be employed.

9.6 Process integration

As the generic process is crafted, the work recommended for the many processes by the several functional departments may be right. Then again, some of these inputs could be unnecessary. Other necessary ones may be missing. The integration agent encouraged is the enterprise integration team (EIT), the enterprise process owner. When first creating this kind of information, the PIT should evaluate all of the task outputs to ensure they are necessary, that someone is using the work products created, and that there are no omissions.

chapter ten

Program organizational structures*

10.1 Overview

The cross-functional integrated product team is advanced as the optimum engine for program work. The teams are identified based on the evolving product structure functionally derived through structured analysis following Louis Sullivan's form ever follows function paradigm. WBS, MBS, DBS, specification tree, and other views are integrated into the system product structure and associated with team responsibilities. How to structure the relationship between the teams and troublesome disciplines is covered. Cross-functional interfaces, one of the principal problems in system development, are highlighted and ways to identify and adjust them are covered.

10.2 The two axes

The matrix has been offered for our enterprise structure despite its well-known flaws because of its recognition of the need to organize in two fundamental ways that are in conflict. The functional structure corresponds to the way we humans and our organization have chosen to partition knowledge of use to the enterprise. These are the fields within which specialists develop domain knowledge. By grouping like specialists together in the functional departments, we encourage enterprise mastery of each of these specialized fields needed to develop products in our chosen product line and customer base.

This functional structure is not well suited for running programs, however. Many companies have made the mistake of permitting functional departments to manage big pieces of programs and found that this approach is badly flawed. The problem is that the system does not commonly follow the organizational structure of the knowledge base. Also, every item development effort is going to require the services of many specialists working together because all of us are narrowly specialized. If we organize on programs by functional disciplines, we make it very hard to communicate across the disciplines.

* The material in this chapter is from Grady, J., *System Engineering Deployment*, CRC Press, Boca Raton, FL, Chapter 7, 2000. Used with permission.

Thus, we require, in addition to the structure for knowledge mastery, a means of organizing for program excellence. The integrated product and process team (IPPT) notion is perfect for this purpose. This structure is overlaid upon the product architecture, and teams thus oriented are staffed by people from the functional departments based on the technology they will appeal to in solving the design problem they are presented with. The matrix, thus, has two axes: the functional department axis providing programs with good people, good tools, and good practices all coordinated around a commonly held view of the development process and a product-oriented team structure.

10.3 Team structures

Figure 10.1 illustrates the recommended enterprise structure showing the two axes reporting to the enterprise executive. In this structure, the

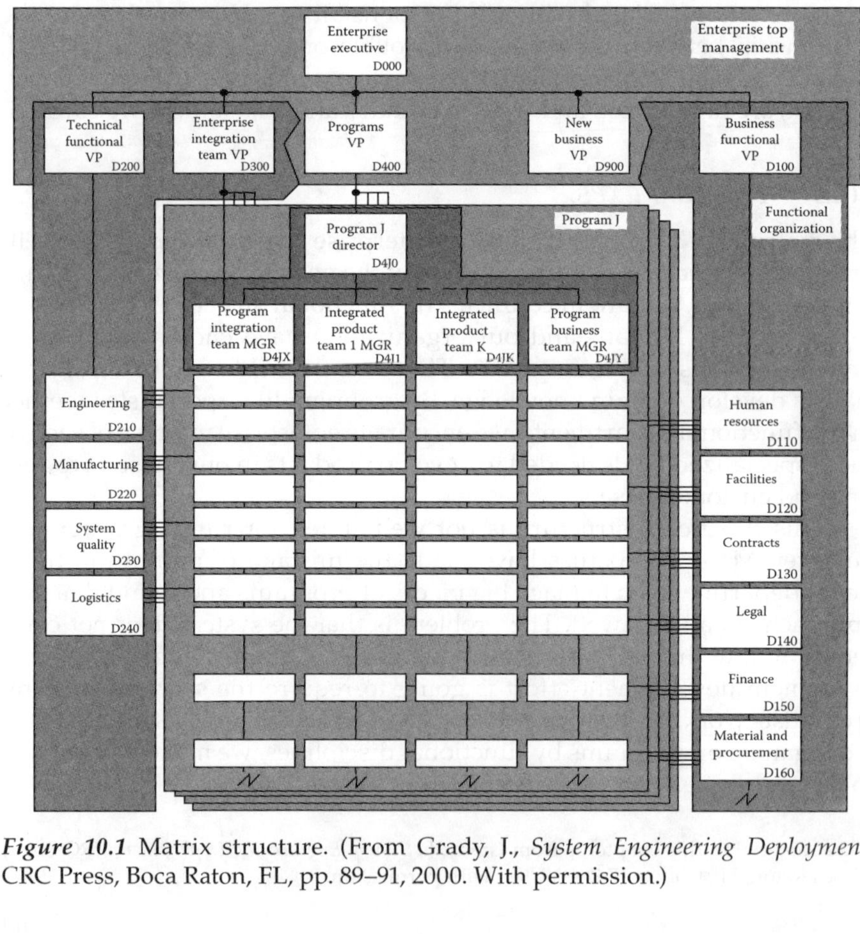

Figure 10.1 Matrix structure. (From Grady, J., *System Engineering Deployment*, CRC Press, Boca Raton, FL, pp. 89–91, 2000. With permission.)

functional departments exist to supply programs with the resources they need to succeed. The program managers are assigned from the program management functional department, and team leaders are selected by the program manager from the people assigned to the program by the other functional departments, or the program manager can request the assignment of particular people by the functional managers to the program to serve as team leaders.

As a program begins in a proposal or study effort, it very likely only requires a business team and one technical team responsible for the whole system to be defined. The author refers to this latter team as the program integration team (PIT). It is staffed from the functional departments with heavy participation by the advanced development department.

10.4 Team formation and staffing on programs

As the PIT identifies the product architecture as a result of a functional analysis that allocates functionality to things synthesized into a physically oriented product architecture hierarchy, it determines how that architecture will be overlaid with team responsibilities. This may require a few teams all at one level or involve two or more layers of teams depending on the complexity of the system. The right mixes of personnel are brought on board the program from functional departments based on the requirements and design concepts associated with the items selected for team coverage. As shown in Figure 10.2, program work is assigned to the teams in especially crafted patterns which, in combination with the nature of the functionality identified for the item, determine the staffing required from the functional organizations.

The first step in forming a team is to identify its leader, presumably by the program manager, though in some companies the team is picked first and allowed to choose their leader. This should be a person with considerable experience with the product line and customer involved. It is not possible to select someone who has mastered every facet of the design problem because of human knowledge span limitations, so we can pick the leader from any discipline but it would be best to select an engineer to lead the early development work, someone from system engineering or the lead design function. The leader should have good interpersonal skills, because this is a leadership position, and have a broad outlook rather than a narrow single-discipline view.

The team leader must then take on the responsibility for creating the team. Figure 10.3 offers a sequence of steps leading to a state of readiness to perform the first responsibility of the team, requirements analysis. The leader can partition all needed personnel into core and extension components. The core people will work full time on the team whereas the extension component is part time. As the core people are identified and

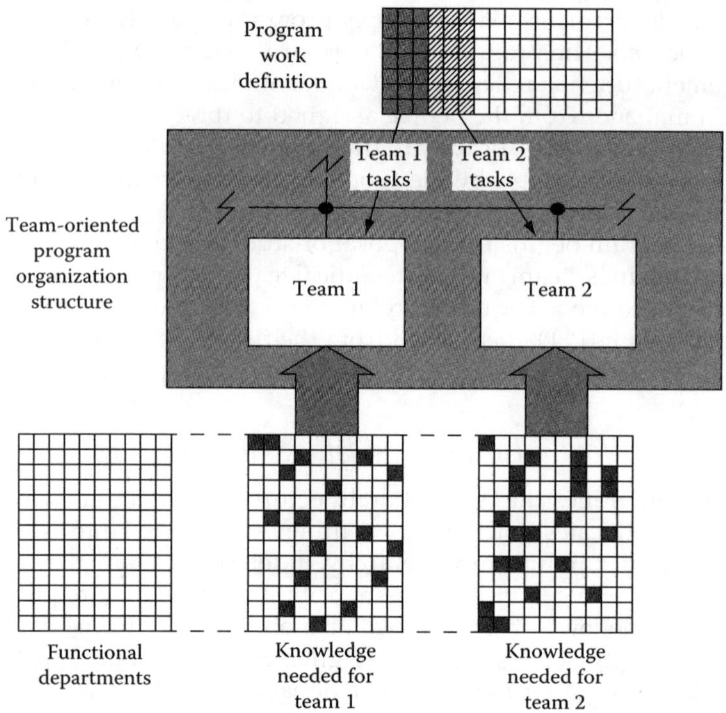

Figure 10.2 Program team formative factors. (From Grady, J., *System Engineering Deployment*, CRC Press, Boca Raton, FL, pp. 89–91, 2000. With permission.)

acquired, the team leader and team members study the team assignment and form a common vision of the task before them.

In support of this first team effort, the PIT should provide the team with a particular collection of information (available team item data in Figure 10.3) as they begin to form:

1. The system specification, a mission analysis report telling what the system must do in a mission context, and a system definition report providing the system requirements analysis products to date (functional analysis diagrams and data, architecture block diagram, and schematic block or n-square diagram).
2. A WBS assignment linked to a particular piece of system architecture. In the ideal case, this should be a single item (this might be a whole aircraft with subteams or the avionics system of the aircraft), but it could involve a contiguous or closely related collection of things as in all of the product software where there are eight software entities in eight different pieces of equipment including two onboard computers and three pieces of support equipment, and three avionics packages.

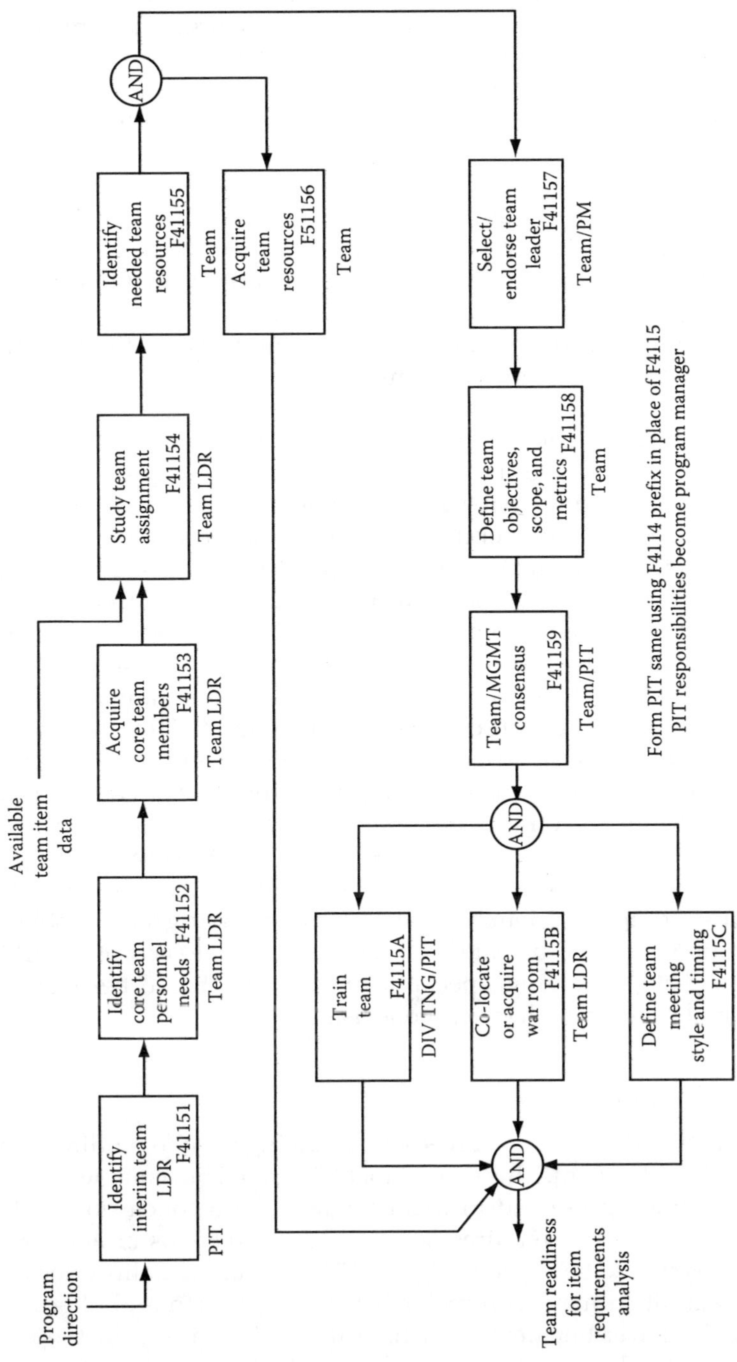

Figure 10.3 Team formation process. (From Grady, J., *System Engineering Deployment*, CRC Press, Boca Raton, FL, pp. 89–91, 2000. With permission.)

3. A contiguous fragment of the program statement of work (SOW), plans, and schedules covering, ideally, the one product architecture item involved in their responsibility. It should be possible to simply lift out the team's planning data from the whole of the program planning data without any complex entanglements with other planning data.
4. A budget tied to the WBS assigned to the team.
5. An item system schematic block diagram showing all of the interfaces of the team item with clear identification of the parties responsible for the other terminal of each of the interfaces that touches the team item.
6. A released or nearly complete specification covering the top-level item for which the team is responsible.

The team must then identify the resources they will require to respond to their responsibilities and undertake action to acquire these resources. Some organizations believe that the team should select the leader, so Figure 10.3 includes an opportunity for the team to endorse the leader selected by program management to get the team started. We are now ready to identify team objectives, the scope of the work to be performed, and any metrics to be used by the team to monitor performance. The program manager should be briefed on the team conclusions (F41157) to make sure that the team's shared vision is in agreement with that of the program manager's. Finally, the team should decide on a meeting style and the timing of meetings and arrange for a war wall or war room upon or within which large sketches can be hung on walls and computer projection used for group access to all information on the network. If the team members have never worked on an IPPT previously, it may be necessary or useful to pass the team through team training covering consensus building, interpersonal relationships, group cohesion, as well as how to avoid group problems such as group think. As the enterprise culture comes to accept the IPPT approach to problem solving, special team training should become less important and eventually the need nonexistent.

10.5 IPPT work

At this point, the IPPT should be ready for applying itself to requirements analysis work for the item. This should include a rapid but complete study of the specification presented to them followed by a conclusion that it is complete and valid or requires one or more corrections or additions. These changes should be approved by the PIT. The functionality allocated to the top team item may require further decomposition and allocation to identify the lower-tier items of which the item consists. In the process, the team should identify those items, see that they are added to the architecture diagram maintained by PIT, and expand the interface definition relative to the item and its parts also maintained by PIT. Depending on the complexity of the item for which they are responsible, the team may have to recommend to the PIT that subteams be formed subordinate to their team.

Should this occur, the IPPT becomes the PIT for those subteams with all of the related responsibilities within the scope of their top-level authority. Alternatively, the team will identify several items subordinate to the top-level item and assign principal design engineers to them from available team members. These principal engineers then continue work as mini team leaders applying the available team personnel resources jointly to their problems. Obviously, conflicts will develop in the joint use of the team members to serve several masters, and this will have to be adjudicated by the team leader from time to time and possibly lead to a request for additional specialists during brief time frames.

Another possibility besides subteams and principal engineers is that the item in question will be procured. In this case, the team must clearly identify the product and process requirements for the item in a procurement specification and a SOW if it is to be procured via a specification. The team may choose to implement the design followed by the procurement of the item to a set of drawings.

As the requirements for lower-tier team items are documented and approved by the team leader, the principal engineers or subteam leaders should be authorized to proceed with preliminary design work, related analyses, and requirements validation activity possibly involving some testing or demonstration. The team should stage a preliminary design review chaired by the PIT followed by the closure of any action items assigned and movement into detailed design probably with an increase in budget and team size. The detailed design should be concluded by a critical design review presented by the IPPT and chaired by the PIT. Upon release of all of the team product design packages, the team task is essentially accomplished except for manufacturing and verification support activity.

At some point, the team can be disbanded with residual work handled by the PIT. Once the whole product is in a manufacturing or procurement status, it may be possible to collapse all of the IPPT back into the PIT with manufacturing facility–oriented teams formed around the way the enterprise has facilitized their manufacturing process possibly called integrated product manufacturing (IPM) teams. In these teams, manufacturing people should provide leadership with engineering support, just the opposite of the IPPT. So, rather than some static team arrangement on programs, the program team structure is a living thing that is born, grows, adjusts to changing conditions, and retracts passing from the scene entirely as part of the produced system life cycle and the life cycle of programs within the enterprise.

10.6 PIT activity

Two teams should exist as long as the program does, and these are the business team and the PIT. The PIT is staffed exactly the same way as the IPPT from the functional departments based on the technologies involved in the system being developed. If a system will appeal primarily to mechanical engineering, we should expect it to be dominated by mechanical engineering functional department people supported by specialists from other departments. If it is mainly electrical in nature, we would expect the electrical engineering department to be heavily involved. Today and in the future, every program with any complexity will involve software people. The PIT will require not just a crowd of system engineers but specialists from many of the other disciplines. The way the PIT can be effective is for the PIT to be staffed to understand the ongoing work in all of the IPPTs so that the results of this work can be assessed from an integration and optimization perspective, and the cross-organizational interfaces between the things the teams are responsible for can be audited for compatibility.

The PIT should have a reliability, availability, and maintainability (RAM) engineer to first develop the RAM models and flow down RAM values to the team level and thereafter audit the work of RAM engineers on IPPT ensuring that everyone is working toward the same RAM models. A mass properties engineer on the PIT should do a similar thing for system end item weight and balance data allocating allowable weight values to each item at the top end of IPPT structures. Life cycle cost is another specialty needed on the PIT, and we may need many others to identify system-level values and allocate values to team product entities.

In some cases, we may only need a particular specialist at the PIT level who allocates requirements throughout the IPPT architecture in each case and monitors team design efforts relative to compliance with these requirements. In other cases, it may make better economic sense to place specialists on the IPPT as well to distribute the work between the teams in a kind of federated structure. In all of these cases, however, the PIT that should own the system models jointly used by PIT and IPPT.

Wherever possible, work should be delegated to the IPPT with PIT auditing and technical oversight of that work. Members of the PIT should participate in all IPT meetings to gain information on the ongoing design effort being alert to conflicts that may emerge between the work on two or more teams without the knowledge of IPPT members who tend to become isolated and focused totally on their area of responsibility within the program. For this reason, the PIT should require a staggering of the IPPT scheduled meeting times throughout the week. This will permit cross-team meeting attendance and the avoidance of multiple simultaneous team meetings.

10.7 The PIT–IPPT team matrix

All of the teams are staffed from the functional departments forming a program/functional matrix at the enterprise level exposed in Figure 10.1, but the functional departments should not be permitted to supervise these people within the context of the program. Functional managers should be encouraged to support their personnel assigned to programs from a discipline perspective but not for the purpose of controlling their work assignments. There is, on a program organized in accordance with the principles covered in this book, another matrix involving work direction from outside the IPPT, however. The PIT is composed of most or all of the same disciplines in the aggregate as all of the IPPT. These PIT member specialists are responsible for these specialist areas for the complete system whereas the IPPT members in those disciplines are responsible for the system subset corresponding to the team responsibility scope. These PIT specialists must interact with their counterparts on the several IPPTs forming another matrix totally within the program organization.

PIT specialists are networked into all of the IPPT for information exchange purposes, but the question remains what authority they should have in directing IPPT member work. The working understanding encouraged is that the PIT specialists are leaders of their specialty on the program and should be the most knowledgeable in their field within the company relative to the people assigned to IPPT. The PIT specialists should cultivate a good working relationship with their counterparts on the teams and organize brief meetings periodically to discuss the evolving work. Generally, they can probably influence the program work in the right direction through peer pressure when things appear to be off the mark on a particular IPPT.

The formal rules of IPPT and PIT team behavior should, however, prevent IPPT members being obligated to take technical direction from their counterparts on the PIT. The IPPT must be allowed to build and sustain a team integrity between its leader and staff. The formal direction of the work should flow from the team leader who is responsible for team cost, schedule, and performance metrics. Therefore, in this environment, a PIT specialist may find that it is not possible to move an IPPT member in that specialty field into agreement on actions to be taken in that discipline in the interest of the system. In this case, the situation should be elevated as an issue to the PIT level and worked out between the teams at that level.

It is true that ego could be the cause of a lot of this kind of hysteresis, but these situations can also represent potentially valid strongly held views by well-qualified engineers who happen to disagree. In this case, the different views should be carefully considered as alternative solutions to a problem and a preferred solution selected with critical comment allowed by all parties to the issue. It is possible in these situations that a particular IPPT happens to have a good idea that is in conflict with the current group think position, and it should be adopted as the norm.

The PIT should manage a formal action item system to track actions required that are identified outside the coverage of the program planning documents. These may cover actions that an IPPT needs the PIT to accomplish as well as the other way round or actions across IPPT boundaries. In addition, the PIT should manage a formal issues list as well. This should be a list of problems not yet well understood in terms of causes, effects, or responsibilities. The issue category focuses the attention of all the parties to items the list during their maturation process whereupon it becomes direction from PIT for a particular IPPT to accomplish or share in the accomplishment of a certain element of work to resolve the issue. The PIT should be working to zero out the action items and issues list content at all times while encouraging a professional job of resolving both kinds of program discontinuities.

part four

Generic planning

chapter eleven

Generic planning strings

11.1 Preamble

In Part five of this book, we will identify program work in a very organized fashion permitting us to catalog all work required on any program in-house. This can be done in any number of ways for any one program. What we desire is one way to do this for all programs in an enterprise allowing for the differences in product and customer from program to program. We will discuss one cataloging method in Part five called the integrated management system that the U.S. Air Force developed, but find that it does not allow for simple enterprise identification of work relative to its own process. The author will offer a cataloging method that embraces the goals of the integrated management system but does allow simple linkage into the enterprise's generic work cataloging system keyed on their generic or common process. A third very simplistic model reflecting what a lot of companies use on smaller programs applies a program-oriented work breakdown structure (WBS) and a statement of work (SOW).

Before we attack program planning, however, we must first have our home court covered. Before we design a program, we should have clear knowledge of the resources available in the enterprise because it is those resources from which the programs are designed. Once again, we intend to apply a generic or common process to every program oriented toward the product entity structure of the program.

11.2 Process component

In Part two, we developed a way to clearly define every component of our process using a process flow diagram. Each process step was named and given an ID in the form FN, where N is an alphanumeric string of numerals and letters using a base 60 system (all of the Arabic numerals, uppercase English letters less O, and lower case English letters less l). Given a process step (also referred to as a function) F141, it was expanded by identifying next lower-tier process steps, linking them together with directed line segments and logic symbols, and assigning F141N strings to those steps where N is one unique base 60 character for each next lower-tier function. We could alternatively have chosen to use a base 10 decimal delimited ID system. The author prefers the base 60 system because it sorts and indexes well in inexpensive databases so long as you do not find it necessary to recognize more than 60 functions subordinate to any immediately higher-tier function and that would not be good diagramming form anyway.

Each block in the common process diagram developed in Part two represented work that had to be accomplished on programs by people from one or more functional departments. Each of these work elements required certain input conditions and resources, operated on them in some fashion, and produced outputs. Each of these steps can be associated with program resources of cost (material resources, numbers of people, and density of their effort) and schedule (start time, end time, total time, etc.).

The proposed process flow diagram is shown in Division 3/Exhibit H/Appendix A. It provides an example of an enterprise process specification. An enterprise could alternatively apply IDEF 0, discussed in Part III of *System Requirements Analysis* (Grady, 2005) or other diagramming techniques also explored in that book, to establish their generic process diagram.

11.3 Responsibility component

In Part three of this book, we established a functional organization structure based on the knowledge our enterprise must master, retain, and continuously improve our command of. Exhibit H/Figure C-1 sheets 1 through 3 offer a graphical depiction of this structure. These are the organizations (functional departments) that we will hold accountable for proficiency on programs in the knowledge areas needed based on our chosen product line and customer base defined in our enterprise vision statement.

We named each department and assigned it a department number such as D1212. We arranged these in a hierarchical structure with all of the technical functional departments reporting to a technical vice president and all of the administrative functional departments reporting to a business vice president, the two of which reported to the enterprise executive. Obviously, there are many alternatives to this structure.

Every piece of work identified in our generic process must have a home in one or more of our functional departments. It is possible that some tasks will require more than one functional department and its associated knowledge space in order to complete it on a program. The higher in the process diagram levels one looks, the more likely this is the case. Obviously, at the enterprise vision statement level, the ultimate enterprise process block, all functional departments map to it. Therefore, we established a way to assign a principal responsibility for each process block to one functional department.

The responsibilities that a principal functional department has for each of its tasks is fourfold:

1. Acquire, train, and husband the necessary number of qualified personnel to do that department's work on all programs requiring those services. Where persons are found to be unfit to perform the work of a department, that home functional department must withdraw and replace that person and find a suitable assignment for the person withdrawn through reassignment, training, or dismissal from the staff.
2. Provide programs with written best practices telling what must be done to accomplish generic process goals and provide reference to sources of information that tell how to do this work. The latter may be in the form of reference to textbooks, videos, courses, or be contained in the department's best practices directly.
3. Provide programs with good tools coordinated with best practices and personnel skills and knowledge.
4. Coordinate with all other departments that have some responsibility under the task in question to ensure that their personnel understand their part in the process, are aware of the aggregate process description involving the principal department's role and that of their own, and understand how to use the supplied tools. Any training that the principal department provides must accommodate the collateral responsibilities of other departments that also have a role in the process step.

11.4 Work responsibility fusion through allocation

In Part three of this book, we applied functional flow diagramming (one could argue that it was actually process flow diagramming without author complaint) to identify the generic steps or tasks that would have to be accomplished on any program in the form of a life cycle flow diagram for any system that we might develop, produce, and deploy. Just as in the case of the use of this modeling environment on a program to develop the product architecture, we applied function allocation in Part three of this book to associate process steps with functional departments responsible. Exhibit H/Table A-2 shows how the process steps in Exhibit H/Figure A-1 could be allocated to the functional departments identified in Exhibit H/Figure C-1.

As noted above, we also had to identify the role that a particular functional department would have in implementing the generic task description and resource base. The department would be either called upon to lead the generic process development and maintenance or cooperate/contribute to its development under the lead of the principally responsible department. Knowing that in every case where there is more than one person or organization responsible for any part of a whole, we established an enterprise integration team (EIT) at the enterprise level that was to act as the enterprise process owner with the responsibility to integrate and optimize the generic process at the enterprise level, leading the functional departments to contribute appropriate content.

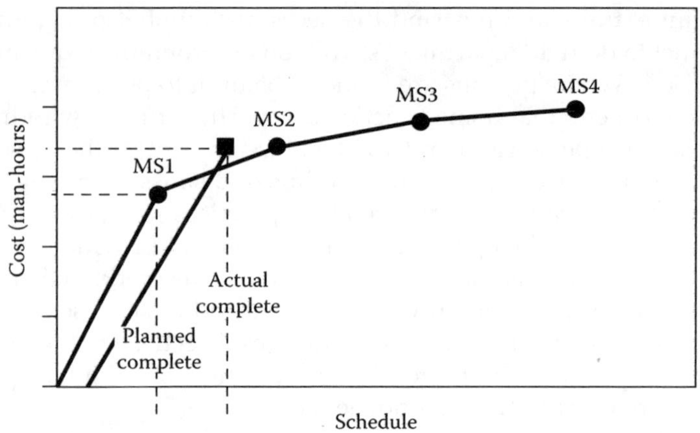

Figure 11.1 Process strings.

11.5 *Functional strings*

All of the work that an enterprise is capable of performing can now be cataloged in the form FY-DZ, where Y identifies a specific task in our common process diagram and Z a particular functional department responsible for accomplishing the related work. This work can be organized at a wide range of granularity of process steps and functional department levels. We need this range of work definition to satisfy our planning needs on programs with a wide range of planning depth needs. On early phase programs, we need fairly high-level identification of work while in later program phases it is necessary to identify work at lower levels of detail.

11.6 *Pull the strings*

The intended generic process structure must connect all of the related resources to the process step IDs such that when a program calls for the application of that process step to some part of the program work, all of the related resources flow into the program with minimum thinking required. As shown in Figure 11.1, when we pull the FN string into the program process flow for the development of a particular item in phase or stage SX, all of the related resources are also pulled into the program.

11.7 *One meaning of lean*

Work is accomplished on a program by people assigned to the program from functional departments and hopefully employed on those programs in accordance with their department charter. All of the work that a given functional department is capable of performing is identified

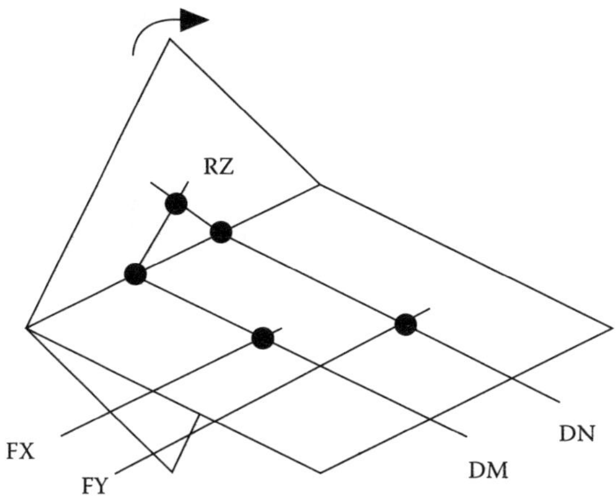

Figure 11.2 Work product transfer.

in the organizational RAS that maps all enterprise functionality to the responsible functional departments. Functional departments synthesize the aggregate of their task responsibilities into a department charter and provide people qualified to do the work in their charter, good practices, and good tools to programs.

Figure 11.2 applies a set of planes used in the author's requirements book to explain traditional structured analysis as it is applied to the development of product systems. The plane in the horizontal is the function-functional department plane sometimes referred to in this book and in *System Requirements Analysis* (Grady, 2005), *System Synthesis: Product and Process Design* (Grady, in press), and *System Verification* (Grady, 2007) as the organizational RAS. It permits the identification of generic program tasks (functions) in combination with the functional department that will supply people on a program to accomplish that work. The vertical plane is an n-square diagram allowing us to link task pairs to show the application of work products created in one task in another task.

In Figure 11.2, function FX will be accomplished by people assigned to the program from department DM yielding an enterprise work catalog element for task FX-DM. In the process of doing the work FX, the people from department DM will produce a work product that must be used by someone for some specific purpose. In Figure 11.2, the work product RZ produced in FX-DM is provided as an input to the task FY-DN. People from department DN use the work product RZ in the process of accomplishing task FY. In so doing, another work product is created.

A particular task may produce one or more work products used by one or more other tasks. The enterprise should not permit that a task may produce no work product or that it produce a work product that has no destination task. This is, after all, the meaning of lean. Every work product must have a useful application.

The task relationships shown such as RZ in Figure 11.2 are the inter-task relationships or interfaces between the program tasks. These can be generically identified as well as the tasks that create and consume them for a given enterprise involved in a particular product line. The enterprise should possess a catalog of these work products keyed to the source and destination tasks as offered in Division 3/Exhibit H/Appendix E.

chapter twelve

Generic task cost and schedule estimating*

12.1 The three principal program planning parameters

Managers focus on managing cost, schedule, and performance on programs. Satisfying all three of these parameters on a program is an evidence of great skill on the part of the program manager. There is another parameter of success that we might add that makes it even harder to be classified as a great program manager. Dr. Marty Wartenberg is a former vice president at Interstate Electronics Corporation and an employee of the University of California, Irvine, at the time this book was being written. Marty also teaches program management at UC, Irvine, and in those classes he likes to ask if any of the managers attending had ever satisfied cost, schedule, and performance requirements on at least one program. He gets a few hands. He then asks if any of these managers succeeded without damaging the company or the people working on those programs and he seldom gets any hands to the second question.

So, you may wish to add this to the three traditional indicators of management excellence. It is postulated here that if you follow the suggestions in this book you have a good chance of succeeding even in this more demanding criteria because the people working on the program will witness and be a part of a well-planned program that proceeds in an uninterrupted pathway to success rather than bounce between many unforeseen constraining conditions each one draining away some of the respect for the program manager people started the program with. Each of these conditions also often damages the company and some people as well. So, plan the work well and work the plan is still good advice for us all.

12.1.1 Margins of various kinds

The performance of the product is defined in the specifications, and the program should have ways of predicting and measuring the evolving design for compliance with these requirements as well as a thorough

* The material in this chapter is from Grady, J., *System Engineering Deployment*, CRC Press, Boca Raton, FL, Chapter 11, 2000. Used with permission.

verification program for proving it. Technical performance measurement (TPM) offers the manager an important tool for tracking product performance for a limited number of key requirements. The requirements that are important enough to manage through a TPM process will often be selected for the assignment of margins to encourage successful management within the context of considerable uncertainty. Margins can be assigned to cost and schedule as well as technical requirements. Cost margins are often collected into a management reserve while schedule margin is often referred to as float.

All margin approaches work essentially the same way. You determine the required value for something and then split it into two components: a target value plus a margin. You then require people to respond to the target value and withhold the margin as a management reserve to use when problems develop in the program. The overall effect of margins is to place more stress on the designers and other workers. In many cases, they will be able to handle the stress performing the task in less time or budget than planned or designing an item better than needed. In the inevitable case where some part of the program gets into trouble, the manager can use these margins to overcome the problem by doling them out carefully to encourage a solution to the problem.

Given that all margins involve taking something away from innocent people in the beginning and giving it away to guilty persons unable to satisfy their requirements, some jaded system engineers claim that margins follow a typical management technique of identifying the guilty and punishing the innocent. Well, that it may be, but it is a very effective technique and made to work by the tremendous skill and creativity of the majority of people currently working in industry in the development of grand systems.

12.1.2 Cost/schedule control systems

Cost/schedule control systems (C/SCS), more often called earned value systems today, are often used to coordinate budget and schedule needs and resources on programs. These systems have to be computerized to be effective on large programs, because there is so much data involved and it must be changed from time to time to adjust to changing conditions. These computer systems should capture the planning strings we developed in Chapter eleven and will extend in Part five of this book. This system should then associate with them man-hour or cost figures corresponding to available budget and time figures telling when the tasks must begin and end to fit in with other work required on the program.

As we begin a particular task on time with the planned budget, our progress is tracked as a result of everyone working on the task turning in their time against the tasks they are working. These inputs are used, of course, to compute the payroll problem for the workers, but these

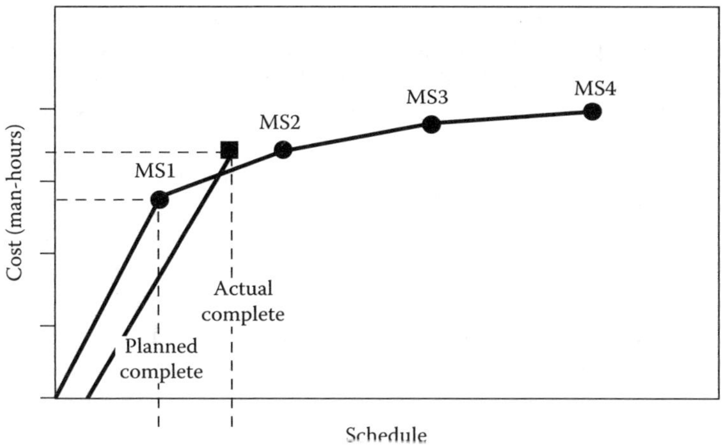

Figure 12.1 C/SCS application.

inputs are also charged against the budget in the C/SCS. Over time, these charges are accumulated and reports made to show progress on the task. Figure 12.1 illustrates the machinery of these systems in the form of a cumulative track of man-hours over a planned period of time. Milestone 1 was intended to be achieved at a planned number of man-hours and on a particular date. In this case, the milestone was actually achieved later and at more cost than planned. When the data collected reveals this kind of situation, a C/SCS will generally generate a variance report requiring the person responsible for this task to answer some penetrating questions about why this task is overrun in both cost and schedule. The person responsible in this case might claim that he or she started late because of difficulty obtaining staff, and the staff was not prepared for the difficulty of the task. In the next cycle leading to milestone 2, this task might be more carefully watched by program management in an effort to cause it in the end to achieve its goals within planned cost and schedule.

The milestones indicated in this figure are mini events of which there may be many between two major program events used in the planning work discussed in prior chapters. In each case, we should have recorded in the planning data a criterion whereby it can be determined when the milestones have been satisfied and they can be claimed.

The C/SCS should also include program evaluation and review technique (PERT) or critical path method (CPM) capabilities permitting the identification of the critical path of work through the program. The diagrams that these computer systems create are often called program networks. These networks involve tasks and events strung together by directed line segments. Both of these systems started out as activity-on-line depictions. All of the tasks we have identified in program planning become lines on the network with vertexes at related events. Activity-on-node systems have also been developed. While these tools have many powerful features, none is more impressive than their ability to display the critical path between any two nodes. Given two or more parallel work paths between two nodes, the critical path is that string of work that requires the most time to accomplish and is characterized by a zero schedule float or margin.

Ideally, we should be able to manipulate the C/SCS to search for sound management policies for situations that arise on the program. Some people call this playing what-if games with the data. Such changes as we might make in the process should not necessarily become a part of the formal record. Only when we have deduced an advantageous change and it has been approved, should we modify the formal plan.

12.2 Task data sheets

The generic planning data should include information about nominal cost and nominal task time span, and this data is updated based on actual program performance. Figure 12.2 offers an example of a task definition form that might be used to capture this information. In this case, the task is a reliability allocation activity performed by someone from department D216-4 performing work on function F41114471. The form includes an estimate of the number of heads and man-hours required to accomplish the task on a nominal program. If, when we planned this particular program, we had concluded that it was a nominal program we would have estimated the job in accordance with the figures on this sheet. If we concluded that the task was less or more demanding, we would have adjusted the estimate accordingly and documented the rationale.

12.3 Estimate improvement

Estimates appearing in generic planning sheets must at some point in time all be estimates analytically formed, but over time the accuracy of these estimates should improve because you readjust them based on actual performance on programs as suggested in Figure 12.3. At some level of indenture, the program estimates for tasks should be compared with actuals, and assessments made of what any difference between these numbers implies. This is a matter of a search for truth. If, on a particular program, the task is badly overrun, does it mean that the task is much harder than estimated, because it was accomplished by people unskilled in the task, or because it was badly managed? The enterprise needs to apply the equivalent of a failure reporting and corrective action system (FRACAS) used to determine the causes of product failures in manufacturing and testing to determine the cause of these kinds of failures in program performance.

Task MID	F41114471
Task name	Reliability allocations (Task 202)
Dept. number	2164
Dept. name	RAM Engineering
Procedure ref.	Company Procedure 42-8; MIL-STD-756B (tailored); *Reliability Engineering*, ARINC, Prentice Hall, Chapter 6.
Objective	Define a quantitative reliability figure for each system entity
Description	The analyst studies the planned design concepts provided by the IPT designer(s) and allocates the system or lower-tier failure rate to subelements from top to bottom in each product entity layer and branch based on available historical data or predictions determined by the item composition.
Significant accomplishment	Reliability figures assigned to all system items to a depth defined for the program.
Completion criteria	The task is complete when every configuration item, procured component, and other items as defined in the program have been assigned a reliability figure and the complete set of data has been checked for internal consistency, customer requirements compliance, and been approved by the PIT.
Events	Model structure complete at SDR, allocations loaded by PDR, and predictions loaded by CDR.
Resources	
Tools	Five networked International Business Machines (IBM) compatible computers each running RAMEASY, Microsoft Word 5.0, and Microsoft Excel for first two quarters. Thereafter, two machines are adequate. Network server memory space 50mb.
Personnel	At peak work load, six experienced reliability analysts skilled in failure rate prediction and reliability math and familiar with equipment used in missile systems.
Facilities	Each analyst requires a standard company engineer's workstation. Personnel will be assigned to teams with two assigned to PIT and four to product teams, one per team during the first two quarters of the program. Thereafter, two analysts on PIT only.
Inputs	1. Design concept information from item designer 2. System architecture definition
Outputs	1. Reliability model complete with a quantitative reliability number for each configuration item and procured item. Data users include IPT as a source of reliability requirements, availability analyst as a source of reliability data for computing availability

Budget EST per quarter heads	1	2	3	4	5	6	7	8	9
	2	5	5	5	2	2	2	2	2
Man-hours	1008	2520	2520	2520	1008	1008	1008	1008	504

Figure 12.2 Sample functional task definition form. (From Grady, J., *System Engineering Deployment*, CRC Press, Boca Raton, FL, pp. 178–220, 2000. With permission.)

Alternatively, we could simply collect the actuals by task and probabilistically deal with them without knowledge of the causes. The mean of the actuals is probably close to where the estimated value should be. Obviously, this could all be done by a computer into which the task estimates and task actuals by the program has been entered. The next

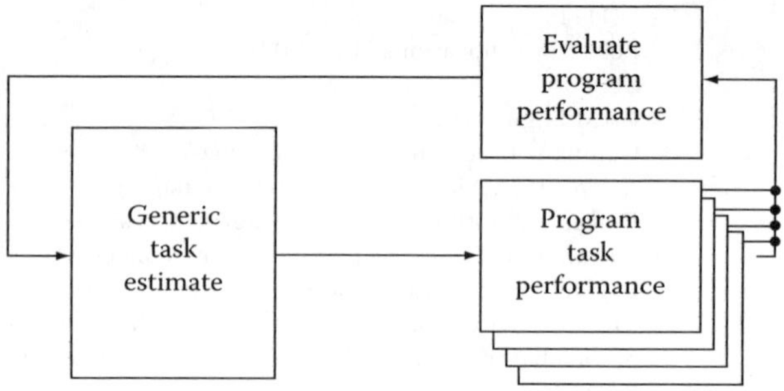

Figure 12.3 Generic estimate improvement. (From Grady, J., *System Engineering Deployment*, CRC Press, Boca Raton, FL, pp. 178–220, 2000. With permission.)

program estimate could be adjusted for the mean of all prior task implementations. Where estimate increments are trending up, we should get a warning signal to search for the cause because our performance should get better each time as a function of repetition with some variance for program difficulty.

We should be seeking predictability in our program planning work. What we estimate during the planning process we should see occurring in actual program performance. Here is where we start to see the big value in generic planning strings. Every program is planned based on the same generic process and planning data, and we can easily capture statistics that have a lasting meaning. If an enterprise plans each program from a different process perspective, it will be very difficult to correlate the results with a planning basis. It will be very difficult to establish a servo loop effect which we can use the differences between planned and actual performance as a basis for improvement where performance does not measure up to expectations.

As important as it is, it is not enough, of course, just to be able to estimate accurately what a task will require. We should also be working to reduce the cost and schedule demands for tasks while improving the quality of the work performed. An analogy with rifle target shooting applies here. Two rifle shooters fire 10 rounds at say 500 yards. One of the shooters observes that his shot pattern is all over the target. The other shooter sees all 10 rounds within a 10 in. radius on the left side of the target. The second shooter will be able to adjust his windage and move the center of the shot group to the bull's eye. The first shooter cannot make any adjustments to improve performance on the second 10 rounds because there is no clear pattern.

So, given that we can measure performance in association with a plan and that our performance shows compliance with estimates, we should inquire if there is some way to purposely change the plan and performance so that customers will receive more value. Functional management and the EIT should be constantly reviewing program performance and reaching conclusions about process steps that could be improved. The EIT should be very careful in selecting improvement actions and do so from a systems perspective. Some improvements may have little effect in reducing program cycle time because the associated tasks do not generally appear on a program critical path. The EIT should manage this process and encourage coordinated improvements implemented by the functional department managers in the resources they provide programs.

chapter thirteen

How-to knowledge access*

13.1 Knowledge depth differences

The normal content of company system engineering practices documents is limited to defining what must be done, the sequence in which it must be done, and possibly why it is done. Few of these manuals include detailed content about how to do this work, at least not to the degree of detail that a textbook on the subject might cover. The system engineer, however, needs access to the how-to information that fills in the gaps in their practices documentation. If the enterprise makes no effort to define the how-to knowledge, then this valuable resource is by default locked into the corporate memory of the current staff. As the staff with experience moves on to retirement or departs for other reasons, this corporate memory drifts out of the company.

It is important that the enterprise identify what must be done in the system engineering field on its programs, but it is not enough. You must move on to the next level of detail and provide access to the how-to information and find some way to make it available to your employees. This book focuses on the system engineering field, but a company can, and should, apply this same model to all fields.

13.2 Sources

The how-to knowledge can reside in several different places and not necessarily the same place for all of this information needed by a company. These places include external standards, textbooks, the company's system engineering manual (SEM), training courses, and human memory.

13.2.1 External standards

There are several standards available in the system engineering field: IEEE 1220, EIA 632, and MIL-STD-499A to name three. The first two are current and maintained while 499A has been discontinued by DoD. All three of these documents contain useful information about the systems approach, but they do not provide in depth coverage; rather a list of things that should be done is provided.

* The material in this chapter is from Grady, J., *System Engineering Deployment*, CRC Press, Boca Raton, FL, Chapter 15, 2000. Used with permission.

13.2.1.1 ANSI/EIA-632

ANSI/EIA-632 lists 33 requirements for system development repeated in Table 13.1. This standard is essentially a specification for an effective development process. It contains little information about how to accomplish any of the work listed. Whatever our system development process, we should be able to show that these things are being accomplished in that process.

Appendix G of ANSI/EIA 632 offers a way to sequence the requirements that the author would describe a little differently. This difference involves the relationships between requirements 14 through 19. The standard encourages that for any one item in the physical architecture, acquirer (user or customer) requirements (14), and other stakeholder requirements (15) combine to form the system technical requirements (16), apparently without benefit from any kind of logical (structured) modeling or analysis (17). Then someone applies logical solution representations as a means to identify derived requirements that together with the system technical requirements drive the physical solution representations (18) and the design solution, which in turn drives the specified requirements (19). On first reading, it looks like the design (18) is driving the requirements (19). The author accepts that requirement 16 represents the system and item performance specifications (using the current U.S. Department of Defense term previously identified as Part I) that should contain design-independent content and be the basis for the design (18) and qualification (30). The author also accepts that requirement 19 can represent the item detail specifications (Part II), which should contain design-dependent content and act as the basis for acceptance (31). Figure 13.1 offers an alternative view of the sequence of requirements showing that requirement 17 should be the basis for all system and performance specification contents.

Many of the requirements in ANSI/EIA 632 map to management practices, and these are simply grouped without sequence in Figure 13.1. The validation and verification requirements are linked to the corresponding tasks in the sequence of requirements and design tasks.

13.2.1.2 IEEE 1220

ANSI/EIA 632 uses requirement numbers to denote the requirements as suggested in Table 13.1. IEEE 1220 is a more normally structured document with paragraph numbers, which are included in Table 13.2 as a way of mapping these requirements to other elements. The document includes additional details expanding upon the indenture listed here. We have a choice in terms of the granularity to which we can map process elements, and the author has chosen first and second indentures for the purposes of this book. This document was initially developed as a commercial system engineering standard, but many companies dealing with Department of Defense have chosen it as the basis for their internal process definition.

Table 13.1 ANSI/EIA 632 Requirements

NBR	Requirement name	Requirement text
1	Product supply	For a system, or portion thereof, supplied to an acquirer, the developer (when acting as the supplier) shall establish and satisfy an agreement with the acquirer
2	Product acquisition	For a system, or portion thereof, acquired from a supplier, the developer (when acting as the acquirer) shall establish an agreement with that supplier
3	Supplier performance	The developer (when acting as the acquirer) shall manage supplier performance to ensure that the technical effort to be accomplished by the supplier provides end products that satisfy the assigned requirements
4	Process implementation strategy	The developer shall define a strategy for implementing the adopted processes of this standard as a basis for project technical planning, and that is in accordance with the agreement
5	Technical effort definition	The developer shall define a technical effort that is in accordance with the process implementation strategy
6	Schedule and organization	The developer shall schedule and organize the defined technical effort
7	Technical plans	The developer shall create technical plans to ensure an integrated and cost-effective technical effort in accordance with the defined schedule and organization
8	Work directives	The developer shall create work directives that implement the planned technical effort
9	Progress against plans and schedules	The developer shall assess the progress of the technical effort against applicable technical plans and schedules
10	Progress against requirements	The developer shall assess the progress of system development by comparing currently defined system characteristics against requirements
11	Technical reviews	The developer shall conduct technical reviews of progress and accomplishments in accordance with appropriate technical plans
12	Outcomes management	The developer shall manage the outcomes of the technical effort

(continued)

Table 13.1 (continued) ANSI/EIA 632 Requirements

NBR	Requirement name	Requirement text
13	Information dissemination	The developer shall ensure that required and requested information is disseminated in accordance with the agreement, project plans, enterprise policies, and enterprise procedures
14	Acquirer requirements	The developer shall define a validated set of acquirer requirements for the system, or portion thereof
15	Other stakeholder requirements	The developer shall define a validated set of other stakeholder requirements for the system, or portion thereof
16	System technical requirements	The developer shall define a validated set of system technical requirements
17	Logical solution representation	The developer shall define one or more validated sets of logical solution representations that conform with the technical requirements of the system
18	Physical solution representations	The developer shall define a preferred set of physical solution representations that agrees with the assigned logical solution representations, derived technical requirements, and system technical requirements
19	Specified requirements	The developer shall specify requirements for the design solution
20	Implementation	The developer shall implement the design solution in accordance with the specified requirements to obtain a verified end product
21	Transition to use	The developer shall transition verified products to the acquirer of the products in accordance with the agreement
22	Effectiveness analysis	The developer shall perform effectiveness analysis to provide a quantitative basis for decision making
23	Trade-off analysis	The developer shall perform trade-off analyses to provide decision makers with recommendations, predictions of the results of alternative decisions, and other appropriate information to allow selection of the best course of action
24	Risk analysis	The developer shall perform risk analyses to develop risk management strategies, support management of risks, and support decision making

Table 13.1 (continued) ANSI/EIA 632 Requirements

NBR	Requirement name	Requirement text
25	Requirement statements validation	The developer shall ensure that technical requirement statements and specified requirement statements, individually and as sets, are stated in an acceptable manner
26	Acquirer requirements validation	The developer shall ensure that the set of defined acquirer requirements agrees with acquirer needs and expectations
27	Other stakeholder requirements validation	The developer shall ensure that the set of defined other stakeholder requirements agrees with other stakeholder needs and expectations with respect to the system
28	System technical requirements validation	The developer shall ensure that the set of defined system technical requirements agrees with the validated acquirer and other stakeholder requirements
29	Logical solution representations validation	The developer shall ensure that each set of logical solution representations agrees with the appropriately assigned subset of system technical requirements
30	Design solution verification	The developer shall verify that each end product defined by the system design solution conforms to the requirements of the selected physical solution representation
31	End product verification	The developer shall verify that an end product to be delivered to an acquirer conforms to its specified requirements
32	Enabling product readiness	The developer shall determine readiness of enabling products for development, production, test, deployment/installation, training, support/maintenance, and retirement or disposal
33	End products validation	The developer shall ensure that an end product, or an aggregation of end products, conforms to its validated acquirer requirements

13.2.1.3 ISO 9001

ISO 9001 from the International Standards Organization is in the quality area (9000) but offers an excellent and brief series of requirements for a sound system development process. These are the absolutes, there is nothing extraneous listed here. At the same time, an organization that does all of these things well and in the right sequence can be assured that they

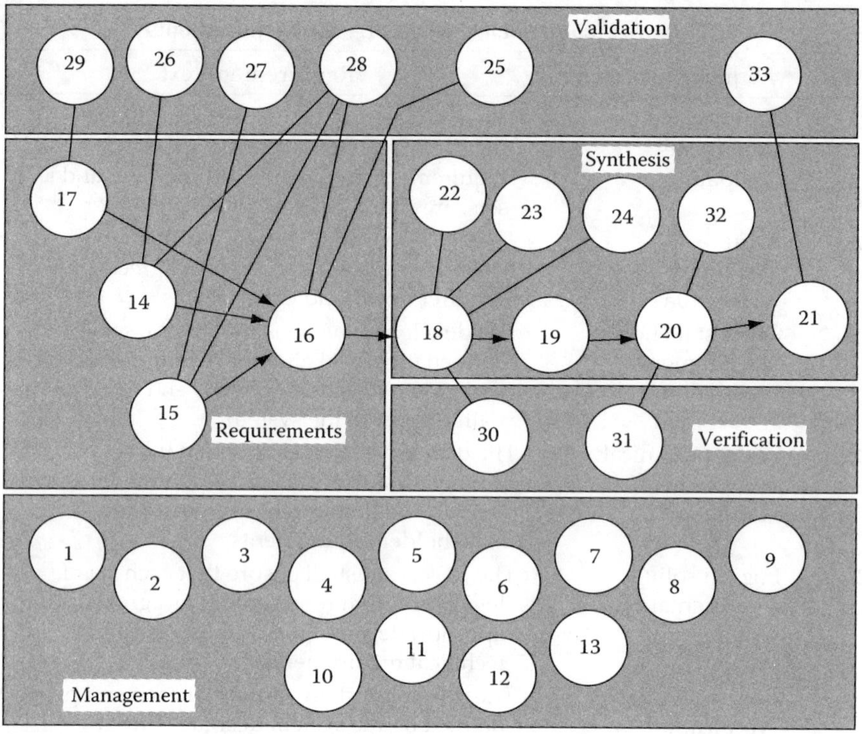

Figure 13.1 Preferred ANSI/EIA 632 requirements sequence.

have an acceptable process. But, that is the problem. Creating a description of a good internal process is not difficult. What is difficult is to cause that process description to be faithfully followed in actually accomplishing enterprise work. In all capability maturity models, we find criteria for following internal documentation.

13.2.1.4 Defense System Management College

Defense System Management College (DSMC) has over the years published a book called *System Management* that is used in a course taught by DSMC for Department of Defense program managers. It provides sufficient detail in some areas to qualify as how-to knowledge. This document has been released in several different versions, and many say that the original one developed from the Lockheed Space and Missiles Division system engineering, manual under the direction of Bernard Morais, then the system engineering director for the company and now the president of Synergistic Applications, was the best of the lot. The current one is not bad, and it can be downloaded from the DSMC Web site.

Table 13.2 IEEE-1220 Content

Para	Topic
1	Overview
2	References
3	Definitions and acronyms
4	General requirements
4.1	Systems engineering process
4.2	Policies and procedures for systems engineering
4.3	Planning the technical effort
4.4	Evolutionary development strategies
4.5	Modeling and prototyping
4.6	Integrated database
4.7	Product and process data package
4.8	Specification tree
4.9	Drawing tree
4.10	System breakdown structure (SBS)
4.11	Integration of the systems engineering effort
4.12	Technical reviews
4.13	Quality management
4.14	Continuing product and process improvement
5	Application of systems engineering throughout the system life cycle
5.1	System definition stage
5.2	Preliminary design stage
5.3	Detailed design phase
5.4	Fabrication, assembly, integration, and test (FAIT) stage
5.5	Production and customer support stages
5.6	Simultaneous engineering of product and services of life cycle processes
6	The systems engineering process
6.1	Requirements analysis
6.2	Requirements validation
6.3	Functional analysis
6.4	Functional verification
6.5	Synthesis
6.6	Physical verification
6.7	Systems analysis
6.8	Control

Source: Grady, J., *IEEE Std 1220™-2005 IEEE Standard for Application and Management of the Systems Engineering Process*, 2005.

13.2.1.5 Other documents

The U.S. Army published a field manual, FM 770–78, in 1979 titled *System Engineering* that went out of use some time ago. It defined a sound systems approach and could be used as the basis for how-to knowledge. Similarly, the Air Force Systems Command years ago published Air Force Systems Command Manual (AFSCM) 375 series that along with FM-770-78 can be downloaded from the INCOSE Web site by people interested in the history of the development of system engineering standards and process descriptions. MIL-STD-499A was used for years in defense industry as noted but is no longer maintained. MIL-STD-499B was not approved as it appeared at a time when DoD was moving toward commercial standards. EIA 731 is a standard for capability maturity rather than intended as a source document for the development of an effective process. Some organizations have used it as the basis for the design of their process, but that was not the intent of the document.

In general, currently available commercial system engineering standards are acceptable as a basis for a practices manual telling what must be done, but generally contain insufficient detail to act directly as a source of how-to knowledge.

13.2.1.6 Multiple standards

An enterprise may be encouraged by multiple customers, each preferring a different standard of performance, to follow their standard as the basis for their process. If an enterprise succumbs to this encouragement, however, they cannot possibly possess a single common process with all of its advantages. The enterprise can have a single common process but can still respond to multiple different standards by tailoring all of them to agree with their common process and mapping the result to their common process description to the content of those external standards.

13.2.2 Textbooks

Several fairly recent books have been published that provide good coverage of the overall systems approach adequate for working system engineers to use as a reference source to find out how to do particular jobs. Some of those are

- B. Blanchard and W. Fabrycky, *Systems Engineering and Analysis*, Prentice Hall, Englewood Cliffs, NJ, 1990.
- B. Blanchard, *System Engineering Management*, 2nd edition, Wiley, 1997.
- George A. Hazelrigg, *Systems Engineering: An Approach To Information-Based Design*, Prentice Hall, Upper Saddle River, NJ, 1996.
- J. Lacy, *Systems Engineering Management*, McGraw-Hill, New York, 1992.

- James N. Martin, *Systems Engineering Guidebook*, CRC Press, Boca Raton, FL, 1997.
- A. Sage, *Systems Engineering*, Wiley, New York, 1992.

Other recent books have been published that provide how-to knowledge about subsets of the overall process. Some of those are

- J. Grady, *System Integration*, CRC Press, Boca Raton, FL, 1994.
- J. Grady, *System Engineering Planning and Enterprise Identity*, CRC Press, Boca Raton, FL, 1995.
- J. Grady, *System Validation and Verification*, CRC Press, Boca Raton, FL, 1997.
- J. Grady, *System Engineering Deployment*, CRC Press, Boca Raton, FL, 1999.
- J. Grady, *System Requirements Analysis*, Elsevier Academic Press, Boston, MA, 2005.
- J. Grady, *System Verification*, Elsevier Academic Press, New York, 2007.
- J. Grady, *System Synthesis: Product and Process Design*, CRC Press, Boca Raton, FL, in press.
- E. Rechtin and M. Maier, *The Art of System Architecting*, CRC Press, Boca Raton, FL, 1997.

In addition, those companies that purchase, or persons who attend, the JOG System Engineering certificate program will possess the four-books covering the complete system engineering process described at the how-to level. This consists of this book and the other three books listed in Table 1.1. plus the student manuals for each course. The latter for the management course includes Division 3/Exhibit E that maps the common process described in Division 3/Exhibit H/Appendix A to the content of these books.

13.2.3 Company system engineering manual

Earlier in this chapter, we mentioned that few SEMs in industry contain sufficient detailed information to act as a how-to source. But, a company could write one that did. The author has copies of old manuals from General Electric, Martin Marietta, and General Dynamics that did provide detailed how-to information but companies preparing manuals today tend to model their practice from IEEE 1220 or EIA 632 that do not provide the detailed coverage. A document that includes the how-to knowledge will invariably be thick, and experience has shown that thickness discourages the use of the document. The enterprise is probably better served by a relatively thin (100 one-sided pages or less) manual backed up with references to how-to knowledge provided externally.

Until a few years ago, the author would have encouraged that a company write a manual with considerable depth. Today the author believes a good manual telling what to do is adequate. This change of position resulted probably because the author has always been able to write easily about our profession and failed earlier to consider how much difficulty many others have in doing so. The author has found a better, easier-to-implement, more flexible way of capturing the how-to knowledge and that is offered in this chapter. If your company has a budding author on the payroll, however, you may be able to satisfy the how-to knowledge need directly in your practices.

13.2.4 Training courses

All of the previous methods of capturing the how-to knowledge require the individual to read and understand the content. Another approach is to use these sources either directly or through suppliers to build one or more training courses that organize the material for more effective learning. These training courses could come from several sources such as

1. In-house courses prepared by company personnel and taught by company personnel.
2. In-house courses purchased from a supplier of system engineering courses traceable to the company process description and presented by company personnel.
3. Traveling short courses such as those provided by JOG System Engineering (the author's company), Synergistic Applications, Center for Systems Management, and Applied Technology Institute (ATI). These courses are commonly offered in various cities with an aerospace population in close proximity and/or on-site at companies by contract or purchase order.
4. Local university extension courses. This may include a complete system engineering certificate program as it does at University of California, San Diego or isolated courses that may satisfy some subset of your needs. Most university extension departments will provide their program on-site at companies. A few universities offer a BS or MS in system engineering, and if your facility happens to be in close proximity to such a campus, that university can, over a sufficient period of time, provide an excellent education for your system engineers.

13.2.5 Registered experts

Your company may have people on the payroll who are respected as experts in one or more fields of system engineering who could act as mentors for those who are not. To take full advantage of this resource, you need to make a list of these people keyed to the areas of expertise. When someone needs to improve their knowledge in one of these areas, they contact the designated expert. In order for this to work in a large company, a funding method must be defined, such as a work order for the expert to charge when mentoring. Where both parties are at the time working on the same program, this cost can very likely be charged to normal work on the program because it is not intertwined with normal work. Over time you will loose these experts and have to consciously refresh your source of experts.

Companies that have an active mentoring program, for college new hires especially, have this arrangement in place and why would not every engineering company have such a program. Managers in the system engineering field commonly are very hostile to hiring recent college graduates, opting in favor of people with some experience in an engineering field in industry. The author while an engineering manager in industry was forced by company policy to hire people right out of college into a system engineering department, and while unhappy about the policy initially, he had to admit that these people turned out as well as experienced people hired from other companies. Their success was in part due to selecting good talent (avoiding the people who could not wait to design circuit boards, for example) and to the way we partitioned the complete system development process into tasks small enough that an intelligent person could master them but large enough to be a whole job and represent a challenge.

13.3 How-to knowledge directory

So, there are several ways by which we can get access to how-to knowledge. Now, the problem is to make an inventory of the knowledge that is required and link it to the source for that knowledge such that all of our system engineers can gain access to it as needed. The needed knowledge is defined by the functional department charters listed in some detail for our sample company in Exhibit H Table D-1 derived from the functional allocations in Exhibit H Table A-1.

13.3.1 Directory format

Table 13.3 lists a sample of the functional tasks from Exhibit H/Table A-1, links them to the responsible department (L for lead in the RESPonsibility column) and contributing departments (C in the RESP column), and identifies the preferred how-to knowledge source. There may be seven acceptable sources of how to do this work but you have done an adequate service if you identify at least one for each task. The contributing departments have been left out in the interest of space except for one example case for

Table 13.3 How-to Knowledge Directory Sample

Task ID	Name	Department number	RESP	How-to reference
F4112L	Item L Requirements Analysis	D2162	L	Book, 4003
F4112L1	Assign/Select Item For Analysis	D2162	L	Book, 4003
F4112L2	Select Item Modeling Application	D2162	L	Book, 4003
F4112L3	Specification Template	D2162	L	Book, 4003, Chapter 5
F4112L4	Item Applicable Documents Analysis	D2162	L	Book, 4003
F4112L5	Change/Revise Specification	D2162	L	Book, 4003
F4112L6	Traditional Structured Analysis	D2162	L	Book, 4003
F4112L61	Functional Analysis Extension	D2162	L	Book, 4003
F4112L62	Performance Requirements Analysis	D2162	L	Book, 4003
F4112L63	Timeline Analysis Extension	D2162	L	Book, 4003
F4112L64	Architecture Synthesis Extension	D2162	L	Book, 4003
F4112L65	Interface Analysis Extension	D2162	L	Book, 4003
F4112L66	Process Analysis Extension	D2162	L	Book, 4003
F4112L67	Specialty Engineering Requirements Analysis	D2162	L	Book, 4003
F4112L68	Service Use Profile Extension	D2162	L	Book, 4003
F4112L69	End Item Zoning Analysis Extension	D2162	L	Book, 4003
F4112L6A	Interface Requirements Analysis	D2162	L	Book, 4003
F4112L6B	Write Requirements Paragraphs	D2162	L	Standard, MIL-STD-961D
F4112L6C	Lead System/Item Requirements Analysis	D2162	L	Book, 4003
F4112L6D	Edit/Update Current Content	D2162	L	Book, 4003

Table 13.3 (continued) How-to Knowledge Directory Sample

Task ID	Name	Department number	RESP	How-to reference
F4112L6E	Integrate Item Requirements	D2162	L	Book, 4003
F4112L6F	Collect SDD Inputs	D2162	L	Book, 4003
F4112L7	Modern Structured Analysis	D2162	L	Book, 4003
F4112L8	Unified Modeling Language			
F4112L9	Database Structured Analysis			
F4112LA	Early Object-Oriented Analysis	D2150	L	Book, 8002
F4112LB	Verification Planning Analysis			
F4112LC	Quality Planning Analysis			
F4112LD	Manufacturing Planning Analysis			
F4112LE	Logistics Planning Analysis			
F4112LF	Material and Procurement Planning Analysis			
F4112LG	Disposal Planning Analysis			
F4112LH	Schedule Planning Analysis			
F4112LI	Deployment Planning Analysis			
F4112LJ	Clone Specification			
F4112LK	Item L Detailed Verification Requirements Analysis			
F4112LL	Study Parent Concept			
F4112LM	Prepare Item L Detail Specification			

Source: Grady, J., *System Engineering Deployment*, CRC Press, Boca Raton, FL, p. 228, 2000. With permission.

task F115J. Also, the function IDs may not agree with the current content of Division 3, Exhibit H, Appendix A content.

Your directory could be improved over that shown in Table 13.3 by including separate columns (fields) for the source type (book, course, or expert), and detailed identification of the source within that type that could include a chapter/section reference in the case of a book (including standards). The reader is by this time probably convinced that it will be impossible to keep track of all of these tables and the information in them. It probably would be if they were all separate tables maintained by typewriter or word processor. All of this information should be in a relational database and the records and fields of the tables set up for maintenance by appropriate assigned personnel. Alternatively, the PIT could be the only party allowed to change any content based on an agreement previously reached. Department numbers come from Exhibit H/Table C-1. As the information is entered or changed in the tables used for maintenance of the data, it appears in all other views of that data.

13.3.2 *Directory development responsibility*

The knowledge directory must be assembled by someone and the question is who? The functional departments should be the knowledge owners for the enterprise, so the managers of these departments or persons they designate should be the correct source of this information. When allocating functionality to departments, we noted that some tasks properly flow to multiple departments at some level of task indenture. In these cases, we have an obligation to identify which of these departments is the principal department and that department should have the responsibility to prepare the first description of the task, identify any tools to be used by all parties to the task, and determine the how-to reference.

The PIT should act as the owner of the directory as they should for the whole set of information defining the enterprise identity supported by the functional departments as a function of the task responsibilities allocated to them.

13.3.3 *How-to knowledge access*

The final issue to resolve is to provide access to the knowledge we have identified in our knowledge directory. If we have a company or engineering department library, this is probably the right place for any textbooks or standards we have identified. There should be a clear policy defined and available to everyone on how to gain access to training course resources identified in the directory. Some companies set aside a training account for all employees and encourage them to independently consume it during the year. In other companies, it may be necessary to gain the approval of one's supervisor. Names and phone numbers or e-mail addresses might be adequate for formally identified experts, but these people may also require a charge number they could use when the demand is time-consuming.

Most engineers spend a lot of time at their computer, so this seems like a good place to deliver the how-to knowledge. The SEM can, of course, be placed online and hyperlinked to more extensive coverage in the form of references to books, more extensive documentation coverage owned by the enterprise, audio and/or video presentation on the topic, or reference to a course in a local university or other sources. It will be necessary to avoid violating any copyright laws in the process of providing access to how-to knowledge.

chapter fourteen

System engineering training program development*

14.1 Training program overview

Training programs provide people with how-to knowledge in an organized framework. We discuss ways to prioritize the parts of the process most in need of improvement and focus energy on providing training for these parts using the low-cost lunchtime seminar as an initial means to satisfy the need. Other alternatives, discussed in the previous chapter, are considered including existing college courses in the area and arrangements that could be made with local universities to create programs that would qualify for company tuition reimbursement. Training solutions of whatever kind should be linked to your how-to knowledge base traceable to the tasks in your generic process, as discussed in Chapter thirteen.

14.2 Process evaluation and training prioritization

Given that your company is in need of system engineering education, and you have a written system engineering practices document that at least tells what must be done on programs from this perspective, you need to decide where to spend your scarce training funds. No matter how the training will be arranged, it will cost something, and industry in general is very frugal with training money (with a few enlightened exceptions like Motorola that allocates training funds to every individual and expects them to be spent each year). To proceed, one should evaluate their process for education priorities. In so doing, you should talk to program chief engineers (by whatever name, the person responsible for program technical matters) and program managers, and find out what system engineering work they think could be done better with the most payoff in terms of program success.

Make a prioritized list of system engineering tasks and start trying to match the items on the top of this list with sources of corresponding knowledge in your organization and in books and standards. The

* The material in this chapter is from Grady, J., *System Engineering Deployment*, CRC Press, Boca Raton, FL, Chapter 16, 2000. Used with permission.

granularity of the beginning task list is important. You should start at the top and, in an indentured outline, refine the work definition down to the level that it would take no more than an hour to cover the topic in a classroom situation. It is at this lower tier that you should select tasks for inclusion on your prioritized list.

14.3 The low-cost lunchtime seminar

If you cannot obtain the funds you need to train your complete staff in system engineering, you have to find an alternative that will pay some dividends. At the other extreme in cost from a complete certificate program, one can create a very effective lunchtime seminar (or brown bag seminar, where people can bring their lunch) series of lectures. The topics are from the top of your prioritized list. The next step is to find someone who is qualified who will volunteer to teach the topic. The list of experts in Chapter thirteen is probably the same list we need here.

The mechanics of the program are very simple. You need to reserve a meeting room of adequate size for each lunchtime you plan to present a course. You should try to get the best room possible to give the program the illusion of management support whether or not this is true. You might begin by offering one class a week always on the same day at lunch. If your company only has a 45 min lunchtime, you may have to reduce the time from 1 h to 45 min, but 50 min is probably a good compromise where the company lunch period is between 45 and 60 min. Those attending could be encouraged to make up any time they lose from work by attending the seminar, especially where they have to drive from one facility to another.

The fact that the lunchtime seminars will not cost your company any money (volunteer instructors from your staff, company facilities used when they would normally be vacant, and employees voluntarily attending on their own time) has advantages, of course. But, there is also a down side. It is probably not legal for you to require employees to attend these training sessions in that they are offered on the employee's own time; so, you can only make a training hit with those who are willing to participate voluntarily. You also probably cannot rank your employees for pay and promotion purposes according to whether or not they attend these classes. If the training is of value, however, one should expect that those who attend will increase their value to the company and their performance improvement on programs should be positively effected as a basis for pay and promotion purposes. You should avoid a direct link between attendance in this kind of program and employee ranking.

You will also encounter hostile signals from people in the management hierarchy. In that attendance is voluntary, it is conceivable that someone from another department, not in system engineering, may choose to attend and that the ideas expressed are in conflict with those of the attendee's management chain. There are also control freaks in organizations who might try to interfere only because they were not consulted. The engineering administration director approached the author when he set up such a program at General Dynamics Space Systems (GDSS) Division and inquired how he was paying for the program. The answer was that it was costing the company nothing. Whereupon the director asked what the employees were charging their time to, and the author responded that they were on their own time and did not need to account for their time any more than if they sat at their desk and ate lunch. The director offered that there was something wrong with the arrangement that he had not yet understood. Well, the author simply continued and expanded the program because he was right. If we wait for approval on activities such as this, we will always find others with reasons that inhibit action.

The author's program included several categories besides system engineering how-to knowledge. In addition, there was a product line series consisting of 1 h classes on the systems in the Atlas Centaur launch vehicle. These were often taught by recent college hires who were thus forced to learn how these systems worked and locate experts in related fields they might not otherwise have met so early in their career. We also offered a graybeard series where very senior or retired engineers were asked to come in and give of their experience. One of these conducted by a former chief engineer from the early Atlas program was attended by several of the engineering department directors. Finally, we offered a series on what the different departments did on programs. The lecturers for these programs were functional department engineering chiefs responsible for these departments. The highlight of this series was when one structural analysis department chief was lecturing and another department chief attending broke into the conversation with the comment that his department did that work on programs. All of these categories provided valuable learning opportunities to the college new hires in the author's system development department.

14.4 Local university cooperation

There are some advantages in acquiring system engineering training from local universities, and chief among them is the shift of this cost from department training funds, always in short supply, to the company's tuition reimbursement funds that a company generally tries to sustain at whatever level demand dictates. If you are successful in making this transfer work, it may cause the top management to reevaluate your company's tuition reimbursement policy, but it is unlikely that the demand that is placed on the program from this source is enough to stimulate a protective response.

14.4.1 University readiness for system engineering training

Few educational institutions are prepared to independently champion the cause of improving system engineering practice in industry by better preparing college students for this work. Few educational institutions today offer students an effective course of instruction in system engineering as practiced in industry. Fewer still recognize the existence of a problem in our unrestrained commitment to specialization. Textbook and reference work publishing houses are organized into very specialized components as well, and this tends to encourage the status quo.

The gestation period today for a system engineer is on the order of several years following graduation with a BS degree from an engineering program. Universities are preparing specialized engineers in a number of engineering fields. Graduates from these programs obtain employment in specialized engineering departments in industry and gain experience in their profession. Over a period of years, these engineers acquire maturity and some demonstrate an interest, aptitude, and ability in generalism that transcends their specialized beginnings. Some of these engineers gravitate to a system engineering organization or responsibility with varying degrees of knowledge and experience.

Engineers trained on the job as system engineers are supplemented in industry by ex-military personnel who gained hands-on experience with complex military systems and, thus, developed an appreciation for and skill in applying thought processes useful in system engineering work.

In some companies, years of progressive engineering specialization (encouraged until fairly recently by specialized government customer agencies and reinforced by specialization in academia), without corresponding improvements in generalism, have yielded islands of specialized, serially and autonomously employed talent. These islands are not easily connected to effective concurrent engineering teams capable of simultaneously creating an integrated product solution consistent with a corresponding production and logistics support capability all optimized for the desired measures of effectiveness (e.g., performance, cost, and operability).

Companies that have been able to build and sustain an effective system engineering capability within this environment are indeed fortunate. If a company does not happen to have an effective system engineering capability, it can be a very difficult task to acquire it. Also, companies with a sound capability are faced with the problem of sustaining it. There are few localities where an infrastructure external to a company exists to support the generation and maintenance of a sound system engineering capability. Companies have been forced to become self-sufficient in igniting and sustaining their own system engineering capability. As a result, knowledge of the process currently resides unevenly in industry and some government agencies.

14.4.2 *Impediments and overcoming them*

Given that the premise is accepted that American industry needs an improved process for educating system engineering talent and improving system engineering capability as a counterbalance to specialization, how do we go about responding to this need? There are many impediments to changing the current condition. But, the consequences of continued movement in the direction of unbalanced specialization are so severe that we should make a serious effort to understand how we might restore a condition of balance between specialization and generalism. Let us list five impediments and then attempt to remove them.

First, companies that have an effective system engineering capability may be reluctant to share their knowledge to help others master the process for purely competitive reasons. Since process knowledge resides largely in industry, it appears necessary for industry to take a lead in changing the current condition of equilibrium through sharing their knowledge. Second, educators now staffing colleges and universities are, for the most part, specialists well qualified to perpetuate the specialized stereotypes and not well qualified to encourage generalism. Third, college curricula in specialized engineering programs are fully loaded now with needed subjects required for accreditation, and it is very difficult to add one or more courses encouraging an appreciation for generalism without lengthening the program. Fourth, there are few good books on system engineering, publishers having succumbed to the same imbalance observed in academia and industry. Finally, the whole educational system is in a state of equilibrium that cannot easily be nudged off the specialization fixation.

Industry must be the principal source of process knowledge in any effort to improve the educational experience for system engineers. This process knowledge is unevenly held in industry. Those who have it have to be convinced that it is in their enlightened self-interest to share their knowledge.

It is true that the management of a company that now has an effective system engineering capability in a geographical area that also includes competitors who do not, may feel protective about their advantage. Those who do not have an effective process and those who share a geographical area with none of their most serious rivals will probably not be concerned about loss of competitive advantage through cooperative development of a local system engineering education capability.

Even in an area where there is one company with an effective process that refuses to join forces initially for the reason of maintaining competitive advantage, that company will eventually join forces once they see that the others are benefiting and eroding the difference between their capabilities. While it is helpful for at least one member of a college system engineering support group to have a sound system engineering process, it is not essential so long as all of those willing to help share a desire to acquire it.

Despite any initial reservations about a competitive position, it is in the interest of all the industries in a given locale to have local resources for system engineering education available for all. Given that system engineering capability is a valuable commodity, the availability of a means of attainment is obviously in the interest of those who do not possess it. But, it is also in the interest of those who now possess it to help give it to others.

There is no company so blessed with system engineering effectiveness that it could not benefit from trying to help others attain it. Any teacher will tell you that the best way to understand something is to try to teach it. It is also true that a fundamentally sound way to improve a process is to expose it to critical comment from as many people as possible. Local support for a system engineering education program can be an effective part of a company's continuous process improvement regimen, because they will find no one more critical of their best efforts than outsiders with different experiences. You can be certain that at least some of these outside ideas will be better than your own.

Every company in a given locality, including those that have a sound system engineering process, will benefit from improving the qualifications of the population pool from which they all draw the majority of their employees. By sharing, we will all become better than we are now. Those with a more effective process at the beginning will retain this advantage if they continue to work at it.

Working system engineers and managers in most companies recognize the benefits from shared access to system engineering education capability and knowledge. Most see a greater benefit from sharing for their own company than from withholding knowledge to protect competitive advantage.

This is the basis for a lot of sharing that does take place now through national organizations, such as the Institute of Electrical and Electronics Engineers (IEEE) and the American Institute of Aeronautics and Astronautics (AIAA). The Defense System Management College (DSMC) in Fort Belvoir, Virginia, offers an excellent system engineering management course open to industry.

These existing efforts are all very commendable, but we need a firestorm of cooperative programs between industry and educational institutions across the country to reinvigorate industry with effective system engineering practices and to break down the growing autonomy in engineering organizations fed by galloping specialization at the very time at which concurrent engineering initiatives offer a hope of rebirth for the system engineering process.

Many practicing system engineers, educators, and government persons familiar with the acquisition process and the intended role of system engineering in that process, banded together in 1990, stimulated by exactly these problems, into what has come to be known as the International Council on Systems Engineering (INCOSE). This organization is a support group for improving the practice of system engineering in industry, academia, and government agencies, and improving the conduits through which talented engineers may pass to become effective, fulfilled system engineers. The charter of INCOSE is as follows:

1. Foster the definition, understanding, and practice of world-class systems engineering in industry, academia, and government.
2. Provide a focal point for the dissemination of systems engineering knowledge.
3. Promote collaboration in systems engineering education and research.
4. Assure the existence of professional standards for integrity in the practice of systems engineering.

INCOSE encourages companies to establish in-house system engineering training programs, changes in college engineering education programs to better prepare graduates for effective contribution in the evolving concurrent work atmosphere in industry, and college extension programs to prepare engineers now in the workplace for improved system engineering skills.

It is industry and government customer agencies that possess the process knowledge and have sensed the need for improvement in the preparation of engineers for system engineering work, and it falls on these sources to provide the ignition energy for these improvements. The fuel to sustain the resultant fires of improvement must be supplied jointly by government agencies, industry, and educational institutions.

14.4.3 UCSD extension case study

The author's purpose here is to expose a pattern by which a local improvement in system engineering education can be realized based on a case study that took place in San Diego, California, in 1990, involving the University of California San Diego (UCSD) Extension program and local industry. In September 1990, very soon after the meeting in Seattle, Washington, that formed National Council On Systems Engineering (NCOSE) (name later changed to INCOSE), the author called on Judy Cottrell, the UCSD Director of Corporate Relations in the Industrial Liaison Office of the Division of Engineering, with a proposal to introduce system engineering courses, such as requirements analysis, into the UCSD engineering curriculum.

After some discussion, we agreed that the UCSD Extension program would be the best place to try out these courses. UCSD actually offered a Master of Science degree in system engineering that was research oriented and focused heavily on control theory at the time.

Cottrell and Professor Dave Sworder, of the Applied Mechanics and Engineering Sciences (AMES) Department, had in August 1989 hosted a workshop, stimulated by the GDSS Division Systems Engineering department, on improving the preparation of graduates for participation in system engineering work in industry. This meeting adjourned with an action item for Professor Brian Mar from the University of Washington to host a second meeting in Seattle in 1990 that became the beginning of NCOSE (now INCOSE).

With Cottrell's encouragement, the GDSS Systems Engineering Director, Dave Clemons, and the author, both founding members of INCOSE, contacted the Director of Business and Engineering at the UCSD Extension, Dr. John Peak, with an offer to help establish an extension program in system engineering subjects. Dr. Peak accepted the offer and astounded his guests by expressing an interest in moving to a certificate program quickly, thereby demonstrating that one of two needed enabling functions were in effect in the San Diego area.

These two enabling functions for creating a local system engineering education capability are (1) a local educational institution with the spirit of adventure and (2) one or more local industries with an effective system engineering capability or interest in building one that will share their knowledge, as discussed earlier. If these two enabling functions are present, the several impediments listed earlier can be dealt with effectively within a cooperative environment.

The initial meeting between representatives from industry and the educational institution cannot take place without some preparation on the part of the industry representatives. The industry representative in this case, GDSS, was formed in 1985 by separating the space and energy functions from the Convair Division of General Dynamics to reform what had been earlier named the Astronautics Division that had produced the Atlas ICBM. The author was called upon to form and manage a systems development department of the new division. He determined that a priority had to be placed on improving its requirements analysis capability and began a program of research into effective methods and tools useful in a distributed, concurrent engineering environment.

A flexible approach was developed featuring three strategies from which a program could select, with the emphasis on structured development. A progressive requirements writing style was developed to match the degree of expected formality with program phase. Commercially, available computer tools were studied for requirements capture capability with the conclusion that there were none available at the time that satisfied the perceived needs. This resulted in an internal effort to develop a tool in a prototyping environment, simultaneous with efforts to define a more effective process. The synergism between these two efforts accelerated both rather than retarding the process, as might be expected.

It was immediately obvious that there would be no way to fund the reeducation needs of the systems development department from the available indirect sources once a more effective process was defined, so the genius of the GD Corporate System Engineering Seminar, which had been offered for several years at that time, was replicated in the form of a Division System Engineering Institute offering lunchtime and after-hours classes to avoid an indirect cost burden. Procedures prepared to define the system engineering process became the textbooks for courses, since there were few good textbooks on system engineering subjects and insufficient budget to purchase them for department members anyway.

As a result of this prior work, when the meeting took place with Dr. Peak at UCSD, GDSS Systems Engineering had been teaching requirements analysis and other system engineering topics to its engineers in-house in its Institute for several years and had evolved, at the expense and hopefully to the betterment of its own population, an effective system engineering teaching capability.

Shortly after the meeting, Dr. Peak formed a System Engineering Advisory Committee, including representatives from General Dynamics (Convair, Space Systems, and Electronics divisions), Scientific Applications International Corporation (SAIC), Thompson Ramo Wooldridge (TRW), General Atomics, and Teledyne Ryan Aeronautical. The combined population of system engineers in these several companies provided a sound student base for the planned courses. The committee met several times to define a certificate program and selected the author to teach the first course, in requirements analysis, in the Spring quarter of 1991. This course has been presented many times in the program at UCSD since the program began, and all of the courses in the program continue to be well attended as this book comes to life in 2010.

There are other examples of this process at work in other parts of the country. In each case, there seems to have been a special relationship between local industry and an educational institution. The relationships between Texas Tech at Lubback, Texas, with Texas Instruments, as well as Boeing in Seattle, Washington, with the University of Washington offer two other examples. Loral Aeronutronic inquired of the University of California, Irvine, about system engineering education opportunities, and while UCI did not at the time have a program in readiness, Mario Vidalon, Extension Director for Engineering and Information, arranged with the author to present a series that has unfortunately floundered since and been terminated. The University of California, Riverside, got into the system engineering certificate program business in response to a visit from a delegation of a local INCOSE chapter representative led by Howard Korman of TRW. This is the perfect cooperative relationship: an INCOSE chapter that took on a chapter project to stimulate a local university to build a system engineering education capability for the area served by the chapter and the university.

Universities, always in need of funds, can, of course, always apply money contributions provided by local industry to encourage particular programs of interest to industry, but the commodity of greater value is access to people and their knowledge of the process within which engineering graduates must function. This need not cost the company a significant amount of money. In the case of GDSS and UCSD, the total cost to GDSS was on the order of no more than 10 man-hours for certificate program meetings. The rest of the time was a matter of individual effort away from the job.

Clearly, system engineering management in industry has a responsibility to provide a means to educate their personnel in system engineering topics, and local educational institutions provide a cost-effective resource that can help managers meet this responsibility. These institutions cannot create effective system engineering education programs out of whole cloth, however, because they do not have the know-how. Industry must provide the ignition source with its knowledge of process and desire for continuous improvement.

It is conceivable that the initial approach could be initiated by an educational institution or by local government representatives responsible for acquisition of large systems and concerned about the state of system engineering education in their geographical area of interest. This initial appeal might be focused on one or more companies that could be encouraged to demonstrate their commitment to civic involvement. However the process is ignited, all three parties are needed to sustain the improvement efforts.

American industry needs to improve its system engineering capability. Your locality probably includes at least one educational institution that would like to participate in helping to satisfy local industry's need, but they have to have industry's help to structure the program and provide knowledgeable instructors. You should consider taking the initiative soon to talk to someone in a local educational institution about your needs and how you can help this institution help you satisfy these needs. The author's motivation in joining with UCSD in the beginning was that he could not get access to department training funds, but there was an almost unlimited availability of college tuition reimbursement funds.

The UCSD program was probably made successful from the university's economic perspective by the work of Charles Zumba, a vice president at SAIC, for he encouraged management to offer SAIC graduates of the certificate program a gift of company stock. As a result, over the years the program has been in existence, nearly half of the class members in courses the author has lectured in have been SAIC employees.

14.5 Purchase of a tailored commercial certificate program

One solution to the enterprise system engineering training program is to simply purchase it from a training provider. In so doing, the company should insist on the content of the courses being mapped to their process definition in the form of a list of your process steps (blocks on your process diagram) in one column and the course content reference in another column of a matrix. In order for this to work, you first must have a process definition, of course.

In some companies where the author has discussed their training needs and provided training, the company representatives have expressed concern that the course material support their written process description only to find that their process description contains only the what-to-do information in such general terms that anything could conceivably map to it. The company may be generally applying a specific set of techniques below the level of process documentation, but it is not captured in writing anywhere and held only in corporate memory. A good compromise in process specificity is to capture the what-to-do knowledge and link it to one or more specific techniques by reference to a book or standard. This makes it unnecessary to essentially write a textbook, while also clearly indicating the preferred technique where several alternatives exist.

For example, the author's company provides a complete system engineering certificate program of four 3 day courses (System Management, System Requirements, System Synthesis, and System Verification) that involves 12 days of training that can be spread over any desired period of time. It can be tuned to any product line through the adaptation of the techniques covered to the company's needs. Each course involves class team projects that can be linked to the company product line, rather than the specific examples used when courses in the program are offered at a university as part of an on-campus certificate program or as a public short course where it is necessary to appeal to a broader product experience base. The courses also cover several alternative ways of accomplishing some tasks, permitting the one preferred by the company to be emphasized. Each course comes with a textbook, a copy of the presentation materials, and access to a collection of templates, data item descriptions, and sample documents. The program can be contracted for directly (and continuing education units awarded to a permanent record if desired), or through a university extension office of choice or a short-course company. This program is also available in four one-day tutorial formats as a refresher or overview program.

14.6 A system engineering help window

Training accomplished closest to a need best serves the need. Most systems engineering training opportunities, where they do occur, are too far removed in time from the need for that training. Commonly, good external training opportunities cost more than we can afford to spend, and are not scheduled for our convenience but rather for broader market demands. The combination of these factors acts to make it very difficult to progress rapidly in improving one's skills and knowledge in this field. This section offers a system engineering organization a way to parlay a prior investment in networking and systems engineering process definition into an affordable solution for this problem. The solution offered also permits the use of a simple generic enterprise systems engineering process manual by providing the detailed how-to information in the form of just-in-time training materials linked to the process manual. The author believes this to be the ultimate in-house system engineering training solution for the current state of technology. This section is based on a paper delivered at the 1997 INCOSE Symposium.

14.6.1 Current realities

The desktop computer has become, over the past decade, as common as the pen or pencil and paper of years past as the preferred instrument for recording the results of system engineering work. A system engineer may very well spend half of his or her time working at such a machine, recording the results of the work directly in its final media. More and more, the product of our work resides in computer media commonly riding on a network server such that it is, or can be, more or less universally accessible within some domain.

While at the GDSS Division in the early 1990s, the author championed an online specification library that provided desktop access to all program specifications whether the person used a Macintosh or an IBM machine. McDonnell Douglas on the FA-18E/F Program has gone one better by providing company and customer personnel a view of the requirements in an RDD-100 database expressed on their screen through an Interleaf application accessed via an intranet Web site. These and other applications are a wonderful advance compared to paper copies of specifications in document cribs and out-of-date copies in piles on all four corners of the engineer's desktop, but they are only offering people the results of work well done. The content had to first be placed into these documents or databases by people who knew what they were doing at the time they were doing it.

What if you have just been assigned the responsibility for a task that you have never before done? Would you reasonably expect in your current situation that there would be an available course or detailed description of how to do this job such that you could start the job from a position of knowledge rather than learning while doing? Perhaps your company has a written systems engineering practices document, but when you open it you find only a brief paragraph on this subject. Your practice, however, does include a reference to a book that goes into detail on this task. You take the time to go to the company engineering library only to find that all of the books were recently given to a local university library as a cost-cutting move.

Well, all is not lost, for you have a supervisor who knows everything and peers who have done this kind of work before. Unfortunately, your supervisor is working on a program at another facility to reduce overhead expenses and your peers all complain that they are working too many hours now just doing their own job because the company is still running lean after several downsizing rounds. There is a local university systems engineering certificate program, but the course covering the task you have to do requires 3 months of one evening a week and it does not run for another 6 months. There is a 3 day course running next week put on by a short-course company in a town close enough to commute by auto everyday, but there is no department-training budget.

You have no choice but to start doing the work and learning as you go. The work product will likely cost more to create, take longer to create, and be of less value than it would have been if you could have started the job from a position of knowledge about the task. There must be a better way to deliver training just in time to our system engineers.

14.6.2 Enter the help window

Since we spend so much time working at a computer, it is perfectly natural to use that computer to deliver the training experience we are looking for. Your company has probably already made the investment in networking and is committed to maintaining this capability. Perhaps your company created a system engineering practice document by editing EIA 632 or IEEE P1220. As a beginning, we could put the company practice online, but this probably will not fill the need for details. What we need is a system engineering process help window on our desktop computer to expand upon the terse content of our practices document.

This window could be opened by an employee in exactly the situation described a moment ago. Our system engineer, John, must start on this new job tomorrow. This afternoon he opens the systems engineering help window and receives a 30 min pitch on technical performance measurement (TPM). He finds out that he must identify a short list of key parameters and gets some help in identifying those most beneficial to the program. John discovers that each parameter must be assigned to a principal engineer and that each principal engineer should have two charts on their parameter. First, they need a tracking chart that shows the parameter required and demonstrated values in time coordinated with related events. Secondly, they need an action plan telling how they are going to achieve coincidence between these two values and when and why we should believe this will occur. This information will enable John to organize an effective TPM plan that otherwise would require lost time and a less effective implementation.

14.6.3 *Route to the initial window*

We assume that your company has a current systems engineering practices manual or some other process document. If you do not have it, then there is some prerequisite work that must be accomplished prior to being able to take advantage of the suggestions in this chapter. The nonavailability of a computer network is another inhibitor.

The first step we should take is to place our systems engineering practice online such that anyone can call it up and check a particular point on screen or print out a copy or some part thereof. Many companies have placed a note on the cover of their online documents to the effect that a paper copy may not be up to date and to check with a person at a phone number given on the paper copy for the current version. This document should be located on a server generally accessible from all programs in your business unit. For our purposes of discussion, we will call this server storage space allocation SEPLANS.

This is a simple move but we should not trivialize it. The fact that your process definition is universally available in the current version is no small achievement however easily it was attained. The remaining problem is that our process description is not sufficiently comprehensive to transform an engineer reading it into an instant powerhouse of skill. So, we have two choices: (1) change our process manual to include the necessary depth to support help window needs or (2) supplement the manual with additional information keyed to the manual content. The second alternative is encouraged in this chapter, and this supplementary information forms the systems engineering help window.

Your preferred systems engineering process description should, of course, be coordinated with the way your company chooses to do the systems engineering work on programs. It is not adequate simply to have a process description and ignore it. The reason for having a process description is to guide people in performing work on programs and to encourage repetition in applying that process to the end, so that the performance of the people doing the work will improve over time. We should insist on continuous incremental improvement of this process rather than periodic revolutionary changes, because we wish to maximize the common portion of the process between any two applications.

Our manual is online. Our manual is maintained consistent with the way we wish to perform systems engineering work. Our programs are applying our standard process described in our generic systems engineering process manual. Given this baseline, there is hope for us in implementing the help window.

It is possible, of course, that we could have a terrible mess as a systems engineering process description. One way to ensure that we are close to quality in our process description is to map what we think is our customer's preferred process description to ours. If we have three primary customers and each has a different view of systems engineering perfection, we may have to make three maps, and to the extent that they do not all describe the same process, we may have to tailor one or more of them to a common process that we describe in our internal process description. Out of this experience will come a conclusion that you have accounted for all of your customers' needs in the area of systems engineering, while providing your people with a single common process description that can be used on all programs.

To pick up the story of the window, we must decompose the process described in our manual into components or modules about which we can prepare real-time instructional material. These components should be no larger than can be described in half an hour using an application like Microsoft PowerPoint, and they must map cleanly to your process description. Ideally, this will involve simple references to a single paragraph rather than bits and pieces spread throughout.

We should make a list of these modules and prioritize them based on some criteria such as (a) which ones are we most in need of improvement, (b) which ones are the easiest to implement giving us early successes, or (c) some combination of these factors. Somewhere in our organization, we should be able to find people who know enough about these topics that they could, within a month or two, even with all of the other demands on their time, come up with a set of self-teaching charts covering an assigned topic.

It remains to manage this whole activity to closure, resulting in a series of computer files, each providing the how-to instructions on a part of our generic systems engineering process description. The training materials should be reviewed by someone and approved for inclusion in the growing library of systems engineering knowledge available to your engineers.

These files are then placed in our SEPLANS directory/folder using file names that are self-evident of their content. For example, we could call the file on technical performance measurement simply TPM. Other files may present a more difficult naming chore, so we may wish to simply number them in accordance with the paragraphing structure of our systems engineering process document, thus gaining good traceability, something we should have regardless of how we name the files, in the process.

The understanding is, at this point, that we only have some charts containing graphics and text that collectively convey to an intelligent engineer how to perform the related task. Clearly, this is not the most powerful training environment in that it does not appeal to the full range of senses of the person being trained. But, it is a beginning.

14.6.4 The external and mixed alternatives

If your company does not have the available talent to develop these training materials, it is possible to bring in an outside agent to develop them. A more likely alternative is that your company can develop some of the modules while others would be more cost-effectively developed by an outside agent.

The materials developed by an external source should be identified with respect to their source. Some materials will come from your own company. Others will come from the external source and may be copyrighted, resulting in the need to pay a license fee for a finite period of time. Other materials may be created by the external agent while employed to do so, and these should be your intellectual property.

14.6.5 Preplanned or incremental window improvements

We could begin our attack on systems engineering ignorance as described above without a lot of serious thought about the future. As is generally the case, a grand plan should come to a better result, however. The only case where we might want to proceed without a plan is where we perceive a planning stalemate, either for political reasons or due to lack of ideas about long-range needs. In this case, movement is preferable to stagnation. This one exception aside, we should recognize our help window's initial operating capability as a beginning with a planned or incremental growth path.

Once the foundation, previously described, is in place, we will find many ways to improve upon it. The way we have described the window so far, it is assumed that the user would have little trouble navigating through the material to find the part of immediate interest. If our window content becomes sufficiently complex, we may need to employ a hypermedia concept to encourage easy navigation. In this arrangement, the user clicks on something of interest, possibly even a paragraph in our practices manual, expanding the view on that topic. This is an alternative to the hierarchical structure of the data implied in the earlier discussion.

While the graphical window data suggested can be very effective as a just-in-time training source, it can be made more powerful by including voice clips briefly explaining particular points. Microsoft PowerPoint and other presentation packages have this capability. You will need your delivery workstations equipped for sound in order to deliver the complete message in this event.

Video clips can also be added to the charts in many presentation graphics packages (including Microsoft PowerPoint). Some points can be communicated much more powerfully via video than by a single graphic or even using the build function of a graphics package where you can link several charts together in a sequence giving the illusion of motion or change in time.

The ultimate application of the systems engineering help window would be implemented through a database application. Even inexpensive databases such as Microsoft Access and Borland's dBase V For Windows are compatible with graphical, audio, and even video clip field data.

The records would correspond to the charts already developed. Each chart becomes a graphic in the database, supplemented with administrative and configuration management data as well as text, voice, and video clips as appropriate. These records should be organized into modules of closely related information as previously discussed, identified by key fields defining the broad subject area (systems engineering), perhaps a secondary organizational layer like requirements analysis, and a module such as TPM.

Table 14.1 offers a way to structure the data. We have assigned 01 for the systems engineering subject area, possibly 02 for design, 03 for manufacturing, and so forth. Topic 01 within the systems area is assigned to requirements analysis, and 18 within requirements analysis has been identified for TPM. Finally, we now get to database records with real content. The record identified by the four ID field values, 01-01-18-05, would

Table 14.1 Suggested Data Organization

ID	Topic
01	SYSTEM ENGINEERING
01-01	REQUIREMENTS ANALYSIS
01-01-18	TPM
01-01-18-05	TPM action plan

Source: Grady, J., *System Engineering Deployment*, CRC Press, Boca Raton, FL, p. 246, 2000. With permission.

contain a chart showing an example of a TPM action plan and a voice clip saying, "The parameter principal engineer should maintain an action plan telling what actions are being and will be accomplished in the future to cause the design to attain the required value. In this case, we see that Bill, the Principal engineer, is intending to perform an analysis of new data derived from a development evaluation test to be completed in two weeks to see what the new demonstrated value is." A video clip could then come up in a window showing Bill talking to a test engineer with some apparatus in the background.

The necessary code can be written to link these records together into a stream of images of value in preparing an engineer for performing this work in the near term. You will also need the code to control the whole help system, making it possible to locate and call the desired material. This can be done through a hierarchical menu system or through a hypermedia application as desired.

14.6.6 Work tool linkage

So far, we have been discussing the systems engineering help window as an isolated utility available to the engineer. The optimum way to deliver this service would be through the very tools that the engineer uses to do work. True, many of these tools have a help function giving support for tool-related considerations. The suggestion is that the toolmakers consider providing links to a company's process help system. Alternatively, this may be a fertile area of expansion for toolmakers, providing a process-oriented how-to help window.

14.6.7 Help window conclusions

The computer will continue to change our lives at an increasing pace. Sometimes these changes will be compatible with a more enjoyable as well as effective work environment for us humans. The systems engineering help window is one of these changes. Ultimately, we should cause these machines to support our needs so that we can reasonably effortlessly apply our skill and knowledge to the problems we are most effective in solving, the difficult ones requiring the application of judgment and interpersonal interaction.

The more of the burden for the mundane that we can remove from management and each other, the more the time and energy that will be available to ply our profession and serve our customers.

It will be a long road from the existing rudimentary applications of the kind discussed in this chapter to a really effective capability useful to new hires from college thrown into the fight, but it is a road that will come to a useful result.

part five

Program planning

chapter fifteen

Product system definition*

15.1 System development process overview

Program planning for a specific program requires an understanding of the product entity structure of the system to be created, because it is the things in this structure about which teams will be formed. These things in the system should be determined through an organized and structured process to avoid omitting needed functionalities or including any unnecessary functionality that adds cost without benefit. Therefore, some system development work must take place before a detailed plan of action for the program can be crafted, and that work will have to be planned at a very high level relative to the things in the system. In fact, that planning can very likely be done based on assigning only a program integration team (PIT) rather than a series of integrated product and process teams (IPPT) as well. During that early phase, it will be the responsibility of the PIT to accomplish the front-end decomposition of the customer need and related mission analysis to define what the top-level items should be that will become the purpose for identifying IPPT on a subsequent phase.

To review, the first cycle of the systems approach entails the following principal steps:

1. Understand the customer's need; that is the ultimate system function, for it tells what the customer wants. It defines the problem that must be solved at a fairly high level.
2. Expand the need into a critical mass of information necessary to trigger a more detailed analysis of a system that will satisfy that need. This may entail customer surveys, focus groups, question and answer sessions, or mission analysis in combination with a high-level functional analysis that exposes needed functionality, system concept design entailing the allocation of the top-level functionality to things placed in a system physical architecture, and the definition of requirements for the complete system (commonly captured in a system specification). Through this process, one will determine to what extent and for what element solutions shall be attempted through hardware and software. This process may require hard decisions

* The material in this chapter is from Grady, J., *System Engineering Deployment*, CRC Press, Boca Raton, FL, Chapter 9, 2000. Used with permission.

fueled by imperfect information calling for the use of the trade study approach discussed in *System Requirements Analysis* (Grady, 2005) and *System Synthesis: Product and Process Design* (Grady, in press).

3. Further decompose the need, which represents a complex problem, within the context of an evolving system concept, into a series of related smaller problems each described in terms of a set of requirements that must be satisfied by the solutions to the smaller problems.
4. Prior to design work, it is sometimes necessary or desirable to improve one's confidence in their understanding of the requirements or to prove that it is physically possible to produce a design that is compliant. This is commonly accomplished through special tests and analyses, the results of which indicate the strength of the currently available related knowledge and technology base and the current state of readiness to accomplish design work using available knowledge.
5. Application of the creative genius of design engineers and market knowledge of procurement experts within the context of a supporting cast of specialized engineers and analysts to develop alternative solutions to the requirements for lower-level problems. Selections are made from the alternatives studied based on knowledge to identify the preferred design concept that will be compliant with the requirements.
6. Integration, testing, and analysis activities are applied to designs, special test articles, preproduction articles, and initial production articles that prove that the designs actually do satisfy the requirements.

We wish, in this chapter, to focus on step 2 where we come to understand what the system shall consist of. It is the physical things in the system that will require development and about which work will be accomplished by IPPT. Conventionally, these things are arranged in a hierarchical family tree structure called an product entity structure diagram. This chapter tells how one can identify the things that should populate the system architecture. This is also the basis for the work breakdown structure (WBS) important in program planning. In this chapter, we will use a construct called a product entity ID as well as WBS notation, but in Chapter nineteen the proposition will be made that since we have aligned the two in our development work, only one is necessary, and the author will propose that the product entity ID becomes the WBS.

15.2 System requirements analysis

This whole development process is held together, especially during the initial development stage, by requirements. These requirements are statements that define the needed characteristics of a solution to an engineering problem prior to the development of the solution to the problem.

Requirements should be identified as a prerequisite to accomplishing design work because of the nature of complex problems and the need for many people, none of whom understand the whole problem or its solution, to work together to develop systems to solve those complex problems. All of the requirements for one item are collected into a specification for that item. The process of determining what the content should be for a specification is called requirements analysis.

15.2.1 *Requirements analysis background*

In the English-speaking world, requirements are phrased in English sentences that cannot be distinguished structurally from English language sentences constructed for any other purpose. It is relatively easy to write requirements once you know what to write them about. A requirement is simply a statement in the chosen language that clearly tells an expectation placed on the design process prior to implementing the creative design process. Requirements are intended to constrain the solution space to solutions that will encourage small problem solutions that synergistically work together to solve the large problem. Requirements are formed from the words and symbols of the chosen language. They include all of the standard language components arranged in the way that a good course in that language, commonly studied in the lower grades in school, specifies. Those who have difficulty writing requirements experience one of the three problems: (1) fundamental problems expressing themselves in the language of choice that can be corrected by studying the language, (2) technical knowledge deficiency that makes it difficult to understand the technical aspects of the subject about which the requirements are phrased, which can be corrected through the study of the related technologies, and (3) difficulty in deciding what to write requirements about.

The solution to the last problem is the possession, on a personal, company, and program basis, of a complete tool box of effective requirements analysis tools, the knowledge and experience to use the tools effectively, and an enlightened management that insists on the application of structured analysis. Providing the toolbox is the function of the system definition process (tasks F41 of Figure A-1, Sheet 4 in Division 3/Exhibit H/Appendix A). This toolbox should encourage the identification of product characteristics that must be controlled and the selection of numerical values appropriate to those characteristics. So, this toolbox must help us to understand what to write requirements about. It gives us a list of characteristics that we should control to encourage synergism within the evolving system definition. If the analyst had a list of characteristics and a way to value them for specific items, language knowledge, and command of the related technologies, he or she would be in a very good position to write a specification.

One approach to a solution to this problem on what to write requirements about is boilerplate. This is essentially what specification standards such as MIL-STD-490A provided for many years and its replacement in MIL-STD-961E is likely to do for many more within the context of military systems. IEEE and other standards bodies provide similar standards for commercial purposes. In this approach, you have a list of paragraph numbers and titles, and you attempt to define how each title relates to the item of interest. This results in a complete requirement statement that becomes the text for the paragraph stating that requirement. The problem with this approach is that there are many kinds of requirements that cannot be specifically identified in such a generic listing. One could create a performance requirements boilerplate that covered every conceivable performance requirement category with some difficulty only to find that it was more difficult to weed out those categories that did not apply to a specific item than it would have been to have determined the appropriate categories from scratch. This is why one would find no lower level of detail in a specification standard than performance even though there may evolve 50 pages of performance requirements during the analysis. Interfaces are another area where boilerplate is not effective at a lower level of detail. Often, environmental and some specialty engineering requirements can be handled partially or completely through standards.

Many design engineers complain that their system engineers fail to flowdown the customer requirements to their level of detail in a timely way. Their solution is to begin designing at their level of detail without requirements, because their management has imposed a drawing release schedule they dare not violate. These engineers are often right about the failure of their system engineers but wrong to proceed in a vacuum as if they know what they are doing. The most common failure in requirements analysis is the gap between system requirements defined in a system specification given to the contractor and the component-level requirements in the performance requirements area. In these organizations in trouble, the goal in the engineering organization is often stated as getting the system performance requirements flowed down to the design level or deriving design requirements from system requirements. Unfortunately, no one in the company seems to know quite how to do that and avoidable errors creep into the point design solutions in that void that develops.

Cloning is one requirements analysis strategy. It just is not a completely effective tool across all of the kinds of requirements that must be defined and particularly not very effective for performance requirements. It consists of three subset strategies:

1. Parent item flowdown bases the content of a child specification on what is included in the parent specification. This is a very effective technique where the child requirement has the same units as the

parent requirement, and we partition the parent value into child values in accordance with some mathematical rule. Other names for this technique are flowdown, allocation, partitioning, or apportionment. Where the child requirement has different units than the parent requirement, the value may have to be derived from one or more parent requirements through parametric analysis, the use of mathematical models, or the use of simulation techniques.
2. Like item cloning uses the content of a similar item specification to determine content for another specification.
3. A component standard can be developed containing the preferred outline and some context that is common to a broad set of specifications.

The problem in the cloning strategy is that you must have a set of parent requirements in order to perform requirements analysis. One may ask, where did the parent requirements come from?

It is possible to sit down at your word processor and write the great American specification using what the author likes to call free style, unencumbered by all controlling influences. Some would also call this the ad hoc approach. If any good content happens to appear in this specification, we will have to thank God because man probably did not put it there using this strategy.

Another strategy, often used in combination with free style or any of the other strategies, is the question and answer game where the analyst tries to think up the right questions to ask customer representatives. This is an important step in the development of the system specification but you cannot depend on coming up with all of the right questions without aid in the form of structured analysis. Structured analysis is recommended, and it is the only strategy further developed in this chapter. Refer to *System Requirements Analysis* (Grady, 2005) for full coverage of this process.

This is not to say that the other strategies do not have a place at the table. Flowdown or value allocation is very useful within the specialty engineering disciplines as in the weights statement and reliability math model. Standards properly play a key role in environmental and specialty engineering requirements identification. The Q&A process should be depended upon heavily in the early development process in combination with structured analysis to help generate good questions. But the principal tool that ought to be brought to bear is structured analysis.

Our toolbox should help us understand what to write requirements about, what characteristics we should seek to control, and what characteristics we need more information on from the customer. Once we have this list of characteristics, we must pair each with an appropriate numerical value and weave words around them to make the meaning very clear. We need to quantify our requirements because we will want to be able to determine whether or not a particular design solution satisfies the requirement. This is very hard to do when the requirement is not quantified. As we shall see, this is the reason why the author encourages simultaneous definition of requirements and the companion test and analysis, or verification, requirements for proving the solution satisfies them.

Our system requirements analysis process should clearly tell us what kinds of characteristics we should seek to write requirements about and provide a set of tools to help us add characteristics to our list. We also need to have the means to quantify them. We seek a list of characteristics that is complete, in that it does not omit any necessary characteristics, and minimized, since we wish to provide the maximum possible solution space for the designer. We would like to have a more specific understanding of how this requirements analysis solution is characterized by completeness and minimization, but there is no easy answer. The best prescription the author can offer is to apply an organized approach that connects specific tools up with specific requirements categories and apply the tools with skills based on knowledge derived in practice or through training.

Figure 15.1 offers an overall taxonomy of every kind of requirement that one would ever have to write in the form of a three-dimensional Venn diagram. The top layer, Part I in MIL-STD-490A or the performance specifications in the context of MIL-STD-961E, corresponds to development requirements, often called design-to requirements that must be clearly understood before design. The lower layer, Part II in 490A context or detail specifications under 961E, corresponds to product requirements, commonly called build-to requirements. The portion of the construct below the heavy middle line is for product specifications. The requirements

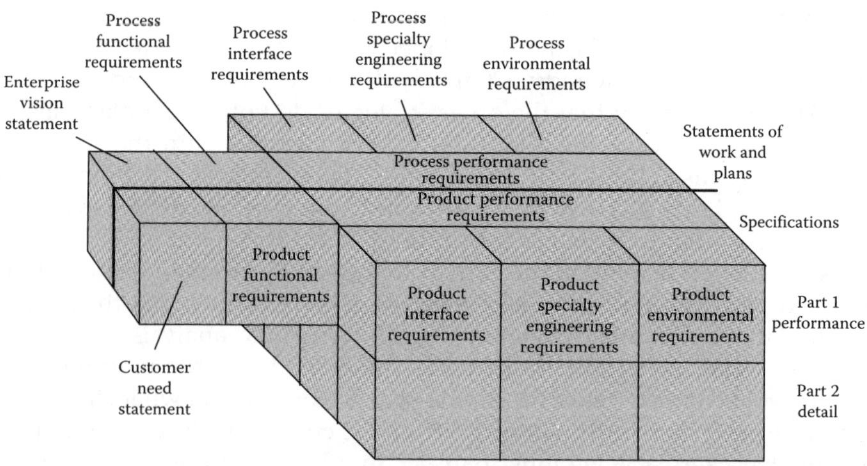

Figure 15.1 Requirements taxonomy. (From Grady, J., *System Engineering Deployment*, CRC Press, Boca Raton, FL, pp. 118–142, 2000. With permission.)

above this line correspond to process requirements captured in statements of work and plans. Many of the same techniques useful for identifying product-oriented specification content apply to process requirements as we will see in this book.

Development requirements can be further categorized as shown in Figure 15.1 as performance requirements and design constraints of three different kinds. Performance requirements tell what an item must do and how well it must do it. Constraints form boundary conditions within which the designer must remain while satisfying the performance requirements. The toolbox offered in this book evolves all performance requirements from the customer need statement, the ultimate system function, through a product functional analysis process that identifies functional requirements. These functional requirements are allocated to things in the evolving system physical model, commonly a hierarchical product structure, and expanded into performance requirements for that item. Functional analysis provides one set of tools. We will identify three other sets for the three kinds of constraints.

As the reader can see from Figure 15.1, verification requirements are not illustrated, and that is because they are assumed to be paired with the item requirements in our imagination. One could construct another dimension for design and verification requirements, but it would be very hard to envision and draw in two dimensions. This method of verification requirement inclusion corresponds to the fundamental notion that flows through this whole chapter that a design requirement should always be paired with a verification requirement at the time the design requirement is defined. The reason for this is that it results in the definition of much better design requirements when you are forced to tell how you will determine whether or not the resultant design has satisfied that design requirement. If nothing else, it will encourage quantifying requirements because you will find it very difficult to tell how it will be verified without a numerical value against which you can measure product characteristics.

Before moving on to the toolbox for the other categories illustrated in Figure 15.1, we must agree on the right timing for requirements analysis. Some designers reacted to the rebirth of the concurrent design approach with acceptance, because they believed it meant that it was finally okay to simultaneously develop requirements and designs. That is not the meaning at all, of course. The systems approach, and the concurrent development approach that has added some new vocabulary to an old idea, seeks to develop all of the appropriate requirements for an item prior to the commitment to design work. We team the many specialists together to first understand the problem phrased in a list of requirements. We then team together to create a solution to that problem in the form of a product design and coordinated process (material, manufacturing, quality, test, and logistics) design. The concurrency relates to the activity within each fundamental step in the systematic development of product accomplished in three serial steps: requirements, design, and verification. Even in the spiral sequence, we rotate between these steps serially repeatedly.

It is true that we may be forced by circumstances to pursue design solutions prior to fully developing the requirements (identifying the problem) as discussed under development environments above but requirements before design is the intent. A case where this is appropriate is when we wish to solve a problem before we have sufficient knowledge about the problem, and it is too hard to define the requirements in our ignorance. A physical model created in a rapid prototyping activity can provide great insights into appropriate requirements that would be very difficult to recognize through a pure structured approach.

15.2.2 The need and its expansion

The very first requirement for every system is the customer's need that is a simple statement describing the customer's expectations for the new, modified, or reengineered system. Unfortunately, the need is seldom preserved once a system specification has been prepared, and thus system engineers fail to appreciate a great truth about their profession. The truth is that the development process should be characterized by logical continuity from the need throughout all of the details of the system.

The tools discussed in this section are designed to expand upon the need statement to fully characterize the problem space. The exposed functionality is allocated to things and translated into performance requirements for those things identified in the physical plane. We then define design constraints appropriate to those things. All of our verification actions will map back to these requirements.

The need statement seldom provides sufficient information by itself to ignite this analytical process so may be preceded by efforts to understand the customer's need and the related mission or desired action or condition. In military systems, this may take the form of a mission analysis. In commercial product developments, we may seek to understand customer needs through surveys, focus groups, and conversations with selected customer representatives.

This early work is focused on two goals. First, we seek to gain expanded knowledge of the need. In the process, we seek to ignite the functional decomposition process targeted on understanding the problem represented by the need and making decisions about the top-level elements of the solution by allocating top-level functionality to major items in the architecture. Some system engineers refer to this process as requirements analysis, but it is more properly thought of as the beginning of the system development process involving some functional analysis, some requirements analysis, and some synthesis and integration. Second, we seek to validate that it is possible to solve the problem stated in the need. We are capable of thinking of needs that are not possible to satisfy. The problem we conceive may be completely impossible to solve, or it may only be impossible with the current state of the art. In the latter case, our study may be accomplished in parallel with one or more technology demonstrations that will remove technological boundaries permitting development to proceed beyond the study.

This early activity should include some form of functional analysis as a precursor to the identification of the principal system items. These top-level allocations may require one or more trade studies to make a knowledge-driven selection of the preferred description of the problem. This process begins with the recognition that the ultimate function is the need statement and that we can create the ultimate functional flow diagram (one of several decomposition techniques) by simply drawing one block for this function F that circumscribes a simplified version of the need statement. The next step is a creative one that cannot be easily prescribed. Given this ultimate function, we seek to identify a list of subordinate functions such that the accomplishment of the related activities in some particular sequence assures that the need is satisfied. We seek to decompose the grand problem represented by the need into a series of smaller, related problems and determine what kinds of resources could be applied to solving these smaller problems in a way that their synergism solves the greater problem.

If our need is to transport assigned payloads to low earth orbit and we assign this to the use system function F47, we may identify such lower-order problems, or functions, as F471 = integrate and prepare for transport, F472 = launch, F473 = transport, and F474 = release payload. This particular sequence of lower-tier functions is keyed to the use of some kind of rocket launched from the Earth's surface. If we chose to think in terms of shooting payloads into the orbit from a huge cannon, our choice of lower-tier functionality would be a little different, as it would be if we elected the Orbital Science Corporation's Pegasus solution of airborne launch.

If our need was the movement of 500,000 cubic yards of naturally deployed soil per day, we might think in terms of past point design solutions involving a steam shovel and dump trucks or open up our minds to other alternatives thus inventing the earth mover (quite some time ago, of course).

So, the major functions subordinate to the need are interrelated with the mission scenario, and we cannot think of the two independently. The thought process that is effective here is to brainstorm several mission scenarios and develop corresponding top-level functional flow diagrams (called master flow diagrams or zero-level diagrams by some) at the highest level for these scenarios. We may then determine ways of selecting the preferred method of stating the problem by identifying key parameters that can be used to examine each scenario. This is essentially a requirements analysis activity to define quantified figures of merit useful in making selections between alternative scenarios.

As just noted, we may have to run trade studies to select the most effective problem statement as a precursor to the allocation of top-level functionality to things. We may also have to accomplish trade studies as a means to derive a solution based on facts for the most appropriate allocation of particular functions. Finally, we may have to trade the design concept for particular things conceived through the allocation process. Trade studies are but organized ways to reach a conclusion on a very difficult question that will not yield to a simple decision based on the known facts. All too often in engineering, as in life in general, we are forced to make decisions based on incomplete knowledge. When forced to do so, we should seek out a framework that encourages a thorough, nonsystematic, evaluation of the possibilities, and the trade study process offers that.

The customer's need need not be focused on a new unprecedented problem as suggested in the preceding comments. The customer's need may express an alteration of a problem previously solved through earlier delivery of a system. The early study of this kind of problem may involve determining to what extent the elements of existing systems can be applied and to what extent they must be modified or elements replaced or upgraded. This is not a radically different situation. It only truncates the development responsibilities by including within the system product structure things that do not require new development. Those items that do require new development will respond to the structured approach expressed in this chapter at their own level of difficulty.

This early analytical process may start with no more than an expression of the customer need in a paragraph of text. The products with which we should conclude our early mission and functional analysis work include

1. A master functional flow diagram (or equivalent diagrammatic treatment) coordinated with a planned mission scenario briefly described in text and simple cartoon-like graphics.
2. A product entity block diagram defining the physical model at the top level.
3. A top-level schematic block diagram or n-square diagram that defines what interfaces must exist between the items illustrated on the architecture block diagram.
4. Design concept sketches for the major items in the system depicted on the product entity block diagram.
5. A record of the decision making process that led to the selection of the final system design concept.
6. A list of quantified system requirement statements. These may be simply listed on a single sheet of paper or captured in a more formal system or operational requirements document.

So, this mission analysis activity is but the front end of the system development process using all of the tools used throughout the development downstroke. It simply starts the ball rolling from a complete stop and develops the beginning of the system documentation. It also serves as the first step in the requirements validation activity. Through the accomplishment of this work, we either gain confidence that it is possible to satisfy this need or conclude that the need cannot be satisfied with the available resources at our disposal. We may conclude, based on our experience, that we should proceed with development, await successful efforts to acquire the necessary resources, focus on enabling technology development for a period of months or years, or move on to other pursuits that show promise. In the case of commercial products, this decision process may focus on marketing possibilities based on estimates of consumer demand and the cost of production and distribution.

15.2.3 Structured decomposition

Structured decomposition is a technique for decomposing large complex problems into a series of smaller related problems. We seek to do this for the reasons discussed earlier. We are interested in an organized or systematic approach to doing this because we wish to make sure we solve the right problem and solve it completely. We wish to avoid, late in the development effort, finding that we failed to account for part of the problem that forces us to spend additional time and money to correct and brings into question the validity of our current solution. We wish to avoid avoidable errors because they cost so much in time, money, and credibility. This cost rises sharply the further into the development process we proceed.

The understanding is that the problems we seek to solve are very complex and that their solution will require many people each specialized in a particular technical discipline. Further, we understand that we must encourage these specialists to work together to attain a condition of synergism of their knowledge and skill and apply that to the solution of the complex problem. This is not a field of play for rugged individuals except in the leadership of these bands of specialists. They need skilled and forceful leadership by a person possessed of great knowledge applicable to the development work and able to make sound decisions when offered the best evidence available.

During the development of several intercontinental ballistic missile (ICBM) systems by the U.S. Air Force, a very organized process called functional analysis came to be used as a means to thoroughly understand the problem, reach a solution that posed the maximum threat to our enemies consistent with the maximum safety for America, and make the best possible choices in the use of available resources. We could argue whether this process was optimum and successful in terms of the money spent and public safety, but we would have difficulty arguing with the results following the demise of the Soviet Union as a threat to our Nation's future.

At the time these systems were in development, computer and software technology were also in a state of development. The government evolved a very organized method for accumulating the information upon which development decisions were made involving computer capture and reporting of this information. Specifications were prepared on the organized systems that contractors were required to respect for preparing this information. Generally, these systems were conceived by people with a paper mindset within an environment of immature computer capability. Paper forms were used based on the Hollerith 80 column card and intended as data entry forms. These completed forms went to a keypunch operator. The computer generated poorly crafted report forms. But, this was a beginning and very successful in its final results.

This process included at its heart a tool called functional flow diagramming discussed briefly above. The technique uses a simple graphical image created from blocks, lines, and combinatorial flow symbols to model or illustrate needed functionality. It was no chance choice of a graphical approach to do this. It has long been well known that we humans can gain a lot more understanding from pictures than we can from text. It is true that a picture is worth 10 to the third words. Imagine for a moment the amount of information we take in from a road map glanced at while driving in traffic and the data rate involved. All of the structured decomposition techniques employ graphical methods that encourage analytical completeness as well as minimizing the time required to achieve the end result.

While functional flow diagramming was an early technique useful in association with the most general kind of problem, computer software analysis has contributed many more recent variations better suited to the narrower characteristics of software. U.S. Air Force ICBM Programs required adherence to a system requirements analysis standard and delivery of data prepared in accordance with a data item description for functional flow diagramming. While functional flow diagramming is still a very effective technique for grand systems and hardware, it is not as effective for computer software analysis as other techniques developed specifically for it. So, our toolbox of analytical techniques should have several tools in it including functional flow diagramming and several others.

15.2.4 *Grand systems and hardware approaches*

Functional flow diagramming is the author's preferred approach for grand systems because of its simplicity and generality. Its simplicity is also its greatest fault as the defenders of more complex modeling techniques would be quick to point out. This process starts with the need as function F that is expanded into a set of next-tier functions, which are all

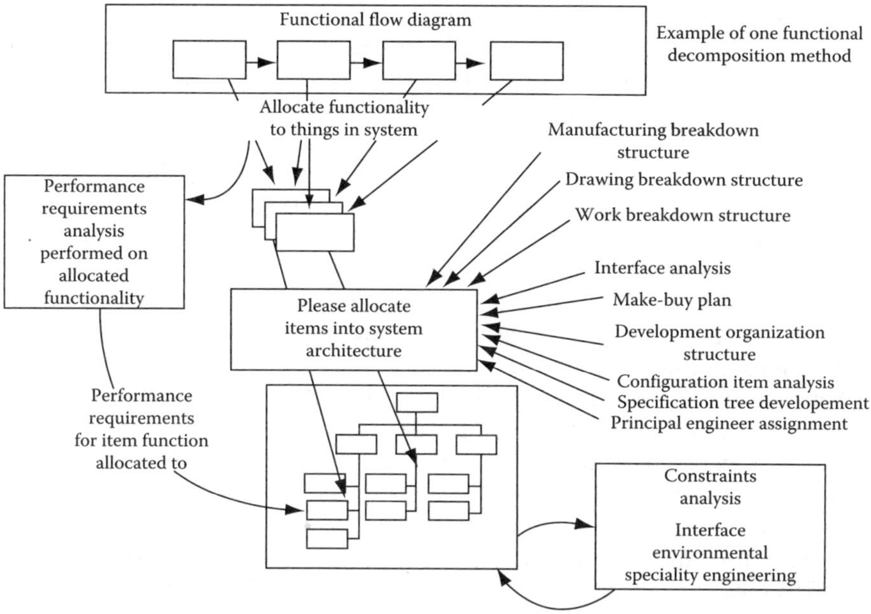

Figure 15.2 Structured decomposition for grand systems and hardware. (From Grady, J., *System Engineering Deployment*, CRC Press, Boca Raton, FL, pp. 118–142, 2000. With permission.)

things that have to happen in a prescribed sequence (serial, parallel, or some combination) to result in function F being accomplished. One draws a block for each lower-tier activity and links them together in a sequence using directed line segments to show sequence. Logical OR and AND symbols are used on the connecting lines to indicate combinatorial possibilities that must be respected. This process continues to expand each function, represented by a block, into lower-tier functions. Figure 15.2 sketches this overall process for discussion.

A function statement begins with an action verb that acts on a noun term. The functions exposed in this process are expanded into functional requirement statements that numerically define how well the function must be performed. This step can be accomplished before the allocation of the function statement to a thing in the physical system architecture or after. But, in any case, the allocation of the function or functional requirement obligates the analyst to write one or more performance requirements based on the allocated function for the thing to which it is allocated. This is the reason for the power of all decomposition techniques. They are exhaustively complete when done well by experienced practitioners. It is less likely that we will have missed anything compared to an ad hoc approach.

This process begins with the need and ends when the lowest tier of all items in the physical architecture in each branch satisfy one of these criteria: (1) the item will be purchased from another entity at that level or (2) the developing organization has confidence that it will surrender to detailed design by a small team within the company.

There are two extreme theories on the pacing of the allocation relative to the functional decomposition work. Some system engineers prefer to remain on the problem plane as long as possible to ensure a complete understanding of the problem. This may result in a seven-tier functional flow diagram before anything is allocated. In the other extreme, the analyst expands a higher-tier function into a lower-tier diagram and immediately allocates the exposed functionality. This selection is more a matter of art and experience than science but the author believes that a happy medium between the extremes noted above is optimum.

If we accumulate too much functional information before allocation, we run the risk of a disconnect between lower-tier functions and the design concepts associated with the higher order product entities that results. If, for example, we allocate a master function for transport dirt to an earth mover, we may have difficulty allocating lower-tier functionality related to moving the digging and loading device (which, in our high-level design concept, is integrated with the moving device). Allocation accomplished too rapidly can lead to instability in the design concept development process because of continuing changes in the higher-tier functional definition.

The ideal pacing involves progressive allocation. We accumulate exposed functionality to a depth that permits us to thoroughly analyze system performance at a high level, possibly even run simulations or models of system behavior under different situations with different functionality combinations and performance figures of merit values in search of the optimum configuration. The allocation of high-order functionality prior to completing these studies is premature and will generally result in a less than optimum system and many changes that ripple from the analysis process to the product entity synthesis and concept development work. Throughout this period, we have to deal with functional requirements rather than raw function statements; so, when we do allocate this higher-order functionality it will be as functional requirements rather than as raw function names. Before allocating lower-tier functionality, we should allocate this higher-order functionality and validate them with preliminary design concepts. These design concepts should then be fed back into the lower-tier functional analysis to tune it to the current reality. Subsequent allocations can often be made using the raw functions followed by the expansion of them into performance requirements after allocation.

Some purists would claim that this is a prescription for point designs in the lower tiers. There is some danger from that, and the team must be encouraged to think through its lower-tier design concepts for innovative alternatives to the status quo. The big advantage, however, to progressive tuning of the functional flow diagram though concept feedback is that at the lower tiers the functional flow diagram takes on the characteristics of a process diagram where the blocks map very neatly to the physical situation that the logistics support people must continue to analyze. This prevents the development of a separate logistics process diagram with possible undesirable differences from the functional flow diagram. Once again, we are maintaining process continuity.

The author believes that the difference between functional flow and process diagrams is that the former is a sequence of things that must happen whereas the latter is a model of physical reality. When we are applying the functional flow diagram to problem solving, we do not necessarily know what the physical situation is nor of what items the system shall consist. The blocks of the process diagram represent actual physical situations and resources.

The U.S. Air Force developed a variation of the functional flow or process diagram called the IDEF diagram. IDEF was originally a compound acronym with the meaning ICAM definition, where ICAM = integrated computer-aided manufacturing analysis. In addition to the horizontal inputs and outputs that reflect sequence, these diagrams also have inputs at the top and bottom edges that reflect controlling influences and resources required for the steps. This diagrammatic approach was developed from an earlier SADT diagramming method developed for software analysis and applied to the development of contractor manufacturing process analysis. It does permit analysis of a more complex situation than process diagramming, but the diagram developed runs the risk of totally confusing the user with the profusion of lines. Many of the advantages claimed for IDEF can be satisfied through the use of a simpler functional or process flow diagram teamed with a dictionary. These diagrams present a simple view that the eye and mind can use to acquire understanding of complex relationships, and the dictionary presents details related to the blocks that would confuse the eye if included on the diagram. The IDEF technique has evolved into an IDEF 0 for process analysis, IDEF 1X for relational data analysis, IDEF 2 for dynamic analysis, IDEF 3 for process description, and many others few of them being well documented at the time this book was being written.

Some system engineers, particularly in the avionics field, have found it useful to apply what can be called hierarchical functional analysis. In this technique, the analyst makes a list of the needed lower-tier functionality in support of a parent function. These functions are thought of as subordinate functions in a hierarchy rather than a sequence of functions as in flow diagramming. They are allocated to things in the evolving architecture generally in a simple one-to-one relationship.

The concern with this approach is that it tends to support a leap to point design solutions familiar to the analyst. It can offer a very quick approach in a fairly standardized product line involving modular electronics equipment as a way to encourage completeness of the analysis. This technique also does not support time line analysis as does functional flow diagramming since there is no sequence notion in the functional hierarchy.

Whatever techniques we use to expose the needed functionality, we have to collect the allocations of that functionality into a hierarchical architecture block diagram reflecting the progressive decomposition of the problem into a synthesis of the preferred solution. The peak of this hierarchy is the block titled system, which is the solution for the problem (function) identified as the need. Subordinate elements, identified through the allocation of lower-tier functionality, form branches and tiers beneath the system block. The architecture should be assembled recognizing several overlays to ensure that everyone on the program is recognizing the same architecture: WBS (finance), manufacturing breakdown structure (assembly sequence), engineering drawing structure, specification tree, configuration or end-item identification, make-buy map, and development responsibility.

As the product entity structure is assembled, the needed interfaces between these items must be examined and defined as a prerequisite to defining their requirements. These interfaces will have been predetermined by the way that we have allocated functionality to things and modified as a function of how we have organized the things in the architecture and the design concepts for those things. During the product entity synthesis and initial concept development work, the interfaces must be defined for the physical model using schematic block or n-square diagrams.

Ascent Logic popularized another technique called behavioral diagramming that combines the functional flow diagram arranged in a vertical orientation on paper with a data and interface flow arranged in the horizontal orientation. The strength of this technique is that we are forced to evaluate needed process and data needs simultaneously rather than as two separate and, possibly disconnected, analyses. The inclusion of product entity synthesis and interface analysis under the functional analysis umbrella does mitigate the differences somewhat. The tool RDD-100 used this analysis model leading to the capability to simulate system operation and output the system functionality in several different views including functional flow, IDEF 0, or n-square diagrams.

Behavioral diagramming was suggested by, or independently evolved along the same lines as, a software model called input-process-output or IPO that was an early attempt to merge computer processing and data analysis as the object-oriented modeling technique more recently.

15.2.5 Computer software approaches

It is not possible for a functioning system to exist that is entirely computer software, for software requires a machine medium within which to function. Systems that include software will always include hardware, a computing instrument as a minimum, and most often will involve people in some way. Software is to the machine as our thoughts, ideas, and reasoning are to the gray matter making up our mind. While some people firmly believe in out-of-body experiences for people, few would accept a similar situation for software. A particular business entity may be responsible for creating only the software element of a system and, to them, what they are developing could be construed as a system, but their product can never be an operating reality by itself. This is part of the difficulty in the development of software; it has no physical reality. It is no wonder then that we might turn to a graphical and symbolic expression as a means to capture its essence.

We face the same problem in software as hardware in the beginning. We tend to understand our problems first in the broadest sense. We need some way to capture our thoughts about what the software must be capable of accomplishing and to retain that information while we seek to expand upon the growing knowledge base. We have developed many techniques to accomplish this end over the period of 40–50 years during which software has been a recognized system component.

The earliest software analytical tool was flow diagramming, which lays out a stream of processing steps similar to a functional flow diagram (commonly in a vertical rather than horizontal orientation probably because of the relative ease of printing them on line printers) where the blocks are very specialized functions called computer processes. Few analysts apply flow diagramming today, having surrendered to data flow diagramming (DFD), the Hatley–Pirbhai extension of this technique, or object-oriented analysis. Alternative techniques have been developed that focus on the data that the computer processes. The reasonable adherents of the process and data orientation schools of software analysis would today accept that both are required and some have made efforts to bridge this gap. The most recent round of tool development focuses on a merger of these two orientations in what is called object-oriented software development. Earlier attempts included IPO that was the basis for behavioral diagramming used by Mack Alfred in the development of requirements-driven design (RDD), and the integration of entity-relationship (ER) diagramming with structured analysis.

All software analysis tools (and hardware-oriented ones as well) involve some kind of graphical symbols (bubbles or boxes) representing data or process entities connected by lines, generally directed ones. Most of these processes begin with a context diagram formed by a bubble representing the complete software entity connected to a ring of blocks that correspond to external interfaces that provide or receive data. This master bubble corresponds to the need, or ultimate function, in functional analysis, and its allocation is to the thing called a system. The most traditional description of this model was developed principally by Yourdon, DeMarco, and Constantine. It involves the expansion of the context diagram bubble into lower-tier processing bubbles that represent subprocesses just as in functional analysis. These bubbles are connected by lines indicating data that must pass from one to the other. Store symbols are used to indicate a need to temporarily store a data element for subsequent use. These stores are also connected to bubbles by lines to show source and destination of the data. Since the directed lines represent a flow of data between computer processing entities (bubbles), the central diagram in this technique is often referred to as a data flow diagram.

In all software analysis techniques, there is a connection between the symbols used on the diagrammatic portrayal to text information that characterizes the requirements for the illustrated processes and data needs. In the traditional line-and-bubble analysis approach, referred to as DFD, one writes a process specification for each lowest-tier bubble on the complete set of diagrams and provides a line entry in a data dictionary for each line and store on all diagrams. Other diagrams are often used in the process specification to explain the need for controlling influences on the data and the needed data relationships. All of this information taken together becomes the specification for the design work that involves the selection of a specific machine upon which the software will run, a language or languages that will be used, and an organization of the exposed functionality into "physical" modules that will subsequently be implemented in the selected language through programming work. A good general reference for process and data-oriented software analysis methods is Yourdon's *Modern Structured Analysis* (1989). Tom DeMarco's *Structured Analysis and System Specification* (1979) is another excellent reference for these techniques.

Much of the early software analysis tool work focused on information batch processing because central processors, in the form of large mainframe computers, were in vogue. More recently, distributed processing on networks and software embedded in systems has played a more prominent role revealing that some of the earlier analysis techniques were limited in their utility to expose the analyst to needed characteristics. Derek Hatley and the late Imtiaz Pirbhai offer an extension of the traditional approach in *Strategies for Real-Time System Specification* (1988) to account for the special difficulties encountered in embedded, real-time software development. They differentiate between dataflow needs and control flow needs and provide a very organized environment for the allocation of exposed requirements model content to an architecture model. The specification consists of information derived from the analytical work supporting both of these models.

Unfortunately, a lot of people do not realize that Hatley and Pirbhai extended modern structured analysis in another fashion that makes it a powerful model capable of being used in a universal fashion at the system level as well as for parts of the problem space hat will be implemented in hardware and software. They did this by allowing the bubbles to represent the processing of energy and material as well as data and the lines to represent the flow of energy and material as well as data.

Fred McFadden and Jeffrey Hoffer have written an excellent book on the development of software for relational databases in general and client-server systems specifically titled *Modern Database Management* (1999). With this title, it is understandable that they would apply a data-oriented approach involving ER diagrams and a variation on IDEF 1X. The latter is explained well in a Department of Commerce Federal Information Processing Standards Publication (FIPS PUB) 184. McFadden and Hoffer also explain a merger between IDEF 1X and object-oriented analysis.

The schism between process-oriented analysis and data-oriented analysis, which had to be patched together in earlier analysis methods, has been joined together more effectively in object-oriented analysis about which many books have been written. A series that is useful and readable is by Coad and Yourdon (1990, 1991) and by Coad and Nicola (1993). Two others are by James Rumbaugh et al. (1990) and by Grady Booch (1993).

At the time this book was written, the software modeling work had been overtaken by unified modeling language (UML) that had also been extended to provide a system modeling language (SysML) intended for system and hardware analysis using many of the same modeling artifacts and techniques effective in UML. Sometime in the 2010 decade the author believes that a universal modeling approach will be formed using the core formed by UML//SysML but prior to that time frame a program will have to deal with a pair of modeling approaches through which the teams gain an understanding of the problem to be solved.

15.2.6 Performance requirements analysis

Performance requirements define what the system or item must do and how well it must do those things. Precursors of performance requirements take the form of function statements or functional requirements (quantified function statements). These should be determined as a result of a structured analysis process that decomposes the customer need as noted above using an appropriate hardware or software technique.

Many organizations find that they fail to develop the requirements needed by the design community in a timely way. They keep repeating the same cycle on each program and fail to understand their problem. This cycle consists of the receipt of the customer's requirements or the approval of their requirements in a specification created by the contractor followed by a phony war on requirements where the systems people revert to documentation specialists and the design community creates a drawing release schedule in response to management demand for progress. As the design becomes firm, the design people prepare an in-house requirements document that essentially characterizes the preexisting design. Commonly, the managers in these organizations express this problem as, "We have difficulty flowing down system requirements to the designers."

The problem is that the flowdown strategy is only effective for some specialty engineering and environmental design constraints. It is not a good strategy for interface and performance requirements. It is no wonder these companies have difficulty. There is no one magic bullet for requirements analysis. One needs the whole toolbox described in this chapter. Performance requirements are best exposed and defined through the application of a structured process for exposing needed functionality and the allocation of the exposed functionality to things in the architecture. You need not stop at the system level in applying this technique. It is useful throughout the hierarchy of the system.

Performance requirements are traceable to (and thus flow from) the process from which they are exposed much more effectively than in a vertical sense through the product architecture. They trace to the problem or functional plane very easily. Constraints are generally traceable within the product or solution plane. This point is lost on many engineers and managers, and thus they find themselves repeating failed practices indefinitely.

Given that we have an effective method for identifying valid performance requirements as described under functional analysis above, we must have a way to associate them with quantitative values. In cases where flowdown is effective, within a single requirement category, such as weight, reliability, or cost, a lower-tier requirement value can be determined by allocating the parent item value to all its child items in the product architecture. This process can be followed in each discipline creating a value model for the discipline. Mass properties and reliability math models are examples. In the case of performance requirements, we commonly do not have this clean relationship, so allocation is seldom effective in the same way.

Often the values for several requirements are linked into a best compromise and to understand a good combination we must evaluate several combinations and observe the effect on selected system figures of merit like cost, maximum flight speed in an aircraft, automobile operating economy, or land fighting vehicle survivability. This process can best and most quickly be accomplished through a simulation of system performance where we are allowed to control certain independent parameters and observe the effects on dependent variables used to base a selection upon. We select the combination of values of the independent variables that produces the best combination of effects in the dependent variables.

Budget accounts can also be used effectively to help establish sound values for performance requirements. For example, given a need to communicate across 150 miles between the Earth's surface and a satellite in low Earth orbit, we may allocate gain (and loss) across this distance, the antenna systems on both ends, connecting cables, receiver, and transmitter. Thus the transmitter power output requirement and the receiver sensitivity requirement are determined through a system-level study of the complete communications link.

This work involved in establishing appropriate values for performance requirements is part of the requirements validation process to the extent that it gives us confidence that the selected values are achievable within the time and money limits established for development.

15.2.7 Design constraints analysis

Design constraints are boundary conditions within which the designer must remain while satisfying performance requirements. All of them can be grouped into the three kinds described below. Performance requirements can be defined prior to the identification of the things to which they are ultimately allocated. Design constraints generally must be defined subsequent to the definition of the item to which they apply. Performance requirements provide the bridge between the problem and solution planes through allocation. Once we have established the architecture, we can apply three kinds of constraints analysis to these items. In the case of each constraint category, we need a special tool set to help us understand in some organized way what characteristics we should seek to control.

15.2.7.1 Interface requirements analysis

Systems consist of things. These things in systems must interact in some way to achieve the need. A collection of things that do not in some way interact are a simple collection of things, not a system. An interface is a relationship between two things in a system. This relationship may be completed through many different media, such as wires, plumbing, a mechanical linkage, or a physical bolt pattern. These interfaces are also characterized by a source and a destination, that is, two terminals each of which is associated with one thing in the system. Our challenge in developing systems is to identify the existence of interfaces and then to characterize them, each with a set of requirements mutually agreed upon by those responsible for the two terminals.

Note the unique difference between the requirements for things in the system and interfaces. The things in systems can be clearly assigned to a single person or team for development. Interfaces must have a dual responsibility where the terminals are things with different responsibilities. This complicates the development of systems because the greatest weakness is at the interfaces and accountability for these interfaces can be avoided. The opportunities for accountability avoidance can be reduced by the assignment of teams responsible for development as a function of the product architecture rather than the departments of the functional organization. This results in perfect alignment between the product cross-organizational interfaces (interfaces with different terminal organizational responsibilities) and the development team communication patterns that must take place to develop them. Responsibility and accountability is very clear.

The reader is encouraged to refer to *System Requirements Analysis* (Grady, 2005) or *System Synthesis: Product and Process Design* (Grady, in press) for a thorough discussion of schematic block and n-square diagramming techniques. As a result of having applied these techniques during the product entity synthesis of allocated functionality, the system engineer will have exposed the things about which interface requirements must be written. Once again, the purpose of our tools is to do just this, to help us understand what to write requirements about. The use of organized methods encourages completeness and the avoidance of extraneous requirements that increase the cost out of proportion to their value.

Once it has been determined what interfaces we must respect, it is necessary to determine what technology will be used to implement them, such as electrical wiring, fluid plumbing, or physical contact. Finally, the resultant design in the selected media is constrained by quantified requirements statements appropriate to the technology and media. Each line on a schematic block diagram or marked intersection on an n-square diagram must be translated into one or more interface requirements that must be respected by the persons or teams responsible for the two terminal elements. The development requirements for the two terminal items may be very close to identical such as a specified data rate, the degree of precision, or wire size. The product requirements, however, will often have an opposite nature to them, such as male and female connectors, bolt hole or captive bolt and threaded bore hole, or transmitter and receiver.

15.2.7.2 Environmental requirements analysis

One of the most fundamental questions in system development involves the system boundary. We must be able to unequivocally determine whether any particular item is in the system or not in the system. If it is not in the system, it is in the system environment. If an item is in the system environment, it is either important to the system or not. If it is not, we may disregard it in an effort to simplify the system development. If it is important to the system, we must define the relationship to the system as an environmental influence.

We may categorize all system environmental influences in the five following classes:

1. Natural environment—Space, time, and the natural elements such as atmospheric pressure, temperature, and so forth. This environment is, of course, a function of the locale and can be very different from that with which we are familiar in our immediate surroundings on Earth as in the case of Mars or the Moon.
2. Hostile systems environment—Systems under the control of others that are operated specifically to counter, degrade, or destroy the system under consideration.
3. Noncooperative environment—Systems that are not operated for the purpose of degrading the system under consideration but have that effect unintentionally.
4. Cooperative systems environment—Systems not part of the system under consideration that interact in some planned way. Generally, these influences are actually addressed as interfaces between the systems rather than environmental conditions, because there is a person from the other system with whom we may cooperate with to control the influences.
5. Induced environment—Composed of influences that would not exist but for the presence of the system. These influences are commonly initiated by energy sources within the system that interact with the natural environment to produce new environmental effects.

As noted, cooperative environmental influences can be more successfully treated as system interfaces. Hostile and noncooperative influences can be characterized through the identification of threats to system success and the results joined with the natural environmental effects. The induced environment is best understood through seeking out system energy sources and determining if those sources will interact with the natural environment in ways that could be detrimental to the system.

The natural environment is defined in standards for every conceivable parameter for Earth, space, and some other bodies. The challenge to the system engineer is to isolate on those parameters that are important and those that are not and then to select parameter ranges that are reasonable for those parameters that will have an impact on our system under development. The union of the results of all of these analyses form the system environmental requirements. It is not adequate to stop at this point in the analysis, however.

Systems are composed of many things that we can arrange in a family hierarchy. Items in this hierarchy that are physically integrated in at least some portions of their system operational use, such as an aircraft in an aircraft system or a tank in a ground combat system, can be referred to as end items. We will find that these end items in operational use will have to be used in one or more environments influenced by the system environment but, in some, cases modified by elements of the system. For example, an aircraft will have to be used on the ground, in the air through a wide range of speed and altitude, and in hangers. These are different environments that we can characterize as subsets of the system environment definition. The best way to do this is to first define the system process in terms of some finite number of physical analogs of system operation. We may then map the system end items to these process steps at some level of indenture. Next, we must determine the natural environmental subsets influencing each process step. This forms a set of environmental vectors in three-space. In those cases where a particular end item is used in more than one process, we will have to apply some rule for the determination of the aggregate effect of the environments in those different process steps. The rule may be worst case or some other one. This technique is called environmental use profiling.

The final step in environmental requirements analysis involves the definition of component environmental requirements. These components are installed in end items. The end items can be partitioned into zones of common environmental parameters as a function of the end item structure and energy sources that will change natural environmental influences. We define the zone environments and then map the components into those zones. In the process, we may find that we have to provide an environmental control system in one or more of the zones to reduce the design difficulty of some components. We may also conclude, given that we must have at least one such system, that we should relocate one or more components into the space thus controlled. So, the environmental requirements for a component are predetermined by the zone in which it is located, and we may derive its environmental requirements by copying (cloning) the zone requirements. Component environmental requirements may have to be adjusted for shipping and other noninstalled situations.

In summary, this set of three tools (standards, environmental use profiling, and end item zoning) may be used to fully characterize the environmental requirements for all system items from the system down through its components. In all cases, we must be careful to phrase these requirements in terms that recognize our inability to control the ultimate natural environment.

The prior discussion focuses on the effects of the environment on our system. We must also consider the effects of our system on the natural environment. This is most often referred to as environmental impact analysis. It has not always been so, but today we must be sensitive to a bidirectional nature in our environmental specification. Our efforts in the distant past were very small compared to the tremendous forces of nature. Today, the scope of our access, control, and the application of energy and toxic substances is substantial, and potential damage to local and regional natural environments is a real concern. Environmental impact requirements are commonly defined in laws and regulations. We can determine in what ways the environment will be exposed to system-generated stress by evaluating all system energy sources, toxic substances used, and any exhaust products. The author prefers to classify this topic under the banner of environmental requirements analysis, but it could be thought of as a specific specialty engineering discipline.

These effects must be considered both for the life of the system during its use and, perhaps more importantly, for its disposition following its useful life. There can be no better example of how difficult and dangerous this can be than the case of the nuclear weapons cleanup process seriously begun in the 1990s after the end of the nuclear confrontation between the USSR and the United States lasting 40 years. Throughout the confrontation, both sides were so concerned with survival and the urgency of development that little thought was given to the problems that they may be causing for the future. By the time the problem could no longer be ignored, it was so substantial that a solution was more difficult than the complex scientific and technical problems that had to be solved to create it. One can be very unpopular expressing interest in system disposition during the heady system development times, especially with one's proposal or program manager. But, customers will increasingly be concerned with this matter as the problems we attempt to solve become increasingly more general and their scope more pervasive.

Computer software environmental requirements are limited. The natural environment impinges on the computer hardware within which the software functions but does not in any direct way influence the software itself in the absence of hardware deficiencies that cause computer circuits to fail due to environmental stresses. The context diagram used to expose the boundary between the software and its environment is the best place to start in determining the appropriateness of any environmental influences directly on the software. Commonly, one such requirements category of interest is the user environment describing how the user will interact with the software. These influences could alternatively be covered under the heading of interface requirements. One must first determine whether the operator is within or outside the system, however, and this is no small question.

The principal element in the software environment is in the form of cooperative systems exposed through interface analysis. What other software and hardware systems will supply information or use information created by the system in question? The noncooperative system component may exist in very complex software systems but is hard to characterize in terms of an example. A very common and increasingly important environmental factor is the hostile environment. This matter will more commonly be addressed as the specialty engineering discipline called security engineering, however, than as an environmental consideration.

15.2.7.3 Specialty engineering requirements analysis

The evolution of the systems approach to the development of systems to solve complex problems has its roots in the specialization of the engineering field into a wide range of very specialized disciplines for the very good reasons noted earlier. Our challenge in system engineering is to weld these many specialists together into the equivalent of one all-knowing mind and applying that knowledge base effectively to the definition of appropriate requirements followed by the development of responsive and compliant designs and the assessment of those designs for compliance with the requirements as part of the verification activity.

There are many specialized disciplines recognized including reliability, maintainability, logistics engineering, availability, supportability, survivability and vulnerability, guidance analysis, producibility, system safety, human engineering, system security, aerodynamics, stress, structural dynamics, thermodynamics, and transportability. For any specific development activity, some or all of these disciplines will be needed to supplement the knowledge pool provided by the more general design staff.

Specialty engineers apply two general methods in their requirements analysis efforts. Some of these disciplines use mathematical models of the system as in reliability and maintainability models of failure rates and remove-and-replace or total repair time. The values in these system-level models are extracted from the model into item specifications. Commonly, these models are built in three layers. First, the system value is allocated to progressively lower levels to establish design goals. Next, the specialty engineers assess the design against the allocations and establish predictions. Finally, the specialists establish actual values based on testing results and customer field use of the product.

Another technique applied is an appeal to authority in the form of customer-defined standards and specifications. A requirement using this technique will typically call for a particular parameter to be in accordance with the standard. One of these standards may include a hundred requirements, and they all flow into the program specification through reference to the document unless it is tailored to reflect precisely what is required on the specific program. Specialty engineers must, therefore, be thoroughly knowledgeable about the content of these standards; familiar with their company's product line, development processes, and customer application of that product; and the basis for tailoring standards for equivalence to the company processes and preferred design techniques.

15.3 Product entity synthesis

Return now to Figure 15.2 and note the development of the product structure as a result of the allocation of functionality to things. In building the physical product model, we should take into consideration the integration of several different views. One of these views is the finance perspective often referred to as a WBS. It is important that the system engineer and the finance people come to a common perspective on the architecture. In the past, in the development of systems for DoD, finance community was responsible for the WBS and system engineering for the top-down development of the architecture. These were seldom totally consistent because the WBS on DoD programs was determined by an appendix of a military standard (MIL-STD-881) before the functional analysis had finished to offer a functionally derived product structure. DoD moved to a position that system engineering should be responsible for defining the WBS in MIL-STD-499B, but that standard was never released.

So, we should be able to establish a single notation that serves both as a WBS and as an architecture identification. The author would prefer the architecture identification system suggested in earlier chapters involving a base 60 system of numbers and letters (upper case less O and lower case less l). If it is necessary because of customer requirements to maintain separate identification systems, then we would have to map the two together leading to some inefficiency.

Large customers have gotten themselves into the habit of using a WBS that covers two different classes of things. DoD, NASA, and other large customers recognize product and process components in the WBS. The product WBS is simply an overlay of the architecture (identical or mapped). In this configuration, only the product-oriented component can be defined through functional analysis of the product. The remainder is commonly defined from a boilerplate used on past programs. The process-oriented component is associated with things that are not easy to connect up with the product. The author maintains that this is mistake that causes unnecessary complexity in program planning and management. The premise is that it is possible to hook all cost and schedule parameters to product.

It is especially easy to do this when we accept the IPPT approach. The program planning process is simplified more when we simply use the architecture IDs as a WBS for finance purposes. If there must be financial component complexity, let that complexity fall entirely within the finance world and their computers. Leave the human part of the process as simple as possible.

15.4 Product-only WBS

If we recognize that work can be accomplished against the whole system as well as any element of the system, we shall have a workable model that will enable an efficient transform between generic enterprise work planning and the work needed for a specific program. The DoD customer has, in the past, been so concerned with their need to control cost and schedule that they have forced contractors to apply a planning model that denies them an efficient map or transform between their identity and the needed program work. Therefore, these contractors have generally failed to develop an identity through common written practices to be applied on all programs, sound internal education programs (on-the-job training and mentoring accepted as well as more formal methods) focused on that generic process, and efficient methods for building program planning structures based on their generic process definition.

Figure 15.3 illustrates the proposed relationships between product items, functional departments, generic work commonly required on the

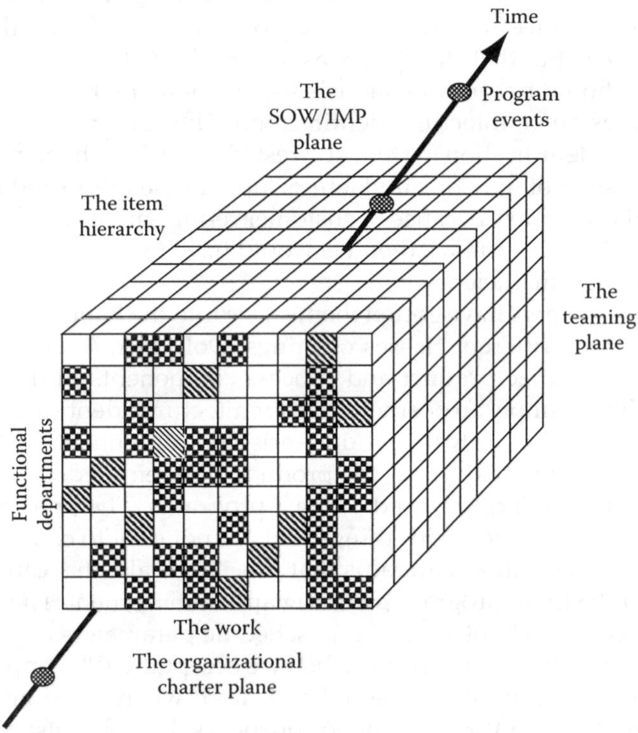

Figure 15.3 Product-only WBS enables good planning. (From Grady, J., *System Engineering Deployment*, CRC Press, Boca Raton, FL, pp. 118–142, 2000. With permission.)

enterprise's programs, and time. The understanding is that the enterprise is managed through a matrix with lean functional departments responsible for providing programs with qualified people skilled in standard practices found to be effective in the company's product line and customer base and good tools matched to the practices and people skills. Each of these functional departments has an approved charter of work they are responsible for performing on programs and which provides the basis for personnel acquisition and training.

The front vertical plane in Figure 15.3 illustrates the aggregate functional department charter by showing the correlation between all functional departments and all possible work they may be called upon to do on programs. The darkened blocks indicate this correlation. The crosshatched block in each column indicates the department that should supply the persons to lead that particular effort on programs in the product development teams assembled about the architecture. The indicated departments must prepare practices supportive of the practice defined by the task lead department (crosshatched block). The task lead department should also be primary in providing tools to do the work for that task on all programs and for providing any training in that activity.

The reader should note that when an enterprise fully projectizes its work force it looses the agent that can accomplish the generic work useful in providing the foundation for all programs. Each program becomes a new process factory. Therefore, there can be no central process, there can be no continuous improvement, there can be no progress, only a never-ending struggle, at best, at the lower fringes of excellence.

In the suggested arrangement, the WBS and the product structure become one since there is no process component of the WBS. When planning a program composed of items identified on the item hierarchy (architecture or WBS) axis in Figure 15.3, we must select the work required from our generic work axis to form the program planning data, which may consist of a general identification of the work at a high level called a statement of work (SOW) and a more detailed expansion of that work definition called an integrated master plan (IMP). This is referred to in Figure 15.3 as the SOW/IMP plane. This work selection leads to a clear identification of the functional departments from which personnel should be selected and the practices that must be applied on the program. Once we have decided the product entity nodes about which we will form cross-functional teams, the teaming plane tells us to which teams these people from the functional departments should be assigned.

The SOW that results from the first level of correlation between the product structure and generic work must then be expanded into a more detailed plan referred to here as an IMP. For each SOW paragraph, we need to identify in a subparagraph in the IMP the work to be accomplished by each contributing department, which should include the cross-functional, co-process integration work to be accomplished by the task lead department. All of the work thus selected must then be placed in a time context by mapping it to major events terminating that work and corresponding to the conduct of major reviews forming the foundation of the program integrated master schedule (IMS). We see the program plan as a series of nodes arranged in a series-parallel network mapped to trios in our matrix illustrated in Figure 15.3. These nodes, in turn, are traceable to a combination of contributing functional departments, product architecture or WBS, and generic work elements.

In that we should form the work teams focused on the product structure or WBS, each team inherits a contiguous block of budget in the WBS and work in the corresponding SOW and IMP section, a clearly correlated IMS expands upon the IMP in the time axis, and a specification at the top level of the hierarchy for which they are responsible is needed. The cross-organizational interfaces, which lead to most program development problems, become exactly correlated with communication patterns that must occur between the teams to develop those interfaces as shown in Figure 15.4. Thus people are clearly accountable for doing this work well.

This product-only WBS has been discussed in terms of a physical product, but most products also require deliverable services such as training, system test and evaluation work, site activation work, and other

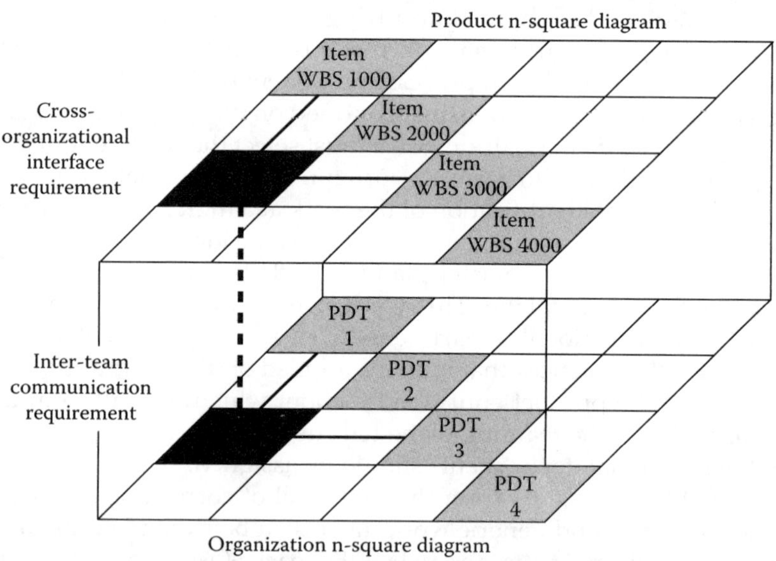

Figure 15.4 Interface development benefits. (From Grady, J., *System Engineering Deployment*, CRC Press, Boca Raton, FL, pp. 118–142, 2000. With permission.)

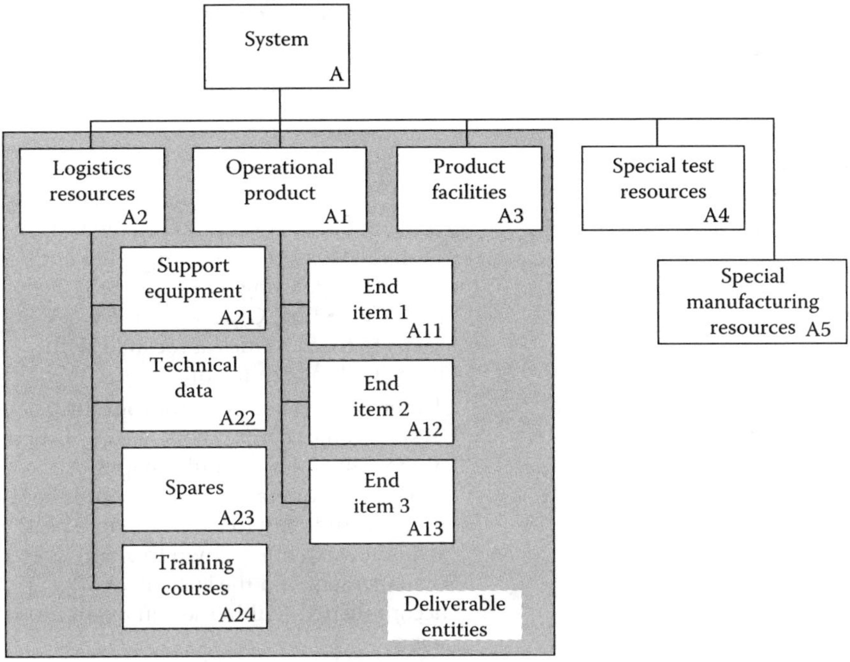

Figure 15.5 The product-only WBS. (From Grady, J., *System Engineering Deployment*, CRC Press, Boca Raton, FL, pp. 118–142, 2000. With permission.)

services. Figure 15.5 suggests that these services can be framed in the context of physical entities some of which may be delivered and others not. The customer training program may entail the services of teaching the courses developed initially or for the whole life cycle. Indeed, the whole system could be operated by the contractor throughout its life and this would clearly be a service. But, these services can be related to the things in the system. The author distinguishes between the services that would be performed on the delivered system and those required to create the system. The latter should all be coordinated with the physical product.

Well, how about a contract that entails only services. The WBS for such a contract can still be formed of services to be delivered, but there remains a distinction between the work that must be done to prepare for performing those services and the performance of those services. The former can be oriented about an entity structure of services being designed. The latter constitutes the product that is delivered. Table 15.1 lists several often troublesome traditional WBS elements and correlates them with the product only blocks of Figure 15.5 to show that this can be done. Note that some of these entities would not normally be delivered like the tooling

Table 15.1 Non-Product WBS Transform

Non-prime item work element	MID	Comment
Training	A24	
Peculiar support equipment	A21	
Systems test and evaluation	A1	A1 elements as appropriate to the test. System testing may be mapped to entities A or A1 depending on the scope of the planned testing. Otherwise, the work maps to the activities associated with items within architecture A1.
Systems engineering/ program management	A	Elements of system A as appropriate. Systems engineering work accomplished at the system level would map to A. Deliverable operational product element work would map to A1. This same pattern would be extended down to team management and the systems work accomplished within the context of those teams.
Data	A22	
Operational site activation	A3	As the work relates to the facilitization and preparation of sites from which the product will be operated and maintained.
Common support equipment	A21	
Industrial facilities	A4, A5	
Initial spares and initial repair parts	A24	

Source: Grady, J., *System Engineering Deployment*, CRC Press, Boca Raton, FL, pp. 118–142, 2000. With permission.

and special test equipment, but the customer may very well own this equipment in that it was developed on the contract.

It is especially important to note that the program management and system engineering services work need not be treated as a service. It can be associated with the system level of the product if nothing else. But by distributing these elements through the architecture where they belong, we distribute the cost and schedule parameters associated with these skills into the teams, PIT and IPPT. Also, we support the notion that system engineering is a process more so than an organization or profession that should be applied by everyone throughout the system architecture hierarchy.

So, at this point, the enterprise has prepared itself by developing an identity composed of an organizational charter plane showing the map between common tasks that must be applied on all programs and their functional department structure. Each functional department has developed written practices telling how to do the tasks mapped to them coordinated with the designated task lead department's practice and tools available to do the work. This how-to information could be in detailed text format in practices manuals or be provided in training materials traceable to the content of an abbreviated practice. This generic identity data is continuously improved based on lessons learned from ongoing programs and is the basis for in-house training.

When a program comes into the house, the work is planned by first performing a structured decomposition and analysis on the customer need that yields a preferred product structure synthesized from the allocated functionality and the aggregate views of engineering (hardware and software influences), manufacturing, procurement and material, finance, logistics, and quality within the context of a PIT. In the simplest program work planning approach, the product structure becomes the WBS (base 60 identification suggested here can be substituted with any other system), which is the basis for development of the program SOW. The SOW is created by the PIT by linking high-level tasks with the WBS elements as discussed earlier.

chapter sixteen

*Interface development in a new world**

The other chapters of this book are focused on the relationships between the elements of the functional organization that will supply the enterprise with the talent and resources with which to create product systems and the teams that will be formed around the product architecture on programs. This chapter focuses on the relationships between the elements in the product system through interfaces and the teams that develop the elements of the product system. There should exist on a program a well-developed correlation between these team relationships and the interfaces between the products for which the teams are responsible.

Interface development is the most difficult part of the system development process because interfaces have two terminals, and all the more difficult when the same party is not responsible for both terminals. Chapter fifteen addressed the need to define the requirements for these interfaces. In this chapter, we will explore the management and documentation aspects appropriate for these interfaces. Interface development has always been a difficult management problem, but rapid changes in ownership of companies common in the 1990s made this an even more difficult task than had been the case previously. This chapter extends common interface development techniques, which are explained in some detail, to the full range of organizational relationships in both static and dynamic business relationship environments.

16.1 Interface fundamentals

16.1.1 Interface defined

A system is a collection of things that interact to satisfy a common objective, goal, or need. These things may be hardware, software, facilities, or people accomplishing procedures. An interface is a relationship between things in a system, between systems, and between systems and their environment. Figure 16.1 illustrates this arrangement along with the system interface concepts. The system is composed of things (not illustrated) internal to

* The material in this chapter is from Grady, J., *System Engineering Deployment*, CRC Press, Boca Raton, FL, Chapter 10, 2000. Used with permission.

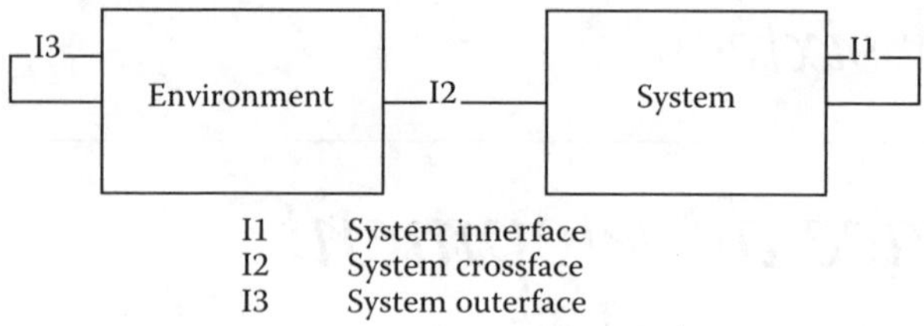

Figure 16.1 Ultimate system abstraction. (From Grady, J., *System Engineering Deployment*, CRC Press, Boca Raton, FL, pp. 147–156, 2000. With permission.)

the system that interact via the interface loop (all of the internal interfaces identified with the modeling ID I1) attached to the system block, and the system interacts with its environment via the interface shown connecting the two blocks, referred to as crossface (with interface MID I2). A third class of interface is possible, identified as I3 and named outerface, that consists of any interfaces within the environment of special interest to the system.

Everything in the Universe that is not in the system is in the system environment. Some of these things are related to the system in some way, while many are not and can be ignored. The environment consists of space, time, and natural forces applied to things throughout the Universe as well as other systems linked by one of the following relationships: cooperative, noncooperative, and hostile. Cooperative systems can best be defined in terms of interfaces between the two systems, because there is someone on each end of the interface to take the responsibility of their end and the relationship can be negotiated. Noncooperative and hostile relationships have to be treated as environmental influences because the relationship cannot be negotiated by persons responsible for the ends.

16.1.2 *Interface characteristics*

All interfaces have three fundamental aspects: two ends, or terminals, and a media through which the relationship is consummated and maintained. The ends touch two things that are in the system in the case of an internal interface, and one thing in the system and one element of the system environment in the case of an external interface. The media of the interface is either entirely within the system, as in the case of a wire harness between two system elements, or in the system environment, as in the case of radio signals through space between two satellites.

16.1.3 *Classes of interface media*

There are many ways to partition all interfaces into subsets relative to the kinds of media employed, as indicated in the following outline:

Electrical
- Power
- Control
- Radio or radar
- Signal bus
- Electrical noise
- Optical
- Visible light
- Infrared (IR)
- Ultraviolet (UV)

Mechanical
- Structural or physical
- Mechanism or linkage
- Human manipulative

Fluid
- Pneumatic (air or other gases)
- Liquids (non-hydraulic)
- Hydraulics

Acoustic
- Sound in air
- Sound in water

The technology associated with these different kinds of interfaces determines the way we would define the requirements for a particular interface in terms of volts, flow rate, forces applied, or intensity. Some organizations prefer to partition all interfaces into the following subsets: functional, physical, and environmental, with a secondary breakdown into the specific media classes noted above.

16.1.4 *Functional versus physical plane identification of interfaces*

The interfaces that exist between the things in a system are predetermined by the way we allocate functionality to the things in the system. Some people claim that they can foresee this pattern on the functional plane prior to the allocation of functionality through the use of functional n-square diagrams (more about these diagrams in a moment), but the author maintains that this only works when one uses a hierarchical functional depiction and is not fruitful when applying a sequence-oriented functional portrayal. The n-square relationships between a sequence of functions are sequential in nature rather than predictive of physical interfaces that will be required. Most people, including the author, have to identify the interfaces within the context of the physical plane and the product architecture based on the way functionality was allocated.

16.2 Interface depiction

Interface relations between things in the system can be exposed to humans through the use of schematic block diagrams or n-square diagrams. Tabular lists could be used featuring pairs of product entity columns to identify the interfaces uniquely, but these are not easy for humans to use commonly. Such a listing is a good idea in the form of an interface dictionary to supplement a diagrammatic portrayal, but not as the sole means of formally identifying interfaces.

16.2.1 Schematic block diagrams

Using this technique, the things in the system architecture are illustrated as blocks connected by lines that signify the interfaces between them. Nondirected lines can be used at the high level where the relationship may entail many individual interfaces from one or more media classes. At the individual interface level, we can use directed lines associated with particular media classes.

Figure 16.2 illustrates one of these diagrams at the detail level where directed lines make sense. We define the way the things shown as blocks in Figure 16.2 are related by defining the interfaces joining them. Items U, V, and Y interact with the system environment, whereas the other items are related only internally.

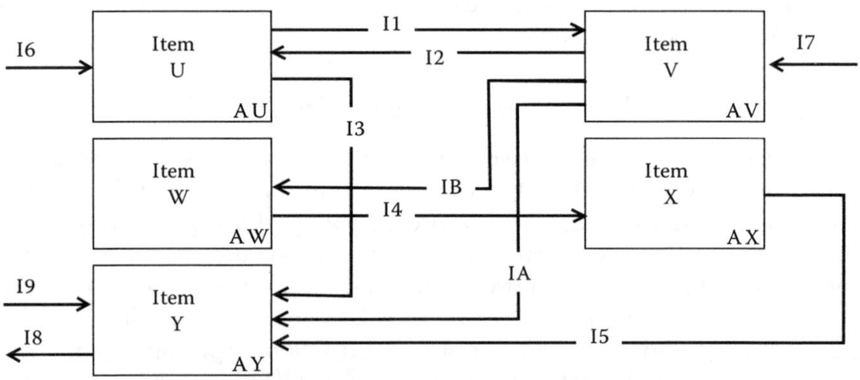

Figure 16.2 Schematic block diagram example. (From Grady, J., *System Engineering Deployment*, CRC Press, Boca Raton, FL, pp. 147–156, 2000. With permission.)

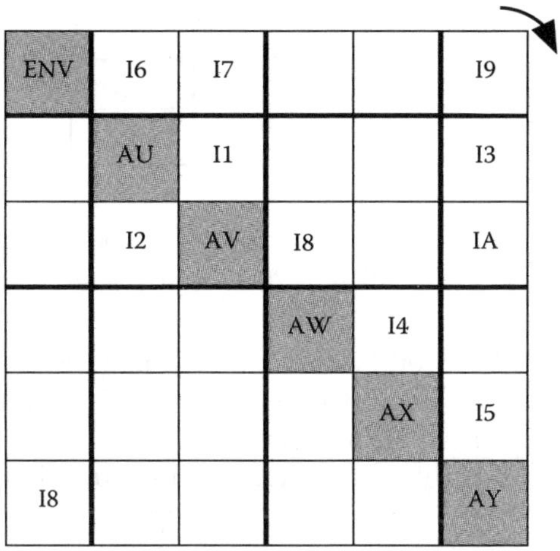

Figure 16.3 n-Square diagram example. (From Grady, J., *System Engineering Deployment*, CRC Press, Boca Raton, FL, pp. 147–156, 2000. With permission.)

The things are identified by names and their product entity IDs, and the interfaces are identified by interface IDs. This diagram should be accompanied by a product entity block diagram and two dictionaries, a product entity, and an interface dictionary, where details can be given and responsibilities defined. The complete set of diagrams and dictionaries should be captured in a system architecture report and maintained during early program development work.

16.2.2 n-Square diagrams

Figure 16.3 illustrates exactly the same architecture and set of relationships as shown in Figure 16.2, but using the n-square diagramming technique. The things in the system are listed down the diagonal along with the external (ENV). Interface IDs are inscribed in the intersections. This diagram should include a directionality arrow, as shown in the upper right-hand corner of the diagram. Interface I9, for example, sends something from the environment to item Y, whereas interface I2 sends something from item V to item U.

16.3 Terminal responsibility

16.3.1 Three faces of interface

The interfaces that appear in systems are looked upon differently by different persons as a function of the number of terminals of an interface that touch the item for which they are responsible. There are three possibilities

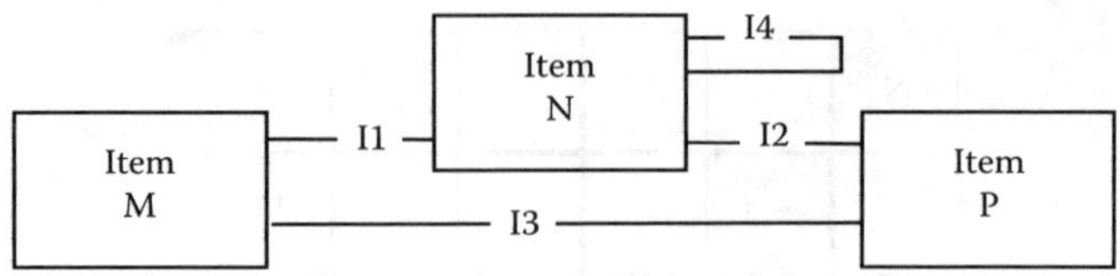

Figure 16.4 Interface viewpoints. (From Grady, J., *System Engineering Deployment*, CRC Press, Boca Raton, FL, pp. 147–156, 2000. With permission.)

obviously: one end, both ends, or neither end touches the architecture item in question. Let us look at these responsibilities in the context of the principal engineer for item N in Figure 16.4.

Note that both terminals of interface I4 touch item N. Clearly, we should expect that this interface can be developed by the principal engineer responsible for item N. This is an example of item N principal innerface. Neither end of interface I3 touches item N, so we can place this interface in item N principal's outerface set, and we should not bother this principal about this interface. Both of the remaining two interfaces in Figure 16.4 have one terminal that touches item N and another that touches a different item. These are all in item N principal's crossface and examples of true interfaces. The true nature of product interfaces can only be exposed in the context of the organizational responsibilities for these interfaces.

16.3.2 Cross-organizational interface

The most difficult problems in the development of systems to satisfy complex needs commonly reside at the interfaces we have described in the previous paragraph as crossfaces. The author would prefer to call these cross-organizational interfaces where the responsibilities fall upon teams or organizations rather than individual principal engineers. It is very important that the system engineer be able to sort out which interfaces fall into this set, because this corresponds with the most serious problems and risks that can occur on programs.

The easiest way to uncover these interfaces is with a schematic block diagram overlaid with responsibility outlines. Wherever the interface lines cross an organizational responsibility line, it identifies a cross-organizational interface. This same effect can be seen when using compound n-square diagrams, as shown in Figure 16.5. Given that each square represents the items under the responsibility of a cross-functional development team at some level of indenture, the interfaces denoted by the lines external to the team squares are cross-organizational interfaces. The interfaces depicted by Xs in each team n-square diagram are innerfaces relative to the team, and we should expect that each team can handle these responsibilities.

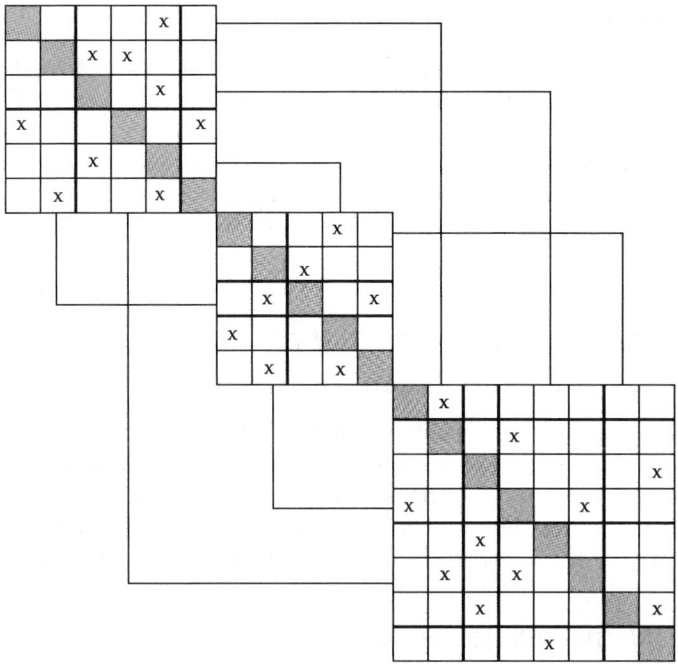

Figure 16.5 Cross-organizational interface. (From Grady, J., *System Engineering Deployment*, CRC Press, Boca Raton, FL, pp. 147–156, 2000. With permission.)

16.4 Interface documentation

16.4.1 Requirements documentation

The development of product designs should be based on predefined requirements captured in specifications focused on the items in the product structure. The specifications generally have a one-to-one correlation with these items. We have several choices when capturing the requirements for the interfaces between these items.

First, we could capture the requirements for each item in the corresponding specification pair. Alternatively, we could create a document that corresponds not to the items but to the interfaces between them, an interface control document (ICD). Where an ICD is used, one should be careful not to double-book the requirements definition. If the ICD is to be used throughout the life of the product, the corresponding specification pair should reference the ICD and not include the requirements. If the ICD is only used as a basis for reaching agreements on the interface requirements to put in the terminal specification pair, they should be deleted from the evolving ICD as agreements are reached and the requirements transferred to the specification pair until the time the ICD is a void when it disappears.

16.4.2 Design documentation

Engineering drawings are commonly used to capture the design solution to the problem defined in specifications. Drawings can, of course, be used to also capture the design of interfaces. These drawings could be released independently and/or included in what is sometimes called a Part II ICD that defines the requirements for the acceptance of the interface terminals. The ICD used during development, in this case, would be called a Part I ICD.

16.5 Traditional management of interface development

The traditional view of interface management is appropriate to a relatively stable world where the ownership of the enterprises involved in a development effort remains essentially unchanged over the life of the development effort. The management activities can be collected into three subsets as a function of the contractual relationships between the parties.

16.5.1 Internal interface management

An internal interface must be developed by the organization responsible. The same enterprise is responsible for both interface terminals. The program should assign interface responsibility to the teams responsible for the development of the things in the system as a function of the way the terminals touch these items, as discussed above in the context of innerface and crossface. All innerface should be the responsibility of the one team whose item is touched by both terminals. The teams should be held accountable for the crossface in pairs, but one of them could be made responsible for coordinating the interface development based on a receiver or sender rule.

This arrangement can be structured through several team levels on a large program, with each team responsible for all items and interfaces at their top item level as well as all interfaces subordinate to that level flowing down to their lower-tier teams held accountable for the subsets through as many layers of teams as necessary.

16.5.2 Supplier interface management

The development work accomplished by suppliers should be controlled through contractual relationships coordinated with an item specification and statement of work. The interface requirements should be clearly defined in the procurement specification and controlled through the contract. This is a case of a contractual relationship between the parties.

16.5.3 *Associate interface management*

It can easily happen on government contracts where the two parties to an interface do not have a contractual relationship between them. This happens when two contractors are under contract to the same customer. This is often called an associate contractor relationship. The common government customer should include interface management clauses in the contracts for each of the contractors, forcing them to cooperate in the development of a compatible interface defined in an ICD.

One of these contractors, or a third contractor also involved in the program, will be assigned the lead interface responsibility of developing an interface management plan and draft ICD. Both these documents should be offered to the parties to the interface being controlled for review and approval. These associate relationships also often involve management through a joint team staffed by representatives from the contractors and the common customer cochaired by the customer and the lead interface contractor. This interface control working group (ICWG) acts to adjudicate differences of opinion on these interfaces and is the final authority for the definition of the interface.

16.5.4 *Cross-program interface management*

Some companies and programs become involved in additional interface development problems because of the use of items in multiple applications. The gray item in the product entity structure illustrated in Figure 16.6 must be compatible with the internal interfaces corresponding to that application. But, if this same item must be used in three other applications as well, the interface development problem becomes significantly more difficult.

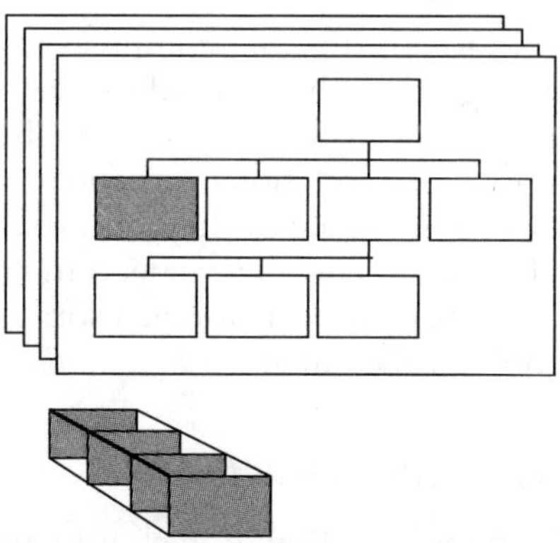

Figure 16.6 Cross-program relationships. (From Grady, J., *System Engineering Deployment*, CRC Press, Boca Raton, FL, pp. 147–156, 2000. With permission.)

In addition to an integration activity within each program to ensure that internal program interfaces are compatible, it is necessary to implement a cross-program integration activity to adjudicate the different demands of the several programs for these interfaces, with the goal being to develop a single interface that is used in the same way in all applications. Where this is not possible, maximum commonality and minimum uniqueness is desirable.

This effect can be realized in a large company characterized by a vertical integration of the product line. One division may provide all of the other divisions with engines or transmissions. To the extent that these units can be identical for similar loads and performance requirements, the aggregate enterprise will benefit from a cost perspective. This same effect can be realized as well when the final assembler wishes to purchase a common design for several products, as in the use of a common engine for several lines of automobiles. The general solution for this problem is an additional integration agent corresponding to the common item with representatives from the several programs involved.

16.6 *Management of interfaces within the context of ownership dynamics*

16.6.1 *A new reality*

In years past, companies tended to remain fairly stable relative to their ownership and employees tended to remain employed by the same company for a considerable period of time. Neither of these are characteristics of aerospace today and may not be for an indefinite period into the future. These factors introduce a new dynamic into the development of systems and, in particular, the development of interfaces within these systems.

The development of the things that compose systems will always be a clear one-to-one arrangement that can adjust to any of these changes fairly easily, but, in that interface development involves a two-to-one relationship and the organizational relationships between the parties responsible for the things can be in a state of flux, and that the development of cross-organizational interfaces has always been the most difficult part of developing systems, this dynamic encourages some concern for a potential need for a methodological change.

16.6.2 *The use of cross-functional teams*

No matter the development situation, one should apply cross-functional teams to the development of systems that are organized as a function of the product structure. Teams simplify shifts in relationships because there are clear relationships between the organizations, no matter how they may change relative to contractual or ownership considerations, and the product elements between which the interfaces connect. The organization of programs using the functional department structure will lead to increasingly difficult management challenges, because the interface relationships between the things for which the organizations are responsible are invariably very complex and difficult to manage.

16.6.3 Team relationship dynamics

For any one business situation, we covered the several kinds of relationships that might exist between two organizational entities during the development of a system. These techniques are effective when implemented in the context of the existence on the program of an overall system agent that clearly assigns interface responsibilities and is able to monitor and audit the performance of all the organizational elements working on the program relative to their interface responsibilities.

In addition, all the organizational entities in the program should accept the responsibility of the integration and optimization of the work and system elements subordinate to their immediate responsibility. Thus, at all levels, one finds effective integration efforts being applied across the cross-organizational interfaces.

Another characteristic of a sound program from this perspective is the existence of a comprehensive model of system interface either in the form of a schematic block diagram or an n-square diagram, where it is easy to see interface responsibilities because of the use of a responsibility overlay.

The relatively new element that creeps into the interface management process is instability in the business relationships between the organizations responsible for the cross-organizational interfaces.

Figure 16.7 identifies all the ways the organizational relationships can change between two organizations during the development of a system due to movement from one business relationship to another. Four kinds of business relationships have been included as described earlier, and the matrix intersections are labeled uniquely keyed to a description of the appropriate response on a program for each one.

In that two of the relationships are supplier oriented (supplier and other division), you might expect that there are only three kinds of responses to all of these possible changes. The key to flexibility here is a clear definition of all the interfaces all the time, no matter the relationships between the various business entities. Whether two teams are in-house, supplier/buyer, or associate teams, they should have a clear understanding between them of what the interfaces are between the

From condition \ To condition	Same division	Different division	Supplier	Associate
Same division		A	B	C
Different division	D		E	F
Supplier	G	H		I
Associate	J	K	L	

A Start treating like a supplier
B Start treating like a supplier
C Start treating like an associate
D Integrate into internal team structure
E Continue treating like a supplier
F Start treating like an associate
G Integrate into internal team structure
H Continue treating like a supplier
I Start treating like an associate
J Integrate into internal team structure
K Start treating like a supplier
L Start treating like a supplier

Figure 16.7 Organizational relationship changes. (From Grady, J., *System Engineering Deployment*, CRC Press, Boca Raton, FL, pp. 147–156, 2000. With permission.)

items for which they are responsible, and manage these interfaces for compatibility.

16.6.4 *Tools and documentation flexibility*

What may change as we move from one business relationship to another is the formality of the documentation of the interfaces unless we look into more effective ways to capture the interface requirements than those currently used. Current requirements database tools are really focused on the things in systems and do not adequately address the interfaces between them. So, one action that would greatly improve interface development amidst business relationship dynamics would be to have available effective interface management tools.

The ideal tool would allow us to capture the interface requirements relative to the interface model captured in the tool and the item requirements about the item model captured in the tool. This could be done in a fairly simple way of linking the requirements to either an architecture ID or an interface ID (see Figure 16.2). In that the tool would also know what the terminal architecture IDs were for any interface, it should be easy for the tool to drop the appropriate interface requirements into the specifications if they were to be documented using the specification pair pattern, or to drop them into an ICD otherwise.

Supported by this kind of tool suite, the business relationships between the enterprises could shift with considerable freedom and a stable model of the interfaces remain under control.

16.7 International effects

The development of systems to satisfy complex needs is very difficult in the best of cases, but it becomes much more difficult when program teams include people with different first languages, where the teams are separated by a considerable distance as can be the case on international programs, and item common interfaces across multiple programs are involved.

A program applying these arrangements must select a principal language and require that all participants master that language and provide for the best possible communications across the distances involved. Many product managers find that they must first travel to the distant place and meet the people involved. Thereafter, they can interact effectively via telephone, e-mail, fax, and videoconferencing. Without such personal contact, communication is much less effective. Development under these conditions can be made even more difficult when a very demanding schedule must be maintained. Difficulty of communications on programs should carry with it additional schedule slack, because it is a risk, rather than less.

Companies that move from domestic ownership to foreign ownership may also face special interface development barriers. The countries involved may have laws on the books that make cooperation difficult across their borders. Whereas a program might have been moving along well prior to the sale of the company, it may stagger to a stop upon the sale of the company.

Interface development is difficult under any circumstances even when the two parties to an interface are in the same company. If we wanted to design a very-difficult-to-develop interface, we would separate the two parties in time and space as well as provide their people different languages and cultures. The more these factors come into play for a particular interface, the more risks the program runs and must mitigate these risks with time and money slack.

16.8 Summary

The standard interface development techniques that have been used for decades are still useful, but contractors must be sensitive to the dynamics of business ownership as a result of downsizing, instability of ownership, and the special difficulties of international cooperation. These enterprise and program relationships will still yield to standard interface development techniques, but may have to be applied in a more flexible fashion than in earlier times.

Key elements in the management of evolving product interfaces during development are the use of cross-functional, product-oriented teams and clear definition of interface responsibilities. All interfaces should be carefully developed based on these clear rules of responsibility.

chapter seventeen

*Program work definition fundamentals**

17.1 Aggregate planning goals

In the aggregate, our enterprise must create a work planning environment that satisfies three important goals as follows:

1. Clearly state the responsibilities of the functional departments in terms of providing coordinated quality resources to programs (people, tools, and practices) that are continuously improved in a way coordinated within the enterprise to maintain.
2. Clearly state the work necessary to perform on a program so as to encourage understanding of the work at all levels on the program and minimize the management difficulty with the result that management intensity can be placed on smoothing current discontinuities and identification, and mitigation of future potential risks.
3. Minimize the program planning transform complexity to the end that everyone can contribute from a position of knowledge in pouring generic planning data into the program framework to create the minimum cost most effective program plan that results in customer best value and enterprise profit.

In order to satisfy these goals, we should clearly identify the entities of interest for functional departments responsible for the generic data and the programs and name them for easiest understanding by the people responsible in each case. If there must be complexity, let it be hidden from the people in computer databases as through mappings between one class of data and another. Let the people refer to things in the simplest possible way they are familiar with.

17.2 Generic planning data structure

On the functional side of the divide, we are interested in generic tasks (the functions of our generic process diagram) that must be accomplished

* The material in this chapter is from Grady, J., *System Engineering Deployment*, CRC Press, Boca Raton, FL, Chapter 11, 2000. Used with permission.

on many things on every program. We then allocated this functionality to functional departments and integrated the sum of those allocations to any one department to craft a set of department charters so that the departments are clear what their responsibilities are.

These functional departments are responsible for acquiring and training qualified people to do the work identified in their charter so they have to know what the aggregate program demand in the future is going to be for these people in order to permit them to match that demand. Thus, the program budgets must reflect back into the functional departments in some intelligible way to the functional managers.

In Figure 17.1, it is possible for the manager of the functional department that is presented with this information to plan on a particular head level for the next six quarters recognizing that this kind of chart always decays toward zero the further out you go in time from the present. What is shown is that the aggregate budget from programs A through E piles up to create an aggregate demand of the number of heads corresponding to the heavy horizontal line in the figure. This is the number of people that the department can support for the near term. If the department has fewer than this number, then the current work force will be overworked or required to work overtime to maintain program schedules. If the functional department has more people than this number, it may have to reassign personnel or in the worst case lay off or furlough some people until the demand climbs back up.

Finally, the functional departments have crafted process knowledge in process manuals such as a system engineering manual and coordinated with the contents of those manuals access to how-to knowledge that people can use to find out how to do the work required on programs.

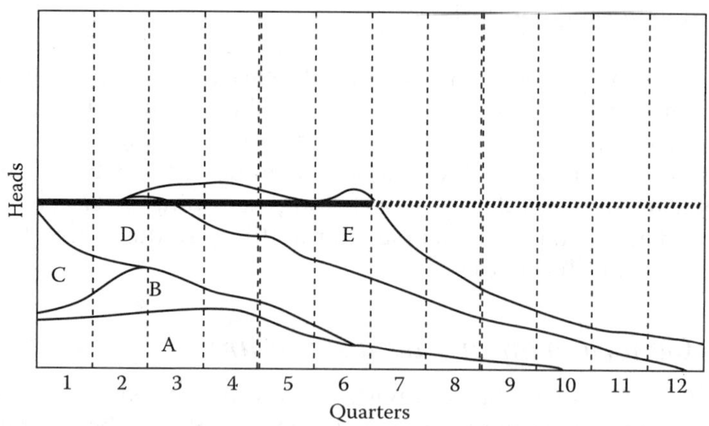

Figure 17.1 Aggregate functional department personnel demand curve. (From Grady, J., *System Engineering Deployment*, CRC Press, Boca Raton, FL, pp. 159–168, 2000. With permission.)

17.3 Program planning structure

17.3.1 Reviewing program management needs

Program management needs some way to identify work in terms of what has to be done, who is responsible for that work, a way to determine what portion if any of a particular work activity has been accomplished at any particular time, and where the program is relative to the use of budget and time. We will coordinate all work with the product architecture as noted in the last chapter using either the term work breakdown structure (WBS) or product entity ID synonymously and link teams to these same entities making a clear connection between work and responsibility. In order to satisfy the other needs, we shall need some way of interrelating time and goals to the work.

Our generic process and planning data will have to be applied to programs in a variety of ways as a function of where we are in a given program. A generic task may have a different flavor early in a program versus later in the same program. Two concepts used to characterize work in a timing sense are program phasing and program events. Phases are major pieces of programs contiguous in time. Events are instantaneous occurrences along the program timeline corresponding to milestones representing important program achievements or goals.

17.3.2 Program phases

Customers of large and complex programs, such as Department of Defense (DoD) and NASA, seek to control risk and the pace of the expenditure of funds by establishing a structure known as phases or stages. A first phase will have very modest goals to define the system functionality and concept, the related mission, top-level system architecture, and demonstrate that the solution posed is feasible with risks identified and controllable. A second phase may involve defining requirements for all of the things in the system and implementing designs, manufacturing and quality planning, and testing to show that the right product has been crafted relative to the requirements. A third phase would involve the manufacture of the product system, customer acceptance of the product, shipment or deployment of the system, and setting up initial operations.

Different customer agencies have developed different ways to break up the overall system life cycle into phases. Figure 17.2 illustrates a three-phase program in the interest of simplicity and coordinates with those phases the implementation of the enterprise generic process. In the first phase, the enterprise is developing concepts applying some of the tasks on the front end of its generic process. In the next phase, some of the work from the first phase must be repeated for two reasons. First, we become smarter as we go

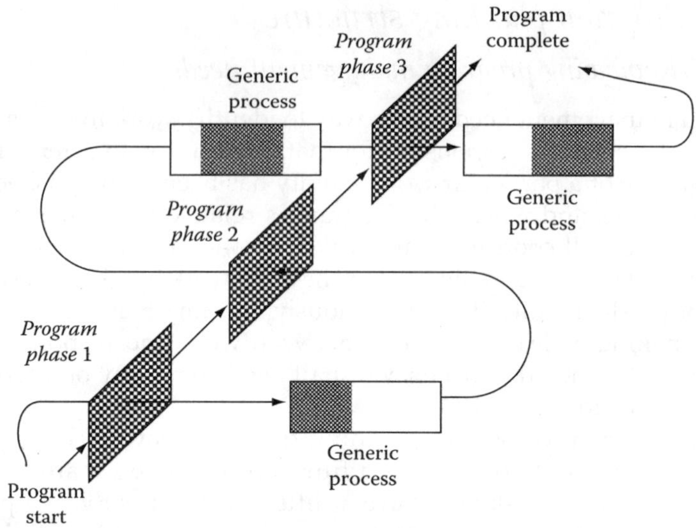

Figure 17.2 Phasing and process correlation. (From Grady, J., *System Engineering Deployment*, CRC Press, Boca Raton, FL, pp. 159–168, 2000. With permission.)

finding that as we know more about our proposed solution we would have done it differently if we had known the problems we have run into. So, we change prior decisions, ripple the results, and go on from that point. Second, in the second phase, we are expanding our definition of the product system downward in detail using the same decomposition and definition techniques from our generic process that we applied to higher-tier elements. So, generic process repetition, or as some people call it, iteration occurs.

Customers of large and complex systems insist on phasing structures because it helps them control risk. The overall problem they seek to solve through a contract with a developer is often too complex to clearly see the preferred solution in all of its details at one time. If a contract were let for development and manufacture of the whole system based on early studies, it would likely be the wrong system that was procured because we simply are not well enough informed about the possibilities and risks at that time. By breaking the whole program into phases or chunks, the customer and contractor are able to lay out clear and specific goals for a collection of work coordinated with a schedule and an amount of money. At the end of that work period, it is possible to clearly establish whether or not the goals were met one of which should have been to clearly define a set of objectives for the next phase. The next phase gets us closer to the overall program goals of a product system in the field performing as required by the customer and the program continues either until that objective is attained, or it has been proven that the system is not feasible or that the current contractor is not the right party to implement it.

The enterprise in applying its generic process to the phases does repeat some activities as noted from phase to phase but over the whole program period, they execute all of the steps in the process moving the process centroid from early activities to later activities as the phases progress.

In planning a specific program, we have to be careful to select work from our generic process that is appropriate to the phase. It would not be appropriate to include detailed design work in an early program phase, for example. Nor would it be appropriate to include mission analysis in a later program phase estimate. If our generic process includes the major reviews that our customer respects, such as system requirements review (SRR), system design review (SDR), preliminary design review (PDR), critical design review (CDR), and so forth, and these major reviews act to terminate the program phases, we can use those major reviews in our generic process to bound the process steps that should be selected for a particular phase. The reader should note that the generic process diagram in Division 3/Exhibit H/Appendix A does include these major reviews that are clearly fed by prerequisite works.

17.3.3 Program events

Each phase may have spaced within them one or more milestones also referred to as events. These are instantaneous happenings corresponding to some planned program activity having been achieved or toward which we are working. Every program work element at every level of indenture should terminate in an event, and it may be appropriate for us to identify intermediate events as well within a particular work element. Given that a phase is an element of work, it too should terminate with an event and may have intermediate events.

Events provide opportunities to status our program progress. It should be very clear how one determines when any work element has been completed, when we can claim the terminating event or milestone. So, every task should have a terminating event and an associated completion criteria by which any reasonable person could judge that the task is complete and about which all reasonable people could agree. In Chapter eighteen, we will use events as one of the program work element identifiers. As noted in Figure 17.3, these events may be terminal (event 2) or intermediate (event 1) types.

We can describe the work illustrated in Figure 17.3 by reference to these event numbers. For example, if this complete work element was identified by the designation 1200-02, we could identify the part of the work between the beginning and Event 1 as 1200-02-01 and the latter component as 1200-02-02 where the latter symbols by convention mean the work element ending at the event defined by the last two numbers

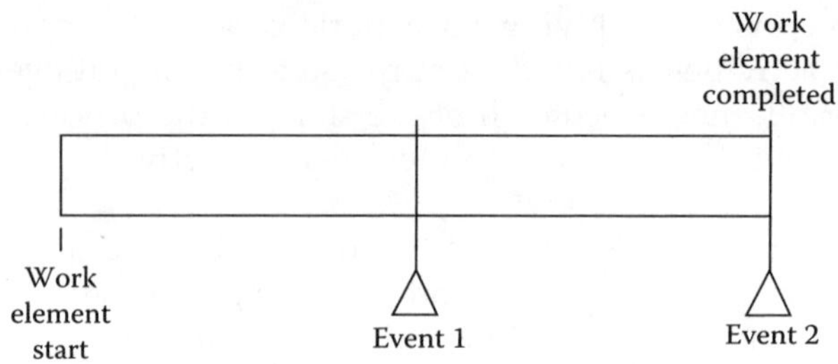

Figure 17.3 Program events. (From Grady, J., *System Engineering Deployment*, CRC Press, Boca Raton, FL, pp. 159–168, 2000. With permission.)

(01 or 02 in this case). By convention, this work element begins at the previous event or the beginning (Event 0) as appropriate.

17.3.4 Teaming phenomena

When developing systems characterized by great complexity and scope, we find that it is not efficient to group everyone working on the grand problem together. It is more efficient to decompose the grand system problem into a series of smaller problems, another example of chunking complexity for compatibility with human intellectual capacity. Our system decomposition process and IPPT teaming concept are driven by this phenomena. There will be system-level work required to accomplish the decomposition activity driving the team identification and each of the teams we establish will have to repeat some of this same work to identify subordinate items. This same process can continue through several levels. Figure 17.4 illustrates this expanding process in the context of Team A1 that will be responsible for item A1 of a system A composed of six major elements each under the responsibility of an IPPT.

Each of these IPPT will accomplish essentially the same work to define their system element and decompose it to lower levels of indenture. If the system is sufficiently complex, it may be appropriate for IPPT A1 to act as an integration team for four subordinate teams (A11, A12, A13, and A14). These teams could involve the development of several lower-tier elements. For example, team A11 might be responsible for developing items A111, A112, and A113 through principal engineers drawing support from specialists on the A11 IPPT. In each of the teams above the principal engineer level, there is both IPPT work and integration work we have placed in the context of a PIT.

The system will very likely involve the procurement of items at all levels of the system architecture. The gray blocks in Figure 17.4 indicate procurement items. Item A4 is being procured, and it is at a very high level in the system. This could be an associate item, a customer furnished, of

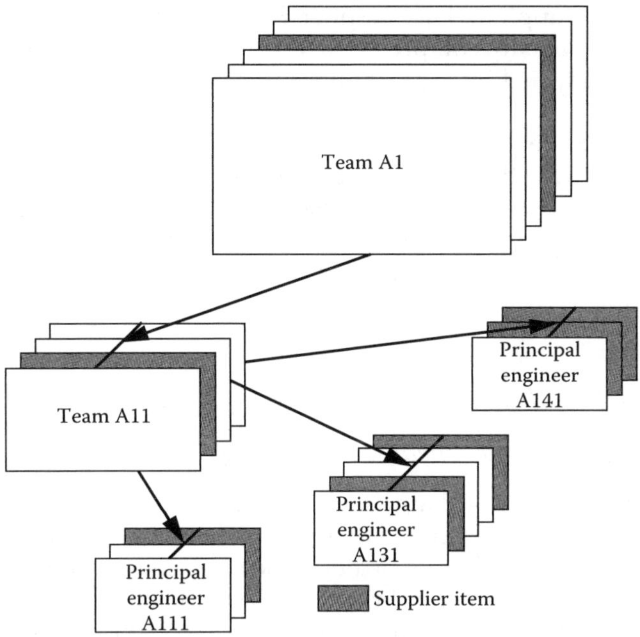

Figure 17.4 Team responsibility structure. (From Grady, J., *System Engineering Deployment*, CRC Press, Boca Raton, FL, pp. 159–168, 2000. With permission.)

a major subcontract item. It may make sense to have a representative of the supplier involved as a member of our PIT to encourage the compatibility of interfaces. We might also establish an IPPT supplier team A4. It is important to note that the way the program manages the teams makes little difference whether they are internal or supplier in nature. In either case, the team will have responsibility for a specific architecture item that has associated with its WBS, planning and scheduling data, specification, and budget components.

The identification of an item as a supplier item significantly effects the continuing decomposition of the subordinate product structure. The subsequent work performed in the company after this identification of a particular item will follow one of three paths generally as a function of the contractual relationship between the company and the supplier.

1. Customer-furnished equipment should not require any development work other than the influences it may have on the product under development. The customer is essentially the responsible team providing the item.
2. A supplier item may be procured to a specification and statement of work (SOW) requiring considerable in-house coordination work

with the supplier. The supplier must design, manufacturer, and test the item to comply with the specification content and the process applied by the supplier must be compliant with the SOW content.

3. A supplier item may be procured to a set of engineering drawings and a SOW. In this case, the supplier must manufacture an item that has already been designed by the buyer.
4. A supplier item may be purchased off the shelf as is and either used in that configuration or altered in accordance with an altered item drawing.

17.4 Cost and schedule estimating

17.4.1 Task cost and schedule estimating

When a new program is planned, we must identify the cost and schedule characteristics of that program whether our customer requires a rigorous identification of these factors or not, unless, of course, money and time have little value for your enterprise and its customers. There are very few situations where this is a valid attitude. The enterprise needs to know what resources will have to be applied to the program.

17.4.1.1 Generic task estimating

In Chapter twelve, we discussed the identification of appropriate cost and schedule figures for each generic task in our process diagram. These numbers could be absolutes or probabilistically stated but generally will have to be adjusted for any program based on program-unique factors. On a large program, we may have to do requirements analysis 43 times based on there being that many things in the architecture that require specifications in several levels. On an actual program, some of this work would be done at the system level by the PIT. Other work would be done at IPPT level. The basic task would be similar in each case but some of these tasks are going to be more difficult to do than others. In some cases, a similar specification already exists that will only have to be adjusted in seven paragraphs. Another specification will cover an item that is pushing the technology in three areas of concern and no one has ever built one of these items before. The requirements analysis job for the latter will take longer and be filled with more serious risks than that in the former case.

Nonetheless, our generic planning data should have established for it cost and schedule figures in some form that are based on actual performance on programs in the past. When you start this process, you may have to estimate the generic estimates, but data should start to accumulate as programs run their course. This is one of the big advantages of a generic process description that is applied on every program. The task work on one program can be compared against the task work on another program yielding real data of value.

People trained in the estimating field, the author is not, can help identify ways to adjust generic estimates for particular program situations. They may encourage the application of different ogives, shape factors, to task estimates based on program-unique conditions. An estimate could be formed by the generic factor acted upon by a program complexity factor and possibly other factors defined by estimating people.

The generic estimates should be contained in a database linked to the functional planning strings (FN-DX combinations). As these functional strings are pulled into program plans, they bring with them their cost estimates. One of the problems in this process is the attitude held by many estimating organizations and their management that denies the working troops any access to historical cost data for proprietary reasons. Perhaps this cannot be shared with everyone in the organization but someone familiar with the technical difficulty is going to have to know. These numbers should not be simply generated by an estimating database with no technical input.

17.4.1.2 Bottom-up program estimates

The total program task–oriented cost estimate can be assembled from bottom-up estimates, and that is the process essentially described above. The total is simply the sum of all of the parts coordinated with each generic task brought into the program. The schedule is simply the critical path through all of the tasks assembled from generic data about the things the system shall consist.

17.4.1.3 Top-down program estimates

Some programs must be planned from the top down. For example, the customer may identify the amount of money they have available within a particular timeframe. These factors probably define the aggregate program cost and schedule directly. From this point, the enterprise must determine what it can afford to do for the customer within those constraints. Limits may be set on the depth of development based on subsequent funding for an additional phase that would complete the job. Limits may be placed on the breadth of the development effort by developing, for example, one of three elements known to be needed.

The money can be allocated down through the architecture like weight to establish programmatic constraints on the development process. In so doing, cost and/or schedule margins should be included for items coordinated with significant risks.

17.4.1.4 Estimating granularity

Throughout our discussion of program planning, we have noted that one must have some understanding of the product architecture because all of the program planning is linked to it. So the question arises, how do you plan a program when you have not yet determined what the architecture is. Perhaps that is the purpose of the program in its first phase. Well, we have choices in the granularity we apply in the estimation of cost and schedule. One extreme condition is thorough knowledge of the architecture down to the component level, the selection of family collections of these things around which to coordinate IPPT, and plan at the team level with team planning extended below that level. The other extreme is at the system level as one pot of money and time. The appropriate level of detail corresponds to what is known. A first phase study contract will have to be done at a high level while an engineering and manufacturing development phase could and should be planned in more detail.

We have many choices in our planning granularity. We can select the product entity level, the generic process functionality level, the functional department level (engineering versus the string of reporting departments or even the next lower tier, for example), as well as the number of events over the program run. Obviously, this is an art form. Probably the biggest mistake made in planning programs using a comprehensive work cataloging systems like that suggested in this book is going into too much detail. This work should be planned from the top down with a halt at a point where the program can be managed with adequate visibility.

17.4.2 Integrated master schedule

The integrated master schedule (IMS) is a time expression of the integrated master plan (IMP). At some level of indenture, planned tasks in the IMP are plotted in time onto a schedule. When the IMP is being constructed, it is important that we not burden those supplying the data with specific dates. That was the reason for the use of program events. Detailed planners need not worry about relative dates of tasks, rather only that they get them into the right buckets of time defined by program events.

A scheduling expert then must take the details of work definition and place them within the event dates determined respecting the task precedents in more detail than those who simply dumped them into the time buckets. Here, the scheduler could fail to understand subtle time relationships between tasks and may have to aided by knowing technical people.

The initial issue of the IMS should be a system-level schedule. Based on this document, the IPPT should develop a draft schedule for their activity supportive of the IMS. Where conflicts arise, they must be dealt with providing adjustments in the IMS or applying management skill to the team schedules to remove those conflicts. Approved IPPT schedules should be combined into the IMS.

As the schedule comes together, scheduling people should identify the critical path and apply schedule slack to this path to the extent possible.

17.4.2.1 Program start and complete framework

The beginning of the IMP rests on program start and complete dates where the development is date driven by other considerations such as a needed initial operating capability (IOC). Otherwise, the customer may recognize that the developing enterprise will have to determine the program time span as they estimate the degree of difficulty and risk involved.

17.4.2.2 IMP task level selection

Select the generic tasks based on the program phase and the degree of knowledge about the application. For front-end jobs, select higher-level tasks. For later phases, select more detailed task levels. The same goes for the functional department level selected in the generic planning strings. Avoid going into any more detail than necessary in this selection.

17.4.2.3 IMP/IMS event selection

Begin by picking major review points as intermediate events. For example, on a DoD program one might pick SRR, SDR, PDR, CDR, and FCA as the events. People supplying the detailed planning data will be able to relate to these time buckets and place within them the right work collections from their functional departments.

If necessary, events can be placed intermediate within the major events but be very careful. The planning cataloging codes expand very rapidly, and once you have started it is hard to get back to something more sensible.

17.4.2.4 IMP import

At the time scheduling begins you have a clean sheet of paper or computer desktop marked with start and end dates. The event dates should be roughly determined and added recognizing that they may have to be adjusted. Every scheduler wishes for a rubber sheet upon which to do this work and available tools accommodate this to some extent. Now the tasks identified by detailed planners from the several functional departments coordinated by the program manager and team leaders (if appointed, otherwise a system engineer) must be imported into the scheduling time frame based on their completion events.

In the process, conflicts will arise and must be coordinated with the contributing parties to remove them. Some system planning talent must be applied to this process because there will be two kinds of errors in the data supplied by the detail people. First, some necessary tasks will not be identified having been missed by the planners. These can be identified by noting the input needs of others and ensuring that they all appear as someone's output. Second, planners will through an error of a careful plan to build an empire insert unneeded tasks in the plan. These can often be located by tracing task outputs to make sure they go somewhere, that someone is planning on using them.

17.4.2.5 IPPT scheduling expansion

The complete program planning job on a large program is too much for one person. The job should be granularized. System-level planning should go down in depth only to the level where IPPT work is identified on Gantt or network charts. Leaders for these teams should be identified and expected to expand the plan through their areas of responsibility. On very large programs, two or three layers of teams might be applied, and this same technique is applied down through the structure with parent team approval up through the chain.

17.4.3 Material cost estimating

There are two cost elements for any program, the task-oriented cost driven primarily by program personnel needs and the material cost based on the things and materials that must be purchased and folded into the product in production. The material cost should be driven by a clear understanding of everything that must be developed including deliverable product and test articles.

A system engineer should be assigned the responsibility for identifying all of the things that must be produced or used in the development of those things produced. Some of these things will properly be funded from capitol while the remainder is a proper program cost element. The product material cost should be based on the architecture of the product at some level of indenture. Design engineers will have to identify preliminary design features that help estimating people to make estimates of materials required.

Special test articles is a common place where cost is not captured leading during implementation to cost overruns. The person assembling the product architecture should include not only the product entities but also all of the elements that will require material inputs. This may include instrumentation equipment as well.

17.5 Making the transform between generic and program models

We have crafted a fine definition of our process and our identity in prior chapters that we have convinced ourselves will work in a real-world program (well, at least the author has been convinced). There are two program characteristics that we have to handle before we should all accept the model as constructed. These realities are customer interest in phasing structures and the effects of teaming as recommended throughout this book. Finally, we will, in this chapter, construct two methods of cataloging work on programs, the first is the United States Air Force (USAF) integrated management system approach and the second is a method evolved by the author referred to as the JOGSE method.

The U.S. Air Force developed a marvelous work identification system in the early 1980s and applied it to several programs, F22 Fighter Aircraft Program among them. They did not include a way to link it to an in-house generic work identification system but the author did so in *System Integration* (Grady, 1994) and *System Engineering Planning and Enterprise Identity* (Grady, 1995). Some of this work is repeated from those books. Some problems with this linkage caused the author to pursue other alternatives over a period of four years and the conclusions from that search are included in Chapter nineteen. A third method is also covered in Chapter nineteen that entails only a WBS and SOW. The reader may come up with a fourth method in keeping with the way that your company captures identifying data and the finance system in use to identify program work units.

17.5.1 The USAF integrated management model

This model identifies six entities for each element of work on a program. The number strings formed by these six identifiers uniquely identify every piece of work in a plan called an IMP and program schedule called an IMS. Table 17.1 offers an example of the structure.

A complete identification sample would then be 031-1100-05-03-03-01. These identifications can be loaded into a database with fields as noted in Table 17.1 and assigned in contiguous blocks to IPPT in that the teams will be assigned as overlays of the WBS that this model is based upon. In previous books, the author has shown how you can link up the significant accomplishments to your internal generic model of work, but it did require an inefficient and redundant identification system. The alternative structure offered covers that linkage and does not appear to violate any aspect of the USAF system.

Under this system, all work related to product entities is defined in the IMP in the numerical sequence noted above, but other work that is difficult to map to product entities is described in narratives in the IMP. As covered elsewhere in this book, the author encourages that narratives not be included in an IMP, rather referenced in the IMP to generic process documentation.

Table 17.1 USAF Model Structure

Identifier	Example	Meaning
WBS prefix	013	A prefix can be used to collect work into different categories of interest like nonrecurring and recurring
WBS number	1100	WBS identification
SOW	05	A paragraph in the SOW under 013-1100 identifying a fairly high-level element of work
Event number	03	A major program milestone/event.
Significant accomplishment	03	One of one or more work elements contributing to work element 013-1100-05-04
Accomplishment criteria	01	Identifies one or more ways one can tell when the work is accomplished in element 013-1100-05-04-03

Source: Grady, J., *System Engineering Deployment*, CRC Press, Boca Raton, FL, pp. 159–168, 2000. With permission.

17.5.2 GSD planning model

The author has developed a method of identifying work using six identifiers as listed and explained in Table 17.2. The model is named the GSD method. The elements of this system are consistent with the content of this book series and involves a minimum of mapping between different identification systems. It preserves the WBS and SOW as intermediate planning structures if one desires to identify them as such. As in the USAF model, the IMP would be created from alphanumeric ordered work identification strings like the one described in Table 17.2.

A full identifier then would be P5-S2-E3-A11-F412111-D2162. The added program and stage identifiers (not included in the USAF model) permits all program planning data to be contained in a single enterprise database, if desired, with each record program-linked by this field. Within the context of a single program, this identifier could be dropped, of course, simplifying the work identification. The prefix letters could also be dropped but are applied in this book to avoid any more confusion than necessary with these long strings.

The reader will note that the identifier does link to the functional department providing the personnel to do the work on the program, and this linkage offers the functional department managers to skim all of the records in an enterprise database linked to his or her department generating a labor demand curve in time looking into the future. Thus, the manager knows what the personnel demands are going to be for his or

Table 17.2 GSD Planning Model Structure

Identifier	Example	Meaning
Program	P5	The fifth program in house using this method
Stage (phase)	S2	The third phase in the customer's life cycle model
Program event	E3	Major program milestone/event as in the USAF model
Product entity/WBS	A11	The product entity ID as explained in Chapter fifteen
Generic function	F41211	Generic work identifier from the Division 3/Exhibit H/Appendix A, Figure A-1 process model
Functional department	D2162	The requirements analysis department will provide the personnel to accomplish this work on the program

Source: Grady, J., *System Engineering Deployment*, CRC Press, Boca Raton, FL, pp. 159–168, 2000. With permission.

her discipline on all enterprise programs. In some cases, the functional manager may be able to work with one or more programs through the PIT to adjust work scheduling to smooth the demand peaks. In other cases, it may be necessary to hire more staff in a timely way or consider ways to trim staff during down periods.

The system also links to the generic work definition as in F41211 (functional analysis) around which the enterprise can have constructed estimating models based on past performance. In Chapter eleven, it was shown how the enterprise can catalog all of its work using functional strings (F-Strings) such as F41211-D2162, which is all work that the requirements analysis department (D2162) is responsible for performing on enterprise programs under task F41211, functional analysis. In the GSD method, we can map F-Strings, already predefined, to program strings such as P5-S2-E-3-A11 that may represent all work done in stage (phase) 2 on program 5 between events 2 and 3 (terminating at event 3) for product entity A11. As program planning strings are assembled, they may be mapped to F-Strings with all functional department resources associated with those strings pulled into the program.

17.5.3 Correlation between models

Table 17.3 shows how the two work identification systems align as well as how they do not. The GSD model is oriented toward the enterprise perspective whereas the USAF model is primarily focused from the

Table 17.3 Model Correlation Matrix

USAF identifier	Example	GSD identifier	Example
NA		Program	P5
NA		Stage (phase)	S2
WBS prefix	013	NA	
WBS number	1100	Entity/WBS	A11
SOW	05	Generic function	F121
Event number	03	Program event	E3
Significant accomplishment	03	NA	
Accomplishment criteria	01	NA	
NA		Functional department	D2162

Source: Grady, J., *System Engineering Deployment*, CRC Press, Boca Raton, FL, pp. 159–168, 2000. With permission.

acquisition agent's perspective. The few differences could be closed fairly easily such that an enterprise could apply either model. We could add a prefix identification between the program identifier and the product entity identifier of the GSD model removing one of the differences. The significant accomplishment of the USAF model is actually satisfied by the functional department field in the GSD model if one followed the author's application of the significant accomplishments identifier to link the program work to functional departments as discussed in Chapter eleven. The GSD model does not include an accomplishment criteria in the work identifier but the database using the identifier as a key field set should include an accomplishment criteria field as well as a task description, name, and other data.

chapter eighteen

Customer life cycle acquisition models

18.1 The general situation

Where an enterprise deals with large customers that insist on managing their acquisition process, whatever method the enterprise may devise to attain a common process, it must permit mapping of that common process to the work elements the customer wishes to identify and through which it intends to manage a program. If the enterprise is a commercial firm not serving large organized customers, then it is free to devise whatever programmatic, phasing, and event infrastructure it wishes. If the enterprise serves a large organized customer with a history of organizing large programs, such as the U.S. Department of Defense (DoD), they will have to subscribe to that customer's methods of phasing programs. If they serve multiple large organized customers, such as DoD, National Aeronautics and Space Administration (NASA), and Federal Aviation Administration (FAA), it may be necessary to evolve the capability to map enterprise common process to multiple external program planning structures. This chapter exposes the current DoD, NASA, and FAA acquisition life cycle models as a preliminary to mapping enterprise common process work to them in Chapter nineteen.

18.2 Department of Defense

Figure 18.1 illustrates the DoD acquisition life cycle model established in the acquisition reform process accomplished in the 1990s. Four major phases have been established, each composed of two or three subphases terminated by a major review that determines to what extent phase goals have been satisfied and the nature of any remaining risks as a prelude to deciding whether to continue as planned, terminate the program, or modify the future plans. Each subphase is identified with a stage ID, such as S31, for Integration. These are the stage (phase) numbers that will be used in the GSD work identification strings.

Figure 18.2 compares the new DoD acquisition life cycle phasing with the previous one and coordinates the major program events between them. The earlier model only addressed systems acquisition and did not

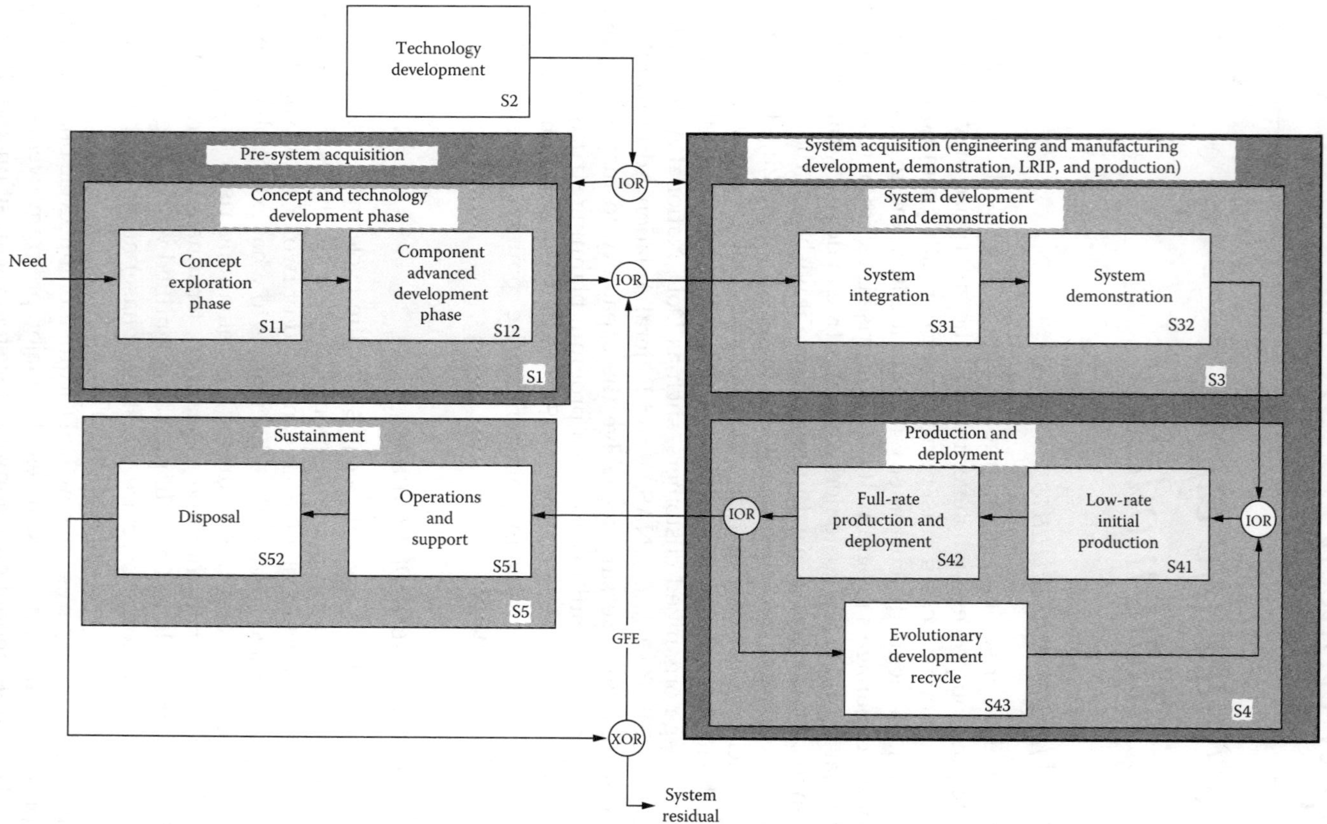

Figure 18.1 DoD acquisition life cycle model.

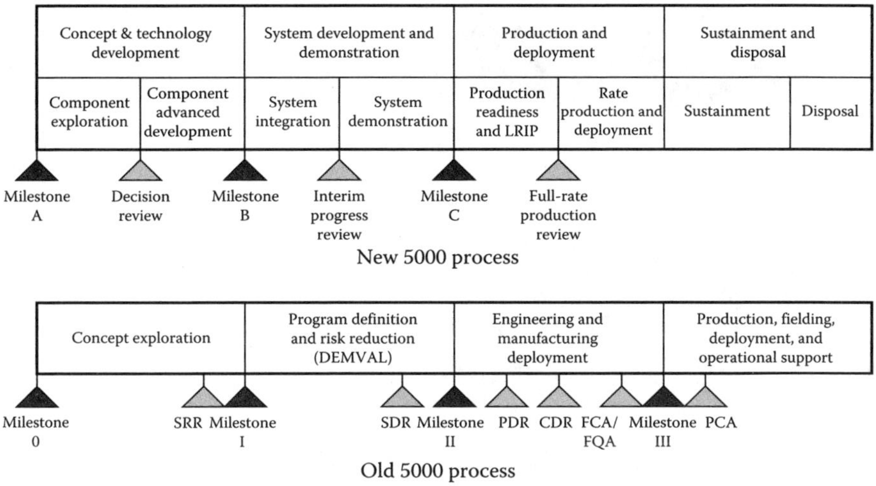

Figure 18.2 Life cycle process comparison.

address the total acquisition system life cycle. The old one treated evolutionary approaches and innovations as "nontraditional" excursions, whereas evolutionary development is a key element in the new one. The earlier approach endorsed tailoring of standards but provided no amplifying guidance to assist acquisition strategy development. Firm decision criteria was also added to each major decision point in the process. The references to 5000 in this figure are from a series of DoD documents that define this process, such as DoD I 5000.2.

18.3 National Aeronautics and Space Administration

Figure 18.3 illustrates the current NASA acquisition life cycle process. As in the DoD case, the author has assigned stage numbers to the steps in the NASA process. Clearly in this case, it is possible to map common process work to these stages, as it is in the case of the DoD model. Major program events can be inserted into the program time span for any stage (phase).

18.4 Federal Aviation Administration

A third large organized customer with a preferred life cycle acquisition process defined is FAA. This process is illustrated in Figure 18.4. The author has numbered the stages again for referencing in program planning strings.

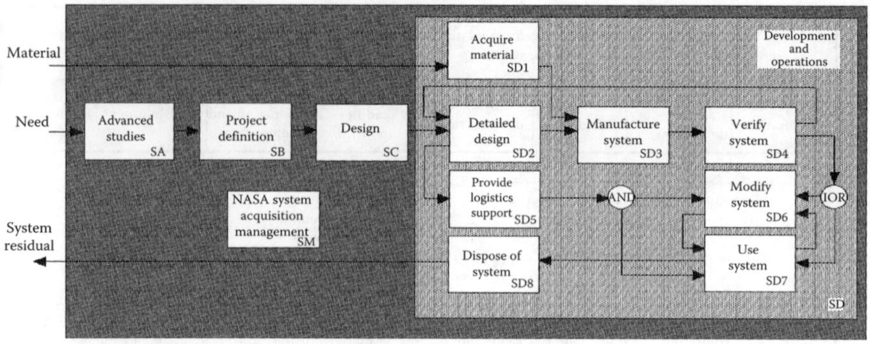

Figure 18.3 NASA acquisition life cycle model.

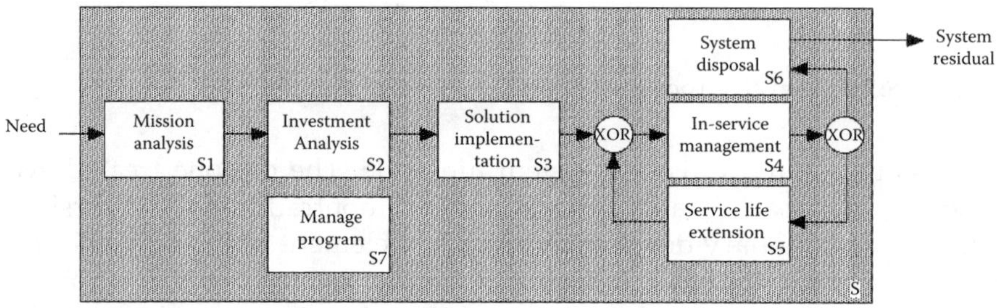

Figure 18.4 FAA acquisition life cycle model.

18.5 Program phasing linkage

The USAF integrated management provides no phase linkage whereas the JOGSE method does. In the later approach, work is identified in a string, such as P5-S3-E2-A-11-F4121-D2160, where S3 identifies a particular program phase. In the three life cycle models covered above, this work would occur in the following stages if the work were done in programs for the indicated customers: (a) on DoD programs, this would be the System Development and Demonstration phase, as illustrated in Figure 18.1; (b) on a NASA program, this would be reidentified as the SC design phase, as shown in Figure 18.3; (c) on an FAA program, it would be the Solution Implementation phase, as seen directly above in the figure. Other large customers, such as the Department of Energy and other governmental agencies, have similar phasing approaches to which enterprise work can be mapped.

The reader can see that these work strings used in the GSD method, such as the one identified in the previous paragraph, are composed of two components. The first one is program oriented and the second one is enterprise oriented. An effective program planning process within an enterprise with a common process defined in FY-DZ strings can be created using the simple three-step process: (1) maintain the enterprise common process work catalog with FY-DZ strings, (2) identify program work in the form PU-SV-EX-AW, and (3) map functional common process strings to program strings. Chapter nineteen explains how to do this.

chapter nineteen

Program planning structures*

19.1 The ultimate requirement and program beginnings

The customer's ultimate requirement is their need, a simple sentence or paragraph that succinctly describes the needed product in terms of its effect on other systems, the environment, or both. This need may be stated by a Department of Defense (DoD) customer as a new threat that must be defeated, an ordnance transportation and/or delivery demand, or an undersea war fighting capability, for example. It may also be phrased by a commercial concern about their hoped-for customers. On the surface, it would appear that the fundamental difference here is that the commercial company must understand its customer base and their needs well enough to know what will sell whereas the military contractor is told what is needed by their customer in a request for proposal (RFP). Actually, every healthy DoD contractor is in the same boat with the commercial companies and must be constantly examining marketing possibilities, developing new ideas that their customer base may find useful, and studying their customer's current situation and potential needs. Also, the DoD contractor must increasingly work to understand all of the stakeholders for a given system, possibly even extending to the population at large. While the planning methods discussed here can be used for any program, the DoD will be used as an example throughout this chapter.

The phrase "customer need" need not scare away those interested in applying this process to commercial practices. In commercial business, it just may be more difficult for the company to find out what the customer's needs are. Success may require good market research, good intuition, lucky guesses, rapid development, and a very fine pencil for cost and schedule control. Whether the system is of a military or commercial nature, no one need respond to the customer needs; however, they become aware of them, with an ineffective, ad hoc, hit-and-miss process. The planning process included in this chapter can be applied to government or commercial development programs. The first step in this organized process is to understand clearly the customer's need.

* The material in this chapter is from Grady, J., *System Engineering Deployment*, CRC Press, Boca Raton, FL, Chapter 12, 2000. Used with permission.

In the structured, top-down development model encouraged in this book, this need is expanded into a system operational requirements document, system requirements document, or system specification through requirements analysis, system modeling and simulation, mission and operations analysis, environmental and environmental impact analyses, and logistics and basing analyses early in program phasing. In addition to the system requirements identified in these early phases, the system architecture is also defined based on the functionality needed to satisfy the need, and it is used to fashion a product work breakdown structure (WBS), a hierarchical arrangement of product material and services cost elements. A DoD program may use some derivative of canceled MIL-STD-881 as a guide for the WBS but the evolving product entity structure based on the allocation of needed functionality should be used as the principal input for the WBS. The customer's needed functionality must drive the system architecture, and therefore the WBS, and not the finance community and their inflexible cost-tracking computer implementations.

The program planning work must begin in the proposal period, or early marketing efforts in a commercial situation, and continue during program execution. Where a proposal is required, the management volume or section should reflect much of the content of the program plan. In some cases, the contractor will be required to provide a formal program plan, system engineering management plan (SEMP), and/or integrated master plan (IMP) with the proposal in draft, preliminary, or final form. In the commercial situation, a new product is a good reason to frame a new development plan or update an old one for new conditions, however simply it may be stated. Granted, there are some commercial product lines richly dependent on serendipity exposed through basic scientific research, an uninhibited creative mind, or good intuition where a degree of nonstructure may be conducive to success. But, even in these situations, once the product possibility reaches the conscious mind, its development and distribution can fit into the pattern described here with some tailoring.

At the time this book was written, the world was unwinding from several decades of fierce competition between the East and West involving an intense system development rivalry to develop the unstoppable weapon system and to counter the other side's systems. This competition energized a search for technology that sometimes resulted in new systems requiring a clean sheet of paper approach with several facets pushing the state of the art simultaneously in several directions. The systems approach matured from this clean sheet of paper environment, but it can be applied effectively on programs entailing mild to massive reengineering or modification of existing systems to accommodate new or changed conditions.

In modification cases, the system engineering products described in this book may or may not have been created when the system was new or they may have been lost or disposed of long since if they were. So, even if a world-class system engineering job had been done on the original system, the results of that work may not be available for use in the changing of that system. In these situations, what can be made clear is the boundaries of the system in terms of its architecture. We can then determine what new functionality is needed and which of the elements of the system architecture must be modified, deleted, added, or replaced to achieve the new goals set for the changed system. Once this is known, we can apply essentially the same system engineering process to the development of the affected items as we would apply to the elements of a new system. It may be a case of working from the middle out rather than from the top down, but it can be done in an orderly process following the tenets of this book.

Several years before this book was written, U.S. Air Force customers had been required to follow MIL-STD-499A in the implementation of a system engineering program. This standard required contractors to develop an SEMP telling how they intended to perform the technical and management activities required by the contract. Some contractors wrote these plans as extensions of the proposal in the same sales language used there rather than as a plan for their own guidance in program execution. This pattern of behavior was a mistake. Whether a customer requires a management plan or not, your company needs one, and it should be written by your company for your company based on a marriage of your customer's needs and your capabilities. This is a fairly complex integration process, linking customer needs and contractor capabilities, composed of many specialized component parts.

19.2 Program plan tree

We would all doubtlessly accept that a specification tree is necessary on a large program to introduce order into the product requirements development effort. What is not as universally accepted is a similar tree for program plans that captures the requirements for the development and production process. All too often programs are implemented allowing planning documentation to be autonomously prepared by the several functional departments (engineering, manufacturing, finance, etc.) contributing work to a program. These plans may be generic company plans or procedure manuals applied to the program or specifically written for the program. Many companies become caught in the trap of trying to be totally responsive to every customer's initially stated process requirements to the extreme that they redesign themselves for each customer in terms of these plans. A future-looking company will apply a continuous process improvement concept to their generic procedures in combination with a rigorous customer procedures tailoring effort on each program to take advantage of the practice–practice–practice notion that world-class athletes use on the road to greatness.

Autonomous, functional department planning, disconnected from program requirements, is the target in this chapter. Program process plans should be architected just as the product system requirements and components should be. Plans should exhibit traceability from the top-level plan down through the lower-tier plans. A structured, top-down planning process will ensure mutual consistency of all of the program plans with a minimum of surprises during program execution. In the process of preparing program plans, it should not be necessary to come up with a new program design for each proposal and program. It should not be necessary to redesign your company for each customer. We need to find out how to apply the practice–practice–practice technique to our work through careful program planning for specific programs and apply continuous process improvement to our methods with a long-term view. Company personnel should be applying the same proven process, incrementally improved in time, to each program and in the process become expert in their specialized disciplines and be able to accomplish needed work at less cost and shorter time frames. This should lead to reduced customer cost and a more competitive position and/or bigger profits.

Very little of the business that a company tries to gain through proposal or marketing efforts involves radically new initiatives. Most of our energy is applied to prospects that are close to our historical product line. Therefore, most of a company's procedures and plans should apply in any new program. This is especially true if your company already does apply an energetic continuous process improvement program.

Given that we are organized in a matrix structure as discussed in Part three, our functional departments should have procedures covering how they perform their function on programs as members of cross-functional teams. The program must knit the functional methods into a coherent process appropriate for the particular product system under development as appropriate to the development phase. In this process, programs should not be allowed to substitute alternative processes creatively without acceptance by functional management because the functional departments should be deploying the very best methods they have developed over time based on continuous improvements fed by lessons learned from prior program experiences.

There are, however, two sound reasons for permitting programs to deviate from the current best practices. First, it is through program implementation that improvements can be developed and tested. A particular program may be asked to experiment with a particular technique in defining interfaces, for example. Perhaps the company's history is to use schematic block diagrams, and the proposition is that program XYZ will use n-square diagrams instead. The second reason is that the customer may have a valid need for a job to be done differently than our current best practices cover. Perhaps the company uses a particular computer program and related procedures to capture logistics support analysis data. The customer may have a big investment in capturing the data in a different computer data structure and have a perfectly valid reason why they need to collect different data than your system will address. You will simply have to adjust your practices to the customer's in this case or run the risk of loosing the competition. In the process of doing so, you may find improvements that can be woven into your preferred practice. But the suggestion is that this should be the exception and the rule should be to follow internal procedures while incrementally improving them.

Given that we all accept that programs should be conducted in accordance with prepared plans for each activity, what plans are needed? At the top of this set of program plans rests the program plan. The program plan can be very simple giving the overall schedule in very broad terms, the customer's need, ground rules and policy, top-level program organization and responsibilities, and reference to other documentation. Figure 17.1 is based on the integrated management system but includes an SEMP and subordinate plans to satisfy the IMP requirement for narrative material covering certain planning areas. Figure 17.1 plan tree illustrates one way to satisfy the requirements of the integrated management approach. The indicated plans provide requirements for the process that will result in the product system. They should be mutually consistent, and this can be demonstrated by establishing traceability between the plans in the patterns suggested by the plan tree.

These plans, in combination with program schedules, tell the humans populating the program what to do, when to do it, and who should do it. Where do these plans come from? What are the right plans to write? Who should prepare them? As mentioned above, we can allow our functional departments to autonomously develop program plans and then try to fit them together during program execution. This is a bottom-up or grass roots approach to planning. Alternatively, we could approach program implementation planning as we should approach product system development—systematically from the top down.

The fundamental difference between past planning processes and the integrated management system is that the integrated management system defines the work to be accomplished very rigorously linked to the program statement of work (SOW) and WBS by a common work identification coding system. In the approach suggested in this chapter, generic functional department manuals become detailed narrative descriptions of how the work defined in the IMP will be accomplished. Also, the integrated management system calls for an integrated master schedule (IMS) rather than the system engineering master schedule (SEMS) called for in MIL-STD-499B. These are essentially the same except in name. The advance made by the integrated management system is to absolutely link all works defined in the IMP with the schedules in the IMS using a common work identification system. This part of the chapter focuses on this approach.

If you are forced by a DoD customer's requirements to plan a program using an approach excluding the integrated management system, the IMP is simply removed from Figure 19.1, the tree collapses to the program plan, and the IMS may be renamed the SEMS. You may conclude from this that the IMP is an extraneous layer of planning, but you are encouraged to finish the chapter before making that conclusion a permanent part of your belief system.

In the integrated management system approach, you would not normally prepare all of the plans subordinate to the IMP. The content of all of those plans could be incorporated as part of the IMP in narrative form. They are separated here for two reasons. First, a company may have to respond to a program not involving an IMP/IMS approach, so the book should offer people in that situation this alternative. Second, the principal approach offered in this chapter recognizes a planning approach that makes best use of the limited time available during proposals (by using the generic planning data as narratives referenced in the IMP rather than physically including them there); focuses proposal work on how these data can be blended into a program-specific IMP; and encourages company use of standard procedures, incrementally improved over time, that provide customers best value. You cannot develop the latter unless you are allowed to practice–practice–practice. If you redesign your company for each new contract (new IMP narratives, for example), you will never realize a single identity, and your work force will not be able to benefit from repetition.

Regardless of whether your common customer base can be expected to call the integrated management system in your RFPs or not, you are

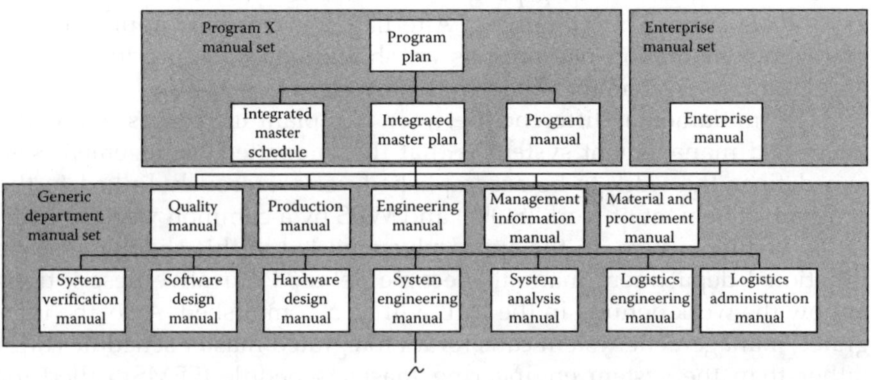

Figure 19.1 Program plan tree example. (From Grady, J., *System Engineering Deployment*, CRC Press, Boca Raton, FL, pp. 174–213, 2000. With permission.)

encouraged to develop one of the organized planning approaches offered in this book. You should, of course, sound out your current and potential customers on this planning approach and get their feedback on acceptability. Generally, you will find them to be delighted with the thoroughness of your approach.

Please understand that the intent is that all of the generic plans subordinate to the IMP in Figure 19.1 are intended to be available as a function of ongoing functional department planning that defines generically how the various tasks in department charters are to be accomplished. On a given proposal and subsequent program work, that basis should already be in place. The proposal work should focus on fashioning a program plan, IMP and IMS appropriate to the customer's needs drawing on your standard planning data as narratives referenced in the IMP, and crafting any program-peculiar planning data required. If you have 10 programs in house, for example, each of the 10 programs would have its own program plan and IMP but each IMP would reference the same set of functional manuals that tell the people on the programs how to do their tasks.

It is true that some customers will be uncomfortable with your generic planning in that you can change it without their approval after a contract has been awarded. Some customers will want review authority on your internal plans and that can be a nightmare when you have multiple customers with very different interests. In such cases, you may have to run a copy of a generic plan and reidentify it for specific use on a program. This program-specific copy could thereafter be changed with customer review without causing chaos in your internal documentation. This is a slippery slope that will force you to deviate to some extent from the practice–practice–practice notion. If all of your customers take this tack, then you may not have achieved very much with a "standard" process. You must understand your customer base and ensure that your standard process will not offend them. You cannot construct your generic process in a vacuum.

The party line in your proposals and in conversations with your customers should be that you are motivated to continuously develop generic procedures that will provide all of your customers with the best possible value. In cases where the needs to satisfy two or more customers are in conflict, you need to consult with those customers as part of the process improvement activity. You should also offer your customers access to your generic internal planning data on the basis that it is proprietary. Let them act as one of the pathways through which you can detect incompatibility using management by exception. One excellent way to do this is to provide internal read-only access to all planning data via computer network throughout your facility and extend this access to your customer base either at their facilities or only at your own.

In any case, the specific work required for each customer's program would be clearly defined in their program-unique and customer-approved program plan, WBS, SOW, IMP, and IMS. These documents would reference the generic planning data that only tells how to do these tasks. The principal obstacle in implementing this approach is, of course, that everyone has not done a fine job of documenting their practices. In the author's opinion, this is not a valid basis for the rejection of the offered planning method, nor does it represent an impossible barrier. It means that you must begin developing these generic practices now and keep improving them over time. Your competition may yet give you time to improve your performance for they are very likely in every bit as much difficulty as you are in this respect.

In the case where a company is organized in a projectized fashion, or only deals with a single product line, you still have to acquire your personnel from the same source as everyone else, from the human race. We are all knowledge limited and are forced to specialize. So, even if you have no matrix and only a project structure in a company with one or more projects, you will have to accomplish program work using specialists whose work patterns should be standardized, one program to another. This standardization can be captured in the kinds of documents shown in Figure 19.1 under the IMP.

19.3 Simple program-centric planning

The first organized planning approach discussed has been used on many programs, the DoD and others. It simply identifies all of the program elements at the level you intend to manage the program in what is called a WBS. The WBS pattern on past DoD programs has included both product and process WBS elements that the author believes to be flawed. The problem is that program product and program process are related orthogonally rather than linearly. The author encourages a product-only WBS. Process should be covered in the enterprise common process description. Enterprise process can then be mapped to product WBS in the planning process.

To implement the WBS/SOW approach, one identifies top-level work elements (WBS) that may be of a product-only nature in a fairly advanced program with a clear architecture or include process WBS elements in early programs heavily dependent on analysis, technology, and science. For each WBS element, a set of SOW paragraphs is completed expanding on the work specificity. In a matrix-managed company, the functional department representatives assisting in program planning can then map work their department should contribute to the SOW paragraphs to form the planning details. Someone familiar with the company and its functional department structure should review the functional inputs in search of omitted and unnecessary elements.

Every task should have associated with it a labor cost figure and total cost compared with planned total budget with any necessary adjustments made to cause equality. Task interdependencies should be identified and work set into the available calendar time in which to accomplish the program. Should the available calendar time be insufficient, an effort must be made to strip out the least necessary work or increase the amount of work accomplished in parallel to encourage on-time completion. Cost reserve and schedule float adjustments should be considered for any work elements thought to offer cost, schedule, or performance risks.

The SOW tasks can be placed on a time axis and start and end points tied to dates forming a program schedule. The problem with this structure is that it seldom supports a closed-loop process where planning information is linked to functional capabilities and collected cost and schedule metrics are tied clearly to the adequacy of the functional department resource base. Thus it is referred to as a program-centric planning method.

19.4 Using the USAF integrated management system

We will next explore the use of an expansion of a U.S. Air Force initiative referred to as integrated management system. It is a variation of the planning process described in MIL-STD-499B (rejected by the DoD prior to approval) that encourages the development of an integrated master plan connected to generic planning data rather than an SEMP. The reader interested in commercial markets should not become disinterested because we are going to use a U.S. Air Force model. The fact is that the Air Force really seized on an excellent planning and management model applicable to almost any development situation.

Unfortunately, the Air Force did not coordinate their model with contractor-generic planning structures, but then why should they have done so and how would they have foreseen a structure that all of their contractors would have been comfortable with? To complete the picture from the contractor's perspective, we will glue onto the Air Force model an internal contractor generic planning capability and continuous process improvement module. We will also merge this with the planning requirements of the integrated management system. This is not optimum in the author's opinion so a third model will be offered after the discussion of the Air Force model.

In applying the resultant planning system, the contractor or commercial company first must understand themselves, their capabilities, and their best practices leading to good customer product value. Next, for each program, they must accomplish a transform between this knowledge of their capability and an expression of the program plan in terms of the customer's need. Then this plan must be well executed. Finally, you must take advantage of lessons learned from each program execution to continuously improve your generic self. This provides an integrated environment within which to attain and maintain a world-class capability in your product line. System integration plays a central role in realizing this goal because we must accomplish our work through the integration of the work products of many specialized people both in the program planning and execution of that planning.

19.4.1 Know thyself through generic program planning data

In the ideal situation, our company will have been involved in a continuous process improvement program for some time, and we will have developed a generic set of good practices coordinated with our tools, personnel knowledge, and skills base. This data should include (1) a generic process flow diagram that hooks these tasks into relative time (Exhibit H/Appendix A coordinated with this book), (2) a functional department charter listing all of the tasks for which each functional department is responsible for maintaining and improving company technology and capability (Exhibit A/Appendix C coordinated with this book), (3) a task planning sheet for each charter task (to be explained later), and (4) an integrated enterprise system engineering manual (SEM) or generic SEMP.

The generic process flow diagram provides a framework into which all of the charter task descriptions fit as pieces into a puzzle. In this model, the process diagram is a simple process flow diagram. It could be expressed as a PERT, CPM, or other network diagram; V diagram; generic schedule diagram; or Gantt chart. The time axis could be in some nominal time measure or the complete period of execution equated to 100%. We can picture this diagram as a rubber sheet being stretched or compressed to satisfy a particular customer's needs with some paths being deleted, and others possibly added, as a function of the contract.

Figure 19.2 offers an example of generic planning sheet for one of the tasks illustrated on the enterprise process flow diagram. It happens to be a reliability task. Much of this information will have to be changed as it is applied to a specific program, in particular the intensity of task application, but it is helpful to have this information as a beginning point. If you have these sheets available as a generic input to the program planning process, you can save time and introduce your company's standards for process quality into the program planning process commonly accomplished during the proposal development period. If you have no enduring generic planning data now, these forms could be filled in initially as part of your next proposal task estimate and later iteratively improved upon in each subsequent proposal process using the incremental continuous process improvement notion.

Some of the charter tasks expressed on the generic process diagram will entail personnel from only one department working alone to complete them. Many tasks will require cooperative effort on the part of personnel from two or more departments as a function of how you are functionally organized and the business in which you are involved. Therefore, the

Funct. task ID	F41114471
Department	D2164, RAM Engineering
Task Name	Reliability Allocations (Task 202)
Procedure ref.	Company Procedure 42-8; MIL-STD-756B (tailored); Reliability Engineering, ARINC, Prentice Hall, Chapter 6.
Objective	Define a quantitative reliability figure for each system item
Description	The analyst studies the planned design concepts provided by the IPPT designer(s) and allocates the system or higher tier failure rate to sub-elements from top to bottom in each architectural layer and branch based on available historical data or sound engineering judgement determined by the item composition.
Significant accomplishment	Reliability figures assigned to all system items to a depth defined for the program.
Completion criteria	The task is complete when every configuration item, procured component, and other items as defined on the program have been assigned a reliability figure and the complete set of data has been checked for internal consistency, customer requirements compliance, and been approved by the PIT.
Events	Model structure complete at SDR, allocations loaded by PDR, and predictions loaded by CDR.
Resources	
Tools	Five networked IBM compatible computers each running RAMEASY, Microsoft Word 5.0, and Microsoft Excel for first two quarters. Thereafter, two machines adequate. Network server memory space 50 mb.
Personnel	At peak work load, 6 experienced reliability analysts skilled in failure rate prediction and reliability math and familiar with equipment used in missile systems.
Facilities	Each analyst requires a standard company engineer's workstation. Personnel will be assigned to teams with 2 assigned to PIT and four to product teams, one per team during the first 2 quarters of the program. Thereafter, two analysts on PIT only.
Inputs	1. Design concept information from item designer 2. System architecture definition
Outputs/ Accomplishment	1. Reliability Model complete with a quantitative reliability number for each configuration item and procured item. Data users include: IPT as a source of reliability requirements, Availability analyst as a source of reliability data for computing availability

Budget EST per quarter heads	1/2	2/5	3/5	4/5	5/2	6/2	7/2	8/2	9/2
Man-hours	1008	2520	2520	2520	1008	1008	1008	1008	504

Figure 19.2 Sample functional task definition form. (From Grady, J., *System Engineering Deployment*, CRC Press, Boca Raton, FL, pp. 174–213, 2000. With permission.)

atomic structure of generic task definition includes a task number and a department (or specialty) number. The generic task described in Figure 19.2 would be referred to as F4111471-D2164. This identification number appears quite long due to the number of process layers included.

With generic planning data in hand, we can apply this data in a tinker toy or erector set fashion to particular customer needs expressed in their RFP. What we need is an efficient transform process between generic planning data and specific program planning data. If you look at the planning data prepared in many companies, it includes a mix of functional department and product-oriented data. Our transform process must map the generic planning data into a program and product context understandable to our customers and valuable to us as a basis for managing the program through product-oriented integrated product development teams.

We have assumed throughout this discussion that the company is organized into a matrix structure. You may not be organized in a matrix, but you will still need to focus on the many specialties the technology base associated with your product line demands. The advantage of the matrix is that the functional organization can provide process continuity and a continuous improvement focus while the program organization runs the programs using resources derived from the functional departments. The danger occurs when the functional organization is allowed to enter the program work supervision path.

19.4.2 Integrated management system overview

During the late 1980s and early 1990s the U.S. Air Force, working with several contractors on several major programs including the F-22 Fighter Aircraft Program, evolved an extremely important acquisition management methodology called integrated management system. One of the beautiful aspects of this Air Force integrated planning initiative is that it strips away a buildup of past confusion and shines a brilliant light on the important things in the program planning process. The fundamental notion is that we should first understand the requirements for the product system and then apply a structured approach to design a program that will produce a system compliant with those product requirements. The program design is captured in a system specification, WBS dictionary, SOW, a list of deliverable data, a set of program plans, and program scheduling data. The content of the plans are requirements for performance of the process that will produce the product defined by a customer system specification. There should be no work planned or performed on a program that does not contribute in a perfectly clear way to satisfying the customer's product needs expressed in their system specification. This shockingly simple concept is, sadly, not always realized in practice.

This methodology recognizes six fundamental documents that collectively contain the requirements for the product system and the program through which the product will be created. Figure 19.3 illustrates the traceability relationship between these documents from the ultimate requirement, the need. The system specification should be driven by the customer's need, and it should recognize a particular top-level architecture that is expanded into a WBS dictionary for the purpose of structuring the program for management purposes. A SOW defines the work that must be accomplished for the product system and each WBS element of the system. The required work must then be planned in an IMP that fits the work elements into a framework of major program events. Finally, the planned work hooked to major program events is scheduled in time in the IMS. The system specification, SOW, and IMP all represent the top ends of trees of specifications, statements of work (including supplier SOWs), and plans, respectively.

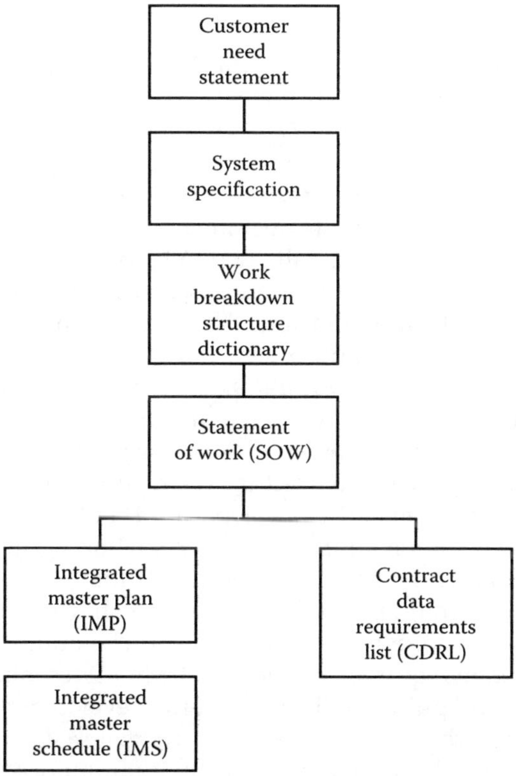

Figure 19.3 Program planning document stream. (From Grady, J., *System Engineering Deployment*, CRC Press, Boca Raton, FL, pp. 174–213, 2000. With permission.)

Some large system customers, like the DoD, use a device called a contract data requirements list (CDRL) to define their requirements for formal delivery of data about the product and the process for creating it. This document lists each item of data that must be delivered and tells what format it must be in (referencing a data item description, or DID), when it must be delivered, to whom it must be delivered, and how many copies are required. These items are commonly tied to paragraphs in the SOW and WBS numbers for management purposes. Some of these information products will be part of the delivered product, such as technical orders that customer personnel will use to understand how to operate and maintain the product system. Other data items, such as cost/schedule control system (C/SCS) reports, meeting minutes, and schedule updates, will be focused on reporting progress in the product system development and manufacturing effort.

19.4.3 Generating the six primary documents

19.4.3.1 The system specification

Depending on the program phase we are discussing, the system specification may be available to us or may not yet exist. Preparing the system specification may be one of the program tasks if the contract involves a very early program phase. More likely, the contractor will receive along with an RFP some kind of requirements document: such as a system requirements document, operational requirements document, or draft system specification from the customer. This input will have to be completed or transformed into a system specification to the mutual satisfaction of the customer and contractor. Commonly, this transform happens during the proposal process with a system requirements review (SRR) scheduled very early in the program execution. Refer to *System Requirements Analysis* (Grady, 2005) for a description of a process that will result in a quality system specification. It often happens that the customer's original need has become lost by the time they are ready to let a contract for the development of their system. Sometimes a need statement is never phrased. But even when it is prepared, it is not uncommon for it to have been so thoroughly digested in the process of developing the system requirements that it simply passes from view. Some people would maintain that once you have the system requirements defined, you do not need one anyway. The author disagrees. The need is the ultimate requirement and, if it truly represents the customer's need, it offers useful guidance throughout the development process.

If you cannot find a need statement in materials provided by the customer, you should ask them for it. If they cannot or will not produce one, write one and get customer acceptance. This may be very difficult, especially where the customer has many faces including one or more users and a procurement agency. But because it is difficult, you should not conclude that it is not needed. Quite the contrary, the harder it is to phrase and gain acceptance of the customer's need, the greater the need to press on to a conclusion. A lot of good work will result that will eliminate many false starts during later program execution. This same policy should be pursued in the definition of system requirements—the harder it is to gain acceptance, the harder you should work to understand the customer's needs and achieve agreement.

All of the requirements in all the lower-tier product requirement documents should be traceable from and to this document. All of the requirements in the system specification should be traceable to the customer's need through a logical process of functional decomposition, allocation, and requirements analysis.

19.4.3.2 The work breakdown structure dictionary

The WBS has, in the past in the DoD market, all too often been crafted by finance people in industry and government based on one of several MIL-STD-881 appendices reflecting different kinds of systems. Even though MIL-STD-881 encouraged that its appendices be used as a guide only, the mindless way it was applied often had a chilling effect on the application of a sound system functional decomposition approach to the development of product systems. Despite the evolution of a conflicting functional system product structure on such a program, the inflexible computer cost management tools used by the customer and contractor finance communities often inhibited alignment of the WBS to the needed functionally derived structure. Under the Air Force initiative that resulted in development of MIL-STD-499B, which was never approved, the government had intended that the WBS be a responsibility of the system engineering community rather than the finance function.

An attempt was made to reissue MIL-STD-881 as a handbook but in 2009 it was restored to standard status because the consequent lack of standardization resulted in (1) poorly constructed WBSs that impede effective program management practices; (2) contractors having to maintain two WBSs; (3) difficulty in understanding and comparing contractor proposals; (4) inconsistent CDRLs; (5) difficulty in reconciling data submissions; and (6) more time-consuming, less accurate data collection and analysis. Hopefully, the return to standard status will not bring back an acceptance of the financial structure taking precedence over that derived through thoughtful engineering modeling.

The WBS Dictionary provides a hierarchical organization of product material and services needed to satisfy the customer's need. The WBS must span the complete system allowing everything in the system to be placed in some WBS category. The content of the WBS must reflect what is needed to satisfy product system functionality and should not be chosen rigidly and arbitrarily based on some financial model. Figure 19.4 lists the partial

Missile system
- Air vehicle
  - Propulsion (Stages 1...*n*, as required)
  - Payload
  - Airframe
  - Reentry system
  - Guidance and control equipment
  - Ordnance initiation set
  - Airborne test equipment
  - Airborne training equipment
  - Auxiliary equipment
  - Integration, assembly, test, and checkout
- Command and launch equipment
- System engineering/management
- Systems test and evaluation
- Training
- Data
- Peculiar support equipment
- Operational/site activation
- Industrial facilities
- Initial spares and repair parts

Figure 19.4 Example of a WBS. (From Grady, J., *System Engineering Deployment*, CRC Press, Boca Raton, FL, pp. 174–213, 2000. With permission.)

content of Appendix C of MIL-STD-881B for missile systems. Only the Air Vehicle, a second-level WBS element, is expanded to the third level.

MIL-STD-881 did not prescribe a numerical coding system for the structure, but one is always assigned. For example, in Figure 19.4, the Air Vehicle might be assigned WBS 1000 and the other second-level items 2000, 3000, and so on. WBS 0000 would be assigned to the whole system in this case. If the system included an Air Vehicle and other things, like a launch site and a final assembly factory, WBS 0000 would be assigned to the complete system and each of these elements, sometimes called segments, would be assigned thousand-level WBS codes. Some programs are so complex that prefixes are assigned to permit cost accumulations in different useful patterns. The WBS 012-1000 might be assigned for nonrecurring development of the Air Vehicle (1000) while 014-1000 is assigned to recurring manufacturing of the Air Vehicle. Costs can be accumulated in WBS 1000 across all WBS prefixes for the complete Air Vehicle cost and within prefix 012 for all development costs.

Lower-tier WBS identification is accomplished by using different numbers in the hundreds and/or tens place of the four-digit WBS code. For example, WBS 1100 might be assigned to the first-stage propulsion system, 1200 to the second-stage propulsion system, and 1210 to the rocket engine of the second-stage propulsion system. We have chosen a four-digit code so far, but there is nothing to prevent the selection of a five- or six-digit code. In a very complex system, a decimal expansion could be added, such as 1210.05, for an engine turbo pump. In this fashion, the WBS can identify everything in the product system in a very organized fashion to any level of indenture desired. It is not necessary to apply a unique WBS number to each product item throughout the system hierarchy, however. The WBS is a management tool and should give managers insight into the system structure. The WBS is actually (or should be) an overlay on the functionally derived system architecture that should be expanded to the component item level (valves, black boxes, mechanical assemblies, etc.). Not all of this detail is needed for the management of the system development. Ideally, the WBS numbering should be the same as the architecture identification numbering eliminating one unique numbering system.

19.4.3.3 The statement of work

The next step in the programmatic requirements development process is to determine what work must be performed to develop, design, manufacture, test, and deploy every product element depicted in the WBS. Every bit of work we perform on a development contract should be included in the SOW at some level of detail and should be traceable to the product requirements in the system specification. The SOW, therefore, will tell what work must be accomplished for each product WBS element at some level of indenture. Only then can the program work be said to flow from the system requirements.

In the past on DoD programs, the SOW has often been prepared by someone in the customer's program office copying customer-created boilerplate SOW material from a similar past program SOW into the new program SOW. The integrated system management approach calls for the contractor to write the SOW based on a WBS derived from the same work that produced the definition of the system contained in system specification. The contractor must decide what work must be done within the context of his plant, personnel base, and product history in order to create a product system that satisfies the provided requirements and organize the work around the breakdown in the WBS.

Two competing contractors may very well offer the customer two very different statements of work with their proposals because they have different plants, product histories and experiences, and personnel mixes. Neither may necessarily be better than the other, only different for these reasons.

Some customer program offices applying the IMP/IMS approach do not call for a SOW, preferring to pass directly from the WBS to the IMP. The author prefers to recognize the SOW as a progressive planning step, but once it is created, it is simply an indenture within the IMP so it could be set aside as a separate document.

If we have properly identified program work, it should be possible to establish traceability between the paragraphs of the SOW and the paragraphs of the system specification. Where the DID CMAN 80008A is called by a DoD customer to define the system specification format, the product-oriented SOW paragraphs can be traced to paragraphs under system specification paragraph 3.7, which captures requirements for major items in the system architecture or WBS. Paragraphs under 3.7 should have been initially conceived from an orderly functional decomposition of the customer need and the WBS developed from that same analytical process.

We can picture all of the supplier statements of work strung out from the system SOW in a tree structure branching from the SOW element of Figure 18.3. Work traceability should exist between the process requirements we accept in the applicable documents referenced in the SOW with our customer and the process requirements we lay upon our suppliers in procurement SOWs. Otherwise, some elements of the complete product delivered by us (containing supplier elements) may not be compliant with our customer's requirements. The Air Force planning initiative brings into clear focus the need for traceability not only through the specification tree but through the SOW tree as well and between a SOW and its companion specification. This means that the complete product and process definition for a system can be unfolded from the customer's need in a structured top-down development effort.

Commonly, the system-level SOW is written to cover all of the work in several WBS indentures. We prepare supplier statements of work for major suppliers but the system SOW commonly provides the only work definition coverage for the prime contractor. The integrated product team (IPT), or concurrent engineering, paradigm suggests an interesting alternative to this arrangement. We could prepare the system-level SOW to cover only the system-level work under the responsibility of the program integration team (PIT) and write internal SOWs for each major system element identified in the WBS and to which an IPT will be assigned. These internal SOWs would each define the work that must be accomplished by one of the teams. The product element would be covered by a specification defining the product requirements and the team SOW would define the process or work requirements that must be accomplished to satisfy the corresponding product requirements.

This arrangement results in product requirements and work definition documents aligned perfectly with the product system elements and the development responsibility definition and should result in great precision in the management of the program. In order to be successful in this approach, you have to have thoroughly studied the customer's need and requirements and decomposed their need into a stable architecture that can be assigned to product teams. Instability in this whole structure can result in a tremendous amount of parasitic work.

The SOW, as the name implies, identifies work that must be accomplished at a high level to provide the items identified in the WBS. Now, how do we identify or describe the work that must be done in detail? A three-step process is suggested.

The first step in transforming generic planning data into the specific program plan for a given program is to map the generic tasks to the WBS. Figure 19.5 illustrates this process. It can be accomplished in a top-down or bottom-up fashion and be accomplished at any level of management desired either by the program team or functional management. Two alternatives for these three planning process components are listed below under headings in the form development direction/planning level/planner, and the reader can imagine other possible combinations:

1. Bottom-up/department chief level/chief—Each functional department chief, or their representative, lists under each WBS, their department tasks that must be performed.
2. Top-down/upper management level/program planning team—A program planning team, composed of people from functional departments, accomplishes the map between the functional department charters and the WBS. The planners shop the functional charters for tasks that satisfy the customer's needs expressed by the WBS Dictionary.

Step 2 is to organize the functional task inputs into a program context within the WBS-driven SOW hierarchy. In the author's preferred scenario,

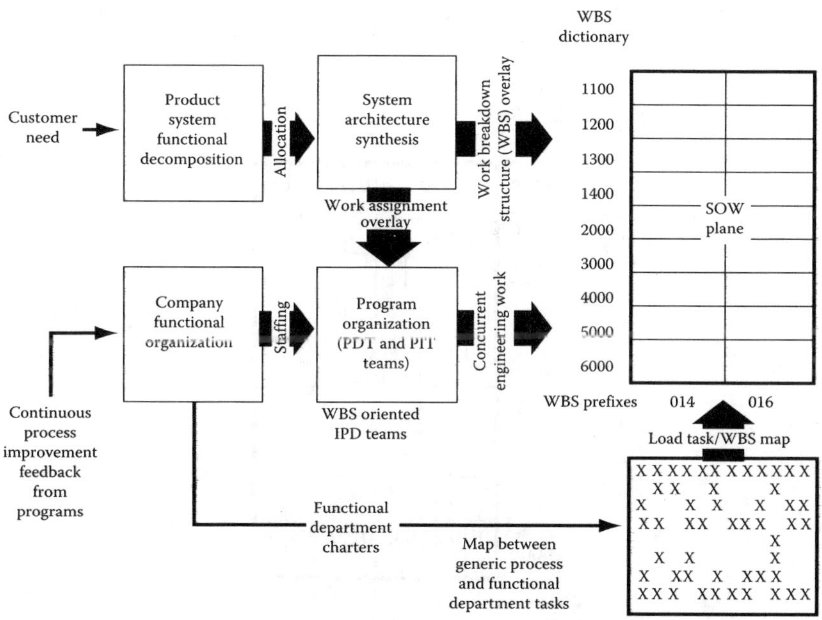

Figure 19.5 Initial SOW task loading. (From Grady, J., *System Engineering Deployment*, CRC Press, Boca Raton, FL, pp. 174–213, 2000. With permission.)

no matter how we orchestrated the first step above, the program should form an integrated planning team composed of people from the functional departments assigned to work on the proposal or program initiation work. Ideally, these same people will later be assigned to work on the actual program from those departments.

The planning work for each WBS should be led by the person who will be responsible during program execution for that WBS, if possible. In the case of product-oriented WBS elements, at some level of indenture this lead person should become an IPT leader. The planning team members for a given WBS must study the functional charter task sheets (see the sample in Figure 19.2) assigned to their WBS and organize, integrate, or synthesize them into a specific set of major program tasks that must be accomplished to satisfy the work element goal or purpose. The aggregate results of this work forms the SOW plane of the IMP Work Space shown in Figure 19.6. We might conclude, as a result of this planning exercise, that all of the work that must be done for WBS 014-1200 (prefixes not necessarily used in a specific situation) can be achieved within the context of major program SOW tasks 014-1200-01 through 014-1200-07, 016-1200-01, and 016-1200-02, for example.

Some of the detailed functional task sheet inputs may be gold plated, even unnecessary empire-building attempts. Some may be understated or even missing. They must be assembled into the fabric of the program by people experienced in the company's product line, organizational

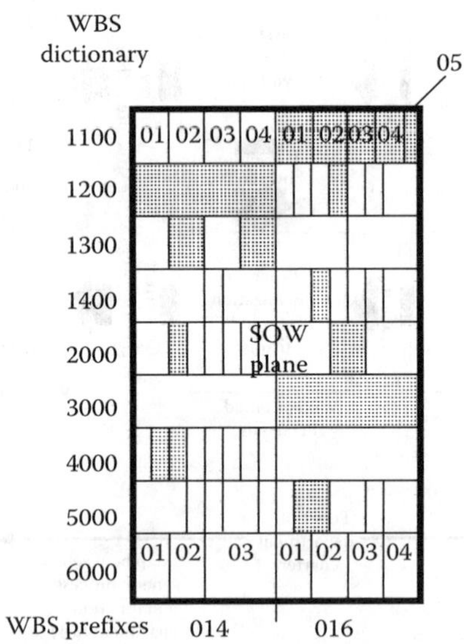

Figure 19.6 The SOW task plane. (From Grady, J., *System Engineering Deployment*, CRC Press, Boca Raton, FL, pp. 174–213, 2000. With permission.)

structure, and functional department charters and the needs of the customer. This is an application of ALL CROSS integration seeking out a condition of minimized completeness.

We have to assure ourselves that we have covered all of the necessary work and have not introduced any unnecessary work. For each planning sheet offered, we have to ask, "What would happen if we did not do this work or did it at a lower level of intensity?" For each WBS element, we have to assemble the complete stack of planning sheets (physically or in our computer) and ask, "Have we covered everything that needs to be done to satisfy this work element?" There is, of course, no substitute for successful program planning experience in this integration process no matter what kind of whiz-bang computer system you are fortunate enough to have.

Prior to kicking off this SOW task planning exercise, the WBS should be well developed and made available to everyone involved in the work planning exercise. It is a great source of confusion when the WBS is being changed twice a day in the middle of the work planning activity. Of course, the WBS may have to change based on new knowledge developed in this process, but the more stability that can be put into the WBS prior to beginning the more detailed SOW planning work, the better.

Very little of the planning information shown in Figure 19.2 need be included in the SOW, but it is needed for the overall planning process including budget determination (cost estimating), detailed program planning, and program facilitization, staffing, and equipping. Planning sheets like that shown in Figure 19.2 received from the detailed task planners form the atomic structure of the detailed planning activity that will result in the IMP and the IMS. These inputs must first be organized within the WBS, as discussed above, and woven into an understandable presentation in the SOW. Figure 19.7 includes a sample SOW fragment framed in the integrated system management style organized not by paragraph numbers but by an expanding work ID number formed from the WBS prefix (if used), WBS number, and SOW task number.

These SOW tasks are fairly generic in nature and are appropriate for a wide range of programs. Note that in the two alternative mapping processes we discussed above you could have used a top-down list like this created by the program prior to the functional department loading of the SOW tasks. The alternative is that you allow the functional departments to load their tasks into the WBS numbers and integrate the result to gain insight into the SOW tasks by combining and unifying. The author prefers the former method but no matter how you gained insight into them, they are the program tasks to which we have mapped all of the functional charter tasks.

Twenty different departments may have mapped a charter task to SOW task 014-1200-01. A system engineering department could have indicated that they will coordinate the requirements development process for the item/team. Fifteen different specialty engineering departments (reliability,

014	Nonrecurring work.
014-1200	Nonrecurring upper stage development work.
014-1200-01	Define upper stage requirements. Conduct analyses to define appropriate requirements for the upper stage based on system requirements and the evolving system concept.
014-1200-02	Review and approve upper stage requirements. Subject the requirements to formal review and approval prior to design authorization.
014-1200-03	Design upper stage. Develop a preliminary and detailed design fully compliant with item requirements.
014-1200-04	Provide design test and analysis support. Perform analyses and tests for the purpose of validating the adequacy of the design concept.
014-1200-05	Review and approve design. Subject the design in preliminary and detailed stages of maturity to a formal review by qualified specialists and approve same with possible redirection.
014-1200-06	Verify design by analysis. Analyze upper stage design for compliance with requirements.
014-1200-07	Verify design by test. Conduct testing to verify that the design complies with item requirements.
014-1200-08	Identify material sources. Define the source of each item of material needed to manufacture the item (make/buy).
014-1200-09	Develop manufacturing process. Define facilitization, tooling and test equipment, and personnel needs.
014-1200-10	Develop inspection process. Define inspection steps necessary in coordination with manufacturing process definition.
016-1200	Recurring upper stage work.
016-1200-01	Acquire and distribute material. Execute material acquisition plans, stock received material, and make available to the production process.
016-1200-02	Manufacture, assemble, and test upper stage. Perform all necessary steps to assemble materials defined in engineering drawings in accordance with manufacturing planning and inspection data. Test upper stage in accordance with approved written instructions for the purpose of assuring item readiness for acceptance by the customer.
016-1200-03	Deliver upper stage. Package and ship item to launch site.

Figure 19.7 Sample SOW fragment. (From Grady, J., *System Engineering Deployment*, CRC Press, Boca Raton, FL, pp. 174–213, 2000. With permission.)

maintainability, mass properties, design-to-cost, etc.) may have indicated that they must develop a particular requirement for the item. Functional charter task F41114471-D2164, displayed in Figure 19.2, could be one of these. A manufacturing engineering organization and logistics department will both cooperate in the development of requirements that are consistent with manufacturing and system operation needs. A test and evaluation department will have to define verification requirements. And finally, an operations analysis department may have to perform a special simulation to define the most cost-effective guidance accuracy allocation for the item. In the case of each line item in this list, we should find one functional department identified with principal responsibility (the bold Xs in Table 19.1).

Table 19.1 Task Responsibility Matrix Fragment

			Team			Functional dept.						
SOW task	Task name	STAT	1	2	3	A	B	C	D	E	F	G
014-1200-01	Define upper-stage requirements	C	X					X	**X**		X	
014-1200-02	Review and approve upper-stage requirements	C	X				**X**					
014-1200-03	Design upper stage	A	X				X		X	X	**X**	X
014-1200-04	Provide design test and analysis support	A	X									
014-1200-05	Review and approve design		X				**X**					
014-1200-06	Verify design by analysis		X					X		**X**		
014-1200-07	Verify design by test		X				X					**X**
014-1200-08	Identify material sources		X								**X**	
014-1200-09	Develop manufacturing process		X					**X**				
014-1200-10	Develop inspection process		X									**X**

Source: Grady, J., *System Engineering Deployment*, CRC Press, Boca Raton, FL, pp. 174–213, 2000. With permission.

In the process of combining all of these functional department charter tasks into SOW task 014-1200-01, we have also defined the specialists who must be brought together in a team environment to concurrently develop the requirements for the item. The WBS manager or IPPT leader who performed the task integration process will have derived knowledge of his or her team composition as a function of the planning experience. Table 19.1 illustrates a fragment of a SOW task responsibility matrix that we could now begin to build. The matrix will be useful in controlling program implementation. Each SOW task is identified by number and name (using a subset of those identified in Figure 19.7) followed by the task status (STAT) column, IPPT responsibilities definition (left blank because we discuss it next), and needed specialized functional department participation. Given that we have progressed in the program to the point that requirements are complete and approved, the "C" in the STAT column would convey that condition. The "A" in the design row indicates that the team may perform design work and supporting analysis and test work.

19.4.3.4 Integrated master plan and schedule

Now that we know how the product system and development process will be organized (WBS) and what work must be done (SOW content expanded by the detailed planning sheets mapped to the SOW tasks), it is necessary to determine who will be responsible (not in the functional organization sense but within the program IPPT structure), how it will be done, and when it will be done. We could establish a tree of plans to define these things like that illustrated in Figure 18.1 topped by a program plan flowing down to an SEMP and sub-plans such as manufacturing plan, procurement plan, and quality assurance plan.

The Air Force integrated management system initiative calls for an IMP, and that is the pattern we will follow here. This does not mean that a program team should overlook all the other plans shown in Figure 19.1. Many of those are still appropriate as IMP sub-plans or integral IMP narrative content. The IMP can replace the program plan and SEMP but may not provide the detail needed by manufacturing, quality, and many other functions. The integrated management system initiative recognizes that all IMP content does not fit neatly into the highly organized structure explained here. It also calls for narrative sections where appropriate. The content of the functional lower-tier plans, if prepared can provide the narrative data encouraged in the IMP.

The author proposes that the enterprise create a generic SEM and the IMP for each program reference this document rather than replicate its content in IMP narratives contained in each program IMP possibly with differences. Other functional department plans can have the same relationship to other IMP narratives. The author's idea of the ideal situation is that the contractors should have a set of functional department plans or manuals that describe how their chapter tasks are to be accomplished. One of these plans would be a generic SEM/SEMP that would tell how the company accomplishes the system engineering process on all contracts. This generic SEM should be mapped to the content of the preferred customer system engineering standard with possible tailoring clearly defined.

Following this pattern for a given contract/program, the company would write a WBS Dictionary/SOW/IMP/IMS document set and in the IMP, where the customer requires narrative materials, refer to the lower-tier functional plans for detailed coverage. This arrangement allows company personnel to follow the same practices on all programs developing and maintaining skills in a single process. Continuous process improvement based on program lessons learned can then be applied to a sound base to move toward the very best possible capability constrained by available resources.

19.4.3.4.1 Program events definition The IMP expands on the SOW telling how and by whom the work will be accomplished and provides a means to determine successful work completion. For each SOW task, at some level of WBS indenture, the IMP must contain planning data that identifies a series of major events through which the development of that product WBS will be accomplished. These events are selected with effective program management in mind. We do not want so many events that we are spending all of our time reporting. Nor do we want so few that we do not know what is going on. This is like selecting test points in a system design that allow rapid identification, isolation, and correction of problems. Too many test points cost money unnecessarily and are tedious to use in isolating faults while too few test points lead to ambiguity in fault isolation. These same extremes apply to program health monitoring systems for management.

The principal event discriminator will be defined by a customer or by your marketing department's determination of a potential date that commercial competitors might be ready to market a competing product. In the DoD, this is called an initial operating capability (IOC) marked by a specific date in the future. This date, or its commercial equivalent, defines a point in time when the program must have completed testing and have produced enough product articles to equip the customer for what they have defined as IOC. This might correspond to two aircraft squadrons, one armored battalion equipped, or the production of the first day's assembly line run of new cars.

Your scheduling experts will then have to fit all program activities in between the program beginning time and IOC with some quality, manufacturing, and logistics events extending beyond IOC. A generic process diagram, drawn on a rubber sheet in our imagination, can be useful as noted in Figure 19.8. The major events on the generic diagram can be

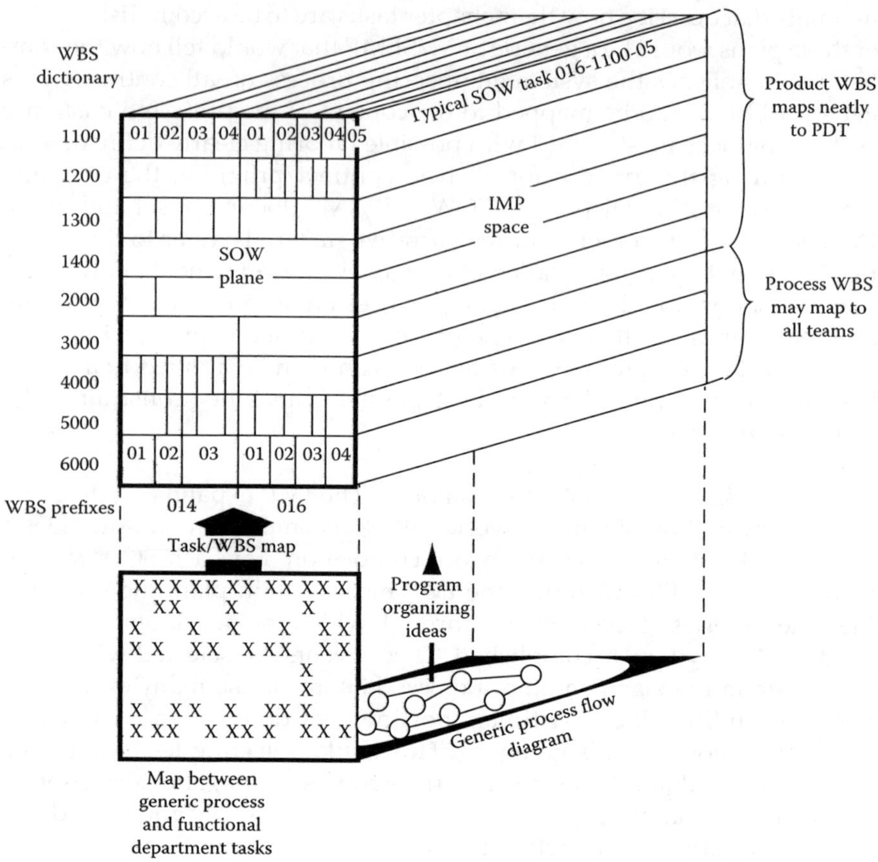

Figure 19.8 Projection of the generic process diagram. (From Grady, J., *System Engineering Deployment*, CRC Press, Boca Raton, FL, pp. 174–213, 2000. With permission.)

pushed and pulled into alignment with the realities uncovered by the scheduling experts.

Depending on the customer, they will insist on a particular stream of major reviews, possibly those included in Table 19.2. Other major events, or milestones, unique to your product line can be added based on the kind of product, the maturity of the product development process, and the degree of intensity with which the customer or the company management wishes to manage the program. The list may include only a single program phase or recognize the full sweep of the development, deployment, and operational steps in the development portion of the system life cycle.

The list of program events in time must now be applied to the IMP space as shown in Figure 19.9. The rubber sheet generic process diagram

Table 19.2 Major Program Event List

ID	Acronym	Event name	Event description
00		Program start	
01	SRR	System requirements review	Joint understanding of requirements reached.
02	SDR	System design review	System requirements validated with risks identified and mitigated to an acceptable level.
03	IRR	Item requirements review	Approval of item requirements as a prerequisite to item design work.
04	PDR	Item preliminary design review	Acceptance of readiness to undertake detailed design.
05	PDR	System preliminary design review	All item PDR complete and acceptance of detailed design entry.
06	CDR	Item critical design review	Design acceptance and authority to initiate manufacturing of test articles or low rate production.
07	CDR	System critical design review	All item CDRs complete.
08	FCA	Functional configuration audit	Proof that design will satisfy requirements.
09	PCA	Physical configuration audit	Proof that product will comply with design, quality and manufacturing planning.
10	IOC	Initial operating capability	Enough resources produced and delivered to provide a complete operating capability at some service level.

Source: Grady, J., *System Engineering Deployment*, CRC Press, Boca Raton, FL, pp. 174–213, 2000. With permission.

now can be stretched and compressed to conform with the program-specific events and be used as an aid in transforming the rough SOW task planning results into the fine structure provided by IMP task planning.

All of the SOW tasks now need to be laid into this time axis. The team responsible for each SOW task should expand the planning detail for their activities by linking up each SOW task with the events. This breaks up each SOW task into a number of segments like the sample illustrated in Figure 19.9. Some of the periods of time between events will correspond to voids for some SOW tasks as indicated by shading in Figure 19.9 meaning no work is planned for those periods and tasks.

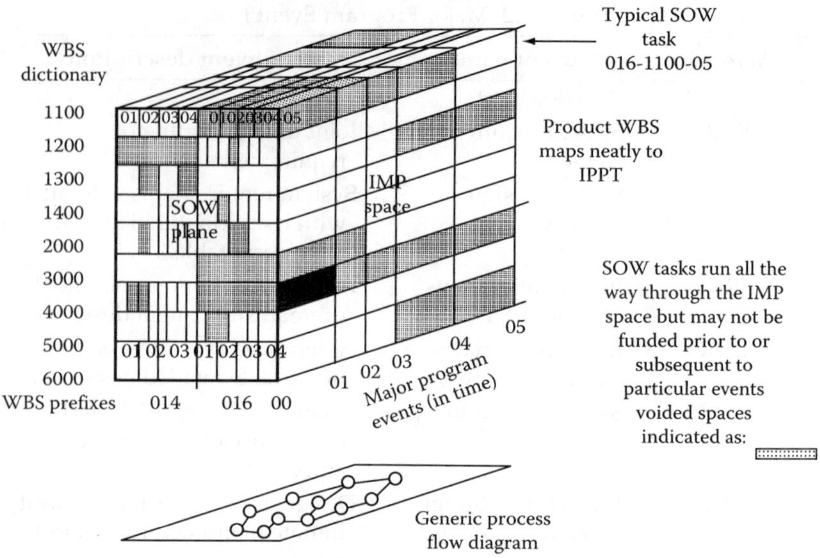

Figure 19.9 IMP space partitioning by event. (From Grady, J., *System Engineering Deployment*, CRC Press, Boca Raton, FL, pp. 174–213, 2000. With permission.)

19.4.3.4.2 Final work definition steps The work components formed by mapping generic functional tasks into the WBS Dictionary, partitioning and combining the work mapped to specific WBS numbers into major SOW tasks, and the time delimitation of those SOW tasks into events provide us with manageable increments of work that are associated with specific WBS numbers, most of which will map to IPPT in an unambiguous fashion if we have organized the teams properly to reflect the product structure. To enhance the management potential for these tasks, the integrated system management initiative calls upon the planner to identify one or more significant accomplishments for each of these IMP tasks and for each accomplishment to identify a measurable accomplishment criteria through which we can objectively determine when the accomplishment can be claimed.

Figure 19.10 illustrates these final IMP task definition parameters. An IMP task is a work increment between two events within a given SOW task. It is composed of one or more subtasks each of which is characterized by a significant accomplishment statement. Each significant accomplishment must have identified for it one or more unambiguous accomplishment criteria.

Table 19.3 illustrates one simple string in this planning process using a dash delimited numbering system. Paragraph 3.7.1.2 of the system specification calls for a space launch vehicle upper stage to accomplish some previously defined functionality. WBS 014-1200 was selected to identify upper-stage development work. The SOW was expanded on

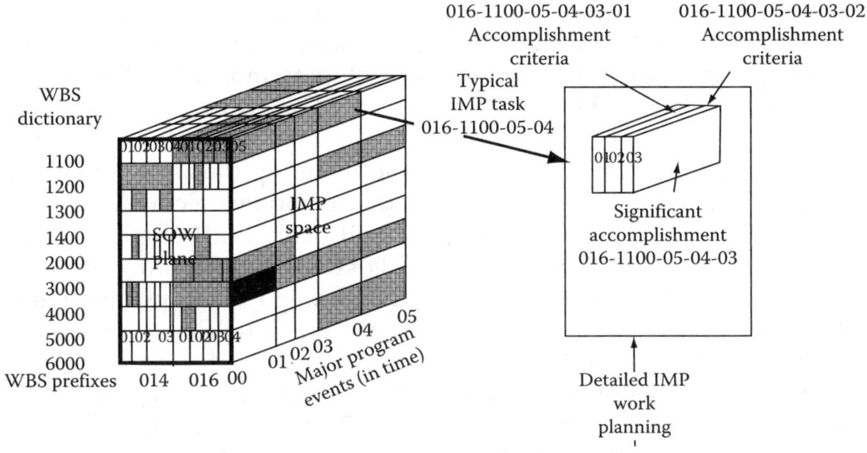

Figure 19.10 IMP task definition. (From Grady, J., *System Engineering Deployment*, CRC Press, Boca Raton, FL, pp. 174–213, 2000. With permission.)

Table 19.3 Single Planning String Example

Document	Reference	Planning content
SYSTEM SPEC	3.7.1.2	Launch vehicle upper stage
WBS	014-1200	Launch vehicle upper-stage development
SOW TASK	014-1200-01	Accomplish design work needed to define the upper-stage configuration.
IMP EVENT	014-1200-01-06	Item CDR
SIGNIFICANT ACCOMPLISHMENT	014-1200-01-06-01	Design complete
ACCOMPLISHMENT CRITERIA	014-1200-01-06-01-01	95% of drawings released
IMS EVENT/DATE	014-1200-01-06	January 20, 2002

Source: Grady, J., *System Engineering Deployment*, CRC Press, Boca Raton, FL, pp. 174–213, 2000. With permission.

WBS 014-1200 to include several work tasks needed to satisfy the WBS 014-1200 requirement. Based on an analysis of the evolving IMS (or SEMS), we selected the completion of design as a critical event through which to manage the program such that we could review design progress before we committed to the manufacture of the flight test vehicles. Traditionally, in the DoD environment, this event was called the critical design review (CDR) for many years, though the term detail design review is becoming popular as this book was in development. We will accept that the launch vehicle design is complete if 95% of all planned engineering drawings have been formally released at that time. Scheduling, with the concurrence of proposal or program management, determines that January 20, 2002 would be the ideal time at which to hold the CDR.

This is only one planning string. It may require thousands of these branching strings to complete the whole planning database for a large program. It quickly becomes obvious that computer technology could be helpful just to capture and organize all of this information. At the time this book was being written, several of the computer tool companies with requirements database tools were beginning to see a potential market for the application of these tools to programmatic as well as product requirements development.

All of these documents have a similar structure of a paragraph, WBS, SOW, or IMP task number; title; and text. These tools could be used to capture the specifications, WBS Dictionary, SOW, and plans and include traceability across the current valleys between them as well as generate the documents, or parts thereof, on demand.

When you have linked up all of the work in strings like the one illustrated in Table 19.3, using manual or computerized methods, you should have a mutually consistent network of management data that provides implementation guidance for people doing the work and management test points for effective program health monitoring.

19.4.3.4.3 Final IMS development As we are developing the IMP content we do not obligate ourselves to define scheduling constraints, only to identify the work that must be done. Once the work is defined, scheduling must arrange the tasks corresponding to the significant accomplishments in time, constrained by the events previously located in time. In the process, we will find that some things cannot be accomplished as planned in the available time, requiring adjustments in event timing and span times between events. This may require several iterations before all of the inconsistencies shake out and the complete set of planning data is ready for use.

19.4.3.4.4 Planning process summary Figure 19.11 combines all of the steps described over the past few pages into a single diagram to better illustrate the intended continuity between the steps outlined. Our approach has been to associate generic functional department work descriptions with WBS numbers defined for the particular product system derived from an understanding of the needed system functionality.

These functional tasks were then combined by the responsible teams into coherent SOW tasks oriented about the product WBS structure. We next determined some major milestones to use in managing program progress and projected these events into a space created by expanding the SOW plane into the time dimension. Finally, we further organized the

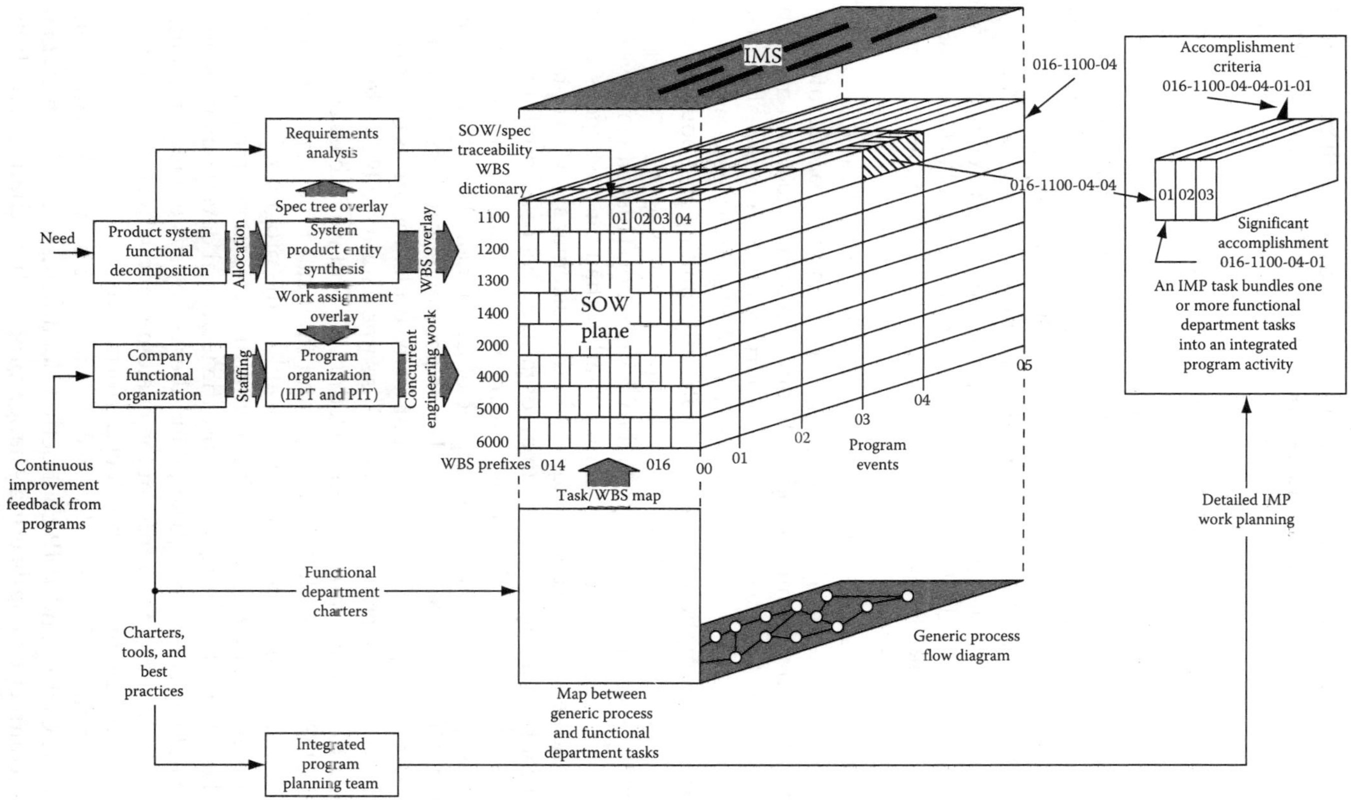

Figure 19.11 Overall program planning scenario. (From Grady, J., *System Engineering Deployment*, CRC Press, Boca Raton, FL, pp. 174–213, 2000. With permission.)

work content of each IMP task defined by a SOW task and a terminal event into a series of specific significant accomplishments each with predefined accomplishment criteria. It should be pointed out that everyone need not see this complete picture out of fear that it will confuse many. One can partition it into parts associated with the responsibilities of individuals. But, someone should understand the whole pattern.

19.4.3.5 Contract data requirements list

The CDRL identifies every item of data the customer expects to have formally delivered. This should be driven by the work that will be accomplished as defined by the SOW. Every task listed in the SOW that will produce a data item of value to the customer in managing the program, or understanding the technical or administrative flow of events, should be referenced there in terms of its CDRL identification and all of these items collected to form the CDRL. The CDRL items should be determined from the SOW tasks and, once identified, should be fed back into the SOW for reference.

Customers sometimes mindlessly boilerplate the CDRL as they do the SOW rather than deciding in an organized fashion what, of the information that the contractor must produce to develop the product and manage the development, they need for the same purposes. Customers realize that formally delivered data costs contractors more than data generated only for internal consumption so there is a naturally limiting cost boundary at work. The customer should be able to acquire data developed on the program anyway whether the item is on the CDRL or not. This data follows the contractor's format and content definition rather than the customer's, as in a CDRL, and requires a special request on the customer's part to obtain it.

The contractor should maintain an organized list of all program data and make it accessible to all contractor team members in a program library. A DoD customer calls this list (exclusive of CDRL items) a data accession list (DAL). This list names all of the data the customer can acquire by specifically requesting it and paying a separate fee under the terms of the contract. The contractor benefits by having this same data available for internal use rather than it becoming lost in desk drawers and waste baskets.

The CDRL must include schedule and financial reporting data that allows the customer to determine the health of the program. In addition, the customer should wish to receive technical performance measurement (TPM) data that offers insight into how the technical development is progressing in terms of meeting a small number of key system requirements. Together these three items inform the customer about evolving program cost, program schedule, and product performance.

The customer will normally wish to make the system specification, WBS, SOW, CDRL, and IMP contractual documents requiring a formal process (contract or engineering change proposal) to change. The IMS, however, should be a CDRL item because it will have to change as a function of showing progress, if for no other reason. All other CDRL items should be selected with care from the work identified in the SOW to provide the customer insight, at reasonable cost, into the evolving product requirements and design synthesis and into the product development, testing, and production processes.

Ideally, the DAL and CDRL items should be in electronic media stored on a computer network server and accessible from all work stations in the contractor's facility. The contractor can then also quickly respond to a customer request to electronically deliver any item listed in the DAL and in the program library. Very little data is produced today by means not involving a computer. Therefore, most items that a customer would be interested in receiving as a CDRL exists in electronic media within the contractor's business. Yet many customers continue to insist on delivery of tons of paper documents that will fill ever-expanding storage space. Alert customers will begin to require delivery of CDRL items in electronic media. This will change the nature of CDRL delivery. Instead of periodic delivery of paper copies of updated documents, the contractor will be required to refresh a customer database at a particular interval or at particular milestones.

The DAL should exist as an electronic data delivery conduit from the program online library to a customer. A customer could be charged a periodic rate plus a fee for each call to gain access to this library section. There are some fine precedents for this kind of service in the form of online databases open to public access like CompuServe. This kind of automated process would result in important savings for large customers like the government. This arrangement will require special provisions in the contract and good discipline on the part of customer personnel to avoid overrunning cost targets.

The company that is able to integrate all of this information product into a central program library has a distinct advantage in concurrent engineering because concurrent engineering is, or at least requires, effective communication of ideas. The people working on the project will also never have to worry about using outdated paper copies of documents. And, incidentally, the facilities people in both customer and contractor ranks will even benefit from fewer file cabinets.

19.4.4 Work responsibility

The IMP must relate program work to work execution responsibility. First, we must decide how we will organize to execute the program. In Part three of this book, we discussed the organizational structure preferred by the author, namely, matrix management characterized by (1) programmatic integrated product development teams and (2) functional departments that provide qualified people as well as proven tools and procedures to programs. Of course, other forms of organizational structures can successfully execute a program. For small companies with little product line differentiation, forms other than matrix may even be an advantage. This book attempts to focus on a situation where a matrix is advantageous for the reasons suggested in Part three.

If we are to follow this pattern, our program work must not be assigned to functional department directors, managers, and chiefs, rather to responsible IPPT. In the process, if we are not careful, we may develop a team structure that is completely unworkable in execution. The team structure must be aligned with the WBS (product architecture) in order to preclude conflicts in the assignment of budgets and tasking to teams.

The product teams must be aligned with the product structure reflected in the WBS because the budget will be aligned with the WBS in order to satisfy the C/SCS criteria required by a DoD customer or an equivalent entity by a non-DoD customer. In 1991, the criteria for DoD contracts was relocated from the DoD Instruction 7000.2 to the newly released DoD Instruction 5000.2 but remained unchanged in content. The C/SCS criteria requires us to manage budget through intersections in a matrix of WBS and functional organizations.

We wish the teams to align with the WBS so the budget for all team tasks can be simply assigned. The larger the number of crossovers that exist between WBS, teams, and tasks, the more complex the program will be to manage. At the same time, the WBS must align with the functional architecture allocated from needed system functionality. This means that the WBS must track the evolving functionally derived architecture reflected in the system specification. This combination will result in the fewest possible crossovers between organizational interfaces and product interfaces, referred to as cross-organizational interfaces, which lead to program problems.

This combination also results in the simplest possible integration task. System integration is a difficult task no matter how expansive. We can do a better job at it if there is less of it to do. The coordination of the organization of the work, product, and performing organization leads in this direction. Figure 19.12 illustrates this point by showing a perfect correlation between a product n-square diagram and an organizational n-square diagram. An n-square diagram displays the relationships between n items by noting these relationships in the square matrix intersections. The n items are listed down the diagonal.

On the product n-square diagram, we see a requirement for an interface between two specific items identified on the diagonal. On the organization n-square diagram, we see a need for the two teams corresponding to the product items joined by this interface to communicate about this interface. If our IPPT are organized about the same structure that the product system uses, the team communication patterns will match perfectly the product item cross-organizational interfaces. It is precisely these cross-organizational interfaces that traditionally lead to development problems.

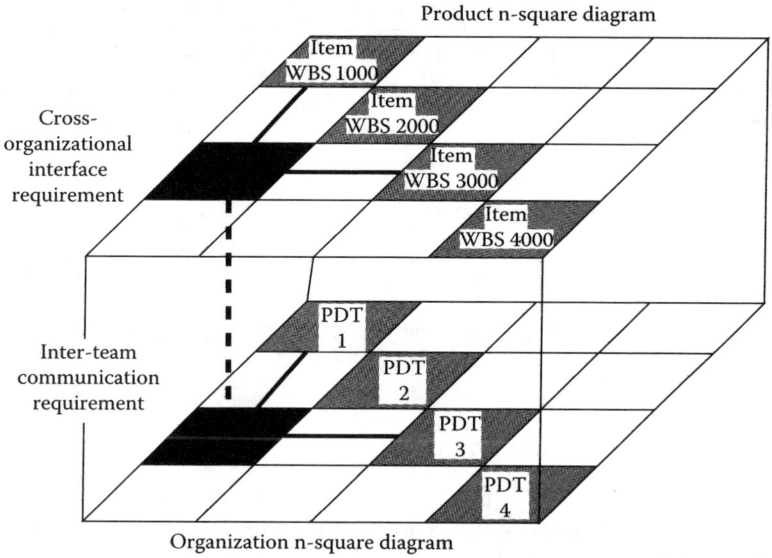

Figure 19.12 Product and teaming n-square alignment. (From Grady, J., *System Engineering Deployment*, CRC Press, Boca Raton, FL, pp. 174–213, 2000. With permission.)

If there is a complex relationship between product development responsibility and product composition, there will be interface development problems leading to unnecessary cost and schedule impacts.

Either the SOW or the IMP can cross-reference the work to the program organization team structure. Since the SOW lists all tasks, you might conclude that it would be a better place to locate the task responsibility matrix than the IMP. The IMP selectively expands on SOW content linking work to specific events and accomplishments corresponding to those events for management purposes. The author encourages that this matrix be placed in the IMP because the SOW tends to be a simple task list for people with green eye shades while the IMP is a plan for us humans. As the product and team structures mature, we must check for the emergence of misalignments between the WBS and the team structure and work toward nulling them out.

Every task identified in the SOW should have some kind of procedural coverage that tells how that task will be performed. This may be in the form of a customer standard (such as MIL-STD-961D), internal company procedures, or commercial standards (such as the American National Standards Institute (ANSI) series). This can easily be defined in a task/procedures matrix or integrated with responsibility assignment into a SOW task/responsibility/procedures matrix such as the fragment illustrated in Table 19.4.

Table 19.4 Task/Responsibility/Procedure Matrix

	Responsible team							
WBS task	1	2	3	4	5	6	7	Procedural reference
1000.01	X							MIL-STD-1422 (Tailored)
1000.02		X						Company Practice 128.23E
1000.03			X					MIL-STD-490A
1000.04	X							Company Practice 153.5
1000.05					X			Program Manual 14.24B

Source: Grady, J., *System Engineering Deployment*, CRC Press, Boca Raton, FL, pp. 174–213, 2000. With permission.

19.4.5 *Who plans the program?*

We have left one thing unanswered up to this point. As a result of our wonderful planning work, we know what must be done, who shall do it, when it shall be done, and how it shall be done. But who should have accomplished this planning work in the first place. Should we let the functional departments engage in bottom-up, grass roots planning integrated by the program? Should the program or proposal team do all of the planning?

The author believes that few proposal teams or program staffs at the time this book was written accomplished their program planning activities in a purposeful integrated fashion, including defining traceability throughout the product and process requirements stream as suggested in Figure 19.2. If the systems approach (or concurrent development) method is useful in developing product systems, why should we not apply it to the program planning process? We, the program team, are after all a system. Let us try it.

Table 19.5 Program Integrated Planning Team Responsibilities

Document	Functional responsibility
SYSTEM SPECIFICATION	Systems engineering
WORK BREAKDOWN STRUCTURE	Systems engineering and finance
STATEMENT OF WORK	Systems engineering and IPPT managers or candidates
INTEGRATED MASTER PLAN	Systems engineering
INTEGRATED MASTER SCHEDULE	Scheduling
CONTRACT DATA RQMT LIST	Data management

Source: Grady, J., *System Engineering Deployment*, CRC Press, Boca Raton, FL, pp. 174–213, 2000. With permission.

First, who should participate? Table 19.5 lists the planning documents discussed above and correlates them with some generic principal functional organization responsibilities. Your company may be organized differently, but these organizational entities are probably fairly widely recognized. While these documents should be developed in approximately the order listed (from top to bottom), we should not permit autonomous work on any of them. There will be insufficient time to sequentially develop these documents during a proposal preparation period of 30–60 days. More importantly, their content must be mutually consistent. Figure 19.13 offers a rough schedule showing the relative timing suggested.

The recommended integrated planning approach involves forming a program planning team with membership by representatives from each functional department noted in Table 19.5. Someone identified by the proposal or program manager should lead the team. All members should be physically collocated in close proximity encouraging easy conversation and a close working relationship. They must have available to them good telephone and computer data communication facilities as well as adequate wall space for posting information for integrated viewing.

The suggested integrated planning approach is not rocket science any more than is the concurrent or integrated product development approach. It simply requires a clear definition of responsibilities, cooperation between the parties, excellent interpersonal and communication skills, and a shared appreciation for the discipline of traceability. The system architecture identified in the system specification must be respected in the WBS. All work must be listed in the SOW and linked to the WBS. All events, accomplishments, and criteria in the IMP must link to the tasks defined in the SOW. All tasks appearing on the IMS must correspond with the SOW and IMP tasks and all events on the IMS must correspond to those respected in the IMP.

Throughout the planning process, the selected IPPT managers must contribute to WBS and SOW development and maintain vigilance for

```
Document                                      Program planning period
-------------------------------------------   -------------------------------
System specification                          <---------->
Work breakdown structure                                <----->
Statement of work                                         <------>
Integrated master plan                                         <---------->
Integrated master schedule                                          <------------->
Contract data RQMT list                                           <----->
```

Figure 19.13 Program planning timeline. (From Grady, J., *System Engineering Deployment*, CRC Press, Boca Raton, FL, pp. 174–213, 2000. With permission.)

crossovers between WBS, team definition, and work responsibilities. They must also work to develop a cost estimate in a proposal situation. Also in a proposal situation, this team will have to coordinate their work with the management volume writing team.

This planning activity probably cannot be accomplished in a straight-line fashion; while preparing the IMP we will get insights into changes in the SOW that may ripple through other documents. All team members must have available to them the full content of all of the evolving documents and be familiar with that content through an almost continuous conversation between team members and access to the actual text on their computer screen and a stickup on a wall in close proximity.

Team members must respect the hierarchy in Figure 19.2, and the content of the documents should be developed in the sequence illustrated there. Each component of each document expands into lower-tier document details. This pattern repeats through the hierarchy. As this expansion is developed, the traceability links should be captured. The information developed by the team members assigned to each document must be constantly checked for traceability and consistency by someone responsible for planning integration.

19.4.6 A generic SEM/SEMP for you

Some people greeted the workup to the release of MIL-STD-499B with screams of, "The sky is falling." You would think that, after many years of contract performance under its predecessor 499A, all DoD contractors would have put in place the fine system engineering activities that they had been describing in the many SEMPs that they had written over the years and submitted with proposals. The SEMPs were never contractual and, sadly, too often were never opened by the contractor subsequent to the contract win. So, the adverse reaction to 499B was often based on not then having in place, despite the fine stories told in SEMPs past, an effective systems approach and a concern for how to acquire an effective process in a reasonable time. Many companies dispatched people to the several MIL-STD-499B short courses offered only to have them return with confirmation that they were in very big trouble.

Whether your company must respond to one standard or another, you really should have in place an effective systems approach because it will result in a product with better value for your customer and more business for your company. If your competitors are able to put in place an effective systems approach and you cannot, you will have great difficulty matching their product cost, schedule, and quality.

The course materials for the author's course coordinated with this book included a copy of a complete generic integrated SEM/SEMP. This document can be used as a basis for a company SEM/SEMP.

It does not follow, of course, that by acquiring a good SEM/SEMP you will overnight become an excellent systems house, though the process of writing one can be quite an education in the right direction. This document must be matched to a current reality within your company. You will need to monitor your performance to this standard and provide the machinery to force compliance on your programs with your own standards. Ideally, in the view of the author, it will be the responsibility of your quality assurance function to verify that the content of the manual is followed on programs. The techniques discussed in this book and elsewhere on continuous process improvement should be applied to your system engineering process to uncover weakness and to understand useful priorities for their correction.

You will also need a way to educate your work force in your system engineering process. This can be done at a low cost through brownbag sessions in your plant for your motivated employees who happen to be the very ones you most wish to retain. The SEM/SEMP can be used as the textbook for these sessions. You should also cooperate with your local International Council on Systems Engineering (INCOSE) chapter and a local college or university to establish or support a system engineering certificate program that qualifies for your company's tuition reimbursement program.

Given that you have a written procedure for performing system engineering, a way to educate/train your work force in performing that procedure, a means to continuously improve upon that procedure, and a way to encourage compliance with your process, you are on the road to success in performing system engineering effectively. You will also have little to fear from the standards any customer may impose upon you.

19.4.7 Rapid identity documentation

It is possible to maintain a current enterprise identity within an environment of a high rate of change through the use of computers. It is time to defend that claim. The reader familiar with database concepts will have already observed, through the previous material in this chapter, the image of a set of databases that could be used for this purpose. Yes, databases rather than word-processed documents. The suggested approach is to apply the same tools used for requirements capture to planning document capture. If you are still using word processors for specification preparation then you will be one step behind the suggested process and a few words of database encouragement are in order.

Several computer applications on the market as requirements tools at the time this book was written were adequate for planning information capture, and several of these tool companies were hard at work more perfectly adapting to enterprise modeling tool sets. Table 19.6 lists several of

Table 19.6 Potential Planning Data Capture Tool List

Product name	Original agent	Supplier in late 2009
RDD-100	Ascent Logic Corp.	Tool not available to public
RTM	Marconi	
DOORS	QSS	
CORE	Vitech	Vitech
Cassets	Rockwell	Tool not available to public
Spec Writer	PRC Division of Black and Decker	Tool not available to public

Source: Grady, J., *System Engineering Deployment*, CRC Press, Boca Raton, FL, pp. 174–213, 2000. With permission.

these tools. These companies have, in the past, been very generous with their time and talent providing classes the author has taught at universities with tool demonstrations and requirements talks. Identifying them here as potential planning tool sources is a small repayment of that generosity. Of those listed below, probably Vitech's CORE is the best selection at the time this book was written.

The advantages of using a database to capture planning data is precisely the same as the advantages of using a database to capture requirements. Modern tools for this purpose will permit not only the storage of and access to requirements or planning data just as documents do (in paper or word processor forms) but also the coordinated capture of other related information that cannot be retained efficiently and economically in paper documents or even in the associated word processor or spreadsheet form. The principal kind of supporting data of interest is called traceability data. But there are some other subtle and powerful benefits that are not so obvious.

You have seen that the SOW content should be traceable to the system specification proving that we are doing no work that is not focused on the product and what it is required to accomplish. Well, how do we prove to ourselves in a fixed price world and show our customer that we have accomplished that goal? The discipline of traceability will go a long way toward satisfying this need. In a modern database of the kinds listed in Table 19.6, traceability fields exist that permit us to link up a paragraph in one document with a paragraph of another document with the meaning that one of these pieces of information was derived from the other or is related to the other. In the process of creating and maintaining traceability data in our database, we will find that certain content is not traceable. This should be a signal that we may have included extraneous material in one document or omitted something important in the other. So, this discipline will allow you to follow the program logic built into your planning documentation subsequent to preparing it but will also encourage the preparation of better planning documentation.

As important as it is, simply providing traceability is not good enough for our purposes. Doing the traceability work will actually slow down the planning process for it is added work. So, how can we accomplish more work faster allowing us to gain the benefit from traceability while completing the planning work more rapidly? If we have captured our generic planning data in a database and our tool set is properly configured, we should be able to generate the program planning data by mapping the generic work to the program work and generate the program plans and estimates from the resulting map. Unfortunately, this capability was not yet fully implemented in the commercially available tools the author was familiar with at the time this was written.

The author has experimented with this approach in his database toolbox called Rosetta Stone with encouraging results. Figure 19.14 illustrates a fragment of this toolbox for discussion purposes. The drums represent database structures. Those with a solid body are program oriented and those with a checkerboard body are generic data systems. The arrows show the flow of work from the customer need through the functional analysis (FUN) tool, the allocation of functionality jointly to the architecture (ARC) and system requirements (REQ), the establishment of the WBS as an overlay of the architecture and SOW content from the WBS with traceability of work elements to system requirements as a cross-check on validity. Each drum is a database and traceability is retained in related databases as paragraph number pairs from the corresponding documents

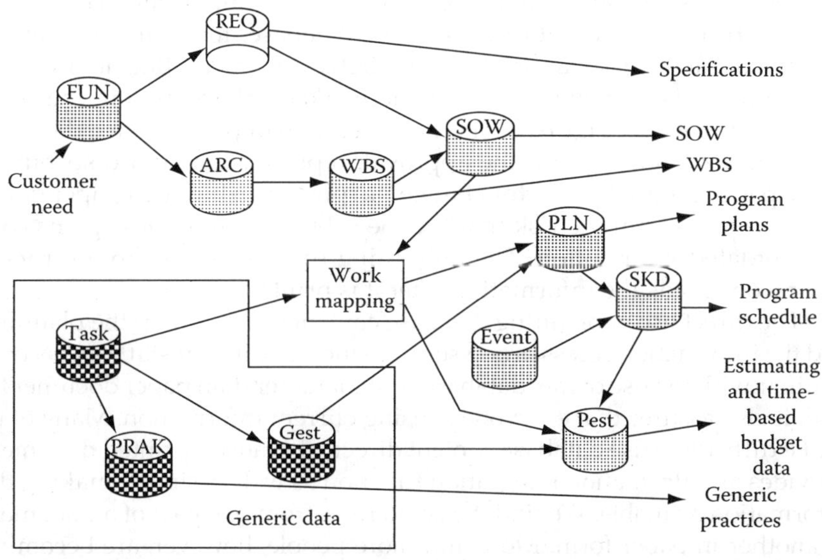

Figure 19.14 Special planning database tool set. (From Grady, J., *System Engineering Deployment*, CRC Press, Boca Raton, FL, pp. 174–213, 2000. With permission.)

(in database format). Commercial tools commonly will permit parsing of paragraphs into subsets for traceability purposes to account for poorly written documents with complex paragraphs.

Generic tasks (TASK) are mapped to practices content (PRAK) and generic estimating data (GEST), and these data do not change over time except for the effects of continuous process improvement. The generic tasks are mapped to the very general SOW tasks to form the detailed planning steps (PLN). Generic estimating data is also mapped to the program estimate (PEST) based on the planned tasks. Planned program tasks (PLN) are hooked to program events (EVENT) and linked to the program schedule (SKD).

The program planning data can be created through this process quite rapidly with contributions from many sources into a common database. The planning work can be accomplished by a team formed from functional department personnel assigned to a program planning team each mapping their charter department tasks to the program SOW tasks. If the integrated management system is applied, the planning strings can be formed in the process of this mapping activity and provide the organizing codes including significant accomplishments and accomplishment criteria drawing explanatory data from the generic data. The SEM/SEMP content would be contained in the PRAK database in this system.

The generic estimating system should be loaded with mean cost data for accomplishing corresponding tasks. As these tasks are mapped to program SOW elements this cost data should be acted upon by schedule considerations and adjusted, if necessary, for program differences.

It is true that the data in the SOW database, for example, may be generated into a very standard printed SOW document. But, at any time, anyone may also directly access the networked information in the database encouraged by Figure 19.14. The generation of paper from a database is not time-consuming but the approval, reproduction, and distribution of the paper originally and for change purposes can be. With paper documents, you also run the risk that the file cabinets full of these plans contain outdated information. The only thing that saves you from errors is that no one reads the information after it is printed.

The direct use of planning database content can speed up the planning and the information access processes tremendously if your staff can become familiar with the use of raw database content rather than paper documentation. It also assures that everyone is using current information. Many of us find it difficult to use database content directly because a printed document provides an information integration function as well as simply making the information available. We find it easy to refer from one part of a document to another in paper form. More and more people, however, are becoming comfortable with direct database use and in time, this will be as easy for many people as you and I might find document use today.

If it is important to have the documents in complete document form rather than using direct database access, the listed tools can generate complete documents that may be loaded into a library hosted on a computer server. These documents will then be available to all via the network. If any one simply cannot use this information efficiently on the screen, they may print one out for their own use.

Most of the tools listed in Table 19.6 offer the direct database user a form with one paragraph or fragment exposed at a time. The document director tool listed in Table 19.6 offers the unique feature of appearing to the user like a document even though the content is in database records. Any of these tools, however, can be used to generate a screen or paper output of a document format from the database content for viewing/reading purposes.

So, it is possible to document our current process, change the documentation in accordance with an effective continuous process improvement program, and quickly generate program planning data based on that generic planning data while incorporating traceability information that ensures completeness and avoids omissions and redundancies.

19.5 JOG system engineering planning model

19.5.1 Goals again

In Chapter seventeen, we stated some important goals to achieve in our planning efforts. These goals were spread across the programs and the functional structures. They were directed at aggregate simplicity for simplicity encourages better management. This is not intended as disrespect for the intelligence of those attracted to management, rather a statement of reality. The simplicity of process encourages more time to focus on real problems where our intellect can be applied to best effect. The mundane will generally take care of itself. Those goals were

1. Clearly state the responsibilities of the functional departments in terms of providing coordinated quality resources to programs (people, tools, and practices) that are continuously improved in a way coordinated within the enterprise to maintain.
2. Clearly state the work necessary to perform on a program so as to encourage understanding of the work at all levels on the program and minimize the management difficulty with the result that management intensity can be placed on smoothing current discontinuities and identification and mitigation of future potential risks.
3. Minimize the program planning transform complexity to the end that everyone can contribute from a position of knowledge in pouring

generic planning data into the program framework to create the minimum cost most effective program plan that results in customer best value and enterprise profit.

19.5.2 Functional structures

The functional organization is responsible for the generic capability of the enterprise. It is from this resource that all of the enterprise's programs spawn. As we have said, this involves people, best practices, and tools. The traditional way of organizing a functional organization, adopted in this book, is by the knowledge sets generally also respected by universities. In Chapter seventeen, we assigned these functional groups department numbers, Figure C-1 in Exhibit H/Appendix C illustrates the structure, and Table C-1 lists the charters for these departments. These department numbers are one of the fundamental organizing markers for the enterprise. The managers of these departments depend on programs to fund the employment of their people so they need information about program demand for these people. Therefore, programs must organize their work in such a way that it is possible to establish traceability between budgets and the functional departments as noted in Chapter seventeen.

These functional departments came into being in our hypothetical enterprise by mapping functionality to them from our generic process diagram (illustrated in Exhibit H/Appendix A). These departments thus formed have responsibilities for supplying work corresponding to these allocations to programs that implement these process steps or tasks. When a program maps work to a functional department as a result of selecting generic tasks from the process diagram, they bring in people with specific skills and knowledge, tools, and best practices created by the principal responsible department and supporting departments.

These two entities, department numbers and generic task identifications, should be included in the program work planning structures in some way. So, the program work identification for functional analysis within a particular IPPT should include F41114-D2162. This covers someone from system requirements department (D2162) performing this function (system and item requirements analysis) on a program.

19.5.3 Program structures

An enterprise will generally have more than one program in the house at any one time so we should have some way of differentiating work for one program from another. Over any reasonable period of time, there would not be more than 99 in the house so we might use the program identifier PXX where XX is the specific program such as P05. By extending this numbering system to the base 60 (all of the numbers, all of the capital letters less O, and all of the lower case letters except l), this system could identify 3600 programs before recycling the symbols. In Chapter seventeen, we added the identifier called stage or phase defined in context with the customer's acquisition life cycle model. Stage S02 might be an early development phase in a DoD program.

In Chapter seventeen, we encouraged joining together the WBS notion with architecture identification and identification of product-only entities within this structure. Under these ground rules, it should be possible to identify all work in association with the product architecture, an example of which might be A12 corresponding to an onboard avionics system that we assign to IPPT 2. The functional analysis work done by this team for this item by department D2162 can be uniquely identified by the string P5-S2-E-3-A12-F4111441-D2162. Note that this string ID would be unique to IPPT 2 because IPPT 2 has been assigned responsibility for item A12, which appears in the string. Team 3 might be assigned responsibility for item A13 for which the same kind of work would be required but its corresponding string would be P5-S2-E-3-A12-F411141-D2162.

Finally, a particular piece of work, especially at the higher levels of indenture, may extend over a long enough period of time to warrant partitioning into a series of smaller and more manageable activities. Let us apply the event numbers covered in Chapter seventeen to associate work with particular time frames intervening between two events. If we set up the events listed below in Table 19.7, we might designate a particular piece of work with the planning string P5-S2-E3-A12-F4111441-D2162. This is functional analysis work performed by someone from D216-2 as a member of IPPT 2 on program 05 in stage (phase) 2 leading up to preliminary design review (PDR) for architecture item A12.

A complete program, P05, may require thousands of these planning strings to define all work, but in so doing we will have achieved the same ends as the USAF method covered in Section 19.4 without the need to cross map so many things to the planning strings. We have used some very fundamental business entities in identifying the work directly. These

Table 19.7 Program Events

Code	Event name
E1	System requirements review
E2	System design review
E3	Preliminary design review
E4	Critical design review
E5	Functional configuration audit

Source: Grady, J., *System Engineering Deployment*, CRC Press, Boca Raton, FL, pp. 174–213, 2000. With permission.

are identification codes that everyone on the program can be very familiar with and not have to think about relationships between one set for program work identification and other sets used within the enterprise.

While this approach is different from that developed by the USAF, it accomplishes the same purposes providing unique identification of all work. It does not provide within the codes themselves the significant achievements or completion criteria of the USAF model but each piece of work identified by a function-department combination should have associated with it a clear statement of achievement expected with completion and a criteria by which one can tell when it is complete. Those two components are not properly part of the work identification string but attributes of a piece of work in the author's view. Another way of saying this is that the parameters captured in the work identification string are all key attributes in a database sense that might be used to capture this planning data and the significant achievements and completion criteria are non-key attributes.

19.5.4 Mapping generic identity to program structures

Figure 19.15 illustrates the flow of work identification using this identification string concept. The enterprise vision is decomposed to identify needed processes or functionality allocated to functional departments

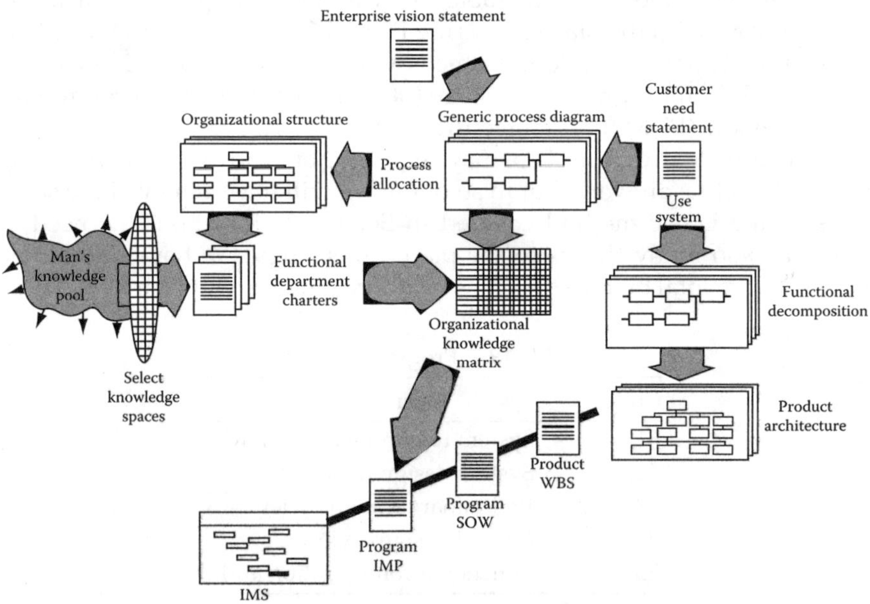

Figure 19.15 Overall work identification flow. (From Grady, J., *System Engineering Deployment*, CRC Press, Boca Raton, FL, pp. 174–213, 2000. With permission.)

forming charters for those departments. When a customer expresses a need, we decompose that need and allocate functionality to things of which the system shall consist. The major items in the product system become the fundamental planning centers for the program.

The program manager determines at what level in the architecture it will manage the program, selects architecture elements at that level to form IPPT about, and builds the program finance system about them as well. The finance people are free to call these entities by the term WBS but they are simply architecture IDs of particular interest from a planning perspective. When we add the program and event identifiers, we have a series of planning entities such as the following examples:

P5-S2-E1-A12	Item A12 Planning Hooked to Event 1
P5-S2-E2-A12	Item A12 Planning Hooked to Event 2
P5-S2-E3-A12	Item A12 Planning Hooked to Event 3
P5-S2-E4-A12	Item A12 Planning Hooked to Event 4

These WBS entities are each expanded into more detailed planning entities by selecting the generic process tasks that relate to the event in question. The kind of work that would properly occur between Event 2 (system design review) and Event 3 (PDR) for item A12 accomplished by IPPT 2 could be characterized as follows:

P5-S2-E3-A12-F4115	Form, charter, and train IPT
P5-S2-E3-A12-F41114	Item requirements analysis
P5-S2-E3-A12 F41131	Review and release specifications
P5-S2-E3-A12-F412	Item preliminary design
P5-S2-E3-A12 F4124	Item PDR
P5-S2-E3-A12-F41133	Validation of product requirements
P5-S2-E3-A12-F41124	Plan qualification verification process

19.5.5 Functional department mapping

The program planning work covered above can be accomplished from the top down by a small program planning team. The remainder could as well no doubt, but it is suggested that the functional departments be brought into the process to map their charter elements into the planning entities listed above. Note that it would be possible to plan the program at any level of detail desired. First, we can decide on the architecture depth selecting either a very low level leading to great complexity in the planning data or plan the program at a high level of architecture. We might choose to plan the work at the end item (A1 air vehicle, for example), subsystem (A12 avionics, for example), or component (A125 onboard computer, for example) level.

Three level	Four level	Five level
F411 Grand system requirements analysis F412 Preliminary design F413 Detailed design	F4111 Define product system requirements and concepts F4112 Define process system requirements and concepts F41113 Manage requirements at PIT level	F41111 Product system concept development F41112 Product system operational environment definition F41113 Perform applicable documents analysis F41114 System and item requirements analysis F41115 Item concept development and review F41116 Verification requirements analysis F41117 Change released specifications

Figure 19.16 Task planning levels. (From Grady, J., *System Engineering Deployment*, CRC Press, Boca Raton, FL, pp. 174–213, 2000. With permission.)

Second, we can also choose the generic process intensity used as the basis for planning. We could choose to use a third-level task identification for task F411 as in the first column of Figure 19.16, a fourth level for task F4111 as in the second column, or the fifth level of task identification as in the third column for task F141111. For simpler programs, we would want to plan at a fairly high level. A more complex program may require more detail in the planning work. Generally, we must fight the urge to plan a program in more detail than is necessary.

The final step, no matter the level at which we choose to do the planning, is to map functional departments to the planning strings crafted above. By way of example, P5-S2-E3-A12-F41114-D2162 represents a complete planning string. All of this planning work would most often be accomplished during the proposal effort. The strings formed up to this point could be crafted from the top down and people from the different departments that will have to perform work on the program brought in to accomplish the detailed estimating and planning. This is accomplished by each functional department representative mapping work to the process steps called out in the preliminary planning strings. These relationships should already have been created in the generic planning so the only real work to do is to estimate how much time will be required to accomplish the work on the program.

19.5.6 The apparent disappearance of the WBS and SOW

In the USAF numbering system, the WBS and SOW structures appeared as intermediate work organizational entities. This was a good idea and has not entirely been dispensed within the numbering system covered in this chapter. The WBS is simply the product entity ID possibly linked to program, phase, and event IDs such as P5-S2-E3-A12. The WBS is expanded into a SOW by adding high-order functional strings to the WBS strings to form P5-S2-E3-A12-F41. This series of strings can be formed from the top down by the PIT. They can then be introduced into the early planning work as a lure for the detailed program planners derived from the functional departments to expand by decomposing them to lower-tier functions and department identifiers thus completing the IMP content.

The WBS and SOW may be used as intermediate program planning strings and published under those names as detailed planning drivers or simply embedded in early IMP content and never really used as separate planning documents.

19.5.7 Integration of inputs into the IMP

All of the inputs to this planning process will not represent the real needs of the program. Some department representatives will fail to identify needed work while others will erroneously or purposely identify work that is not required. All of the inputs have to be reviewed and assembled into aggregate planning string sets and voids and excesses identified and corrected. The total of all of the planning strings arranged in alphanumeric order forms an IMP identifying all of the work required on the program. Obviously, the integration work must be accomplished by people with a system perspective from the PIT.

19.5.8 Integrated master schedule

Every element in the IMP must appear on program schedules at some level of indenture. The scheduling representative on the PIT should craft the top-level IMS showing all IPPT work and lower-tier work in the context of the high-level planning strings noted above but not necessarily detailing that work. The responsibility for expanding the schedule data can either be retained in the PIT or delegated to each team that could be tasked to craft a team schedule subject to PIT review and approval.

19.5.9 GSD planning development process

Figure 19.17 illustrates the overall process described in this section. This work should be performed by a proposal team that could be the precursor of the PIT organized as a function of the volumes to be produced rather than the structure of the product system to be created.

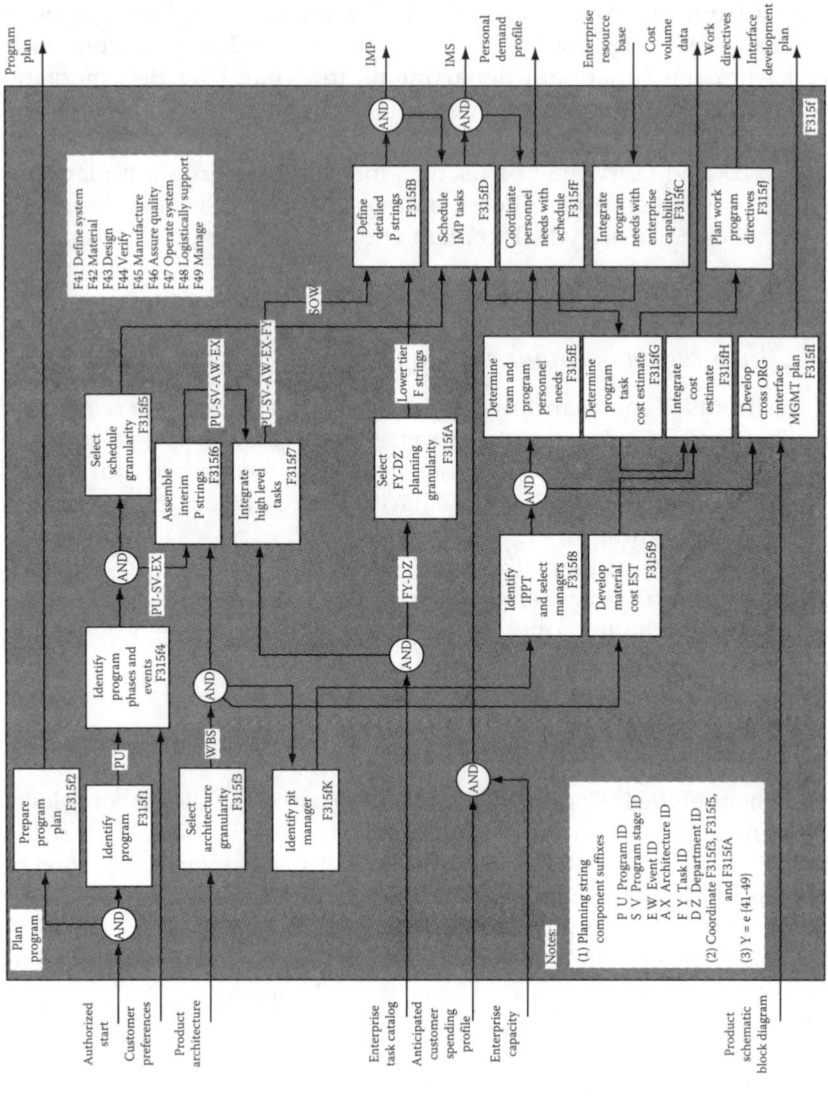

Figure 19.17 GSD program planning process.

part six

Program implementation management

chapter twenty

Proposal and concept development

20.1 New business acquisition

An enterprise cannot long exist without new opportunities of developing systems. A development organization subsists on customer problems for which customers are willing to compensate the developer. Figure 20.1 illustrates several elements of the work that must be done to constantly feed the enterprise new programs.

Customers may ask for information about your enterprise's willingness to participate in a program they are entertaining in the form of a request for information (RFI). The enterprise response should be to respond with a message about its intent to respond (yes or no) to a request for proposal (RFP) and, if it intends to do so, to provide information about its qualifications in response to any detailed information requirements of the customer.

In order to be ready to respond to an RFI and an RFP, an enterprise must be constantly doing its homework, including conducting independent research and development (IRAD) related to the enterprise's product line and customer base, performing precontract work to improve understanding of the customer's needs and continued marketing work to seek out new customers, improving the relationship with existing customers, and seeking information of value in new business activity.

20.2 RFP and proposal process overview

The most intensive work that is undertaken for new business activity is in responding to customer RFPs. Figure 20.2 illustrates this process. The customer creates an RFP describing their need and providing rules for responding often through interaction with potential suppliers. The RFP is submitted to potentially adequate enterprises, each of which must make a decision about whether or not to respond with a proposal. This decision must be based on the chance of winning and the value to the enterprise of a win.

Given a decision to compete, three principal activities must be undertaken: accomplish applicable document analysis, create a program

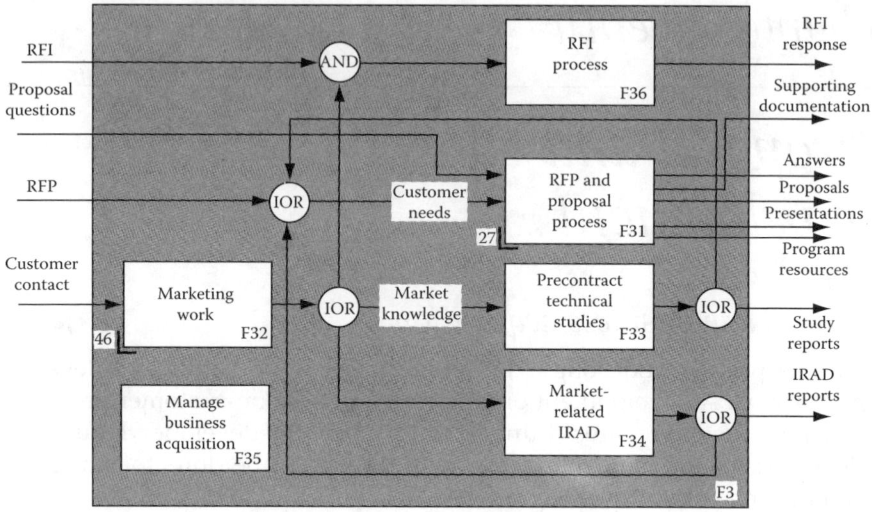

Figure 20.1 New business acquisition process.

infrastructure, and prepare the proposal. The completed proposal is submitted to the customer on time, and a waiting period is entered while the customer reviews the proposals and reaches a selection decision. In the process of so doing, the customer may require the respondents to answer questions about their proposals and respond with a best and final offer (BAFO), generally encouraged by customers to squeeze any final cost concessions out of the contractors. The customer may also require contractors to prepare and deliver an oral presentation either in person or via tape.

Once the selection has been made, the winning contractor and the customer enter and complete contract negotiations, resulting in the actual contract that both sign. If the winning contractor has applied the right amount of resources to creating a program infrastructure, they will be ready to start contract work at the time contract funding becomes available.

20.3 Prepare proposal

Figure 20.3 offers a proposal preparation process expanding on function F315 of Figure 20.2. The primary components of this work involve actually writing the required volumes of the proposal, but considerable effort must go into clearly understanding the customer's need through requirements analysis and initial product architecture identification. The latter is needed in order to describe product concepts and planned contractor work related to these items as well as to develop an appreciation for the things in the system for which program planning strings are assembled.

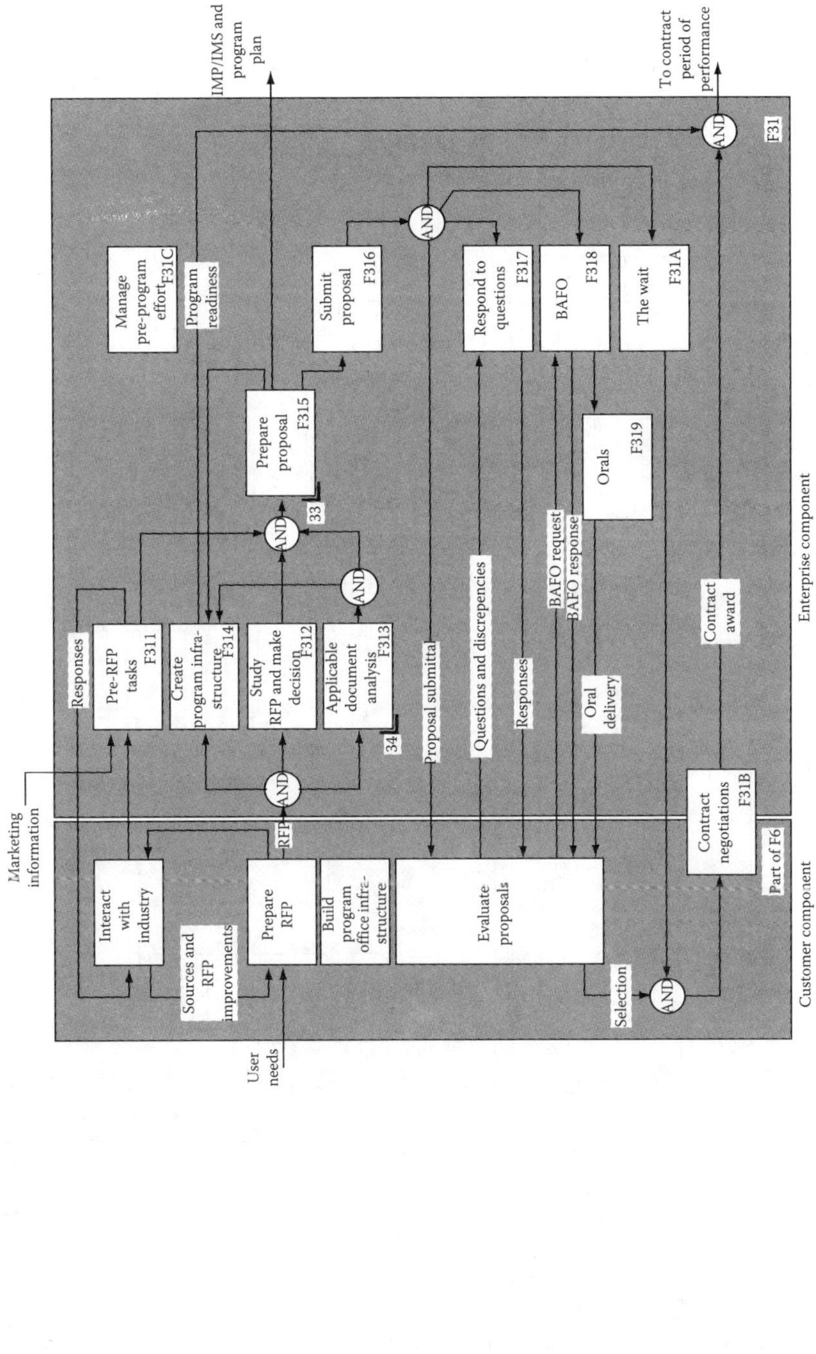

Figure 20.2 RFP and proposal process.

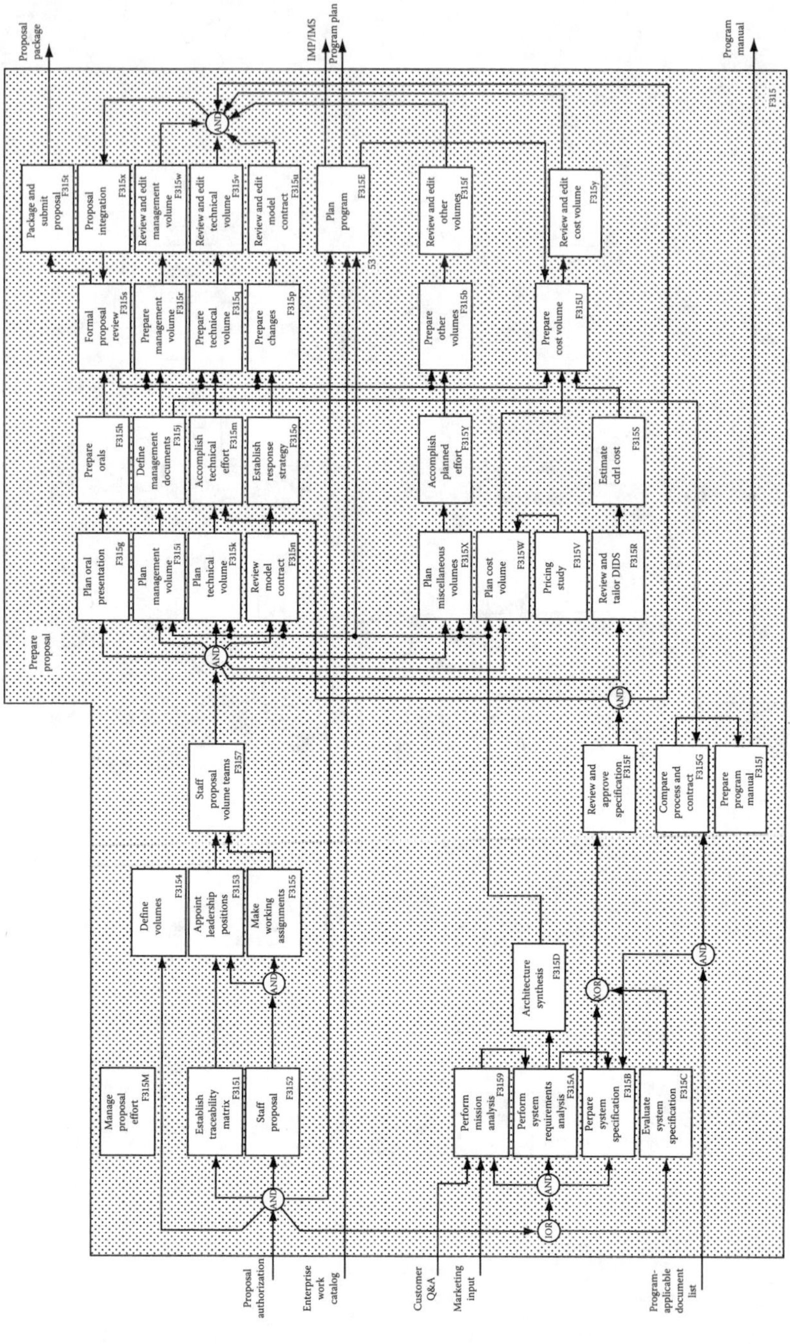

Figure 20.3 Prepare proposal.

The development cost must be determined including the task-oriented cost composed of human labor, the material estimate based on the things identified in the product architecture supplemented by all of the special test articles and any special facilities required in development.

This is the time that the content of Part five of this book must be put into practice to produce the work breakdown structure (WBS), the statement of work (SOW, if required), the integrated master plan (IMP), and the integrated master schedule (IMS).

20.4 *Proposal team structure*

The proposal effort should be initiated by the selection of a proposal manager and the formation of a program integration team (PIT). The initial requirements analysis work must be completed by the PIT to structure the technical, cost, and risk situations. The proposal team should probably be organized about the volumes that must be produced rather than in the form of an IPPT built around the product architecture resulting from the PIT analytical work.

20.5 *Post-proposal activity*

If the proposal is a win, the program must be formed, budgets set up, and teams recruited and facilitized. If the planning data follows the instructions in this book, it will be obvious what resources are needed and where they are coming from in the functional department structure. The program pulls the planning strings and the resources come with them. Actually, getting these resources may be something else and require assertive work on the part of the program manager to do so. This is an area where an enterprise integration team (EIT) can be very helpful in interceding between program and functional structures to balance program needs and functional resources based on what is best for the enterprise rather than what is best for individual program and functional department manager careers. If the proposal effort resulted in a loss, the management should make some effort to understand why and use this information to change practices as needed to encourage future wins.

chapter twenty-one

Resource acquisition and program facilitization

21.1 The major steps

All of the management activity involved in implementing the plan can be cataloged under several headings. We must first get the business through marketing and customer interaction, possibly involving a fairly elaborate proposal telling how we would satisfy the customer requirements, as discussed in Chapter twenty. Second, we must provide the program infrastructure, including personnel, facilities, materials, and practices, as discussed in this chapter. Then, we must implement program management of the three fundamental development steps: define the problem, solve the problem, and prove it, as discussed in Chapters twenty-two, twenty-three, and twenty-four, respectively. The problem to be solved must be defined in a series of specifications that are created through a requirements analysis process, ideally driven by modeling. Teams must be formed and presented with the responsibility to accomplish design synthesis work (design, integration, and optimization) compliant with each of the specifications. Obviously, we also have to acquire and process materials needed in a manufacturing process, and manufacture the product. Then, we must prove through verification work that the resultant design satisfies the specification content. Throughout the program work, we must accomplish four fundamental hallmarks of good management: (1) risk management, as discussed in Chapter twenty-five; (2) work performance management, as discussed in Chapter twenty-six; (3) cost and schedule management, as discussed in Chapter twenty-seven; and (4) configuration and data management, as discussed in Chapter twenty-eight.

A program should be served by an effective set of computer applications that assigned personnel are qualified and experienced in applying. Ideally, these tools would be to some extent coordinated or even connected as the enterprise works toward evolving a model-driven capability, as introduced in Chapter twenty-nine.

Subsequent to the production of some product, we may also be called upon to modify the product in some fashion. Finally, we should make some kind of after-action analysis to determine how well we did on the program and what we can do to improve our capability on future programs. This latter step should be going on throughout the program yielding metrics and lessons learned and, in the author's opinion, should be accomplished under the leadership of the quality assurance function.

Each of these steps involves unique technical and management activities mixed with generic good management techniques. The reader is encouraged to consult a good book on project management for coverage of the generic management activities, such as Harold Kerzner's *Project Management* (2003) in the current edition. This book will focus on system development—related management activities, commonly found necessary in the development of systems to solve complex problems.

21.2 A good plan

A program needs a good plan ideally formed using the techniques covered in Part five of this book. In the best of all worlds, a program will have a sound plan, the people working on the program will implement the plan faithfully, and conditions upon which that plan was based will unfold as foreseen in the plan. Few programs have been accomplished completely under these conditions because the plan crafted was a poor plan, the people failed to follow the prepared plan, and/or the future failed to comply with what was assumed would happen. Religiously following a bad plan is not a good idea either.

If an enterprise follows the structured approach covered in Part five of this book, it is hard to craft a bad plan. If the enterprise has an effective functional management staff responsible for supplying programs with good people, good practices, and good tools, all coordinated by an enterprise integration team (EIT), they should expect good performance in implementing their good program plans. There are some things that little can be done about, however. The future cannot be known with certainty. We can only think about what might be. But this is a great deal better than not thinking at all. Therefore, we should try to identify things that could go wrong with our plan and find ways to reduce the chances of these things occurring and the amount of damage done if they do occur. The art and science of doing this is called risk management, discussed in Chapter twenty-one.

21.3 Program organization and personnel

As the program comes into being with an appointed program manager, a staff must be assembled consisting of the leaders identified in Chapter ten and shown in Figure 21.1 in the darker gray area. This includes a PIT manager, a program business team (PBT) manager, and a manager for each IPPT; only two shown, but there may have to be more and they may be

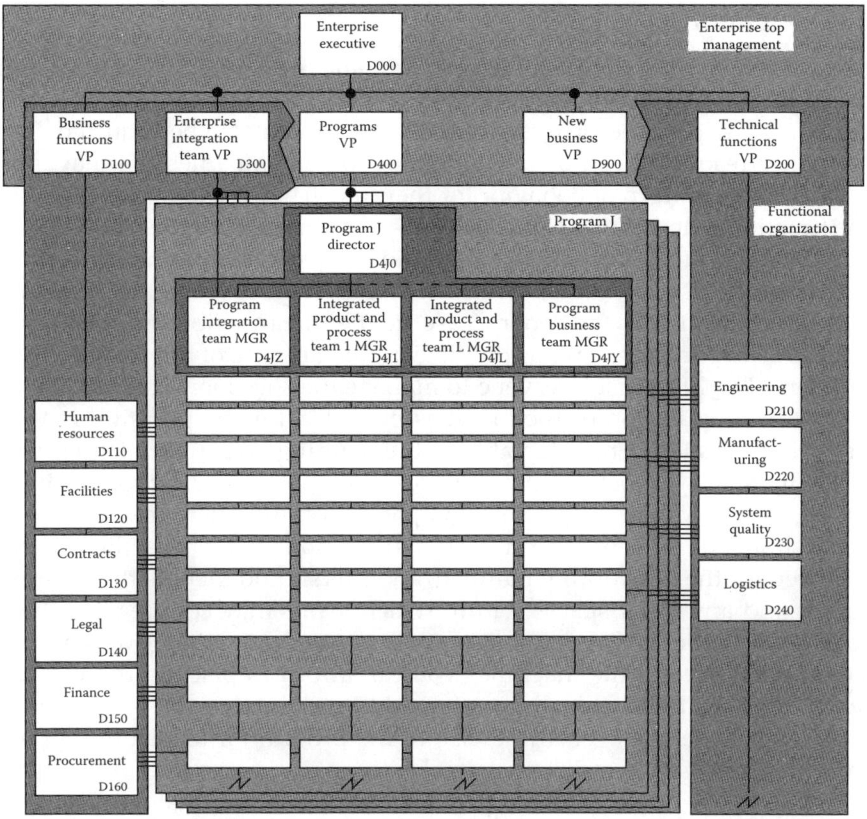

Figure 21.1 Typical program organization structure.

organized into a hierarchy on a very large program. Each team manager must acquire the personnel needed from the functional organizational structure consistent with the work defined in the IMP linked to the functional organization structures automatically by the work-cataloging strings that identify generic processes and responsible functional departments, as covered in Part five. Based on the relative win probabilities among programs awaiting a customer selection, PIT managers on the potential programs must interact with functional managers to line up IPPT staffing. The personnel assigned to a new program cannot all be assigned from new hires. It will probably be necessary to provide a mix of new hires and transfers from existing programs to new ones. If there are several new programs hitting the enterprise at much the same time, this can lead to a great deal of personnel assignment instability that can be injurious to existing program teams, so team leaders must assert special efforts to encourage continuing team success.

21.4 Facilities

In addition to the people, the managers must ensure that the tools and practices are pulled into the program as well. This may require computers networked within the program space, which must also be acquired. The computer resources can be a problem in some companies that have not worked out a pattern of behavior for their use. They could be provided by the functional departments to satisfy the charter services these functional departments must provide. They could be provided by an enterprise information systems department. Other possibilities exist, but these are two common ones. In the general case, neither the functional departments nor the IT department have a complete understanding of how to solve this problem. The EIT must intervene to manage the solution of this problem.

The configuration of the facilities provided for the program is very important. It should be adaptable to the teaming structure such that all of the members of each team are in close proximity, leading to the shortest possible physical communication pathways. Adequate meeting-room space should be present or modifications accomplished. The facility should be wired for the networking planned, and this should include the meeting rooms and any other spaces within which teams may congregate. These rooms should be outfitted with computer projection with network access. The more powerful the machines you can afford to place in these meeting rooms, the better. Ideally, you should be able to project engineering drawings as well as everything else in the program information library, including the use of engineering analysis applications.

Where it is not possible to provide each team with its own meeting room, the furniture should be laid out to permit the use of walls as stand-up meeting places where large sketches can be located and computer projection directed to, without overlapping into heavily traveled passageways interfering with their use for the meeting purpose. Alternatively, temporary partitions can be purchased or made, upon which large sketches, projection screens, and white boards may be placed.

21.5 Work product readiness

An enterprise should enter a program with a certain degree of readiness to produce work products. The work products produced in program tasks form the process interfaces. A later task will frequently require as an input a work product produced in an earlier task. The enterprise should have a list of every work product coordinated with the common process task from which it is derived. Each of these work products should be expressed as a template giving the generic structure of the work product, an item description (for a document, this is expressed as a data item description), and an example of a program work product. With this capability, every program should be able to start the early tasks with a high degree of readiness to produce the work products required to enable other early tasks. Many programs start out in schedule trouble, because they spend too much time during early program work trying to decide how they should report work results. Common process coordinated with responsible functional departments and enterprise management demand that functional departments provide generic work product materials that can be used immediately on a program, goes a long way toward preventing early program schedule drift.

chapter twenty-two

System requirements analysis management

22.1 *Overview*

In Chapter fifteen, we briefly discussed methods for performing requirements analysis and specification generation on a program. In this chapter, we seek to develop the tools to allow us to plan and manage the many tasks needed to make this effort successful. *System Requirements Analysis* (Grady, 2005) covers requirements analysis and management in detail.

The approach covered focuses on using an integrated product and process team (IPPT) concept staffed from a functional organization structure within a matrix. A program integration team (the system IPPT called the PIT) is assumed to be in overall control of the requirements analysis process for a program. That team would accomplish the requirements analysis work necessary to develop and maintain the system specification and the top-level specification for each IPPT. As each IPPT comes on line, the PIT provides the leader with a specification for the team item, budget, and schedule corresponding to the architecture item assigned, and a contiguous piece of the IMP/IMS also corresponding to that item. Team leaders may assign a system engineer the responsibility for coordinating all of the requirements analysis work for the team and the term SRA lead engineer is used in this book to identify that person.

The teams must develop an effective way to manage their development effort, generally in a top-down direction, and regulate the relative progress of the requirements analysis and concept/design development activities such that requirements analysis leads the design effort. A means is needed to assure that the principal engineers and IPPT leaders will respect the requirements-first direction. PIT must oversee the work of the IPPT to ensure that this happens.

The principal tool for managing program specifications preparation and maintenance is the specification tree. It is used to identify the architecture subset that will require the preparation of specifications and as a placeholder against which responsibility for requirements analysis and specifications preparation is identified. We must also have effective methods for the management of a specification baseline on a program. This involves a clear definition of what specifications are required (via the specification tree) and sure knowledge of the current status of each of the documents identified on the tree. Further, the paper and electronic versions of each current baseline document must remain synchronized.

Whether the program employs a paperless approach or not, it is important that a program library be formed to safely retain the electronic media for all specifications in the interest of minimum cost. If this library is computer networked and provided with the proper infrastructure, it can also greatly reduce the number of paper copies of documents reproduced, the use of outdated requirements, document review costs, and document revision costs.

We will complete the process management picture with a replicable pattern of behavior that can be applied as a module of requirements analysis activities for any item. Many of these modules may be fitted together to cover the complete requirements analysis task for a program.

22.2 Total quality management

Too often we find ourselves driven from panic to panic during proposal and program execution phases responding to expressed and implied customer requirements that force us into procedures different from our own internal ones. We tell ourselves that we risk being nonresponsive to a customer's request for proposal (RFP) if we deviate however slightly from the customer's requirements. As a result, we allow our customers to redesign our organization and methods for each program. These program differences prevent us from developing an identity of our own and from working toward perfecting our process to better serve all of our customers and our own interests.

Dwight Stones was a perennial world-class high jumper for many years and holder of the world record several times. He did not become a world champion in his field overnight or easily. But by watching Dwight perform just once you could see the secret to his success. You can apply this same secret to your enterprise's activities beginning today and, over time, become as great as you can be.

As Dwight Stones prepared to make a jump, you could see him mentally going through every step of his jump, reprogramming himself for the jump. During his approach and jump he tried to repeat a pattern of behavior that he had perfected over years of practice, practice, practice. But, he also learned from his mistakes and made small changes to prevent repeating them. Dwight Stones developed a baseline approach that he consciously tried to repeat each time and made small changes based on experience and education. Constant practice against a standard changed only in small increments for sound reasons (as well as a lot of natural athletic ability) elevated Dwight's performance to the peak of human capability in his sport during the period he competed. We can apply this concept in our work as well.

Total quality management (TQM) may be defined as a continuous process of improving an organization's methods resulting in increased customer satisfaction as a result of lower cost and better performance of products produced by the organization. The essential elements of an effective implementation of the TQM concept are

1. Understand the current process your organization uses and write it down in such a way that everyone may gain easy access to a description of their role and everyone else's role in the process.
2. Encourage everyone to be consciously alert while doing their job to the possibility of better ways of doing it that result in less cost and better product quality.
3. Empower someone to play the system engineering role applied to your process and encourage them to focus on seeking out suboptimization at the lower-tier organization level rather than at the overall process level.
4. Provide an organized method to collect input ideas on ways to improve the process and screen and prioritize them for action to make improvements.
5. Develop a minimum but effective set of metrics that measure your performance in critical areas with which you can monitor performance and plot improvements and identify areas most needing improvement.
6. Actively seek out your customer's views on your performance and listen carefully to their inputs.
7. Selectively implement the results of this continuous process study using your metrics to help identify the most costly parts of the process that will leverage the biggest change and change the written process description accordingly.
8. Provide an organized way to train the workforce to implement the process and continuous changes to it.

We should also study our competitors and other organizations we respect in an effort to acquire knowledge of what constitutes a world-class capability and compare our organization's current capability against this benchmark. This knowledge can help to steer and prioritize our improvement efforts. If we repeat this process, continuously taking advantage of lessons learned from each program application, there is a good chance that our organization will improve its performance and achieve increasingly greater success.

But, how can we possibly do this and avoid the risk of being branded nonresponsive in our proposals? DoD has identified this as a problem in which they need to cooperate in solving. In DoD Directive 5000.43 Acquisition Streamlining, they specifically instructed procurement offices to encourage contractors to apply their proven processes on contracts. This position is based on the notion that good contractors will apply TQM principles to their business and products and provide their customers with good value. The DoD Instruction 5000.2 revision issued in 1991 canceled 5000.43, but it includes the same intent and subsequent changes in DoD acquisition regulations have continued this framework.

It is true that DoD agencies at one time imposed many standards on the contractor through system specifications and statements of work. To the extent that this is still being done, it forces contractors to comply with external procedures not tuned to their development and production processes nor their facilities and tooling base. But, DoD Directives have radically curtailed the reference to military specifications especially in manufacturing and management areas. Acquisition guidelines encourage program offices and contractors to jointly tailor (adjust) any remaining applicable documents to permit the contractor to apply their processes wherever possible.

The way a contractor can stay on top of this problem is to have sound internal procedures covering how they perform their work and map these procedures to a set of applicable documents commonly applied by their customer base. Where these customer documents would require the contractor to perform contrary to their preferred process, the contractor should develop tailoring to make the customer document identical to their internal procedure. When a proposal is prepared against an RFP containing applicable documents, those documents, and tailoring offered by the customer, are compared with the generic tailoring previously prepared and the differences used to generate alternative tailoring initiatives offered with the proposal keyed to the contractor's minimum cost, best quality approach.

There will be cases where the customer will insist on their own tailoring version, but generally a convincing case can be made for your version because the result is your least-cost, best-quality approach satisfying their need and giving them best value. In the few cases where the customer cannot deviate from a particular requirement, the contractor makes an adjustment in their internal procedures for a particular program and adjusts the cost estimate accordingly.

So, it is possible to build your own organization's identity and avoid the stigma of noncompliance. If each program can be performed using the same processes, the contractor will benefit from the practice, practice, practice notion through which sports figures become world-class competitors. If a contractor is introspective about their performance on each program and consciously seeks to make small improvements in their process based on lessons learned from programs, they will constantly improve their baseline and practice will improve the performance of their people.

This combination of performance to a standard that is referenced to customer-preferred standards and continuous small improvements of the standard will move the contractor toward a goal of being as good as they can possibly be.

This book offers a comprehensive process for planning, managing, and accomplishing an effective requirements analysis and specification generation process, but it probably is not perfect for any given organization and program. The reader is encouraged to pick components from it and mix them with the good parts of their own existing process. Then you should work continuously to improve your process in the interest of better serving your customer's needs. As always, profit will follow satisfied customers.

22.3 Program initiation

We should not start work on a new program without a plan nor without preparation. The best time to accomplish this planning and preparation is during the proposal period preceding a contract win while all of the great ideas that flow into the proposal are still warm. The activities covered in this section should begin during the proposal period and the product of that work retained for use as a basis upon which to build during the contract period of performance.

22.3.1 Resource identification and acquisition

The lead SRA engineer must identify the resources needed to initiate and maintain the specification development and maintenance program. This begins in the form of an appeal to functional management for qualified personnel for their team consistent with the planned budget and schedule and to functional or program management for furnishings, computer hardware and software, materials, and working space that is consistent with the planned requirements analysis and specification development approach. The lead SRA engineer must then assertively work toward acquisition of these resources in a timely way relative to program events.

22.3.2 Program specification development methods

The PIT must identify any and all specification agents and define and/or negotiate a program specifications development approach with these agents, the program chief engineer, program management, and configuration management. A specifications agent is a person responsible for

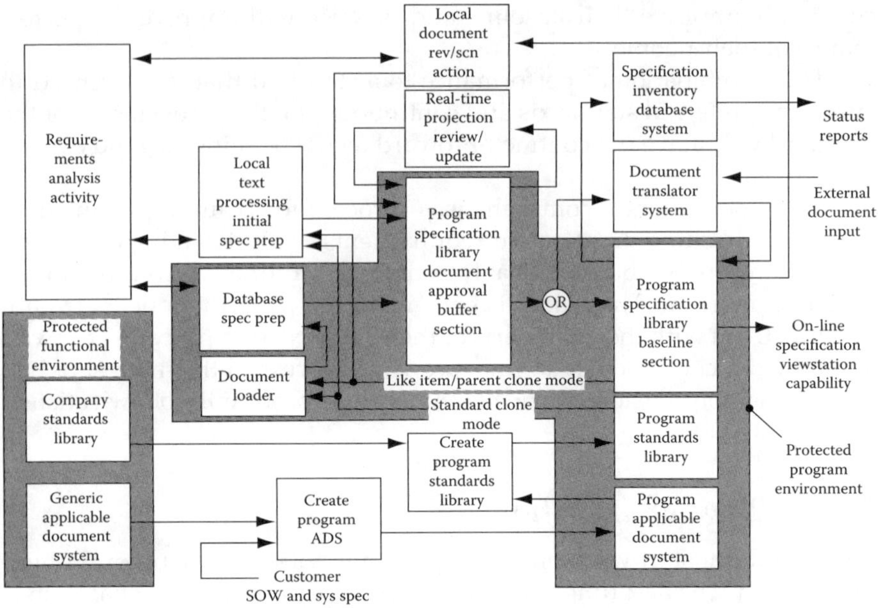

Figure 22.1 Specification development environment.

the development of one or a series of specifications and the configuration management of the electronic media for those specifications.

The recommended specification development environment calls for the use of networked microcomputers containing one or more specification libraries each operated for the specification agent by a specification librarian, in accordance with a published standard. Figure 22.1 illustrates such an environment.

On a program where the cloning strategy is used, one or more of these libraries should contain specification standards that are made available to all principal engineers with a specification responsibility in read-/copy-only mode. The principal engineer copies the standard to his or her local workstation and transforms it into a specification for the item based on the results of the requirements analysis activity they perform and coordinate. This document is made available in a buffer on the network for review and marked up or annotated by reviewers using networked computer markup software. After corrections are accounted for, the principal offers the document for chief engineer review. The released, approved document is loaded by the appropriate librarian into the expanding specification library, where the document resides in read-only mode.

Where the program is equipped for computer video projection in meeting rooms, the document review can be accomplished with real-time correction by the principal engineer in control of the computer driving the video projector. Alternatively, a program properly equipped could hold telephone conference calls between the principal, chief engineer, and reviewers with all of these people viewing the specification at their own desks on networked workstations. Once again, the principal engineer would be the only one allowed to change the text during the review.

Where the program uses a requirements database approach, the data should be created, reviewed, and approved while in the database form. Specifications are then created from the approved database content using a specification generator at baseline events. Only one requirements database should be used on a program unless classified data is involved where a separate database for classified data is appropriate. In Figure 22.1, a real-time computer projection review is illustrated based on a text document. When using a requirements database, this computer projection review should be done directly from the database to prevent the text document from becoming uncoordinated with the database content.

During the planning phase, the program should select one of our three strategies of cloning, structured analysis (preferably using a database approach), or freestyle. As a program matures from one phase to another, it may make sense to change strategies. A system requirements document (SRD) might be created very efficiently for a concept phase program by an experienced system engineer using the freestyle strategy. In the demonstration and validation and early engineering and manufacturing development phases, the program should use structured analysis. Finally, in the production phase, the large set of baseline documents in the electronic library may make it possible to use like-item cloning very effectively.

22.3.3 *Applicable documents assessment*

During the formative program stages the proposal or program team must perform an applicable document selection and tailoring exercise. If the customer provides a detailed list with tailoring as a part of a system specification and statement of work, the task is to evaluate the documents on that list against company practices and offer alternatives that bring the references into alignment with company practices. The customer is interested in the best value for their dollar and company practices should reflect the best you are able to perform in developing a good product for a reasonable cost. *System Requirements Analysis* (Grady, 2005) includes a comprehensive applicable documents approach.

22.3.4 *Program specifications plan*

The specifications plan is a subplan to the system engineering management plan (SEMP) when separately prepared. As an alternative it can be integrated into the SEMP. The plan should address the nine specific items explained below as a minimum.

22.3.4.1 Program-specific document identification

The plan must contain a clear methodology for identifying what specifications will be required on the program. The methodology encouraged is to review the items on the product entity structure, ideally created through an application of structured analysis, and compare the item characteristics against a specification selection checklist. *System Requirements Analysis* (Grady, 2005) provides such a checklist.

22.3.4.2 Responsibility assignment

The plan must define a methodology for the assignment of responsibility for the development and maintenance of specifications identified as noted above. This can be as simple as a program specification list with department numbers and names after each document. In some companies, engineers assigned to program teams from a system engineering organization prepare all specifications making this list a no-brainer. The plan must also offer an effective means to communicate this list to the assigned IPPTs or principal engineers following a formal review of the list by the chief engineer.

This book uses the IPPT acronym to denote a cross-functional team composed of engineers, analysts, and support personnel assigned under a leader to develop a major system item. The term principal engineer is used to identify an individual engineer responsible for the development of a lower-tier item possibly under the cognizance of an IPPT. In the context of this book, principal engineers will generally be dependent on support from engineers and analysts not under his or her supervision who may also be assigned to the parent IPPT and shared by several lower-tier principal engineers. The principal engineer for an IPPT is the IPPT leader.

The preferred method of assigning responsibility for requirements analysis and documentation is for the responsible team leader to gain the support of the chief engineer (PIT leader or principal engineer), program management, and functional design supervision in the identification of principal engineers for product entities and to identify those principals also responsible for the development of the associated requirements document defined on the program specification tree. This responsibility normally can be broken into four levels and mapped to the responsible program organization (PIT or IPPT). The personnel to manage or accomplish the writing of these documents should be obtained by program teams from the functional departments noted in Table 22.1.

Systems engineering could be held responsible for requirements analysis and specifications generation for all items designated by the customer as configuration items. Seldom will any of these items be designated as

Table 22.1 Principal Engineer Levels

Level	Scope	Principal organization
1	System, segment, prime item, facilities	Systems Engineering
2	Subsystem	Design Systems Groups
3	Component	Design Groups
4	Parts, materials, and processes	Parts Engineering, M&P

subsystems but some may be component-level items resulting in a potential conflict with the pattern defined in Table 22.1. These few ambiguities can be decided on a case-by-case basis.

It is understood that all of the work noted in Table 22.1 is to be accomplished in the PIT or IPPTs, and we are only concerned with who shall lead the action. The entities listed after level 1 will all require participation by a wide range of specialized engineers who are members of the PIT commonly. Subordinate items within that product entity branch may then be assigned by the team to lower-tier principal engineers. The same team that is responsible for concurrently developing a product design solution mated to a manufacturing and logistics support concept is tasked to first concurrently develop the requirements for the item.

22.3.4.3 Specification scheduling and statusing

The plan must contain a specification schedule in terms of due dates and a means of keeping track of the status of each document in work at any point in time and how the program team is doing over all in specification development in terms of schedule compliance. The actual schedule data may be included in the IMS. You should negotiate due dates with the document principal engineers that are consistent with overall program schedules for the development of item concepts and designs and to include these dates in the specification list.

Actuals then can be included in this same list and a dual "S" curve generated from these two sets of dates to reflect general program specification development performance for periodic review by the chief engineer. The due dates and actual dates need to be explained in the plan. For example, the due dates could be principal engineer document completion dates, review dates, or release dates. Several dates could be tracked to provide feed forward information.

22.3.4.4 Specification baseline identification

The plan must provide a means to identify program baselines in terms of dates and/or program milestone events that will be generally recognized by the program. These may coincide with customer-defined baselines (functional, allocated, and product) only or include additional internal baselines as well. Clearly, this is a responsibility of a configuration management specialist on the PIT.

22.3.4.5 Baseline definition documentation

The plan must define how the program will define a specification or requirements baseline. The suggested approach is to prepare a program requirements baseline report (RBR). This report includes a list of all of the program-specific specifications with status, revision, and responsibility information and a list of all applicable documents the contractor is required to comply with and the contractually binding tailoring associated with each. The formal specification tree can also be included in this report. All of this data could alternatively be appended to a program specifications plan, but the constant update of the data would make configuration control of the plan (which should not change all that much) a constant problem.

22.3.4.6 The physical baseline

The plan should define what the actual physical specification baseline is. In the past, the normal physical baseline has been a set of signed, released, paper masters located in a vault augmented by any group released masters retained in engineering department repositories. In many organizations today, the masters are actually in electronic media residing on some program computer in a CAD application. As an interim step toward the future, one could locate the network file server containing the released specifications in the same Vault, or under their control, that contains the signed paper masters. Somewhere in the release cycle you must ensure identity between paper and electronic baselines. This could be done in the Vault or by a preceding release checking function.

22.3.4.7 Electronic specification library

The program should operate an electronic library of specifications as the working masters for all specifications. This may comprise a central computer-networked library (under release group control) or a distributed, or federated, set of libraries each with its own librarian. These libraries can provide the program with online, electronic access to all specifications if networked, but as a minimum, they must protect the investment in typing labor to support future revisions.

22.3.4.8 Specification change management

The plan must cover how changes will be handled on the program and how they will be handled differently in different program phases if the process is graduated in some way. This may be as simple as to refer to the change procedures in a company work standard or configuration management manual.

22.3.4.9 Program preparation

Ideally, the manual would call for the appropriate functional department to prepare for program specification success by identifying a preferred specification structure that coordinates with the modeling work planned. For example, the author has devised a universal specification format that is coordinated with the use of a combination of UML, SysML, and a few modeling artifacts from traditional structured analysis. The format will work no matter the nature of the entity in terms of hardware or software. The pattern of use should call for the use of modeling and traceability captured between all requirements and the model from which they were derived.

22.3.5 Program specification standards

If the program intends to apply specification standards as templates or boilerplate for use in developing program specifications, these standards must be controlled. The program/proposal requirements lead engineer should be held responsible for the preparation of a set of program specification preparation standards (boilerplates). The preferred approach for this task is to convert a previously prepared set of generic specification standards available in electronic media into program standards by including the results of (1) an environmental use profile and zoning analysis, (2) an identification of the specialty engineering disciplines needed for the program in the form of a specialty engineering scoping matrix explained in *System Requirements Analysis* (Grady, 2005), and (3) the program applicable documents analysis as amplified below:

1. Customer requirements compliance: Company standards can be derived from a document such as MIL-STD-961E with possible adjustments in style for electronic publishing compatibility. These changes may not be fully acceptable to the customer in which case the program standards will have to undo these changes. Other customer requirements must also guide the adjustment of the company standards.
2. Environmental requirements analysis: It is normally senseless to develop a generic company boilerplate for product environmental requirements, but developing generic program environmental requirements boilerplate is commonly a money-saving action that makes it unnecessary for each principal engineer to repeat essentially the same analysis. The results from an environmental use profile and zoning analysis should be captured in program specification standards. The component-level principal engineers should only have to identify in which zone of a prime item their component is located and pick off the environmental requirements identified for that zone.

3. Applicable documents analysis: Another major piece of information to be added to the program standards is the results from the program applicable documents analysis. This should consist of a list of customer compliance and guidance documents with some tailoring. This list should be included in program standards section 2 as a shopping list. The item principal engineer should exclude any information not referred to in the document. Any other documents not on this list that the principal engineer adds as compliance documents should be carefully considered by the Chief Engineer for need at an internal review of these documents listed.

The specifications plan must tell how the program standards will be made available to the principal engineers (on a network, copied to the principal engineer's blank disk by the lead requirements engineer, given out in paper form, etc.).

While the program must convert the company standards to program standards, it may also be useful to convert the program standards into multiple type standards. For example, given that a company standard for in-house requirements documents has been converted into a Program XYZ standard, the support equipment design function may find it advantageous to create a support equipment standard from the more general program standard. Likewise, if 15 valve specifications are required for outside procurement, it may be advantageous to prepare a valve standard from the program procurement specification standard. Alternatively, a general valve specification could be prepared that is referenced in each specific document.

22.3.6 Program specifications library

22.3.6.1 PSL initiation

A program specifications library (PSL) should initially be set up during the proposal phase and plans laid for full implementation at contract award. The program may have more than one of these libraries each of which includes specifications under the responsibility of one or more departments or all of these documents may be merged into a single library for the program. The aggregate of these libraries are referred to as a PSL. A PSL has the following major components:

1. PSL librarian. A PSL should be operated by a librarian trained in the use of the computer hardware and software used on the PSL, Library procedures, and document configuration control procedures. There should be at least one other person qualified to act as the librarian to provide service during periods of regular librarian sickness and vacation.

2. Library data. A central PSL should have control over (1) the master electronic media for each released specification, (2) all electronic program specification standards masters for system, configuration item, subsystems, components (in-house or procurement items), parts, materials, and process specifications, (3) program-peculiar specification status database information, and (4) an applicable documents electronic database containing a listing of all customer applied compliance and guidance documents with tailoring.
3. PSL workstation. One workstation under the control of the central library agent should be identified as the PSL workstation. This workstation should be capable of (a) reading and editing all PSL documents, (b) the preparation of floppy disk copies of specification standards for principal engineer use or otherwise providing for distribution of the standards, (c) the control of the PSL baseline whether installed on a computer network server, managed on the PSL workstation hard disk, or located on floppy disks under librarian control, and (d) maintaining the status of the electronic baseline.
4. PSL workstation software. Software must be installed on the PSL workstation to provide the capabilities noted in "3" above.
5. Other workstations. The program needs one or more other workstations, as a function of the number of documents that might be simultaneously worked, capable of preparing program specifications within the context of the program specification standards.
6. PSL library standard. The library must be operated in accordance with a company or program procedure. The procedures for all libraries (if more than one) on a program should have some consistency.

22.3.6.2 PSL variations

Libraries can be established with a range of characteristics as a function of program needs. Four specific ways they may vary are

1. PSL variations by responsibility. A system engineering PSL may be created as a centralized repository for all program specifications, it may be structured to retain only system-level documents, or it may capture something in between. Alternatively, the program may elect to have some other department (data management, engineering administration, the Vault, or configuration management, for example) manage a central repository for all program specifications. Two or more specialized libraries could be created, each with a librarian (software, parts engineering, materials and processes, and all other specifications, for example).
2. PSL variations by storage media. The PSL data may be retained in electronic media in any of several ways. The documents may be

stored on floppy disks, removable hard disks, magnetic tape, or a hard disk internal to a stand-alone workstation or network server. In any case, data must be protected against loss of the primary storage media by suitable backup procedures. One could use only paper copies produced on typewriters, but most companies are well beyond this evolutionary stage.

3. PSL variations by connectivity. The PSL may be organized as a stand-alone computer workstation or located on a network server (the preferred configuration). Wherever possible, the documents should be stored for accessibility by the program population through a computer network for the purposes of reading, reference, or under suitable controls, for update. Any updating or revising should be done external to the PSL.
4. Variations by word-processing software. The program may use any of the many good computer word-processing or desktop-publishing software applications as a function of customer preferences, available program computer assets (central word processing center, for example), the preferences of program personnel, or the resources provided by supporting departments. Wherever possible, the program should reach an agreement on minimizing the diversity of applications in the interest of assuring the simplest possible access to documents by the widest population on as many workstation types as possible. Microsoft Word is one word processor that can function on Macintosh and Windows machines equally well. Networking solutions exist that permit a computer network composed of IBM clone and Macintosh machines to jointly share a central repository with the documents appropriately formatted in a word processor like Microsoft Word. If a company has only one kind of machines in use, the solution only becomes easier.
5. Variation by machinery. Much of the prior comments apply to the use of word processing as the principal means of capturing requirements, but the enterprise should be moving as rapidly as possible to the use of requirements modeling and analysis computer applications as resources permit. A networked capability should be available for use by principal engineers and assigned analysts. The requirements would all ideally be derived from these modeling experiences and flow into databases within which traceability and other management activities can be accomplished. In these cases, provisions must be made available for configuration management of the evolving mass of data. As specifications are released following a formal review and approval, the related data should pass over to configuration management control requiring a requirements review process to make changes.

22.3.6.3 *Security*

The data residing in the PSL must be secure from unauthorized change. If located on a network server, it should be protected from change, erasure, or addition in that environment by anyone except the PSL librarian. If located on the hard disk of a stand-alone machine or on a series of floppy disks, the librarian must protect the data using some combination of physical security and file access security. How bad can it get? One company suffered the physical loss of the whole hard disk containing the complete specification library. It was stolen apparently by an employee notified of his termination.

Should the PSL contain classified data (secret or confidential), the data must be handled, stored, and processed in accordance with customer-classified data regulations. In general, it will be prudent to locate classified data on removable computer data storage media that can be physically marked and safeguarded in an appropriate classified data cabinet. Generally, this data must only be processed on a stand-alone computer workstation specifically approved for use with classified data.

All PSL data must be protected from willful or unintentional loss or damage through an effective data backup procedure. Depending on assets available to the program, this may involve the use of a tape drive, floppy disks, CD-ROM, or other storage media regularly used to copy master files. The library equipment and procedures must be so designed as to preclude loss of data due to any single point failure of the system including human users through willful or accidental actions.

22.3.6.4 *Availability*

The specification data in the PSL must be readily available for read-only access by anyone on the program. This requirement can best be served by placing the PSL on a network server protected as read-only. Network service to individual engineering workstations and CAD terminals is suggested, but if resources are unavailable to satisfy this capability, a number of well-placed networked view stations can be provided at minimum cost.

22.3.6.5 *PSL finances*

Each specification agent should be held responsible for their own budget to maintain their PSL. But, if other departments choose to integrate their specifications into a central PSL or they are directed to do so by program management, the lead requirements analysis/specifications person must set up procedures to accommodate these other documents. If the library operating cost is included in the budget for the department responsible for the central library, there will be no problem in allocating cost to users.

Otherwise, a burden-sharing arrangement is required to ensure that all participants are paying their fair share of the cost for library services.

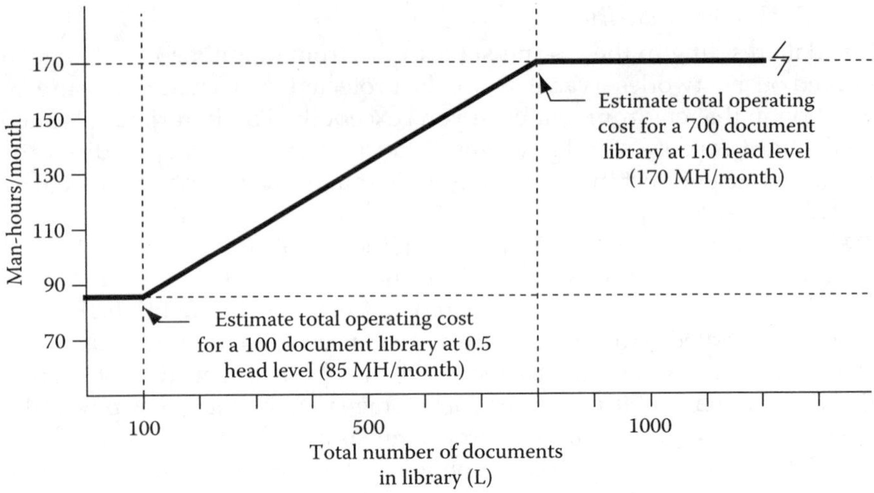

Contributing department monthly cost (Cd)

$$Cd = \begin{cases} 85D/L & L < 100 \\ (0.1417L + 70.83)\,D/L & 100 < L < 700 \\ 170D/L & L > 700 \end{cases}$$

D = Department document count
L = Total library document count

Figure 22.2 Cost-sharing formula.

Figure 22.2 provides an example of a formula for determining cost that a particular department or team should be expected to pay (Cd) in man-hours per month as a function of the total number of documents (L) in a library and the number of documents that the department (D) has in the library. This formula is based on the illustrated inverse relationship between the total number of documents in a library and the cost per document to maintain them in the library. This relationship reflects an expected economy of scale as the library becomes large.

22.3.6.6 *Specification standards loading*

The requirements analysis/specifications lead person should be held responsible for the acquisition of department specification standards and translating them into program specification standards. These standards can be made available to program personnel assigned a specification preparation responsibility either as a floppy disk copy or through network service.

22.3.6.7 *Requirements database interface*

If the program uses a requirements database system, an interface must be provided between the database system and the library function to couple specifications generated from the database into the library. This could be as simple as some kind of desktop publishing application, such as Ventura or Interleaf, permitting database output, boilerplate text, and graphics to be integrated into a single electronic document. The database is set up to output to a particular file name to be located in a particular directory/folder.

22.4 Program implementation

At the time a program begins under contract, the team should have available to them the infrastructure described above and be fully ready to run. This is an uncommon condition, but it is a goal that the proposal team should work hard to achieve. The management of the early requirements analysis activity during the subsequent program implementation can then focus on the technical task at hand without being burdened with the administrative details that should already have been disposed of.

22.4.1 Specification tree development

The requirements lead engineer in the PIT should be held responsible for transforming the evolving system product entity structure into a program specifications tree that identifies what requirements documents must be produced, what type they shall be and assigned principal engineer responsibility. The initial, top-level specification tree should be prepared during the proposal period for inclusion in the proposal. This activity must be coordinated with the WBS, make-buy plan (procurement specifications), and configuration item identification activities (where a customer manages the program in this manner).

The program specification tree must cover all requirements documentation on the program. This includes system, segment, configuration item, and procurement specifications, interface control documents, and internal requirements documentation. On a DoD or NASA contract, the customer may require the preparation and maintenance of a contract specification tree that identifies only the system, segment, and configuration item specifications identified by the customer. This contract tree will be a subset of the complete tree.

The specification tree can be prepared in diagrammatic or indentured list format or a combination of these formats. The diagrammatic format illustrates the breakdown of architecture items that must have their requirements formally documented. Figure 22.3 shows the top end of such a tree. The alternative is to list the documents in product entity order with some method of indicating indenture. A mixed alternative could include a diagram for top-level documents supplemented by an indentured list for lower-tier requirement documents. If the program maintains a graphical product entity structure diagram, this diagram can be annotated for

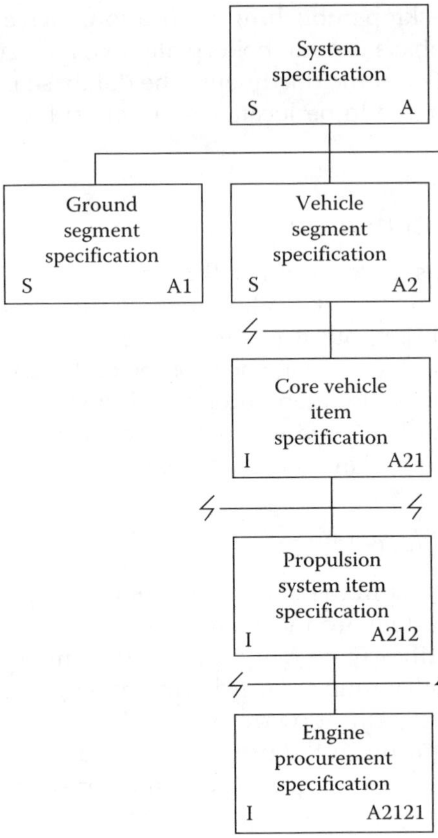

***Figure* 22.3** Typical specification tree.

requirements documentation identification to satisfy the specification tree requirement, but it will clutter the diagram. A tabular list is generally best in that it is easy to maintain and can include additional management data that is distracting on a diagram.

There are essentially three kinds of requirement documents that have to be identified in our tree, and these kinds relate to the contractual relationships represented by the documents. A customer will generally define the items for which they require specifications based on the intensity they wish to employ in managing the program. These specifications are usually identified as system, segment, and configuration item specifications, and they are all contractually binding, authenticated (approved) by the customer, and formally delivered to the customer as contract data requirement lists (CDRLs).

The second kind is created by the contractor for items that are purchased from their suppliers to ensure that suppliers clearly understand needed product characteristics. Finally, there are some elements that will be designed and manufactured by the contractor for which the contractor will choose to prepare requirements documents. The latter are generally items intermediate between system and configuration item specifications or between configuration item and procurement specifications. They are prepared to capture the flowdown of requirements in an orderly fashion across what otherwise would be gaps in the requirements documentation.

22.4.2 *Principal engineer selection, assignment, and training*

The principal engineer is the person who is assigned the responsibility for the identification of the requirements for an item in the system architecture and the capture of these requirements in a document of the appropriate type as noted on the specification tree. As these assignments are made, we must make certain that each principal engineer understands his or her roles and responsibilities. It may be necessary to offer training for one or more principal engineers in their requirements analysis responsibilities, though that can be very difficult to do amidst the actual program implementation. Working requirements analysis luncheons or after-hours classes may provide a means to interlace the work and training needed. Avoid giving the principal engineers training on the whole SRA process. Focus the training on the few elements they need to be effective.

22.4.3 *Regulating the plunge*

Requirements analysis should precede design or procurement activity. It often happens that the engineering organization panics, stimulated by a pressing schedule, and concludes that they cannot waste time on requirements. Management feels that they cannot meet the drawing schedule unless they begin making drawings right away. This is a tragic conclusion that will almost always lead to a very expensive development activity due to the tremendous number of engineering changes that must be made later.

The development effort should not be driven solely by the drawing schedule, as important as that is. The overall program phase schedule should be created based on a balanced appreciation for requirements analysis, concept development (pre-design activity), and design work with the knowledge that later development activity will be easier if the requirements are carefully thought through before the design drawings are initiated.

Figure 22.4 illustrates how the chief engineer can control the movement of the development process downward through the product entity structure,

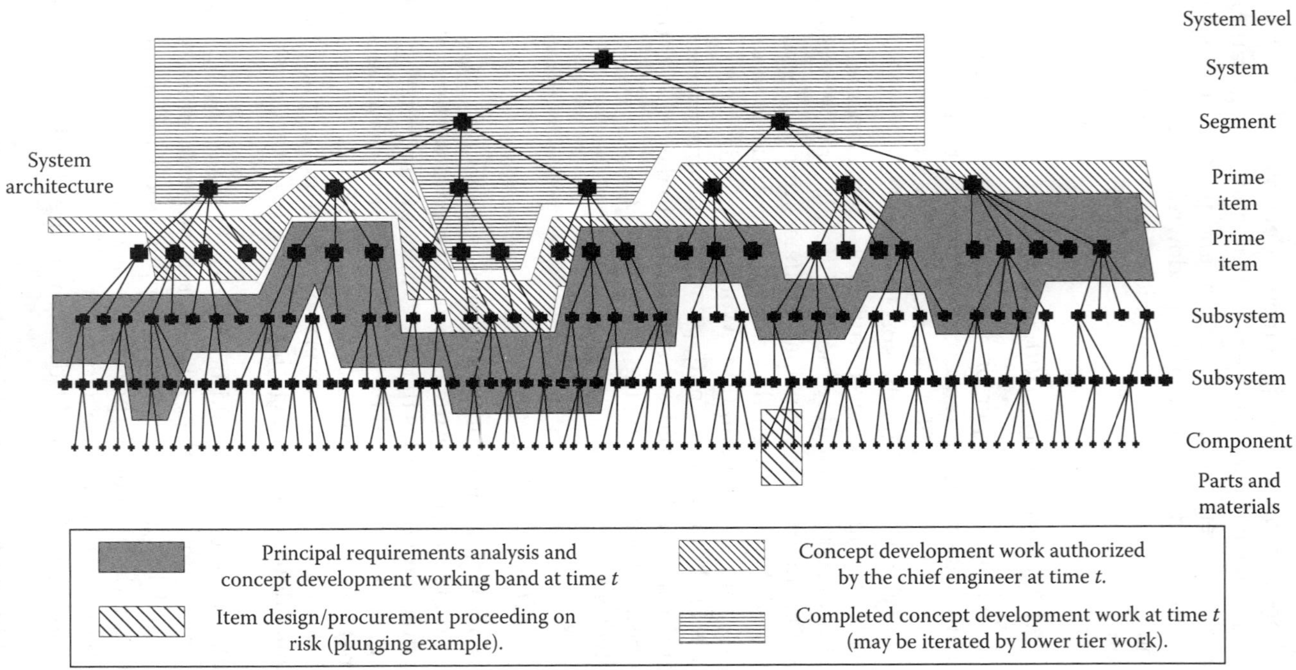

Figure 22.4 The advancing wave. (From Grady, J., *System Integration*, CRC Press, Boca Raton, FL, pp. 139–140, 1994. With permission.)

assuming the program chooses a top-down development direction. This view would have to be modified for a middle-out direction sense. The chief engineer needs a means to communicate his or her authorization for works indicated in Figure 22.4 in the several categories noted in this figure. Table 22.2 suggests one possible means to do this.

Note in Table 22.2 that the left columns identify the item, the responsible contractor (for a case where multiple contractors are involved through teaming, joint venture, or associate arrangements) and principal engineer, document type, and number. Columns are then provided for requirements documentation scheduling and status. The date in the RQMT DOC DUE DATE column should be earlier than the date in the CONCEPT DUE DATE column. The current status carried in the DOC STAT (status) and CONCEPT STAT columns should reflect a more mature status of the requirements than the concept until they are both noted as complete. Additional columns can be provided for preliminary design, detailed design (coordinated with the drawing release schedule), and test-related events.

The requirements in early program phases can be captured in simple requirements lists, rather than rigid specification formats with reams of confusing boilerplate, using the progressive requirements analysis concept covered in the author's *System Requirements Analysis* book (Grady, 2005). The concept data can be captured in an organized set of data generated by the principal engineers and supporting personnel on program microcomputers. This whole set of data provides the principal engineer with briefing data for chief engineer approval review and formal design reviews and the program with a way to communicate the resultant evolving baseline to everyone.

22.4.4 *Selective requirements development*

Selective development is defined as the development of requirements and concepts for system elements the parents for which do not have a set of requirements and concepts developed. It plunges into the architecture ahead of the structured process to develop lower-tier elements without the benefit of the information that flows from the structured process. The small cross-hatched area on the lower fringe of Figure 22.4 identifies such a case.

Selective development offers a third alternative between the extremes of unstructured and top-down structured development. The wholesale application of selective development is probably better termed unstructured development. Selective development should be reserved for specific purposes and not abused. There are several appropriate reasons for engaging in selective development. Three are explored below followed by a discussion about inappropriate applications of the technique.

Table 22.2 Development Control Table

Architecture identification		Principal			Requirements documentation						Concept documentation			
Arch. ID	Item name	Contractor	Dept.	Engineer	Type	Number	Rev.	Due date	EBR NBR	STAT	Due date	EBR NBR	STAT	Remarks
A14	Cosgrove upper stage	GSTS	120-1	Jones	CI	72-1934	—	2/25/1990	51	G	3/10/1990	63	G	
A141	Equipment module	GSTS	120-1	Burns	DRD	72-2430	—	3/10/1990	62	G	4/20/1990	68	G	Platform mounting scheme
A1411	EM structure	GSTS	231-3	Adams	DRD	72-1122	A	4/5/1990	66	G	5/10/1990	80	O	
A1412	Avionics subsystem	GSTS	342-0	Perkins	DRD	72-3323	—	4/10/1990	67	O	5/15/1990	81	R	Chip GIDEP akert
A14121	Guidance and control subsystem	GFG	350-1	Smith	DRD	72-2579	B	5/20/1990	83	R	6/25/1990	—	O	
A141211	Onboard computer	TELEBOND	780-2	Brown	PS	72-2582	—	6/10/1990	91	O	7/10/1990	—	O	
A141212	Intertial navigation set	Reynolds	780-2	Windham	PS	72-2585	A	6/15/1990		O	8/10/1990	—	O	
A141213	Star tracker	GFE	100-0	Fletcher	—	S-1267-3	A	2/10/1990		G	2/10/1990	—	O	Strike scheduled
A144122	Electrical subsystem	GSTS	420-0	Blackmer	DRD	72-9333	—	5/10/1990	82	G	6/10/1990	93	G	
A141221	Battery	RPS	420-1	Gomez	PS	72-9354	C	6/10/1990	90	R	6/15/1990	95	O	
A141222	Wire harness	GSTS	420-2	Chin	—	72-9202	—	3/6/1990			6/20/1990	—		
A141223	Power control unit	ESI	420-3	Johnson	—	72-1211	B	5/10/1990	89	G	7/25/1990	—	O	

Source: Grady, J., *System Integration*, CRC Press, Boca Raton, FL, pp. 139–140, 1994. With permission.

1. Commonly, it is necessary to develop procurement specifications to support a cost proposal bid much earlier than the structured process has developed the requirements appropriate to these elements. Potential suppliers will generally, and understandably, not provide a firm bid without a clear understanding of the requirements for the element they are asked to bid on. This means that they will require a procurement specification. The requirements for these items may have to be selectively developed.
2. Situations sometimes develop where schedule risks are identified in the development of particular system elements that force early development as the most advantageous abatement technique. This may entail parallel development of two or more approaches each of which must be supplied with a requirements set before the parent requirements are defined.
3. The customer or the contractor Program Director may direct that particular system elements be developed with high priority even before the parent requirements are fully defined for reasons other than perceived risk.

Because of our familiarity with our product line, we may conclude, even properly sometimes, that it is necessary to conduct specific concept trade studies far in advance of the requirements development. But selective development should not be allowed to become the norm. The customer schedule not uncommonly panics the engineering community into believing that we must race to drawing development; its every man for himself. The system engineering community must counter these attitudes with factual support for the structured approach.

At the time it is decided to plunge, the principal engineer or the IPPT responsible for the selective development activity must be charged with establishing a set of departure requirements. These should be based on the requirements for the lowest-tier parent element in the same architecture branch as these elements appear. During the period of the selective development activity, the structured process may progress further down that architecture branch providing refinements in the requirements available to the team.

It is important to understand the lowest tier of requirements available in the branch where the selectively developed element resides so that requirements can be made to reflect them as much as possible. It will not be possible to establish traceability connections to these requirements because there will be a gap; that is the nature of selective development. Traceability problems can be minimized by using the available superior element requirements as a guide during selective development events.

When the structured development process arrives at the previously selectively developed requirements, a traceability audit is required. You may very well find upon arrival at the join-up that serious problems exist requiring significant alteration of the requirements and concept for the plunged item or for the higher-level system elements. When we undertake plunging as a means to mitigate risk, we should understand that we are really only trading one kind of risk for another.

22.4.5 Modularization of the schedule

The principal engineer should be shown how his or her work fits into the overall program specifications schedule and encouraged to prepare a detailed schedule for that item. Figure 22.5 illustrates the expansion of a program schedule into an item schedule. The principal engineers have to be brought into the responsibility network for making the program schedule work by making their own schedules work. The chief engineer must manage the overall engineering development activity and hold the principal engineers accountable for their part in schedule performance.

22.4.6 Process controls

The chief engineer and system engineering community need mechanisms to ensure that the planned activity covered on program and item schedules is happening and producing a cost-effective solution to the customer's need. An effective way to focus management energy is to define a development module that can be applied by each principal engineer. These modules can be connected through multiple layers, if necessary, to provide for the development of multileveled architecture items.

22.4.6.1 IPPT meeting structure

Ideally, IPPT should be physically co-located to encourage communication and concurrent interaction. Each team should capture the results of their requirements work in a networked computer database or public folder system such that all program personnel may gain access to the current baseline at any time.

The ideal situation for an IPPT is for all of the team members to be given a space and furnishings arranged to permit easy conversation between team members during their work. A fundamental principal of concurrent development is to encourage practices that minimize the need for communication while maximizing opportunities for communication.

As a minimum means of implementing the concurrent engineering approach during the requirements identification period, the IPPT leader must hold periodic meetings to ensure that team members are kept abreast of the current baseline, suggest changes to it, and solicit input. All team members must be consulted during formal trade studies conducted by the team. Each team member or discipline is responsible for completing their

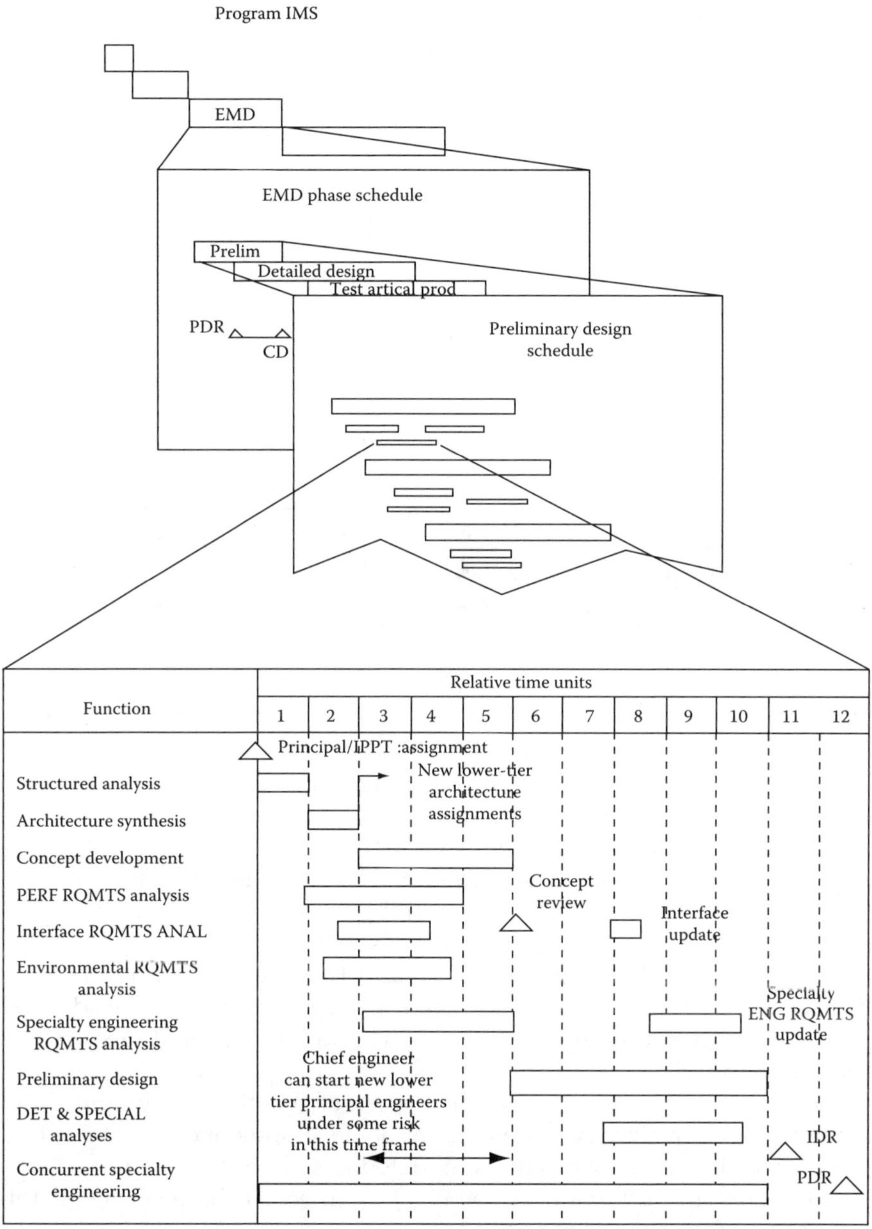

Figure 22.5 Development schedule modularization.

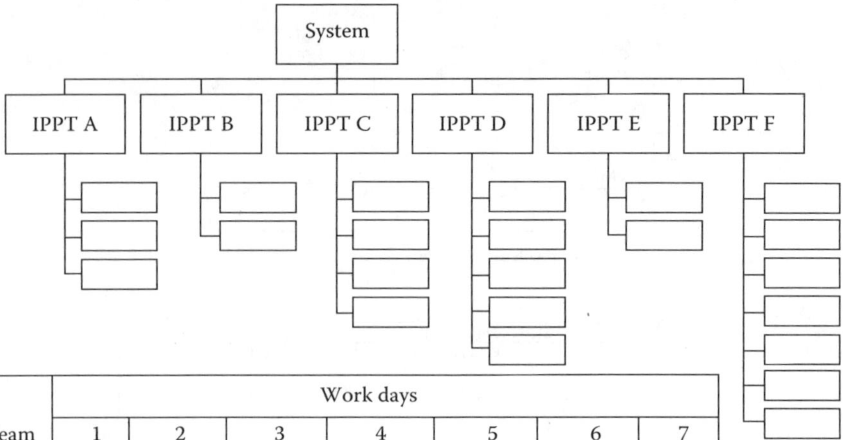

	Work days						
Team	1	2	3	4	5	6	7
A	●		▲	▲		▲	
B		●			▲		▲
C	▲	▲	●		▲	▲	▲
D	▲	▲	▲	●	▲	▲	▲
E		▲		▲	●		
F	▲	▲	▲▲	▲▲	▲	●	▲
SYS							●

Figure 22.6 Sample IPPT meeting cycle.

portion of the requirements set for the item and for understanding the complete requirements set for the item.

Figure 22.6 illustrates a meeting cycle that can be used to review evolving requirements and concepts developed by the integrated development teams. The meetings are obviously arranged in this example in a cyclical fashion with 7 day centers. This cycle span should be selected based on how many IPPTs the program has and the intensity with which the development process must be managed. The cycle time could be adjusted over time with low span time at the start and less frequent meetings as confidence builds.

The chief engineer can use this series of meetings to monitor and manage program activity as defined in the development control table and item schedules. These data can be updated based on the meeting results. Action items and direction by the chief engineer at the meetings is followed up at future meetings.

The circular symbols indicate IPPT meetings chaired by the chief engineer. A system meeting is included in the cycle to provide a time to evaluate the integration and system optimization of the results of the prior round and focus on actions and review items for the next round. The IPPT leaders may need to hold their own brief meetings (triangular symbols) as indicated on major items within their area of responsibility.

22.4.6.2 *Requirements traceability audit*

As the requirements fall out of the process, they should be hooked up to the process through which they were derived and to item parent requirements. This can be done by requiring each principal engineer to do their best to include this data in their requirements analysis data and auditing their work. It may not be possible to audit all of the requirements for traceability. In this case, priorities must be established and a sampling technique instituted. The results of the audit must be to correct flaws in the condition of traceability. Questions about requirements that do not trace to parent requirements must be asked of the offending principal engineers and those questions resolved.

22.4.6.3 *Status tracking*

The scheduling and control measures outlined above provide the framework within which the system engineering community can track the status of the requirements analysis process and compare with planned performance. The resultant error signal is used to identify and correct unsatisfactory performance. The meeting cycle and one-on-one discussions are used to communicate corrective action direction based on observed status problems.

22.4.6.4 *Integration and optimization activity*

The system engineering community must constantly study IPPT products to assure that the evolving requirements and concepts are optimum and balanced. Conditions of suboptimization are identified, alternative system-optimization approaches evaluated, and a course of action selected and implemented. These influences on the architecture items under IPPT development responsibility may cause recycling of some requirements and concepts.

22.4.7 *Technical risk management*

The chief engineer and system engineering community have two excellent tools to monitor and identify program technical risks related to requirements. These are margin management and technical performance measurement (TPM). As risks are identified through the application of these tools, we also need a mechanism for mitigating them.

22.4.7.1 *Margin management*

In Chapter fifteen, we described a method for assigning margins to requirements for the purpose of providing room for maneuver as we find it difficult to satisfy requirements. The chief engineer should identify the parameters he or she wishes to manage by margins and require his or her approval (at the IPPT meetings) for changing them or the target values associated with them. The chief engineer might delegate this authority within each IPPT to the IPPT leader and withhold the authority to make changes across IPPT and parameter boundaries.

Each margin category should be assigned to a specific margin principal engineer whose responsibility is to manage that margin account. Most of these accounts are easy to associate with a particular discipline clearly identified with an engineering department. Examples of some common margin categories illustrate this: reliability, weight, radio/radar link gain, and design to cost (DTC).

22.4.7.2 *Technical performance measurement*

TPM is a methodology that charts the history of the values of a selected set of quantifiable requirements in time for the purpose of clearly illustrating the changes in value in relation to program events and goals for that requirement value. The TPM chart is annotated with major events such as major reviews and events that have caused and will, in the future cause, a significant effect on the requirement value. Figure 22.7 illustrates a typical TPM that will be used as an example in this discussion.

The TPM parameters are key variables carefully selected by program contractor and customer management because their values have a powerful effect on the ability of the system to satisfy its intended function.

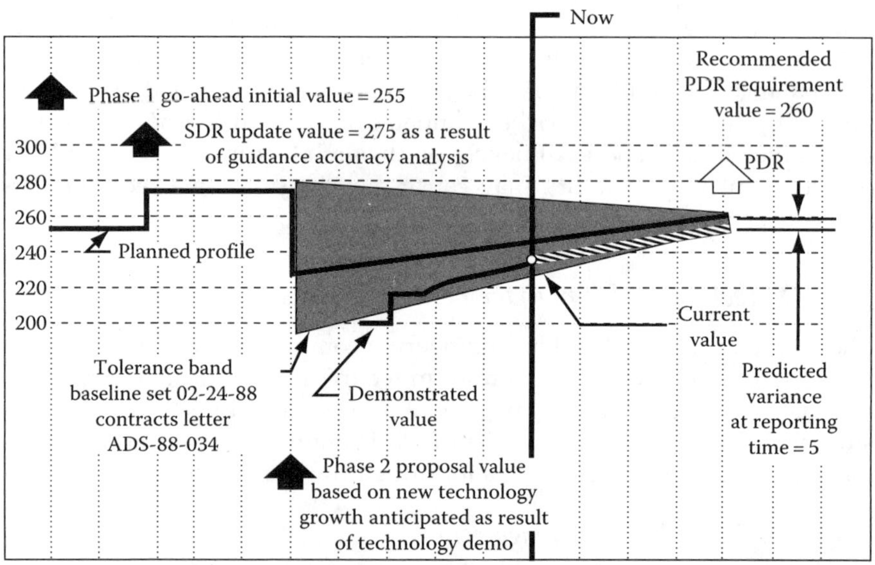

Figure 22.7 Typical TPM chart.

To select too many TPM parameters trivializes management energy by squandering it on nonessentials. To select too few or the wrong parameters runs the risk of management missing important indicators that could give forewarning of pending disasters that will cost much more to correct if detected after the disaster has materialized. The intent behind TPM is to provide management with what is called by some "feed forward" (as opposed to feedback). TPM should provide management with advance information about how the development process is performing with respect to satisfying the system need.

TPM is the technical hand maiden of cost/schedule control techniques that focus on program cost and schedule risk control. And, yes, there is a link between TPM and risk management. Given that TPM offers insights into the evolving capability of the system to perform the intended mission, it provides valuable signals on potential technical risks. At the same time, it is part and parcel of any good technical risk abatement plan because it provides the visibility needed to track how well corrective actions are faring in mitigating the related risk. The TPM chart should be annotated in these cases with remarks about future events related to risk abatement.

Where we identify a near-term or projected problem in satisfying required values, the principal engineer for the parameter (such as the mass properties engineer, for example) is required to develop an action plan to bring the parameter back into the control region by a particular date or program event.

22.4.7.3 Risk mitigation process

TPM or margin problems have to be surfaced by the architecture principal engineers, system engineers, TPM parameter margin category principal engineers, specialty engineers, and the chief engineer. Each of these problems must be evaluated for program technical risk and those that qualify must be subjected to a risk mitigation action. The suggested approach is to identify a risk principal (the margin category or TPM parameter principal engineer) and require that principal engineer to prepare an action plan to mitigate the identified risk. Action plan status is tracked in the system IPPT meeting on a regular basis until the risk has been reduced to an acceptable level.

22.5 Development data package concept

Once engineering drawings start coming off the boards or CAD stations, standard configuration management procedures are very effective in controlling the design baseline. Many organizations find it very difficult, however, to control the evolving requirements and concepts baseline and maintain decision traceability during the sometimes chaotic period leading up to the preliminary design review (PDR). Many engineering organizations have found themselves in the sad predicament at a critical design review (CDR) without the backup data asked for by the customer for a critical decision made months earlier.

The requirements database can provide a means to capture not only program technical requirements but also the rationale, sources, and traceability associated with these requirements. So, the database approach can be used to satisfy the need to retain rationale data for requirements, but may not totally solve the design concept decision capture problem external to the requirements information component. We seek a solution that embraces both requirements and concept information and ways to manage the evolution of this information resource in early program phases.

Many system engineering organizations have evolved over a period of years a universal view of all of the information of interest in the early phases of system development and defined a particular organizing structure for that information. One such concept, called a development data package (DDP), was advanced by logistics engineers at General Dynamics Space Systems Division in the early 1990s. Figure 22.8 illustrates one way this information package can be organized for a project. The matrix horizontal axis lists the sections of the DDP. The vertical matrix axis lists the organizations participating on an integrated product development team focused on developing a particular item that happened to be an avionics box.

The understanding is that there must be a place in the DDP for every member of the team to place all of their information product during the life cycle of the DDP. This includes the kind of information that many organizations expect engineers to place in their engineering notebooks, journals, or logs. The DDP captures all information of interest to the team and makes it available to all team members. Later we will see how computer technology can satisfy the availability requirement. First let us discuss DDP organization and life cycle.

Under this concept, each IPPT leader and principal engineer must create and maintain a DDP for their item beginning when the chief engineer authorized the team to begin development work. The DDP provides a means to capture development information from all IPPT members in a common format between initiation of team activities and completion of the PDR. It must provide a place for every concurrent engineering team member to place their information product. Between PDR and CDR the content of the DDP should flow out to formal documentation destinations such as specifications, engineering drawings, and planning data and DDP maintenance should be discontinued as each section makes the transition. Table 22.3 defines the formal documentation destinations of each DDP section (noted in the vertical axis of Figure 22.8) subsequent to PDR. Generally, the team should be allowed the time between PDR and some time prior to CDR to complete the conversion between DDP content and the formal data destinations.

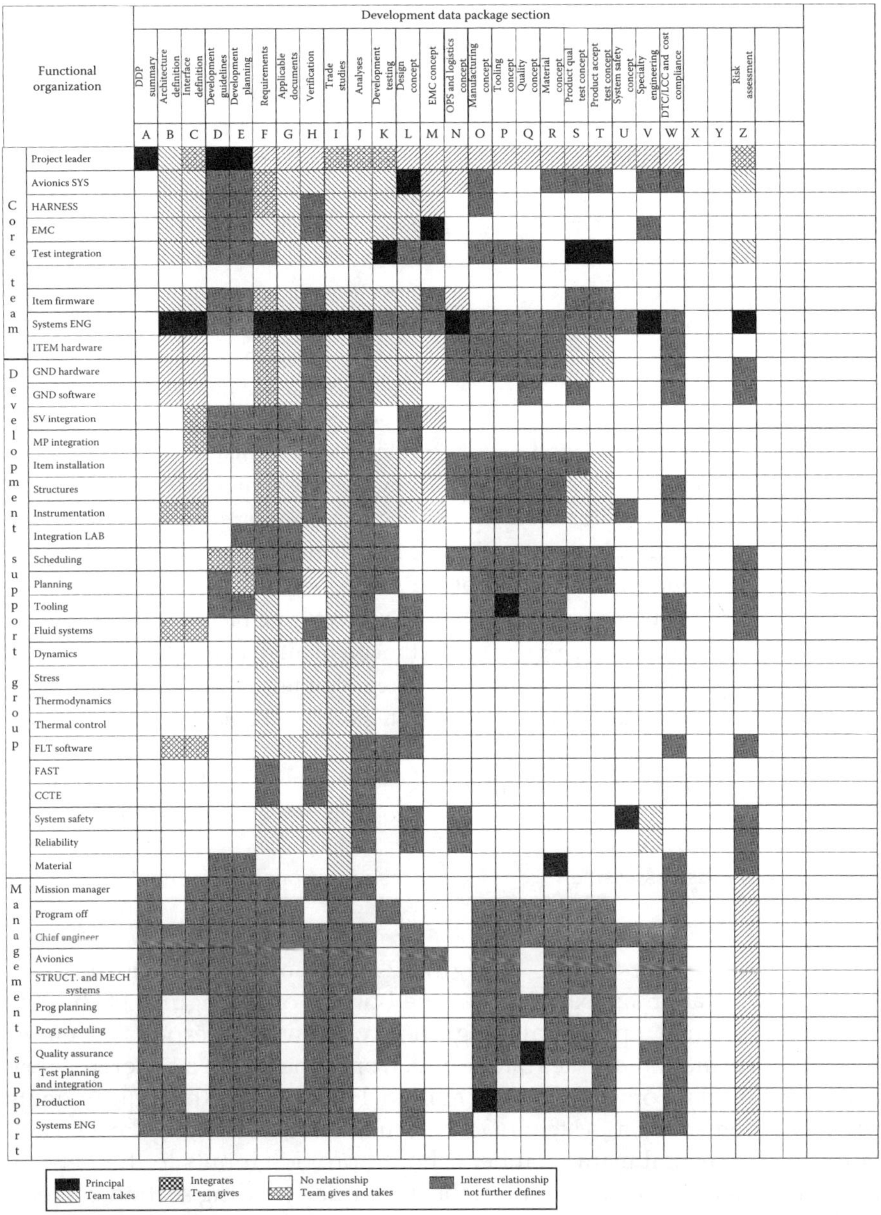

Figure 22.8 DDP responsibility matrix.

Table **22.3** DDP Data Destinations

DDP section ID	Title	Formal documentation destination
A	Architecture	Specifications, specification tree, specialty engineering models
B	Interface	Specifications, engineering drawings
C	DEV guidelines	Program planning, supplier SOWs
D	DEV planning	Program planning, supplier SOWs
E	Requirements	Specifications
F	Applicable DOC	Specifications
G	Verification	Specifications, program test planning
H	Trade studies	Design rationale traceability documentation
I	Analyses	Design rationale traceability documentation
J	Development test	Integrated test plan
K	Design concept	Engineering drawings
L	OPS/logistic concept	Logistics support plan
M	Manufacturing concept	Manufacturing plan, facilitization
N	Tooling and STE concept	Procurement documents
O	Quality concept	Manufacturing planning documents
P	Material concept	Procurement documents
Q	Product quality testing	Integrated test plan, test procedures
R	Product acceptance testing	Integrated test plan, test procedures
S	System safety concept	System safety plan, hazard reporting
T	Cost compliance assurance	Program planning, specifications
U	Risk assessment	Design rationale documentation

The matrix intersections of Figure 22.8 are annotated with responsibility information as explained at the bottom of the figure. Each DDP section is assigned a principal integration agent who is responsible either to input information or integrate inputs from others into a coherent story consistent with all other DDP information. For example, the requirements section is owned by a Systems Development function in Figure 22.8. As you can see, several team members should provide inputs to this section and others should be held accountable for understanding section content. Still others are identified as interested parties with no obligation to interact over section content.

If the program uses computer word processors to prepare specifications rather than a computer database system, the content of the DDP requirements section can be initially provided using the word processor of choice in the primitive list style. As the item definition matures, these primitive statements may be expanded into full specification text in time for publication as the item specification in the required format (configuration item responsive to the customer data item description, procurement specification, or in-house requirements document).

Where the program uses a computer database to capture item requirements, the DDP requirements section may simply reference the database content or be used as a baseline repository for the most recently approved snapshot of database content while the working database content continues to mature.

Whether word processing or database technology is employed, the team responsible for the particular DDP would apply a sound requirements analysis process, such as that covered in this book, to identify the content of the requirements set for the item.

The DDP could be assembled in paper media using typewriter or standalone computer technology, but the most powerful application of the concept requires networked microcomputers tied into the DDP located on a network server. The server is set up with (1) a set of templates for each section that requires a specific application program (such as dBase, Excel, PowerPoint), a drop box, and a working baseline consisting of all of the work completed to date.

Someone is appointed to manage the database. That person makes sure that everyone on the team understands how to gain access to the templates and how to use them to create their product. Each person with a DDP input responsibility, as defined in some variation of Figure 22.8, copies the appropriate template to their local workstation and proceeds to enter their work. At points in time defined by the DDP data manager, each contributor drops a copy of their section in the drop box on the server. Periodically, the DDP data manager looks in the drop box for input. New inputs are placed in the working master by the data manager, the only one with the password to change working master content.

At any time, anyone on the program may gain read-only access to anything in the working master. If the program is equipped with meeting room computer network access and video projection capability, periodic concurrent engineering team meetings may be accomplished by projecting directly from DDP resources. Periodically, even on a daily basis, an approved copy of the working master (complete or a subset) could be transferred to customer access. In fact, in early program phases, this would be a much more effective CDRL item than the piles of reports commonly delivered to customer file cabinet resting places.

As useful as the DDP is in solving the information communication and integration problem during early program phases, the DDP concept is probably not the most efficient long-term solution to a company's information needs. But, given that a company does not now know what its aggregate system development information needs are or how those needs relate to long-term information needs, the DDP concept can provide a manageable growth path from ignorance to understanding. At the terminal end of this path, after applying the DDP concept on several programs, the company will have an excellent understanding of its needs and be able to phrase requirements for their information system builders.

22.6 Tailoring the development intensity

This book provides a systematic requirements analysis methodology supportive of a DoD weapons system development activity requiring adherence to acquisition process controls and procedures. This chapter in combination with Chapter fifteen, or *System Requirements Analysis* (Grady, 2005) could be used as the requirements analysis process description for such a program referenced in the program SEMP.

However, all programs will not require every component part of the process included in this book nor the level of intensity suggested here. People responsible for engineering management on each program must decide the right intensity for each of these system engineering activities for the particular program phase based on the customer statement of work and company standard practices. A good way to organize this approach is to define a system engineering process criteria (SEPC) (including within it a requirements analysis criteria) for your company that is composed of a series of simple statements that define the minimum system engineering process features to be applied on each program. This generic criteria can be included in the company's generic system engineering manual or management plan.

Then, make it the responsibility of each program to define in the program SEMP how they will comply with the company's generic criteria subject to the approval of company functional engineering management.

A DoD weapons system may have to use the full content of this book and *System Requirements Analysis* (Grady, 2005), *System Synthesis: Product and Process Design* (Grady, in press), and *System Verification* (Grady, 2007) to comply with customer requirements. A skunk works operation can get by with much less but even here there are limits beyond which you court disaster. We close this chapter with a sample criteria for early program phases where requirements analysis work is most intense. It may be instructive for you to write down how your program or company normally complies with each of these criteria. Then you might ask yourself how your company should respond. If there is a difference between the two positions, you may have identified a basis for improving your company's capability.

Sample SEPC for activity through PDR

1. A clear statement of system need developed by or agreed to by the customer.
2. A set of requirements expanded from the need to define the mission, brief definition of major elements in system, system performance goals/requirements, external interface definition, system environment definition, and a minimum of specialty engineering constraints in the simplest format acceptable to the customer.
3. An organized means to decompose the customer's need into a system architecture consisting of hardware, computer software, facilities, materials, and personnel that can be related to the customer's definition of cost accounts.
4. An organized method for assigning responsibility for the development of the system elements in the architecture to personnel and/or teams that is mutually consistent with customer's definition of work breakdown (in support of effective cost collection) and the enterprise development team concept applied on the program.
5. A means to define and enforce the definition and approval of requirements for items prior to design work.
6. A means to capture all requirements correlated with their architecture item and to make them available for reference by all program personnel.
7. Traceability of requirements to sources, rationale, other requirements, and the development process.
8. An organized way to develop compatible internal and external interfaces between elements depicted on the system architecture.
9. An organized way to capture and retain the results of development team product and process concept and design work so as to be easily accessible at any time by all other program personnel and available for iterative improvement by responsible engineers and analysts.
10. An organized means to schedule work performed on development teams and to monitor progress to those plans. This is also a C/SCSC criteria.
11. An organized method for gaining insight into the results of development team work on the part of management coupled with a means to direct the work of teams based on these insights.
12. An organized means to integrate the work of separate product development teams and to optimize at the system level across design, manufacturing, test, operations, and logistics support to overcome problems resulting from suboptimization at the individual team level.
13. A means to define what the system and its components must do in the customer's environment to maintain needed readiness and

operational capability and to attain mission success. This method must provide for development and evaluation of the system maintenance and logistic support concepts and coordination of those concepts with the evolving system design concept and a mechanism for gaining insight into needed procedural and human task-oriented requirements. This must include a way to feed back to the design evolution the results of analyses into these matters.

14. A concurrent method for the evaluation of the system and its components for reliable and safe operation. This must include a way to feed back to the design evolution the results of analyses into these matters.
15. Provide a means to reach sound decisions where current work plans have come to discontinuous points in the development process requiring direction.
16. A means to capture design rationale and traceability of the decision process that led to the preferred system concept.
17. A means to clearly define the current system requirements and concept baseline during the period leading up to PDR. Subsequent to PDR the baseline is clearly captured in a set of specifications, reports, and engineering drawings listed in an engineering drawing management system and rigorously controlled by configuration management.
18. A means to concurrently perform specialty engineering work and to efficiently integrate the results of that work into the ongoing development process and product definition.

chapter twenty-three

System synthesis management

23.1 Program sequencing

A program can be planned based on several patterns of behavior. If we have good knowledge of the problem to be solved, can plan the work in some detail, and are fairly certain that we have control over the unknowns and risks, the whole program can be outlined in a grand, detailed plan prior to any program work being accomplished. Alternatively, if there are significant unknowns because the program involves unprecedented problems or designs, we may be better advised to bite off only a piece of the problem. Subsequently, when we know more about the application, we can plan another piece of work and, eventually, complete the program incrementally. There are other reasons why we might prefer to accomplish the program incrementally rather than in one single effort, but in general the rationale is bound up in risk.

23.1.1 Grand plan development

Under this development sequence, the intent is to move from need to deliverable design solution and manufactured product in one continuous stream of work. This is not to say that phases cannot be used to mitigate risk to some extent.

23.1.2 Incremental development

In incremental development, the intent is to achieve some intermediate condition short of the ultimate result. In the process of reaching this goal, new information is acquired that removes some of the concerns about moving closer to the ultimate desires of the customer. In this form of development, there are some very demanding requirements that have to be satisfied but a shortfall in technology relative to satisfying them. The customer accepts an initial capability that can satisfy current needs and can later be upgraded to satisfy ultimate goals in one or more program increments.

The DoD often follows this pattern by plan or necessity. The B52 bomber has provided the U.S. Air Force a very long and distinguished service by undergoing major modifications to adapt to new needs and deal with new threats that were not apparent when the original design was created. Other systems, including aircraft, land warfare vehicles, and ships, have passed through a similar history. A policy that appears to work in these cases is to acquire a sound structure and propulsion system and adapt this basically sound design to constantly improving command, control, communications, and computing capabilities.

The incremental approach is often followed in software procurement where the software is delivered in builds. In some systems, these deliveries are made each year with each delivery adding capabilities, removing bugs from prior builds, and, no doubt, adding new bugs to be dealt with.

23.2 Integration

Integration is at once a very simple concept involving combining two or more entities, and a very complex notion. The author was once asked to define it in a way in which it could be taught to people, leading to a stunned silence and 2 years of effort to understand its essence, which is hopefully presented here.

23.2.1 Integration components, spaces, and cells

23.2.1.1 Setting the stage for integration decomposition

It would be convenient if we could describe all of the facets of system integration in terms of a single entity. The author is convinced, after years of work, observation, and study in this field, that many of us accept incorrectly that integration is one activity. Few of us are able to describe how this single activity is performed, however. Many people assign the term "system integration" a mystical quality. Whatever it is, it is the answer to every system engineering problem, and seemingly just connecting the term in the same sentence with the problem is sufficient to define an approach to the concept in our proposals and conversations.

We will see in this chapter that the integration process can be decomposed into several parts and these parts explained in an uncomplicated way. System integration then becomes the sum of these parts. We must, of course, be watchful that we do not lose contact with important information about the whole when thinking in terms of the parts.

We begin with an acceptance that an engineering organization that must deal with multiple system development activities should organize in a matrix structure. A matrix has the advantage of focusing day-to-day work on specific program problems while providing a good environment within which to improve the organization's skills, methods, tools, and knowledge. We should also accept the good sense that we cannot beat the odds on specialization. Our engineering organization must select its personnel from the same pool as everyone else, humanity. We have seen that we are limited in knowledge, and we solve this problem through specialization.

Therefore, we will organize our personnel into functional specialties led by functional department chiefs. These chiefs will be responsible for providing all of our programs with qualified personnel, skilled in using a particular tool set and following the standard department procedures proven effective on past programs. We insist on standards that are continuously improved, because we wish to take advantage of the practice–practice–practice template used by great athletes.

Our product will be organized into sub-elements that can be worked on by two or more integrated product and process teams (IPPTs), and we will assign our personnel to these teams that will form the principal personnel supervisory structure. The work of all the IPPTs will be coordinated by a program integration team (PIT).

We will organize all program work into process steps linked to product entities under the responsibility of one or more IPPTs. Each process step will have a set of goals and a simple task description. We will map our IPPT responsibilities to the processes and identify leaders for each process. All of the processes will be laid into our integrated schedule with clear start/stop dates and budgets that reflect back into the IPPT structure. Each process will have clearly identified information and/or material product outputs that are needed in other processes as inputs.

23.2.1.2 Integration components

These assumptions and selections leave us with the problem of combining, or integrating, the work of many people in different functional disciplines, working on different product system components, in many different process steps over time. We define these three fundamental integration components as function, product, and process, respectively. Be very careful that you understand that we have used the word "function" to mean functional organization and not product system function.

In each of these components, we have to concern ourselves with two fundamental kinds of integration: co-component and cross-component integration. In addition, we have to account for the possibility of one or two of the components not being involved. This means that we have a three-valued situation: co, cross, and null values. With three variables, each with three values, it is obvious that we are working with 27 (3 cubed) different integration possibilities. It is helpful to have a picture of this problem to better understand all of these possibilities.

Unfortunately, we cannot easily illustrate the three-valued relationship for each of our three components, so our visual model will disregard the null possibility for each component. Figure 23.1 illustrates the three components as a three-dimensional system. We can imagine that we can assign positions on the function axis to discrete organizations in our

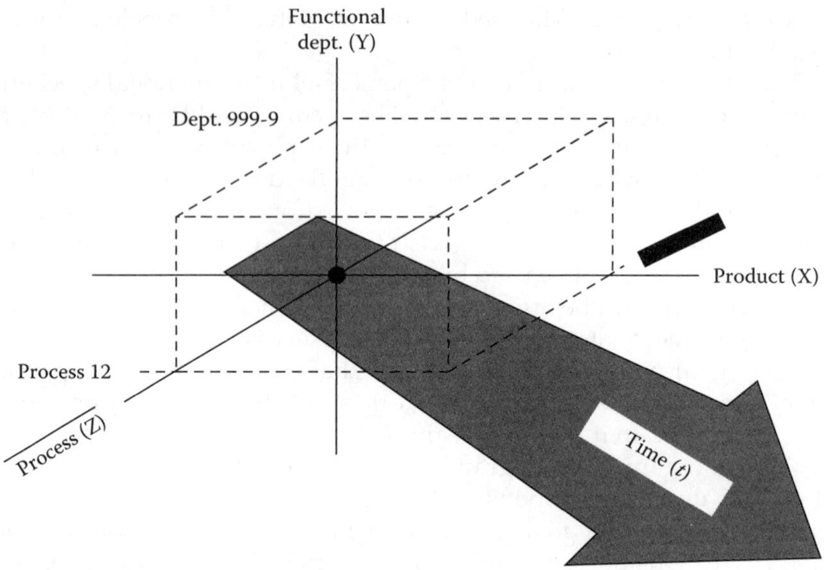

Figure 23.1 Integration components. (From Grady, J., *System Integration,* CRC Press, Boca Raton, FL, pp. 43–228, 1994. With permission.)

functional organization, such as reliability, structural design, and quality assurance. Similarly, we can assign positions on the product axis to the elements of the system (avionics system, onboard computer, circuit card, transistor, etc.) in a hierarchical fashion. Finally, the third axis can have positions marked out corresponding to the processes on our program process diagram, such as identify item X requirements, design item X, and test item X.

For a given functional organization, all the work that a specialty discipline does can be thought of as being within the plane passing perpendicular to the function axis at that function position. The task of integrating all the work on this plane is called co-function integration, the integration of work accomplished by two or more specialists in that one functional discipline.

Similar planes can be constructed perpendicular to the other two axes at different points on these axes to capture all of the co-process and co-product integration work. The careful reader will observe that every point in the three-space, then, corresponds to some combination of co- and cross-component integration for the three components.

Let us now define the three integration components in terms of their possible values. The three combinations are nulls, meaning no integration for that component: null-function, null-process, and null-product integration. The other six possibilities are co and cross combinations with each of the three components. Let us take each of these six cases in turn assuming, in each case, that the other two components have a null value for the moment.

Co-function integration coordinates the work of two or more persons from the same functional department (specialty discipline) to ensure they are all using the same tools, techniques, and procedures in an appropriate fashion and that their results are consistent with other work on one or more programs or systems. This task is the responsibility of the senior program functional specialist for the project or the functional supervisor, in the case where all company projects are the target of the integration work.

Cross-function integration coordinates the work of persons from two or more functional disciplines in search of suboptimal design and specialty engineering solutions needing rebalancing, mutual conflicts between specialty requirements or the corresponding design solutions, available unused margin to be repossessed and applied more effectively, and wayward interpretations of the project or product requirements that may lead to conflict. This is a program responsibility that may fall upon an IPPT leader, a task principal, or a person from the PIT as a function of the relationship of the work to process or product and the integration level.

Co-process integration coordinates the work of two or more persons working on the same program process to ensure that they all are focusing on the same process needs within available budget and schedule constraints. This task is accomplished by the assigned task leader who is responsible to achieve the task goals on time and on budget.

Cross-process integration coordinates the work of two or more task teams working on different processes. It seeks to ensure that the work of these two or more teams is mutually consistent and driven by the same program goals. This integration task is the responsibility of a program person.

Co-product integration coordinates the work of two or more persons developing the design solution for a particular product item. This may be to develop an integrated set of product item requirements; evolve an optimum synthesis of these requirements in terms of a design concept; ensure that all of the cooperating specialty views are respected in the design solution; or integrate the test, analysis, and design work associated with this product item. Where an item requires a special test article, such as a flow bench for a fluid system, this work should also ensure that the test article properly reflects the same requirements and design embodied in the product item.

Cross-product integration is the commonly conceived integration component that most people would first think of in response to the word "integration." It focuses primarily on the integration of interfaces between product items to ensure that all interface terminals and media are compatible both physically and functionally. This work could be as simple

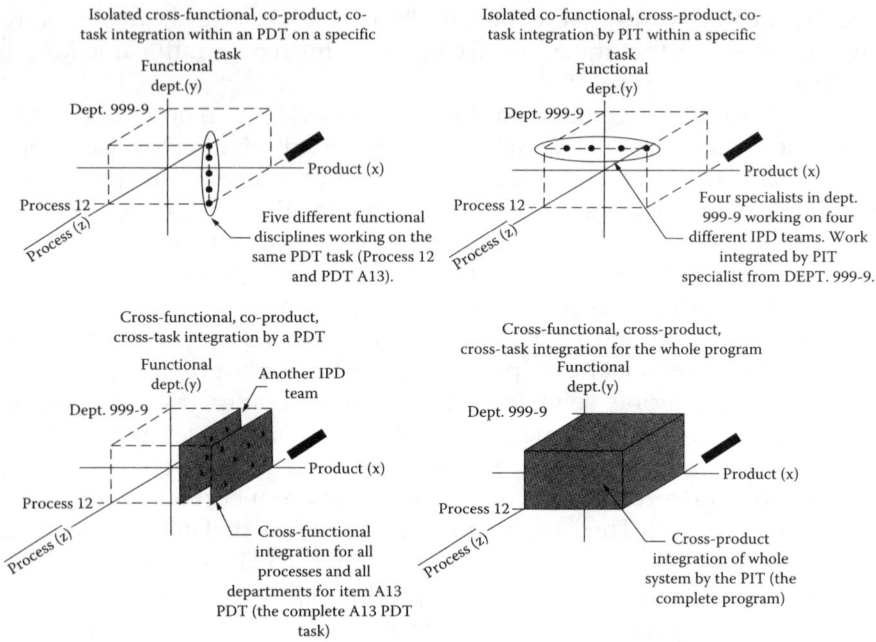

Figure 23.2 Integration space examples. (From Grady, J., *System Integration*, CRC Press, Boca Raton, FL, pp. 43–228, 1994. With permission.)

also as ensuring that all items are painted in the correct color, have satisfied a particular specialty requirement, as in being maintainable defined by an ability to remove and replace in 10 min. It is difficult to talk about this form of integration without linking other integration components.

23.2.1.3 Integration spaces

Integration work seldom falls into one of these pure cases. Generally, we have to deal with more complicated situations involving combinations of two or three components with mixes of cross, co, and null values. Figure 23.2 offers four particular examples of these possible combinations, or spaces, for further discussion. Once again, we disregard the null-value case to enable a simple graphical portrayal in three dimensions on a two-dimensional paper.

In Table 23.1, we use a simple tertiary counting scheme to ensure that we have not omitted any combinations of the three variables. The reader should understand that there can be no other forms of integration work than those listed in Table 23.1 given that we have included every appropriate integration component. It would be possible to add one or more components to our list in addition to function, process, and product. This

Table 23.1 Integration Space Identification

ID	FUNCTION	PROCESS	PRODUCT	INTEGRATION TYPE NAME
0	NULL	NULL	NULL	INDIVIDUAL Effort
1	NULL	NULL	CO	ISOLATED CO-PRODUCT
2	NULL	NULL	CROSS	ISOLATED CROSS-PRODUCT
3	NULL	CO	NULL	ISOLATED CO-PROCESS
4	NULL	CO	CO	CO-PROCESS & PRODUCT
5	NULL	CO	CROSS	CO-PROCESS/ CROSS-PRODUCT
6	NULL	CROSS	NULL	ISOLATED CROSS-PROCESS
7	NULL	CROSS	CO	CROSS-PROCESS/ CO-PRODUCT
8	NULL	CROSS	CROSS	CROSS-PROCESS & PRODUCT
9	CO	NULL	NULL	ISOLATED CO-FUNCTION
10	CO	NULL	CO	CO-FUNCTION & PRODUCT
11	CO	NULL	CROSS	CO-FUNCTION/ CROSS-PRODUCT
12	CO	CO	NULL	CO-FUNCTION & PROCESS
13	CO	CO	CO	ALL CO
14	CO	CO	CROSS	FOURTEEN
15	CO	CROSS	NULL	CO-FUNCTION/ CROSS-PROCESS
16	CO	CROSS	CO	SIXTEEN
17	CO	CROSS	CROSS	SEVENTEEN
18	CROSS	NULL	NULL	ISOLATED CROSS-FUNCTION
19	CROSS	NULL	CO	CROSS-FUNCTION/ CO-PRODUCT
20	CROSS	NULL	CROSS	CROSS-FUNCTION & PRODUCT
21	CROSS	CO	NULL	CROSS-FUNCTION/ CO-PROCESS

(*continued*)

Table 23.1 (continued) Integration Space Identification

ID	FUNCTION	PROCESS	PRODUCT	INTEGRATION TYPE NAME
22	CROSS	CO	CO	TWENTY TWO
23	CROSS	CO	CROSS	TWENTY THREE
24	CROSS	CROSS	NULL	CROSS-FUNCTION & PROCESS
25	CROSS	CROSS	CO	TWENTY FIVE
26	CROSS	CROSS	CROSS	ALL CROSS

Source: Grady, J., *System Integration*, CRC Press, Boca Raton, FL, pp. 43–228, 1994. With permission.

would require us to multiply the number of different integration spaces. The number of integration spaces (*S*) is, of course, predictable as follows:

$$S = C^n,$$

where

C is the number of integration components
n is the number of values for each variable

If, for example, we concluded that there should be five components instead of the three covered in this book, each having three values (co, cross, and null), we would have $5^3 = 125$ integration spaces. At the end of the last chapter, we speculated about recognizing program as an additional component and noted that it, in combination with the three covered here, would require 64 integration spaces (4^3). These 64 spaces would include co, cross, and null program integration as well as all the new combinations with the other components. In the context of the new ideas you have just been exposed to, the author elected to set this component at a fixed value of co-program and focus on a single program. As you can imagine, the identification of additional components could get out of hand very rapidly, making the description of integration more complex than the work itself. Three components and three values appeared to the author to be a good compromise between completeness and understandability for the purposes of this book.

We will use the INTEGRATION TYPE NAME in Table 23.1 to refer to these combinations of integration components, referred to as integration spaces. The integration space ID numbers are used in a database designed by the author to experiment with system engineering information relationships, and are included here only to illustrate the simple tertiary counting scheme for ensuring completeness. Each integration space involves a merger of a unique grouping of three particular values for the components and represents one integration mode useful in one or more particular situations.

Of the 27 integration spaces, we can dispense with INDIVIDUAL EFFORT integration immediately. We must assume that a single individual is fully capable in their field of specialized knowledge and able to effectively apply this knowledge to a specific product item and in a particular single process. This is why we specialized, after all, to create a human task that is within the power of a normal, single, specialized individual to master. We assume that each specialist can carry on an internalized conversation with themselves and use the power of their specialized discipline to solve the small problems we have tried to frame in our decomposition efforts.

Similarly, the ALL CO integration space is usually not very interesting to a system engineer, since it involves people from the same functional department performing work in the same process step, for the same system element. True, on a very large program, we could imagine how this could become quite a challenge. For example, we might assign seven technical writers to the job of writing a single technical manual for a single item of equipment. Integrating the work of these seven writers would be a case of ALL CO integration.

One other fairly simple integration space to explain, though the most complex of them all in practice, is ALL CROSS integration involving cross everything. When a member of the PIT integrates the work of several members of two IPPTs developing the design of two different product elements and the work of a facilities engineer responsible for the factory that will assemble these two elements and a tooling designer responsible for the manufacturing equipment that will hold the elements during mating, we have an example of this kind of integration. The reader will be able to imagine several other cases of this integration space.

We have given examples now for 9 of the 27 integration spaces, including the 6 isolated integration cases, INDIVIDUAL EFFORT (or ALL NULL), ALL CO, and ALL CROSS. This leaves 18 remaining to be explained with at least one example. Some of these 18, you can see from Table 23.1, are very difficult to name, so we will simply use the ID number for a name. Examples for most of these other 18 spaces will be stated in the remaining chapters of this book.

23.2.1.4 Integration cells

In almost any given system development work situation, we will find it necessary for some combination of the 27 integration spaces to be applied. Very little system development work can be accomplished in total autonomy today. This phenomenon appears because we have had to specialize very finely to master enough of the available knowledge base to be competitive. The number of these combinations is finite but quite large for a large development program.

We have many options in grouping all program work into unique combinations of integration spaces, but the best way to do this is probably driven by the program tasks that would appear on a program task network. Each program task has associated with it some set of functional disciplines performing work on some particular combination of product elements. Many of these tasks, possibly most, will require some form of integration in the context of some combination of the 27 integration spaces defined above. Let us call any one of these combinations of task, product, and functional organization an integration cell.

The more finely we divide the overall program into tasks, the more unique integration cells we will have. The more finely we assign IPPTs to develop the system, the larger the number of integration cells. The more functional disciplines that must be assigned to the program, the more integration cells that will be necessary. At the same time, the larger the number in any of these integration components, the more simply we can describe each integration cell because these will consist of a less complex combination of integration spaces.

23.2.1.5 Program world line

So, the integration process is more complex than we might have first imagined, composed of a finer structure than we might have thought. But the complexity does not stop here. We must not only apply each of these integration spaces well within the context of the integration cells defined for the program, but we must apply them in a pattern coordinated with the program schedule. Figure 23.3 illustrates this need by placing the integration spaces coordinate system on a program world line. Throughout program passage on its world line (network or schedule), appropriate specialists must apply the appropriate integration spaces to the planned integration cells in accordance with an integrated plan. To the extent that we can reduce aggregate program work by early identification and resolution of inconsistencies between product and process elements, we encourage success in system integration. Ideally, we should be able to do all work error-free on the first pass.

Now we can answer the question, "What is system integration?" It is the rich mixture of three integration components applied in combinations defined by the resultant integration spaces to the work confined to a finite number of integration cells across the program world line that actually comprises system integration on any given program. Throughout the remainder of this book, we will consider some of these integration situations as they relate to process and product integration.

23.2.1.6 This may be a little mad

Before leaving this subject, we should question one of our assumptions as a way of cementing our understanding of the integration spaces concept.

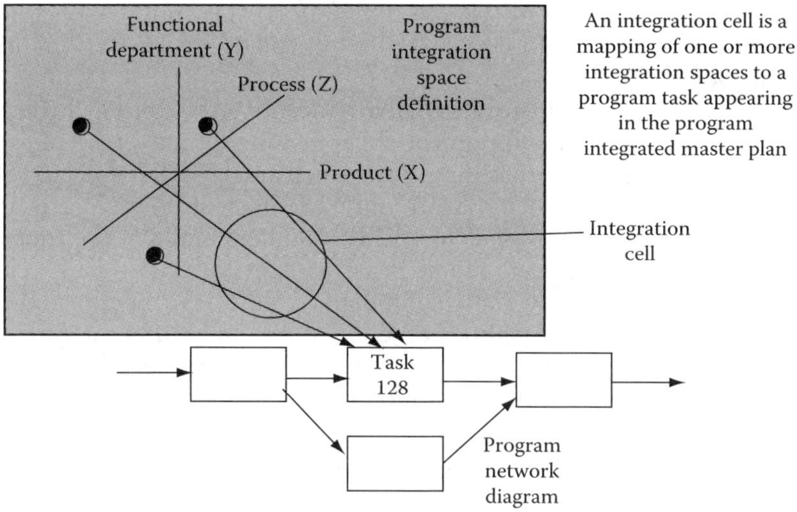

Figure 23.3 System development world line. (From Grady, J., *System Integration*, CRC Press, Boca Raton, FL, pp. 43–228, 1994. With permission.)

Perhaps, we were being premature in dispensing with the INDIVIDUAL EFFORT space. Suppose that company XYZ has acquired an engineering staff composed of 20 multiple personality individuals and that each of these persons has 20 personalities.

Let us say that it will require knowledge of 20 specialized disciplines on program ABC. Let us also accept that a person with multiple personalities enjoys inter-personality communication of knowledge and that all personalities can be simultaneously functioning, that is, anything known by personality 1 is available to personality 2 and time sharing of the senses and mental processes is not required. We have to accept that speech is a single channel function for these personalities, however. Finally, the members of the XYZ development staff all graduated from the MultiMind University where they have perfected the technique of educating each personality in a different engineering specialty.

Might we not have to rethink the nature of the INDIVIDUAL EFFORT integration space in company XYZ? One person could completely populate an IPPT if the 20 specialties they had mastered happened to match the 20 disciplines needed for program ABC. Let us assign two engineers to staff the PIT leaving the other 18 to work on 18 simultaneous IPPTs. Perhaps this will allow company XYZ to take advantage of the ultimate downsizing opportunity, while also accomplishing the work necessary to develop the customer's product.

We have said that the need for integration is driven by the need for humans to specialize. In this case, the specialization process has been internalized, and while external integration may not be necessary, there must be a means to integrate the information each personality is working with.

Exactly how does integration take place? It takes place through a human interface called communication, appealing to our senses of touch, sight, and hearing, and, to a lesser extent, smell and taste depending on the product. We specialists must share a common information base permitting communication of our ideas and an ability to understand the consequences of information communicated to us in association with our own specialty.

If the MultiMind University had fouled up the class schedule and educated each of the 20 personalities of Joan Doakes in a different language, Joan would be hard pressed to take advantage of her formidable internal integration possibilities. A common language, logic, and mathematics form the principal integration instruments at the most primitive level.

Whether we are dealing with graduates of the MultiMind University or Georgia Tech, these people must enjoy a common basis of communication and reasoning through which they can exchange ideas that spring from applying their specialized knowledge base to a problem they share in solving.

23.2.2 *Interface integration*

The principal residence of system engineers is the interface between the things in a system and the interfaces between the team and different specialists. It is important in system development to organize the system, the organization that will develop it, and the relationship between them such that it is very clearly understood to everyone on the program where and how these two structures come together. The ideal way to do this is to organize the people into cross-functional IPPTs and assign these teams to things in the product architecture, as discussed in Part five of this book.

The technical job of interface integration involves auditing the two sides of the interface from the perspectives of the parties responsible for the two terminals and comparing these perspectives. To the extent they differ, problems are exposed that must be resolved. In doing this work, we should recognize that everyone does not look at the interfaces that exist in quite the same way.

23.2.2.1 *Three views of interface*

Interfaces exist between objects of systems for the purpose of introducing synergism between the objects. A system is said to be a collection of two or more objects that interact in purposeful ways to achieve an objective that no subset of the objects could otherwise achieve, and it is through the synergism provided by interfaces between the system objects that this takes place. Interfaces communicate information from one object to another that the other would have no way of obtaining through its own unaided capacity. They provide energy with which to function, as in the case of electrical or hydraulic power. Interfaces act to physically tie a system together, where objects are bolted or otherwise joined together.

An interface element has two fundamental characteristics: (1) a terminal at each end and (2) a media of communication. The terminal will always be an element of the system depicted on the system architectural block diagram or an element of the system environment. This includes the elements of the system of interest, possibly some cooperative and hostile systems within the system environment, natural environmental elements, and no other possibilities. Likewise, the media of the interface element will be provided by some element of the system of interest, or the system environment (natural environment or cooperative systems). Wire harnesses, fluid plumbing, and the space through which radio signals pass are examples of interface media. The first two of these media are provided by the system architecture and the last is an example of an element completed by the system environment.

We take three different views of the system interface here as a function of the perspective of the observer. These three views take into account the variable intensity of interest and the responsibility for interface by the design community. The basis of these differing views is the relationship between the terminals and media of an interface element and the responsible design organization. It is a great systems management truth that the interfaces that will result in the greatest system problems are those where different organizations are responsible for two or three of the following: source terminal, destination terminal, and media. Relatively little difficulty can be expected where the same organization is responsible for all three. Where differences occur, the interface is a member of the cross-organizational interface set upon which system engineering must focus its energy in interface integration work.

The traditional view of interface is that an interface exists where two different contractor organizations are responsible for the design of the two terminals of the interface. We take a more general view here, where an interface is any means of relating one system element to another at any level of system indenture. An interface thus imagined may be partitioned into subsets of interest as a function of organizational responsibility by establishing three classes as follows.

1. OUTERFACE: Interface elements with neither terminal nor media under the design responsibility of the organizational element X are in the organization X outerface class. Organization X has no immediate interest in this class of interface and is not responsible in any way for its proper design.

2. INNERFACE: Interface elements with both terminals and the media under the design responsibility of organization X are in the organization X innerface class. Only organization X is immediately concerned with this class and is completely responsible for its proper design.
3. CROSSFACE: Interface elements with some subset (other than the null set) of two terminals and media under the design responsibility of organization X and the remaining subset (not the null set) under the design responsibility of a different organization are in the organization X crossface class. Organization X and some other organizations are jointly responsible for the design of this class. This is the interface class where system engineering energy and budget must be focused, because it is where system problems will develop. These are the cracks between the specialized knowledge and experience of the engineering organizational elements. They are inevitable in any system and are determined by how the organization is structured with respect to the structuring of the system architecture. It is important for the system engineering community to be capable of sorting all of this class of interface from the whole and assigning responsibility for its development.

23.2.2.2 Interface responsibility model

How shall we assign responsibility for interface development? It is clear that there is a fundamental difference between architecture development responsibility and interface development responsibility. The former can be completely assigned to a single development agent with little difficulty. Interface on the other hand has two, and with the interface media possibly three, agents that should be interested in any one interface element. We need a way to make interface responsibility unambiguous.

The previous discussion of the three classes of interface as a function of your outlook offers a solution. First, let us create a hypothetical schematic block diagram in Figure 23.4. It includes seven components in three subsystems. Our challenge now is to determine who, among the seven principal engineers at the component level and the three principal engineers at the subsystem level, is responsible for each of the interfaces illustrated here.

The system illustrated in Figure 23.4 is composed of a set of architecture consisting of A = {A11, A12, A21, A22, A31, A32, A33} arranged into three subsystems (A1, A2, A3) and a set of interfaces defined by I = {I11, I12, I13, I21, I22, I23, I31, I32, I33, I411, I412, I42, I431, I432, I441, I442, I511, I512, I52, I53, I54, I55, I6, I71, I72, I7311, I7312, I7313, I732}. Let us now assign some human beings as principal engineers for these elements.

It is easy to assign architecture responsibility. Let us assume that A1 is a mechanical subsystem with Bob, the principal engineer, supported by

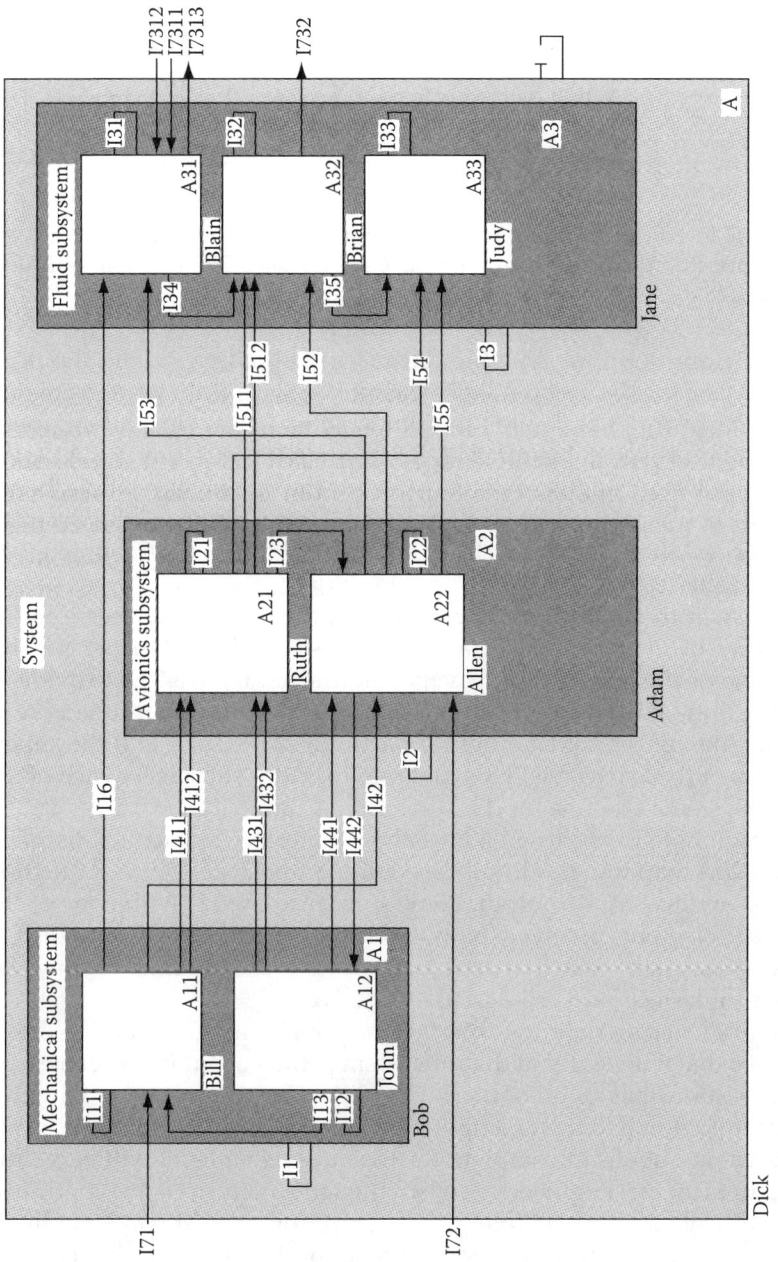

Figure 23.4 Interface responsibility model. (From Grady, J., *System Integration*, CRC Press, Boca Raton, FL, pp. 43–228, 1994. With permission.)

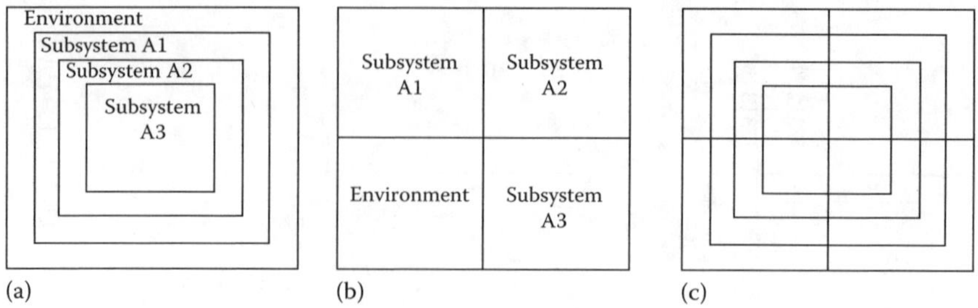

Figure 23.5 Interface partitions: (a) input partition, (b) output partition, and (c) superimposition. (From Grady, J., *System Integration*, CRC Press, Boca Raton, FL, pp. 43–228, 1994. With permission.)

Bill for A11 and John for A12. Similarly, we will assign Adam, the principal engineer for the avionics subsystem, A2, with Ruth responsible for A21 and Allen for A22. Jane is the fluids subsystem principal engineer, A3, with Blain responsible for A31, Brian for A32, and Judy for A33. We must have a system engineer responsible for the system as a whole, hence assigning Dick to this task. Now, how do these people share the development responsibility for the interfaces between the elements for which they are responsible?

To answer this question, let us partition the set of interfaces into the three views of innerface, outerface, and crossface from the perspective of each one of the subsystem principal engineers. Figure 23.5 provides a Venn diagram of all the interface possibilities for the system based on an input partition of the set I, an output partition of the set I, and the superimposition of these two sets. This diagram includes subsets for each of the three subsystems and one for the system environment.

You will note in Figure 23.6 how the 29 interface elements illustrated in Figure 23.4 map into the 16 subsets defined in Figure 23.5 as a function of subsystem level input/output. Some subsets are voids with none of the 29 interface elements involved. Now, let us, in Figure 23.6, see how each of the three subsystem principal engineers views these interfaces from their innerface, outerface, and crossface perspectives.

The final step in defining interface responsibility is at hand. Note in Figure 23.6 that it is clear that the subsystem principal engineers can easily be made responsible for all of their innerface. They may wish to flow this responsibility down to their component principal engineers, and the same method we are about to complete for the subsystem level will serve the subsystem principal engineer's needs. This same pattern can be continued through all layers of system interface in a top-down development of interface in step with the top-down development of the system architecture.

The problem remains to decide which of the subsystem principal engineers should be responsible for interfaces identified as crossfaces in

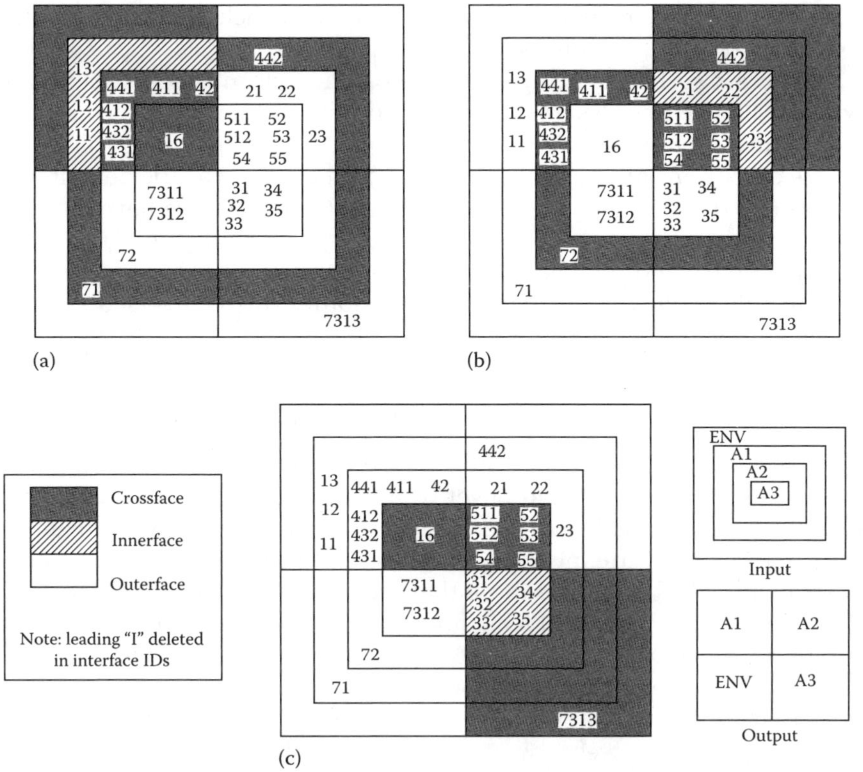

Figure 23.6 Subsystem principal engineer views: (a) subsystem A1 perspective, (b) subsystem A2 perspective, and (c) subsystem A3 perspective. (From Grady, J., *System Integration*, CRC Press, Boca Raton, FL, pp. 43–228, 1994. With permission.)

Figure 23.6. We have three choices: (1) make the architecture principal engineer on the source-end responsible, (2) make the principal engineer on the destination end responsible, or (3) let them share the responsibility. Clearly, they have to work together, but one of them should be made responsible for the proper development of each interface. It is possible to select either rule 1 or rule 2 above to avoid interface responsibility ambiguity. In this era of interest in total quality management, it may be popular to select rule 2, since the destination principal engineer can be thought of as the customer for that interface.

Whichever rule you select, you will find many good reasons for exceptions based on (1) architecture elements that are grandfathered into the system as customer-furnished equipment derived from previous systems, (2) existing equipment designs that are only modified for this application, and (3) biased engineering judgment. Since it is so difficult to pick a foolproof interface responsibility rule, you may conclude that it is foolish to try, but try to resist this feeling.

Where a system is being developed from a clean sheet of paper and will use no existing designs (a very uncommon situation), we could pick rule 1 or rule 2 without fear of compromise. In a real-world situation, where exceptions are possible, the rule could be 1 or 2 as the default with exceptions decided on a case-by-case basis. Given that we have an interface dictionary, we can include interface responsibility information in this dictionary and remove all doubt about who is responsible. Either the default rule applies or the default responsibility rule is reversed as noted in the dictionary.

23.2.3 The special need for external interface development

Interfaces between elements that are both under the same company's design responsibility should never be a problem in a development activity, but they commonly are even where different design departments of the same company are involved. Interfaces between elements a contractor is developing in-house and those that are procured from an outside vendor should never be a problem, since the contractor controls the vendor through a subcontract or purchase order. Interface definition is a common area of dispute between contractors and their suppliers, however. If interfaces that are fully under the control of a contractor are difficult to develop flawlessly, those between two contractors that have no contractual relationship could be nearly impossible. Because it is so difficult to develop these interfaces, we need some special controls at these very important interface planes.

Interface development is an engineering management and technical approach to the development of interfaces where the contractors responsible for the design of the elements at the two terminals are different and not linked through a contractual arrangement. Commonly, these contractors are called associates, and each has an independent contract with a common contracting agency. This management approach seeks to introduce discipline and precision into the technical communications between the contractors and to formally resolve problems across the interface between their products.

A customer will commonly require two interfacing associate contractors to reach an agreement on how they will jointly work on a mutual interface and to jointly sign a memorandum of agreement stating the means by which they will cooperate. They may also be required to develop an interface management plan to expand on the memorandum of agreement and both sign this plan. One of the associates will be required to prepare an interface control document (ICD), and this associate and the common customer will manage the development of the interface through this ICD.

An interface control working group (ICWG) will be formed with the customer chairing its meetings. The ICWG will meet periodically and take up interface issues. Over time, the interface definition evolves between the two elements managed in this way. Interface issues will be resolved by joint action and approved as evidenced by approving signatures on the ICD and revisions (interface change notices, or ICNs) by ICWG members.

Two possibilities exist for this ICD. One possibility is that it will only be used as a means to an end, that being to provide each associate with the same interface definition for use in their design work. In this case, each associate uses the ICD as the source of the interface requirements for the specification on the element on their side of the interface plan. When these two specifications are authenticated (signed) by the common customer into the program baseline, the interface can thereafter be managed though engineering change proposals (ECPs) against the two specifications. The ICD makes its exit at this point having served its purpose.

The other alternative is that the ICD remains a living document throughout the program. In this case, the interface requirements contained in it may never be copied over into the specifications at the terminals. Each terminal specification will reference the ICD for the interface requirements.

There is a third alternative that is nothing but trouble, and this is the case of a living ICD and individual contractor attempts to capture the interface requirements in their terminal specifications. This results in a triple set of books that will almost certainly get out of synchronism.

23.3 System optimization

The handmaiden of system decomposition and interface integration is system optimization. We carve the system functionality into an architecture assured that the aggregate of the elements will satisfy the system need in an optimum fashion, entailing maximum capability for minimum cost. We may not have allocated the needed functionality so as to yield a well-optimized system. It is also possible, during the design evolution, that one or more IPPT may develop a design solution, within the boundaries established by their requirements, that places the overall system in a condition of nonoptimization. The design solution in one team may force a solution on another team that is unnecessarily costly or force a severe weight penalty on another team. Each team should be forced to carry a reasonable share of the design difficulty, if possible.

It is relatively easy to detect the symptoms of nonoptimization during the design process. One or more teams will be struggling to meet their requirements while one or more others are having no trouble satisfying their milestones, possibly underrunning team budget and schedule allowance in the process. Mathematical models developed to solve requirements and design problems are also excellent tools to help identify suboptimization as well as to look for optimizing alternatives within the bounds for which the models are appropriate. The PIT must be alert to detect these symptoms and react to them, to understand the driving forces, and bring about a balanced condition between the teams, if possible.

It may be possible to rebalance the requirements for reliability, mass properties, design to cost, or other parameters such that the teams are equally sharing the burden and a more balanced system will result. If this cannot be done, it may be necessary for one or more teams to alter their design concept or even the technologies to which they are appealing. These actions can ripple into the risk assessment and reshuffle these priorities.

On the other hand, it may not be possible to rebalance the pain across the teams, but it may be possible to identify areas of growth potential, corresponding to the overachieving team areas, that the customer may choose to exploit while providing additional funding. It is also possible to reduce the budget and schedule allowance for one or more overachieving teams, leading to increased margin (or profit) and schedule slack that may have to be spent at a later date to cure a discontinuity. The point is that we should be actively seeking out these conditions rather than letting the corresponding opportunities slip through our fingers from indifference. The alternative to an active program of optimization is that suboptimized conditions will materialize in a time of panic and have to be solved by throwing money at the corresponding problem.

23.4 Other PIT actions during design

In addition to the top-level interface integration and optimization work for which the PIT must accept full responsibility (task 14111 in Figure 23.7), the PIT must also provide all of the integration services noted in Figure 23.7. These include test-planning integration (tasks 14113 and 14114), system-level analyses (task 14112), the risk management program (task 14113), and the several nonengineering integration functions noted.

The PIT must be constantly alert for opportunities to integrate the design work between product and process. The production process member(s) of the PIT must remain aware of the unfolding production solutions from the IPPT and integrate them into a system production solution. At the same time, they must remain alert to possible opportunities to better balance the design solution between the product, operational employment, and production.

This same attitude must be present in the material procurement members, quality members, integrated logistics members, and others. Each team member must remain attuned to what their counterparts are doing on all IPPTs and integrating these solutions into the system solution while working as a team at the system level to constantly question whether they

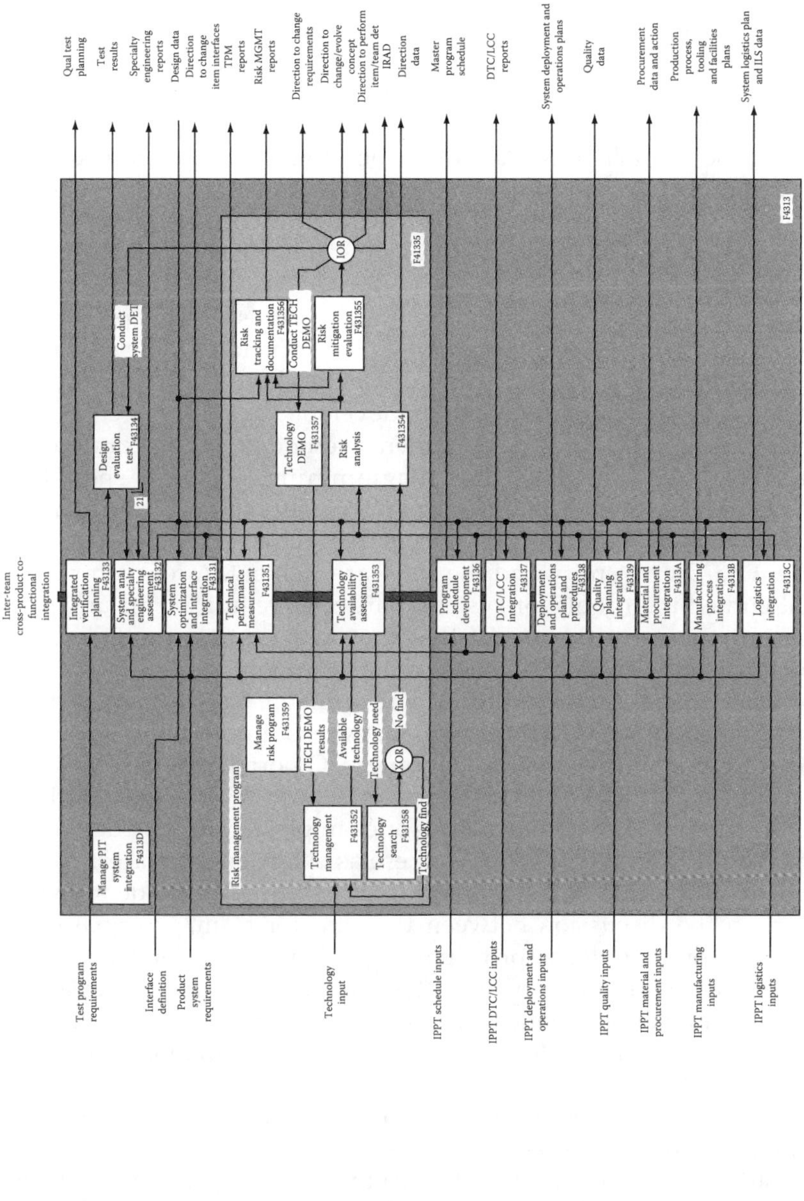

Figure 23.7 System integration process. (From Grady, J., *System Integration*, CRC Press, Boca Raton, FL, pp. 43–228, 1994. With permission.)

have achieved the best condition of balance between the evolving designs for the product, the production process, and operational and logistics deployment considerations.

23.5 Special hardware–software integration needs

All of the previous content of this chapter applies to computer software as well as hardware, but software does impose some special concerns and opens up some special opportunities. Commonly, a software entity can be properly assigned to a specialized IPPT focused on that one item. This team and the PIT must adhere to the same prescription discussed earlier, where the cross-organization interfaces are clearly defined as are all the other requirements that the software entity must satisfy. A software person might call these requirements collectively their customer requirements.

A program or customer that mindlessly requires a single prescribed functional decomposition methodology, like functional flow diagramming, will immediately run into difficulty at this juncture. The computer software engineering community has developed several effective specialized decomposition and development methods that the community would prefer over functional flow diagramming, and they should be allowed to use these techniques. One of the big problems at this interface is the difficulty of requirements traceability across the abyss between the parent hardware entity and the software element. A few companies have integrated their tools for system/hardware requirements and software requirements to the extent that they can capture the flowdown relationships between them in one unique location.

Software interfaces are a little more difficult to characterize than hardware. The software community may need information about an interface that extends beyond the computer I/O port. It may be necessary to characterize the delays throughout a one-line circuit diagram all the way to the ultimate load (an actuator, for example) to properly define the software needs. The author refers to these as extended interfaces. A simple reading of the interface responsibilities rules may exclude software people from the discussion between the actuator team/principal engineer and the computer team/principal engineer who may jointly feel that their interfaces are in the software outerface set. This is true as far as the n-square diagram or schematic block diagram might show, but we have to be attuned to special needs for extended interface information on the part of the software teams.

There is a feeling among system and hardware people that software offers a safety valve for mistakes in other parts of the system and this sometimes results in late changes to software that are very difficult to make. This attitude is driven by knowledge that the software-manufacturing process is embedded within the software development process, and software changes are focused less broadly across an organization than in the case of hardware changes late in the development process. Hardware changes commonly drive manufacturing changes (possibly extending to tooling and test equipment), procurement changes (with significant cost and schedule impacts), test plans procedures changes, and quality inspection changes. Computer software changes, on the other hand, offer an irresistible alternative completely addressed with the software IPPT. Development teams should consciously recognize the attraction for late software changes and carefully weigh the real relative benefits with accurate inputs on the software impact of the needed changes.

Effective software development work requires adequate testing and simulation capability as do some forms of hardware development. There may be a tendency in the software and hardware communities to isolate on their immediate needs in this area, resulting in specialized resources that may duplicate some functionality and expand cost unnecessarily. For example, systems involving guidance and control functions, and computer software implementation of these functions, are good candidates for simulation and test resources shared by hardware and software communities.

The hardware and software communities also respect vocabulary differences that can lead to communication problems across their interfaces. It is extremely important that engineers involved in discussions across these interfaces challenge each other to ensure understanding. The phrase, "I understand you to mean…," should often be heard in these discussions. In the next chapter, we will discuss one of these vocabulary differences that can lead to misunderstandings, but there are many.

The greatest problem in hardware/software integration, however, is a separation between hardware and software people that sadly is becoming a tradition. We said earlier that it is not uncommon for engineers at the cross-organizational interfaces to withdraw rather than plunge across these interfaces. This is nowhere more fatal than when the interface is between hardware and software. The PIT on a program and the leadership of each IPPT must be sensitive throughout the development activity to any symptoms of separation between these parties and encourage in every way possible a close working relationship between them.

Software developments commonly include in-process reviews by several different names. The IPPT with software responsibilities should invite related hardware people to these reviews to educate them and to get useful feedback from them. Likewise, hardware-dominated teams should encourage participation in their meetings by software people even though there may be no software under their responsibilities. Communications is the machinery of integration.

chapter twenty-four

System verification management*

24.1 Verification classes

Three specific verification classes are encouraged and discussed in this chapter in a management context. The details of the work managed are covered in *System Verification* (Grady, 2007). The first of the three classes is item qualification, where it is proven whether or not the design solution to the problem defined in the corresponding performance or development specification has satisfied the requirements in that specification and, therefore, whether or not the item is qualified for its application. This class is accomplished at the item level. The second class involves aggregating items into end items that satisfy an end-use function and subjecting these entities to verification actions at the system level. Often this class entails system test and evaluation. In the case of an aircraft, this would involve a flight test. The third class subjects the first production article to an acceptance process, captures the results of that process, and compares the results with requirements to reach a conclusion of whether the production process produces a product complying with engineering and manufacturing planning data and therefore acceptable to the customer. This class is then extended throughout the production run to verify that individual articles are satisfactory for delivery to the customer.

24.2 Item qualification verification

24.2.1 Setting the management stage

This chapter covers the overall management of the item qualification verification process using the integrated verification plan (IVP), the integrated verification report (IVR), and the verification management report (VMR) introduced in earlier chapters. All of the work grouped together under qualification is focused on a major review at the downstream end of the qualification process called by the Department of Defense (DoD) the functional configuration audit (FCA). At this review, the customer is presented with the evidence accumulated through the qualification program to prove that the design satisfies the requirements in what would now be called in the DoD the performance specifications. In past years, these specifications

* The material in this chapter is from Grady, J., *System Integration*, CRC Press, Boca Raton, FL, Chapters 5 and 6, 1994. Used with permission.

would have been called the development specifications or the Part I specifications, but MIL-STD-961E changed the terms for the DoD in 1995.

Given that we have captured the correct data and organized it well for implementation, the remaining problem is one of doing the planned work and management of that work. Management involves bringing the right resources to bear on the problems we have defined in accordance with the schedule, also predefined, within the context of limited economic resources consistent with the planned work. We desire to accomplish the planned work that will result in the evidence of a condition of compliance. Where this evidence does not support the conclusion that the product satisfies its requirements, we must take management action to bring ourselves to a condition of compliance. The possible actions we can take entail changing the requirements, changing the verification planning data and reverifying, changing the product design and reverifying it to possibly new planning data, or obtaining a temporary waiver from the customer for the noncompliant design features.

Management skill was required throughout the preparation for verification. This included insisting on solid verification requirements as well as sound planning and procedural data, following the development of needed verification resources and providing guidance and direction where needed, and arranging for availability of product to test in accordance with the planning data. As the product, in its several representations (engineering drawings, parts lists, physical product, and lines of code), becomes available for test, analysis, demonstration, and examination, the verification management focus must shift to taking action to cause planned work to occur as scheduled, maintaining status of planned work, overcoming conditions that cause work stoppages, and taking action to prevent or correct for budget and cost overrun conditions.

As the item qualification verification work begins to come to a close, verification management must focus on communicating the results to the customer through a formal review of verification data and achieving customer acceptance of a successful item qualification verification effort. The prize at the end of this run is the authorization to engage in system-level verification activities potentially involving actions that are hazardous to life and property, such that occur in the flight or ground testing of weapons systems. Progress and incentive payments may also be tagged to the completion of item qualification.

24.2.2 Documentation overview

Table 24.1 shows an example of a specification verification traceability matrix. It parses the requirements into fragments (FRAG) if necessary and maps item requirements to verification requirements and methods to be applied in the verification. The requirement in paragraph 3.2.1.5 of

Table 24.1 Sample Verification Traceability Matrix

SEC 3 PARA	FRAG	Title	Methods T	A	E	D	Level	SEC 4 PARA
3.2.1.5	1	Guidance accuracy	X				ITEM	4.2.37
	1			X			ITEM	4.2.38
3.2.1.6	1	Clock accuracy	X				SYSTEM	4.2.41

Table 24.2 Sample Verification Compliance Matrix Fragment

SEC 3 PARA	Title	Methods T	A	E	D	LEV	SEC 4 PARA	VSN	VTN
3.2.1.5	Guidance accuracy	X				ITEM	4.2.37	0032	F412314
			X			PARENT	4.2.38	0033	F412822
3.2.1.6	Clock accuracy	X				ITEM	4.2.41	0102	F412314

this specification has only one fragment, but will require two different methods applied to verify compliance. The FRAG field is needed if the paragraph includes multiple requirements. As an alternative to the use of fragments, we could use requirement IDs assigned in a database system to each unique requirement, but humans will relate more easily to paragraph numbers. The line items in this matrix in each specification form the atomic structure of the verification process called verification strings.

Table 24.2 illustrates a fragment from a verification compliance matrix composed of the union of all verification traceability matrices, and it maps every verification string to a verification task number (VTN) that relates to the specific tasks in the qualification program. This is the matrix that the verification engineer should use to design the verification process, and this takes place by mapping, assigning, or allocating each string to a task and collecting all of the verification strings under each task as the basis for preparing a plan and procedure to accomplish this task.

As shown in Figure 24.1, the content of the compliance matrix is filtered into the verification task–tracking matrix that has one line item for each VTN. This is the matrix that the program manager should use to manage the program and track status. If any program had to maintain all of these matrices as independent entities, it would tax available resources. The only way to maintain this data is through the use of relational tables in a database. The verification plans and procedures are focused on the VTNs and collect in the sections of the IVP, as outlined earlier. The verification coordinator uses the overall verification plan, schedule, and verification task–tracking matrix to determine when to assign and authorize

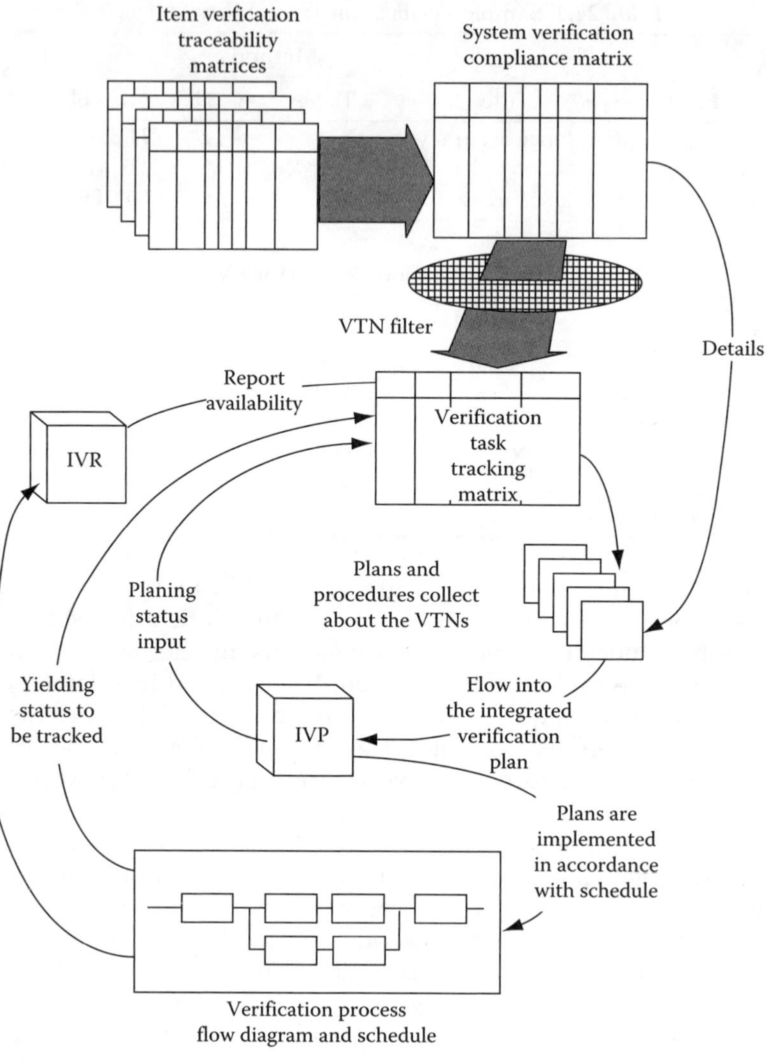

Figure 24.1 The verification task–tracking matrix. (From Grady, J., *System Validation and Verification*, CRC Press, Boca Raton, FL, pp. 213–275, 1997. With permission.)

tasks, and tracks these tasks in work as they mature and complete their cycle in accordance with the plan content.

Figure 24.1 illustrates the general management pattern during the item qualification process. Those responsible for verification tasks are stimulated to implement planned work identified in the verification task–tracking matrix. Corresponding planning data is implemented by assigned personnel in accordance with the schedule and budget, yielding results documented in the corresponding integrated verification task report. The tracking matrix is updated and the whole program continues to follow this pattern toward completion of all planned verification work.

This chapter discusses the implementation of IVPs, procedures, and reports. It should be noted that this has probably never been done successfully on a large program. More often, individual plan, procedures, and reports are prepared by principal engineers and separately released.

The line items in the verification task–tracking matrix are VTNs derived from the detailed listing of the verification compliance matrix that is the union of all the verification traceability matrices in all the program specifications.

24.2.3 The management agent

Many enterprises have strong test and evaluation functional departments associated with all test activities, and interest in the remainder of the methods is spread across engineers from many functional departments on a program. The T&E community is often very good at what they do and insist on controlling all of the testing work, but show no interest in the other methods. An enterprise would be well served to rename this organization the Verification Department and make them responsible for managing the whole verification process on programs. In the author's view, this would place a senior engineer from this department in the PIT with overall responsibility, perhaps named the verification coordinator.

Within a program that applies the cross-functional team organizational approach with staffing from lean functional organizations, the overall verification management responsibility should be assigned to the PIT, by whatever name. All item verification tasks should be assigned to one of the product teams with a clear definition of resources and timing in the published planning data, as discussed in previous chapters. As the time planned for these tasks comes into view, the PIT verification coordinator should ensure that the responsible team is moving toward readiness and that the necessary resources are being provided by supporting personnel. An effective means should be in place for all verification actions in progress to encourage persons responsible for these actions to make available to the PIT accurate status information. The PIT must review this information periodically and take action where appropriate to maintain the verification process on track.

The system verification management agent (the PIT in this book) must take direct responsibility for all qualification verification actions accomplished at the system level including any tasks focused at a higher level than the product teams, which may include tasks below the system level depending on how the product teams are assigned. The responsibility for a particular task could be delegated to a team, of course.

All verification work in terms of VTNs is assembled from verification strings and oriented about items in the system product entity structure by a principal engineer assigned to each task. Each task should have associated with it a task plan and a task procedure captured within the IVP and should result, when implemented, in a component of the IVR. The tasks provide a sound basis for verification management, but there are groupings of these tasks that also prove useful.

The sum total of all VTNs associated with a particular system item is referred to as an item verification task group, and the sum total of all the item verification task groups assigned to a single person, group, or team is a responsibility verification task group. There will be one or more item verification task groups subordinate to every responsible verification task group. If a team is responsible for verification work for two items, its responsibility task group will consist of two item task groups. These verification task groups consist of all the verification tasks, of potentially all four methods (test, analysis, demonstration, and examination), associated with the items involved.

The person from the PIT leading the item verification process should form a virtual organization of all the persons responsible for these verification task groups, which may involve tiers of groups and individuals rather than a simple single-layer fan structure. This responsibility is that of a project engineer working across organizational structures to achieve overall program goals in this specific area. This is an integration responsibility not strictly a management responsibility. But, this is not an unusual condition, as it is the role of most people in a system-level team. The system reliability engineer is looking for the optimum reliability condition across the system. The system-level mass properties engineer is similarly interested in prime item mass properties rather than that of the items upon which the teams are focused.

Figure 24.2 offers a way of illustrating this grouping of individual verification tasks into groups oriented toward architecture items and teams. Given that integrated product and process team (IPPT) 1 is responsible for architecture items A11 and A12, the responsibility verification task group for this team includes all of the verification tasks in all of the item verification task groups mapped to this team. In the verification task management matrix, all of the tasks indicated under team 1 would show a responsibility assignment to this team or someone reporting to this team.

We now have the makings of a manageable situation. We have identified a comprehensive series of tasks clearly defined in plans and procedures, driven by a clear set of requirements, hooked into budget and schedule figures, and assigned to specific persons or organizations. The remaining problem is to track the progress of the responsible persons in satisfying task goals within assigned cost and schedule figures and take such actions as required to encourage the planned outcome.

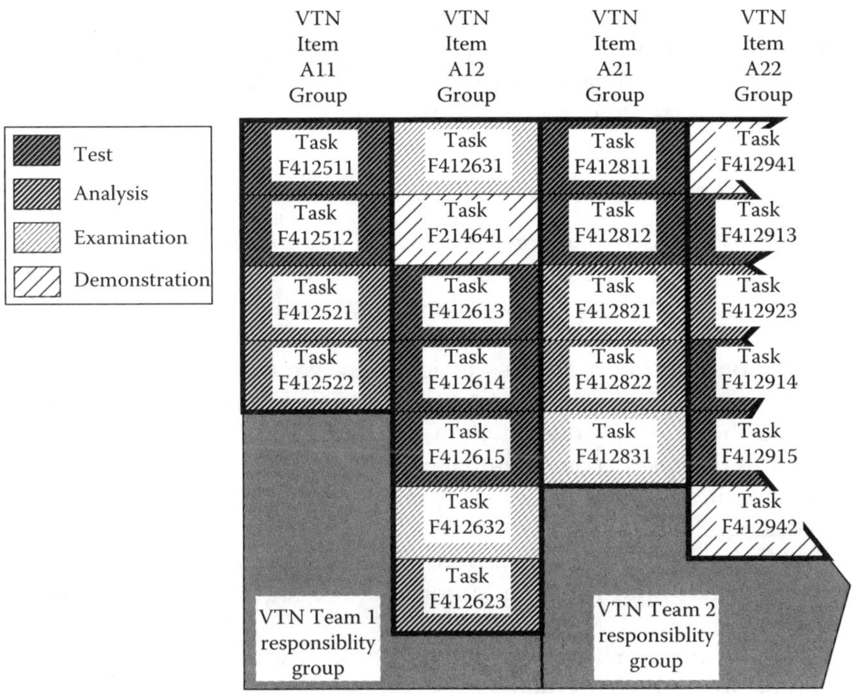

Figure 24.2 VTN groups. (From Grady, J., *System Validation and Verification*, CRC Press, Boca Raton, FL, pp. 213–275, 1997. With permission.)

24.2.4 Task integration and scheduling

As the atomic components of the verification process become available from teams or individuals in the form of specific verification tasks of several kinds (test, analysis, examination, and demonstration), they must be coordinated with respect to the place of implementation, the time frame within which they will be accomplished, and other tasks upon which they depend and to which they must supply results. Ideally, this will be possible within the requirements tool applied on the program, but if a truly comprehensive system is not available, a good alternative for this part of the process is offered by the many good network tools capable of PERT or CPM. Microsoft Project, Mac Project, and Primavera are some examples. The tasks can be entered and linked to time and resources in these tools and schedules generated for use by those who must implement the qualification tasks. The critical path is also clearly identified in these tools.

The tasks (VTNs) do not stand alone, generally. They are interconnected by information dependencies. These dependencies should be noted in the planning data and used to help determine the impact of problems in completing particular tasks on schedule. Where a task has no downstream dependencies, it will be less of a problem to allow it to overrun schedule-wise. Where it has many downstream dependencies, it will commonly appear on the critical path and signal a need to pour in sufficient resources to expedite completion.

24.2.5 *Task readiness review and task authorization*

In the case of tests, there is a generally accepted notion of reviewing readiness to execute the test as a prerequisite to beginning the test. This reflects an appreciation of the cost and schedule consequences of implementing a test prior to a condition of readiness. In any action, if we begin work before a condition of readiness is achieved, we may have to repeat the work. This could be called a first-degree risk, where the spent money and time characterizes the extent of the risk. In some qualification tests, the item under test becomes damaged as a result of the stresses applied, and this unit thereafter is not acceptable without refurbishment for the subsequent retest when it is discovered that a condition of readiness was not in effect and a retest is required. This results in a second-degree risk involving funds and time in addition to those required to simply repeat the work.

As a minimum, the verification task readiness review concept should be applied where there is a second-degree cost and schedule risk. Alternatively, we could require a review prior to every verification task, recognizing that most of these would require no more than 10–15 min by the principal engineer and the team leader. Each VTN listed in the verification task–tracking matrix should be annotated to indicate the kind of readiness review required from a set of possibilities such as the following: (a) none, (b), team leader, or (c) program verification coordinator.

In the case of verification tests on high-dollar items involving large cost and schedule increments, the customer may wish to be present and voice an opinion on readiness. The matrix should also note these cases so that there will be no mishaps where the customer is not notified of such meetings.

24.2.6 *Evaluation and refinement of evidence*

The final review of the qualification verification evidence will take place at the FCA. But this evidence should have received a thorough going over long before then by several parties. It does happen in some organizations that considerable money is spent on planning for and accomplishing qualification tests only to have the results spark no interest. This is especially true for supplier test data, but also occurs for one's own test reports. Verification actions should be undertaken for good reasons and the information produced as a result is important deserving of careful scrutiny. If the work is not important enough to read the report, it probably was not important enough to have done the work in the first place.

24.2.6.1 *In-house review of in-house verification reports*

The verification task–tracking matrix should identify who in plant must review and approve the results of each verification task. This may be the principal engineer, the product team leader, or the program chief engineer and leader of the PIT. In some cases, we may wish to have several people to collectively review a report and meet together to reach a consensus on the results. Where one concludes that the report does not provide clear evidence of compliance, the problem should be elevated to a management level where a decision can be made about what to do.

Each verification task listed in the verification task–tracking matrix should have an assigned review responsibility and indicate the status of this review. The verification management agent must ensure that these reviews take place and that the results of these reviews are acted upon.

24.2.6.2 *In-house review of supplier verification reports*

A large company working on a complex product will procure many parts of the system from other companies. Where the product is purchased off the shelf, the buyer must simply accept it or appeal under a warranty or guarantee where it is defective. Where you thereafter alter the item, you have to accept the responsibility to verify that the resultant product satisfies your requirements. Where you procure an item built to your blueprints, you should do the qualification testing yourself since they are your requirements, in which case the verification work simply falls under your qualification program. Alternatively, you could require the supplier to test it to your requirements, possibly with your witnessing of the tests. In any case, you should expect some kind of written report of findings, and it should be reviewed for findings and the veracity of these findings.

Where you procure an item developed in accordance with your procurement specification and statement of work, the verification work is best done by the supplier. Your subcontract should call for review and approval of the qualification plans and reports. On a large program with many suppliers, the amount of evidence to be reviewed and retained for future reference can be almost staggering. General Dynamics Space Systems (GDSS) Division (prior to sale to Lockheed Martin) accumulated so much supplier data by 1993 for the Titan Centaur Program that the file cabinets within which this paper data was stored became so numerous that data management had to move people to other office space to make room for them. Later in the program, data management brought in a system provided by Hershey Technologies that largely solved the storage capacity problem and went a long way toward solving some other problems in routing the data for review and approval.

This system included a scanner to convert paper documents received from suppliers into bitmapped images that were temporarily stored on a server where they were available for review by persons contacted by data management. As a result, it was no longer necessary to run copies of these documents and physically distribute them to a list of persons. The people notified by e-mail that they had to review these documents could open them up during a scheduled review window and respond with comments. When a sufficient body of data was available on the server, data management loaded it onto a CD-ROM using a CD-ROM maker built into the workstation. This CD-ROM was then stored in a file cabinet. Subsequently, an engineer could call data management and request that a particular document be uploaded to the server from a CD-ROM, and the engineer could make a paper copy of the parts needed or view it on the screen of his or her workstation. In a company so equipped, this document could be called up across the network to a computer projection unit in a meeting where the content of the document could be discussed. This is a tremendously powerful capability that can repay the small cost of implementation many times over in efficiency or productivity.

Other companies have come up with alternative ways of making supplier data available for review and approval as well as use. One could put the CD-ROMs in a carrousel or disk storage bank to avoid the manual upload problem, but there is a cost-effectiveness trade that should be run here. If the data will seldom be requested after initial review and approval activity, it can be more cost effectively provided using the verbal request method discussed above. The problem with scanned data is that it is very consuming of computer memory even where special compression algorithms are used. You may conclude that it would be better to scan the incoming data as text and graphics to reduce the memory space and also to provide the opportunity to edit it subsequent to receipt. You should be aware that even with the very best scanner and optical character reader software, there will be errors in the resultant data. There will be no errors in a bitmapped image. Further, we should not modify data received from our suppliers even if we could. This is formally delivered data across the contractual boundary, and changing it internally after receipt could lead to many ethical and legal problems.

The workstation for supplier data receipt should also include a modem for receiving data submitted by electronic data interchange (EDI). Over time, this pathway will increase as more and more companies become EDI capable. The paper submission method will become less popular but will likely never become a complete void. One step beyond the paper and e-mail or fax delivery of data, we find database delivery. Every company today creates its documentation on a computer, and eventually more and more companies will create their data in relational databases. This data could be delivered to customers by refreshing the content of a customer version of the database in accordance with a schedule or at specific milestones. Internet, ever evolving, will also offer useful data delivery options in the future.

24.2.6.3 *Associate interaction and interface verification resolution*

Interfaces pose one of the most serious development challenges on a program. Where the two sides of the interface are represented by different companies with no contractual relationship, the contractors are referred to as associate contractors. This happens where a common customer separately contracts for several segments of a system. Each of the contractors has a contract with the common customer, but no contract with the other associates. The customer solves this problem by requiring each contractor to submit a memorandum of understanding, citing that they will cooperate with other associates in accordance with the associate contractor clause in their contract and jointly agree to work in accordance with an interface development plan and interface control document (ICD) that one of them will be called upon to create and maintain as the integrating contractor.

A team is formed called an interface control working group (ICWG) cochaired by the common customer and the integrating contractor and including representatives from all the associate contractors as members. This group is responsible for developing an interface development plan and developing compatible interfaces across all the associate interfaces in accordance with this plan. The first step in this process is to identify each necessary interface, define interface requirements for each of them, and capture them in an ICD.

An ICD can be used in one of two ways. First, it may be a living ICD that remains active throughout the life of the program. Alternatively, it can be used only as a means to define the requirements that are moved to the two specifications for the two interfacing elements. In the latter case, the ICD passes from the scene when the interface is mature and defined compatibly in the two terminal specifications.

In the case of the living ICD, the verification requirements should be captured in the ICD Section 4. The verification traceability matrix in Section 4 should list all of the requirements and identify the methods to be applied in verification, the level at which they are to be verified, and the responsibility for the verification work. An interface plan and procedure prepared by the ICWG should provide all of the necessary data in support of the verification work to be accomplished. When this work has been completed, the responsible party should be required to prepare a report of results referenced in the interface compliance matrix prepared and maintained by the ICWG. An interface FCA could be held or the results could be audited at the system FCA.

If the passing ICD is applied on the program, it is consistent to require the associate contractors to prepare interface verification planning and procedures in their IVP with traceability to the ICD. If this work is done well, the ICD verification traceability matrix tells responsibility and points to the responsible contractor for each requirement. The interface verification data developed on the two sides of the interface should be reviewed in-process by the ICWG. A joint FCA could be called or the two sides of the interface audited separately by the same group of people close together in time.

24.2.6.4 IV&V witnessing and review

The customer may not have assigned to its program office the personnel required to interact technically with the contractor(s). To satisfy this need, they may contract with another company to provide technical, independent validation and verification (IV&V), or system engineering and integration (SE&I) services. Four of several companies that have done this kind of work in the past are Aerospace Corporation, Scientific Applications International Corporation (SAIC), MITRE, and Teledyne Brown Engineering. The customer may require the contractor supplying the product to invite the technical services contractor to in-plant meetings for qualification test planning and test readiness purposes. In these meetings, the technical services company represents the customer technically but cannot direct that such changes be made that change the contractual scope or alter previously-agreed-upon formal baselines.

The agent may have to be invited to witness all or specific qualification tests or be provided with the results of qualification testing for review. The agent may interact with your company over technical details of the test and results and with the customer about their conclusions on product suitability for the application as exposed in the qualification test.

Some development organizations are offended when their customer calls in an IV&V contractor, because they feel that the customer does not trust them. This is a very immature attitude. The facts are that very complex systems are not easy to create and an independent view is a healthy thing. The responsible contractor cannot be totally removed from the thoughts and emotions involved in the development of such a system. The GDSS Division decided to produce space launch vehicles for the commercial market as the NASA and military launch market weakened in the late 1980s.

Some people in the GDSS were relieved when it was no longer necessary to deal with the IV&V contractors that NASA and the U.S. Air Force brought into their programs. It was not long, however, before the GDSS realized that it was a healthy thing to have an IV&V contractor reviewing its development work and hired a firm to provide this service for its commercial programs.

24.2.6.5 Customer coordination

As verification reports become available, they should be made available to the customer incrementally for review. Ideally, the customer should have access to updates of the tracking matrix, which shows the building status of this evidence, and should be able to obtain computer access to the evidence directly. In the best of all worlds, the appropriate members of the customer organization will remain tuned in to the availability of this evidence, review it as it becomes available, and notify the appropriate person in the contractor's ranks about their conclusions. To ensure that this does happen, the principal engineers identified for each verification task or the team leaders responsible for the accumulating evidence should talk to their counterparts in the customer's organization and encourage them to review newly available reports and discuss their findings with them. There should be some way for the contractor to capture in a formal way the results of a customer review of the reports, but at least a verbal comment should be requested.

One of the possibilities is that the customer will be convinced that the requirements were satisfied based on reading the one or more verification reports related to the item. In this case, the contractor should be able to move on to other things of importance and consider that the item in question is ready for the FCA. It is also possible that the customer will not be convinced by the evidence. In this case, a mutual understanding must be reached between the customer and the contractor to take one of the following several actions.

1. Agree to disagree until the FCA where management action will be taken to resolve the dispute. This course of action may have schedule risk problems in that if it is finally concluded that the design is deficient, it will require some time to correct the problem and verify the fix, possibly adversely affecting other schedules, such as the beginning of the flight test.
2. Agree that the evidence does not support the conclusion that the item satisfies its requirements and take one of the following actions.
   a. Change the design and reverify.
   b. Request a waiver that would authorize the use of the current deficient design for a limited period of time or limited number of articles until such time that the problem can be corrected. This will require formal work through the contract to achieve and should be complemented with the necessary work to understand how the waiver condition can be lifted and when this can be done.
   c. Change the requirements to reflect the level that was satisfied in the verification work. This course of action may entail the

contractor giving the customer consideration for permitting a relaxation of the requirements. This consideration may be in the form of accepting a lower price for the product, agreeing to fix some other problem that is out of scope but desired by the customer, or paying back an amount of money to the customer.

3. Reach an agreement that the evidence really does support the conclusion that the product design satisfies the requirements.

If this kind of work is not done during the qualification process, it will all pile up at the FCA leading to a lengthy meeting and numerous action items dragging out the final qualification approval. The FCA should be a relatively short meeting where agreements are quickly arrived at between the company and the customer regarding the evidence of compliance, because both parties are familiar with the item requirements and the verification evidence. It should be possible for the cochairs from the customer and the contractor to move the agenda quickly through the items being reviewed. There may be a few items where real disagreement remains, and these few actions can be the subject of splinter sessions yielding outbriefs and direction from the cochairs on a course of action.

The tracking matrix should include information on who the customer contact is and the status of their review of the report. The corresponding contractor representative, the VTN leader, the item principal engineer, or the corresponding team leader should recognize it as one of his or her tasks to coordinate with this customer to encourage them to obtain and read the report, send it to them if necessary, and discuss their conclusions after having read the report. The results of this conversation should then find their way into the tracking report. If the customer reviewer finds a cause for concern about the report, these concerns should be considered by the contractor and a mutual agreement derived for action to remove such concerns. In the worst case, this could mean rerunning some parts of the qualification work to improve the evidence of compliance. By the time the FCA takes place, this matrix should be full of okays for customer review and acceptance.

24.2.7 Staging the FCA

24.2.7.1 Single or multiple audits

All major reviews, including the FCA, are intended to be held at the configuration or end-item level. Where a contract only involves a single item, a single FCA is clearly called for. Where the system entails several end items, a separate FCA may have to be held for each item. Alternatively, all of these audits could be combined into a single FCA. Where two or more separate end-item FCAs are held, they should be followed by a system FCA, as indicated in Figure 24.3.

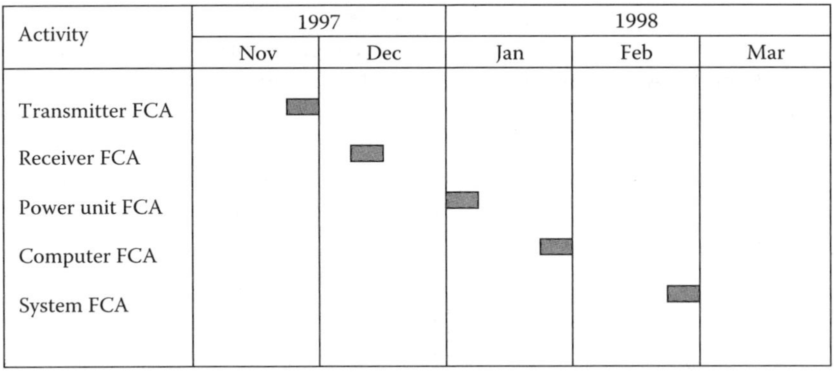

Figure 24.3 Multiple FCA schedule. (From Grady, J., *System Validation and Verification*, CRC Press, Boca Raton, FL, pp. 213–275, 1997. With permission.)

24.2.7.2 Audit preparation

24.2.7.2.1 Completion criteria Generally, the FCA will be scheduled for a specific date, but the customer and contract may respect an event-driven schedule where the audit is not held until the contractor is completely ready to stage it. In the latter case, all of the tasks that have to be complete in order to support the FCA will have clearly defined completion criteria. This may extend to all of the qualification test, analysis, examination, and demonstration tasks, including publication of the results in each case. As we begin to prepare for the FCA, we should also clearly understand the completion criteria for the FCA. The customer should have clearly defined these conditions, but if they have not done so, the contractor should ask for them or offer a closure criterion for consideration. This criterion could include the following conditions:

1. All verification tasks have reports available describing the results. This could be confirmed by the customer reviewing the task-tracking matrix and noting the references to reports and their status, and further confirmed by spot-checking the availability of several.
2. Customer review of the task reports for some subset (possibly all) of the tasks with a conclusion that the evidence supports the conclusion that the item satisfies its requirements in each case.
3. All action items stemming from the FCA closed. This may entail rerunning some tests and reporting the results, or simply responding to specific questions.

MIL-STD-973, configuration management, includes instructions for FCA for DoD programs and offers a set of forms for formally closing out the FCA. The reason this is so important is that a successful closure will enable important program activities, such as system testing, and it may very well enable one or more progress payments to the contractor. If the closure status of the FCA simply drifts along without decisive action, it will also result in inefficient use of personnel and possible program loss of key personnel to other programs within the contractor's organization. If failure to close out the audit precludes forward motion on the program, then the program must reduce personnel to prevent a cost overrun, since many of these people cannot possibly be efficiently employed. If the program drops some critical persons for lack of budget support, the program may never get them back when sufficient funding is made available again.

24.2.7.2.2 Audit agenda The contractor should prepare an initial agenda for the audit and submit it to the customer for review and comment. The contract may require that this agenda be submitted as a contract data requirements list (CDRL) item. Figure 24.4 offers such an agenda based on the assumption that we will only be auditing one item.

The task audits could be reviewed in sequence at a single meeting (as suggested in Figure 24.4) or broken up into subsets for auditing by teams composed of the appropriate disciplines from company and customer staffs. The latter will permit more work per unit of time, but require a final integrating session (not included in Figure 24.4) where each audit team reports back to the committee of the whole.

If the audit will cover multiple items, then the agenda should reflect this by including an indenture within Section 4 for items. Item 4.1 would then be Item 1 audit, and items 4.1.1–4.1.8 would become 4.1.1.1–4.1.1.8.

24.2.7.2.3 Agenda responsibilities assignment The customer may respond with some recommendations for agenda revision, and these changes must be worked out with company program management resulting in a final agenda. Each of the agenda items must be assigned to a presenter and these people asked to prepare their presentation. The responsibility for Sections 1, 5, and 6 of the outline is obvious. Section 2 should be presented by the principal engineer or team leader for the item. Section 3 should be presented by the program verification manager or one of his or her people. Each task in Section 4 should be audited under the company responsibility of the engineer named in the task-tracking matrix as the person responsible for that task. This person should make the necessary documentation available for the task audit, guide the discussion during the task audit, and report the results to the whole if the parallel auditing approach is used. Each task audit may also have assigned a customer cochair where parallel meetings are employed.

This principal presenter need not necessarily be from the contractor. Where the customer and major suppliers have been active team members throughout the verification process, these other parties may be the best

1	AUDIT OPENING	20 min
1.1	Host Co-Chair Welcome	
1.2	Customer Co-Chair Comments	
1.3	Host Chief Engineer Discussion of the Agenda and Audit Plan	
2	ITEM OVERVIEW	40 min
2.1	Item Requirements Discussion	
2.2	Item Design Features Summary	
2.3	Item Risk Coverage	
3	VERIFICATION OVERVIEW	60 min
3.1	Verification Traceability Matrix	
3.2	Verification Requirements Coverage	
3.3	Verification Planning and Procedures	
3.4	Summary of Verification Tasks	
3.5	Item Verification Schedule and Process Flow	
3.6	Verification Documentation Cross Reference	
4	VERIFICATION TASK AUDITS	60×N min
4.1	Verification Task 1	
4.1.1	Task Objectives	
4.1.2	Review of Task Resources	
4.1.3	Task Personnel and Qualifications	
4.1.4	Requirements To Be Audited	
4.1.5	Exposition of the Evidence of Compliance	
4.1.6	Recommendation	
4.1.7	Discussion	
4.1.8	Conclusion	
4.N	Verification Task N	
5	EXECUTIVE SESSION	120 min
5.1	Disposition of Action Items	
5.2	Closure Criteria Review	
5.3	Closure Plan	
6	AUDIT FORMAL CLOSING	60 min
5.1	Summary of Action Items	
5.2	Review of FCA Closure rules	
5.3	Planned Closure Schedule	
5.4	Customer Co-Chair Remarks	
5.5	Host Co-Chair Remarks	

Figure 24.4 FCA agenda. (From Grady, J., *System Validation and Verification*, CRC Press, Boca Raton, FL, pp. 213–275, 1997. With permission.)

qualified to make the presentation in some areas and should not be denied this opportunity based on contractual boundaries. This depends on the customer, of course, and their tolerance for these ideas. Coordinate with each responsible party and let them know what is expected and negotiate the period of time they have to present their topic. Let them know what presentation resources are available (overhead or slide projection, computer projection, video, film, audio, etc.) and inquire if these will be adequate for their needs.

24.2.7.2.4 Audit venues and resources You should then plan for the necessary resources including

1. Main meeting room and smaller caucus rooms
2. Presentation equipment
3. Presentation materials generation, coordination, printing, and distribution
4. Meeting administrative services for attendee check-in, attendee telephone and computer use, meals (if provided) and refreshments, and incoming telephone calls for attendees

It may be possible to decrease travel costs by holding the conference with videoconferencing equipment, but you will have to provide for tighter discipline in the meeting rooms and brief local chairmen that they must brief their participants on the need for more formal identification of speakers and questioners than is necessary when all are physically meeting at one place. A simple meeting etiquette sheet would be useful to this end. You should determine whether there is any benefit in recording the meeting and, if so, notify the participants of the decision to tape. You should recognize that taping will probably change the character of the meeting. Generally, it will amplify the normal differences between introvert and extrovert attendees. Introverts will tend to be more quiet, and extroverts will tend to be more active. As a result, the chairman will have to make an extra effort to bring out the opinions of the demure and control the energy of the others.

Presenters may also find it useful to bring the reviewers to a particular test cell or location in close proximity to the review site in order to convey in some detail the significance of the ongoing system verification work or to emphasize the validity of past testing. You can show slide after slide on equations and environmental parameters at the review to little effect on the attitudes of some doubters, but they may be convinced completely by a brief trip to the test cell where the item is put through a representative round of tests involving thermal, vibration, shock, rain, and salt-spray cycling. Thereafter, the briefing materials will likely be more believable to everyone. Still photos and video can also be very effectively used for the same purpose where the test site is geographically removed from the review site.

24.2.7.2.5 The presentation materials The biggest job in preparing for the audit will be the acquisition of the presentation materials from the presenters, the review and approval of these materials, and the integration of them into a final presentation handout for attendees. Each presenter should be given a style guide in electronic media, a page-count maximum, a clear goal for their segment, and a date when their contribution must be turned in for internal dry run and approval. If this material is all created in electronic media, you should give some thought to a computer projection presentation. Otherwise, the materials will have to be converted into overhead transparencies or slides only to be thrown away after changes are made in these materials in preparation for the final review materials.

24.2.7.2.6 Audit dry run You may choose to dry-run the audit to make certain that everything is in order for the real thing. You need not take as much time as the planned audit. Rather, the first three sections can be discussed to the extent that the program manager has confidence in the readiness of presenters. Then each task leader can briefly preview his or her plan and recommendations. In some cases, the program manager may want a full dry run of the evidence.

24.2.7.2.7 Audit data reproduction Paper copies of Sections 1 through 3 should be provided to customer personnel in the form of a paper handout. You will need to make at least one paper copy of the Section 4 data that will be reviewed. It is unreasonable, however, to make a copy of every document for every member of the auditing team. The amount of documentation can be very voluminous and, therefore, very costly to reproduce. The ideal approach, where all of this data is online on a server, is to provide a workstation and a local printer in the area of the audit such that if anyone wants a copy of a particular document, they can call it up and print it. Each room used for the audit should be equipped with computer projection capability such that any of the data may be projected in the room for all to see. This combination will dramatically reduce the amount of data reproduction required in support of the meeting. A few paper copies of the report data should be made available in the meeting room for use by anyone not computer literate or to use in combination with data viewed on the screen for comparison purposes.

24.2.8 *Presentation of the qualification results, the audit*

Refer to Figure 24.4 for a list of audit topics for each task to be reviewed. As noted above, all of the data should be available for computer projection from a central verification library consisting of many text and graphics documents or the content of a relational database covering the whole process from requirements in specifications through verification reports. In some cases, the customer representatives will have read the related task reports and accepted the conclusion that the item design satisfies its requirements. If the company principal engineer has maintained contact with his or her customer counterpart and assertively, but politely, encouraged customer readership as the reports became available, this is a more likely result than if the FCA is the first time the customer has seen the documentation.

The contractor and customer representatives may be able to agree at the outset of a specific task audit that everything is acceptable except one or two particular areas where the conversation can focus avoiding wasting time discussing things that both parties agree on completely.

Where the customer representatives are reluctant to accept that the contractor proved that the item satisfied its requirements, the contractor task lead may find it useful to take the customer representatives to a close-by test lab to see the test apparatus or go over the simulation or modeling software and ways they have validated them. The discussion about the physical resources used to verify the requirements may be more convincing than the data produced as a result.

From the contractor's perspective, the audit should provide convincing evidence of the following:

1. Test procedures comply with specification content.
2. Procedures were run as documented.
3. Results reflect that the design complies with the requirements.
4. All requirements were verified (completeness).

The evidence may not be convincing to this end. If there were areas where verification action was not taken, could not be taken, or was taken and the results were either inconclusive or proved that requirements had not been satisfied, these facts need to be presented along with a recommended action. This action may entail retesting, suggested changes to the requirements, design changes followed by retesting, or a deviation approved by the customer permitting the delivery of some number of products that are less capable than required while the problem is being corrected.

Unless clear evidence has been uncovered during the audit that the item failed to satisfy its requirements, the company representative should make a recommendation that the customer accept that the requirements have been verified for qualification purposes. Before leaving the audit session, a clear statement of the customer's conclusions should be solicited if it is not forthcoming. If the two parties cannot come to a common conclusion, they may have to jointly brief the complete audit team of their different opinions. The cochairs will have to agree upon a way to finally resolve the dispute.

24.2.9 Post-review action

Subsequent to closure of the review, the contractor must work off all the action items and track the status of this work as it is driven to null. Some of these action items may entail additional qualification work to close out customer concerns expressed during the review. During the review, the customer could finally grasp that they were receiving a product that does not satisfy their current wants. The product may perfectly satisfy the current requirements as expressed in the customer approved baseline, but the customer has discovered that this item could, but does not currently, satisfy other requirements not identified earlier in any formal way. This may trigger engineering change proposal (ECP) action that will define new requirements and design changes, leading to fabrication of the changes and qualification of these changes. Any ECP action should not be allowed to interfere with closure of the current FCA. The changes can be the subject of a delta FCA or regression-testing FCA at a later date.

The contractor should keep track of each specific task that must be completed to gain FCA closure, assign someone responsible for closure of each item, and manage the closure to completion. This is especially true of all the action items assigned at the audit.

24.2.10 Information resources

As the reader has observed, this process is very documentation intensive, for it seeks to capture formal evidence of compliance in the form of test reports and analysis reports. Therefore, the management process must rely very heavily on sound information management techniques and systems. Not only must we ensure that the reports are created and reviewed, but we must be able to access any of these reports on short notice whenever we or the customer want them.

Engineers on programs that were not well run from a verification aspect have experienced the stomach-churning realization that the documentation for particular verification tasks was not available during or shortly before the FCA. Some programs permit the principal engineer for a verification task to retain the corresponding documentation providing copies as required on the program. It has happened on such programs that one or more of these engineers has retired, died, or left the company for other reasons and the contents of his or her desk and files were disposed of in the trash or to the retiree's garage in a pile of boxes never to be opened again. Sometimes you might be lucky enough to locate the box in the engineer's garage in one of these cases and be able to retrieve the only copy of the report in time for the FCA, but do not bet on it.

These goals can be achieved through the use of verification documentation captured with word processors with good discipline, but the optimum way to do this is to use a relational database. Some commercially available computer tools like DOORS include many good features supporting the verification process, and the schema of many of these tools can be adjusted by the buyer to improve and tailor the capability for specific enterprise needs. The ideal tool you select or build for your own purposes should capture the requirements completely in a relational database from which a specification can be printed and within which traceability can be established within and among the requirements and between the requirements and the verification process information.

The verification requirements should be captured in the same relational database keyed to the same specification they relate to with traceability between the requirements section (usually 3) and the verification requirements section (usually 4). Further, it should be possible in this database system to also assign these verification requirements to a method and collect them into predetermined verification tasks. The IVP content should be created in this same database system with its paragraph numbers, titles, and text also included in records just like the requirements. The planning content should be traceable to the verification requirements from which they were derived.

Finally, the verification reports should also be database-captured with traceability back to the planning and procedures. With this kind of structure, all of the planning and reporting data is fully traceable and specific documents can be called up or printed in whole or in part on demand. It should also be possible to status the whole process in some automated fashion based on the content of the database. Figure 24.5 illustrates this structure in an IDEF-1X format explained in the Department of Commerce FIPS Publication 184.

24.3 System qualification verification

24.3.1 System verification overview

The qualification process properly occurs at the end-item level, possibly supported by lower-tier component tests and analyses. These activities are accomplished within the context of the item verification qualification process, as discussed in Section 24.2.

Most customers of complex systems will also require system-level tests and analyses to show that the end items will interact as planned to achieve system requirements. This is primarily accomplished through the use of the system in accordance with its planned-use profile. If the prime item is an aircraft, we call it flight test. Land vehicles may be tested through on- or off-road ground tests depending on the kind of vehicle

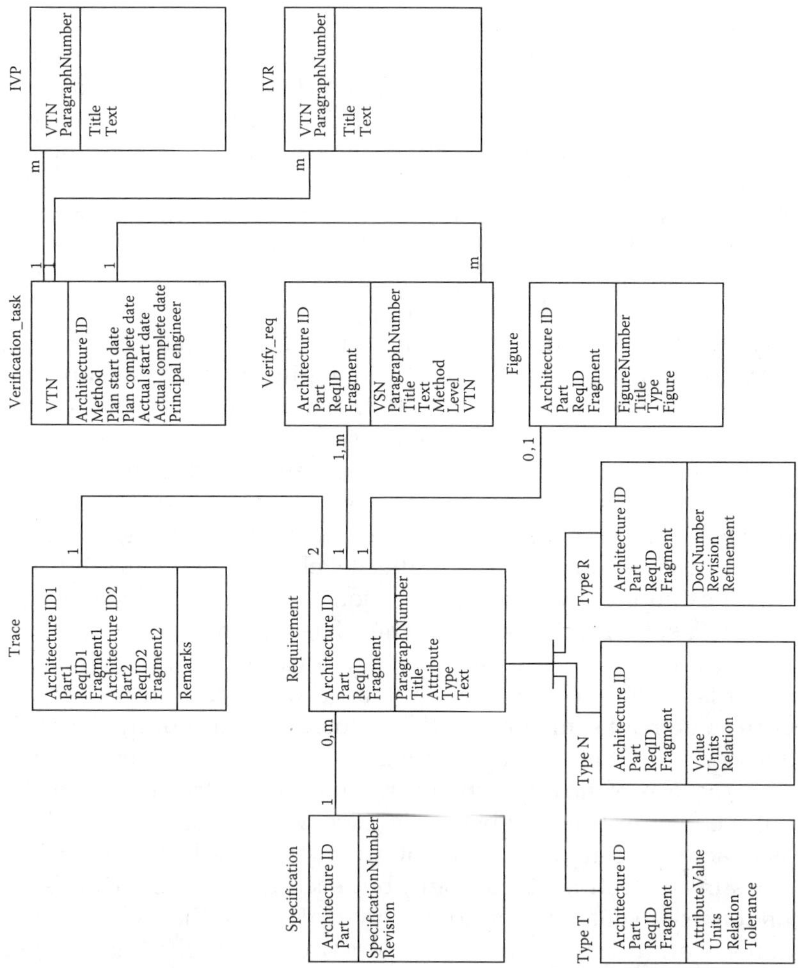

Figure 24.5 Specifications and verification support database structure. (From Grady, J., *System Validation and Verification*, CRC Press, Boca Raton, FL, pp. 213–275, 1997. With permission.)

involved. A missile system will undergo test firings under a range of conditions anticipated in the actual use against targets that perform like the planned targets. This system testing is not thought of as part of the qualification process that culminates at the FCA, because qualification is a component activity that is a prerequisite to the operation of the complete end items in which they are installed.

If we have had to conduct a segmented FCA process because of the complexity of the system and the availability of end items distributed in time, the customer may also require a system-level audit to go over the results of all of the item audits and ensure that all prior action items stemming from item FCAs have been closed. This system audit acts to link together all of the individual FCAs into an integrated activity. It is uncommon to coordinate the system FCA with the end of any system testing that is required, since the system testing may require a very long time, years on a new fighter aircraft, for example, and both the customer and the contractor have used the FCA process only to ensure that the end or configuration items are qualified for the application.

The system FCA, like the sensible use of multilevel hierarchical testing, illustrates a point about which system engineers must remain consciously aware because of their perfectly normal and useful mania for decomposition of complex problems into a series of related smaller problems. In the opening chapter, we noted that because we decompose large problems into small ones in the interest of focusing our energy onto problems consistent with our technical and managerial capacities, we must also integrate and optimize at the higher levels. It is necessary but not sufficient, for example, to verify the operation of five items independently that will form a subsystem in the end item. There are very likely higher-level subtleties that cannot be exposed by lower-level work. In space programs, it is normal to test at every level of indenture in the system architecture. The reader may recall the unfortunate case of the Hubbell Space Telescope program where an integrated test of the optics was not performed because of facility scheduling and cost difficulties and the belief that the item testing had been adequate. With systems that will be more easily accessible after deployment, we may properly be more tolerant of holes in the testing string logic, but each gap in the string is an invitation for risk to enter the program. As in all cases, the technical risk of not doing work must be balanced against cost and schedule risks of doing it possibly unnecessarily.

We have several goals in testing the complete system and combining the results of the end-item verification results. First, we wish to establish the readiness of the complete system for deployment or delivery and operational employment. We also hope to validate the models and simulations used in the development so that they may be subsequently used to predict performance in later studies and change analyses. In a commercial situation, we may also be interested in proving that the product is safe and reasonably free of flaws that might stimulate legal action.

There are many product lines where government regulations require new products to pass the scrutiny of special testing laboratories or criteria. The FAA requires new passenger aircraft to be certified for revenue-producing flight. Road vehicles must pass or be prepared to pass tests focused on safety regulations. New drugs must wind their way through lengthy evaluation before being allowed on the shelves of pharmacies. Many other products must pass similar hurdles in route to public sale or distribution. In some cases, the responsible agency will accept your test results for review. In other cases, the agency insists on their own tests at their own facility or conducted by a contracted licensed test lab.

The system verification process depends on both the results of the item verifications and on unique system-level activities that cover requirements that could not be verified at the end-item level. The latter consists of all the verification tasks that were identified in the compliance matrix for verification at the system level. These tests and analyses must be characterized by progressively in order to incrementally approach conditions where the maximum capability consistent with minimum risk to life and property is realized. We also need to focus on comprehensiveness or completeness over the full range of system operational and logistics actions. Actual completeness may entail an infinite space, and thus be practically impossible to achieve in a sufficiently complex system, so system testing is also a great compromise requiring careful consideration by experienced engineers to determine what subset of completeness will be sufficient.

24.3.2 System verification audit

Where item verification audits are held, it is useful to terminate the verification audit process with a system review even if there are no system-level verification activities. The system audit in this case acts to tie together any loose strings that were identified at the end-item audits and ensure that all action items identified at these audits have been properly addressed and been closed out. Where the system includes both hardware and software elements, as is often the case, the system verification audit also permits an integrated consideration of the isolated results in these two areas.

If only a system audit is held, where the system includes two or more complex items, this audit must accomplish the intent of the item and system audits. The system audit should first be broken down into a series of items audits (following the pattern discussed in paragraph 3.6.6.2), followed by an audit of system verification in accordance with guidance offered in this chapter.

24.3.2.1 System audit planning

Prepare an agenda in cooperation with the customer and other interested parties. Determine who is chairing the audit. Is it the customer, an IV&V agent of the customer, or some third party? Decide on some goals for the audit, which might include the following elements:

1. Review and approve the actions taken at all previous item audits.
2. Ensure that all past action items have been addressed and closed.
3. Review the results of any system-level verification actions that have occurred, and place earlier item audit results in context with them. The system test work will commonly not be complete when the system functional audit takes place.
4. Review the effects of any engineering or manufacturing process changes that have occurred since the completion of item qualification work, and ensure that any and all effects on previous qualification conclusions have been considered and, where appropriate, have resulted in additional verification work, which has also been the subject of FCA action either independently or within the context of the system FCA.
5. Formally conclude the qualification process with a formal notice of completion.

The latter may be a conditional conclusion subject to the completion of a small number of remaining actions, but there should be closure on this issue rather than allowing it to remain a lingering sore for an indefinite duration.

24.3.2.2 The audit

If there have been no significant changes in the product since item FCAs completed, the system FCA should only focus on the results of those FCAs and action items that lingered after meeting close. So, there should be an agenda section for each item FCA. If there have been significant changes since item FCA closures, the customer could have elected to stage delta FCAs for all or some of the engineering changes where the differences were subject to review. These delta FCAs simply become add-ons to the corresponding item FCA discussion.

There may be changes where delta FCAs were not held with the understanding that the matter would be taken up at the system FCA. In this case, the agenda should include a full-blown audit of the changes rather than a review of the prior audit results.

24.3.2.2.1 Item verification summaries and integration The current status of all item audits should be reviewed with special focus on any retesting that had to be accomplished subsequent to the item audit. ECPs that have been identified and processed subsequent to the item audit should be discussed in terms of their effects on the design and the results of the qualification process for the item. It should be agreed that subsequent to closure of the system verification audit, any change proposals should carry with them the responsibility and costs for any re-qualification action that may be necessary.

There may be some item requirements that were not verified wholly at the item level, rather at a lower or higher level. In all these cases, we should connect the verification actions focused on items and review the adequacy of the collected work for each item. Ideally, this should have been done for each item audit where promotion or demotion occurred, but some of these actions may not have been complete at the time of the item audit in question. For example, if we completed the verification work for item A123 and held an item audit on February 21st and some of the requirements were going to be verified at the parent level (item A12) in testing that was not scheduled to take place until June 13th, our A123 FCA would be incomplete. In this case, there should have been an action item assigned to report the results of A12 testing related to item A123 and offer a recommendation regarding closure of the A123 audit. In cases where there are many promotions and demotions, it is possible that not all of the intended integration actions occurred. At the system audit, all of these strings should be traced and proper closure assured.

24.3.2.2.2 Audit action item closeout status It is likely that some action items were assigned at each of the item verification audits. It is hard to imagine that all item audits were held without any action items having been assigned. So, any open action items should be reviewed for status and ways found to close them out. The customer and the contractor should jointly review those closed out previously and concur that they were completely addressed in their closing action. MIL-STD-973 provides forms for action items and verification certification that can be used as a guide even with a nonmilitary product.

24.3.2.2.3 Interface verification audit When you arrive at the system audit, item interfaces had probably previously been verified at the item specification pair level. In this case, we should compare the verification results for both ends of each interface and ensure that the results were satisfactory and complementary. At higher levels in the system, some interface requirements may be documented in ICDs and have to be verified with respect to this documentation. The only difference in interface verification results is that they relate to an interface rather than to an item in the system. The evidence is developed in the same way in test, analysis, demonstration, and examination reports. The review of system-level interfaces could be viewed as part of the system-level work, but is singled out to preclude it being overlooked.

The interfaces may have to be physically verified quite late in a program. Space systems often first come together completely only at the launch site. So, many interfaces have to be verified as the first vehicle is built up in preparation for launch. As a result, adequate time has to be built into the schedule for this work that is in addition to normal launch processing. Many of the problems that can occur in this work can be eliminated long before this time through the use of matched tooling for mechanical-fit checks at the factory of one or both mating items and special test equipment reflecting the design of the mating item.

24.3.2.2.4 System testing results If the system FCA covers system testing as well as the integration of the item FCAs, we should check to ensure that the system-level test and analysis work has produced a comprehensive body of knowledge covering the whole system. This work will have consisted of a finite number of tasks accomplished at the system level, leading up to the system review. Each of these verification tasks should have had associated with them one or more goals or purposes and specific actions defined in a verification plan and procedure driven by requirements in a specification. Most of them will be based on the system specification, of course, but lower-tier requirements may be included that were to be verified at a higher level. The review must determine if the results of the testing and analysis have proven that the requirements have been satisfied. Each system verification task should be reviewed to determine if the results cover the planned verification space and that the whole leaves no planned verification work undone.

24.3.2.2.5 Audit follow-up and closeout The contractor should create a final report of the system-level audit noting any decisions arrived at, action items assigned, and any remaining steps necessary to complete the whole qualification verification process. Any remaining action items should be assertively worked to completion and submitted to the person or organization requesting them. If actions were assigned to the customer, then associate contractors, a prime contractor (if you are a major subcontractor), or team members will require a lot of telephone and e-mail contact to encourage closure on time. As soon as all remaining actions have been completed, this fact should be communicated to the customer, sponsor, or IV&V contractor identified by the customer and formal closeout action requested.

The final action in this long and difficult process is that the contractor should encourage the customer to send them some kind of letter or form, formally indicating successful completion of the qualification verification requirement. On many programs, completion of the verification process is required to begin any system-level testing that may entail risks to damage or loss of scarce development resources (like flight test aircraft) and death or serious injury of personnel involved in the testing. It is a good idea for contractual and legal purposes to acquire some kind of formal agreement that the program was ready to move to this stage before doing so.

24.3.3 System test planning

System test planning should start way back with the development of the system specification. The V diagram illustrates this very well. As noted in Figure 24.6, the content of the system specification should drive system test requirements crafted at the same time as the specification is written. Ideally, these requirements will be included in the verification section (4 in a DoD specification) of the specification. Some organizations place these requirements elsewhere or fail to go through this step, and the result is a poorer specification. If you have to write the verification requirement as the product requirements are being written, you will write better requirements. There will not be any requirements statements like "… must work well and last a long time…" in the specification, because one cannot craft a way to test or analyze whether or not the final product satisfies this kind of requirement.

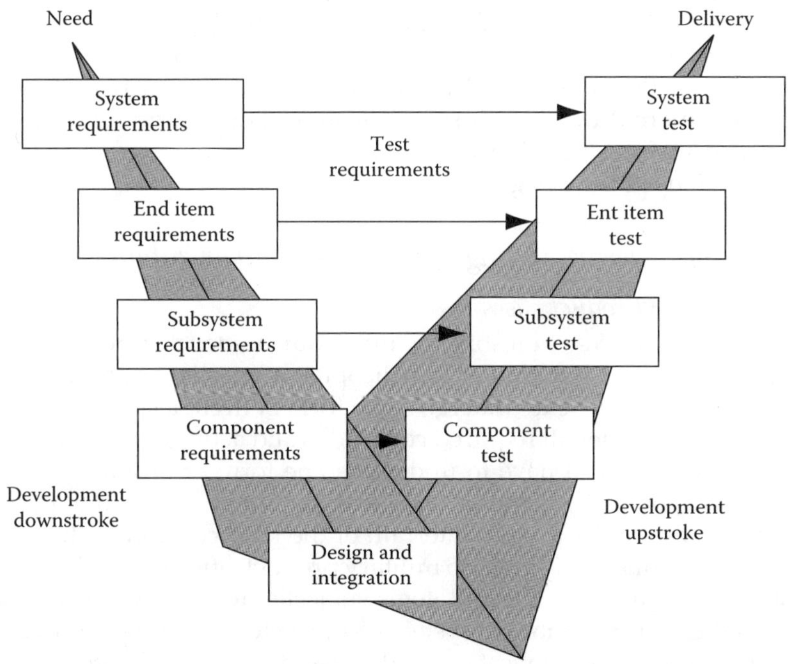

Figure 24.6 The V model encourages the right timing. (From Grady, J., *System Validation and Verification*, CRC Press, Boca Raton, FL, pp. 213–275, 1997. With permission.)

All of those system requirements mapped to test in the system specification traceability matrix should come to rest in the hands of the person responsible for creating the system test plan. These may be augmented by verification requirements, or verification string numbers in the context of item verification work, promoted from subordinate items to parent-level verification. As described for items, the system-test-planning process must determine how to least expensively and most effectively verify that the system satisfies these requirements.

The first step is to synthesize the test requirements into a set of test tasks. On a simple product, this may only require a single test task. On a new aircraft, this can entail many flights over a long period of time preceded by many ground tests and taxi tests. The first flight might be a takeoff, followed by an immediate landing, followed by a takeoff and several touch and go landings with all flight within visible distance from the flight test base or under surveillance from chase planes. Based on the results of these tests and analysis of any instrumentation data recorded during them, engineering should provide guidance on subsequently planned flights. It may be necessary to alter initial flight plans restricting certain flight regimes or loading conditions, or moving them into later flights until data can be further analyzed.

Generally, it is a sound practice to test from the simple toward the complex, from the small toward the large, from the specific toward the general, from the safe toward the more hazardous, and from the easy toward the more difficult. Our system testing should progressively build confidence in future testing by proving the soundness of increasingly larger portions of the design.

24.3.4 System-level testing and analysis implementation

24.3.4.1 Product cases

The nature of the system testing required during development is a function of the product. An aircraft, a helicopter, a missile, or an unmanned aircraft will likely have to undergo some form of flight test to prove that it can actually fly under all required conditions and achieve mission success. A ground vehicle will have to undergo some form of ground testing. A tank, a truck, or any other kind of vehicle should be required to perform over real or especially prepared terrain of the kind covered in the system specification while verifying, in a military product, offensive and defensive capabilities. In all cases, the end items subjected to system test should be composed of components, subsystems, and systems that have previously been qualified for the application, as discussed in earlier chapters.

Because we have tested all of the parts, there may be a temptation to conclude that it is not necessary to test the whole. This conclusion is sometimes correct but more often not. The problem with complex systems is that it is very hard to intellectualize everything. This is true especially in space systems. Many readers will recall the problem with the Hubbell Space Telescope that could not focus properly when deployed because the optics were not correctly configured. Cost, schedule, and security problems during development with respect to the use of facilities capable of adequately testing the telescope resulted in the rationalization that major elements tested separately would work properly together. Since the telescope was placed in a low Earth orbit, it was possible to modify it while in orbit using the Space Shuttle but only at tremendous cost.

Space systems are commonly tested throughout the product hierarchy starting at the bottom and working toward the whole up through each family structure. This is true during development to qualify the system as well as when assembling a particular deliverable end item for acceptance. The logic for this is pretty simple. Shuttle repair of the Hubbell notwithstanding, it is not possible to routinely solve design problems that were not observed in testing during development after launch. We simply do not have the same flexibility in space systems that we have in ground, aeronautical, or oceangoing systems, systems with which we can interact with relative ease after they have departed their manufacturing site.

The intensity with which we will choose to pursue system testing is a function of the planned use of the product, its complexity, and available resources. This should not be a matter of chance or personal preference. One of the common failures in system test planning is to select those things that can be tested based on the current design. System testing should be based on system requirements. We should be trying to prove that the system performs as required in its system specification within the context of prior successful component and end-item test results. During the design development, based on specification content, we should be concurrently designing the system test. Where it is very difficult to accomplish a particular verification action at the system level, and it cannot be accomplished at a lower level, we should consider a design change to make it possible. This is, of course, the value of concurrent development. If we can catch these problems early enough, it is economically feasible to develop the best-possible system that is also consistent with the right test, manufacturing, logistics, and use processes.

There are relatively few good written sources for the testing of purely hardware systems or the hardware elements of grand systems. Those that the author has found useful are military standards. One of these offers a good superset for testing guidance within which most test situations will be a subset. This is MIL-STD-1540C, Test Requirements For Space Vehicles. MIL-STD-810D, Environmental Test Methods and Engineering Guidelines, and MIL-STD-1441, Environmental Compatibility Requirements For Space Systems, also provide useful content relative to environmental testing, a significant portion of the test problem for systems and lower-tier elements.

There is a lot of good literature on the testing of computer software in the form of books written by Boris Beizer and William Perry, to name two respected authors in this field. Software cannot, of course, form a complete system in that it requires a medium within which to run, a computer of some kind. Given that the machine, its compiler, and the selected language are fully compatible, the software can be treated as an isolated system, however. The software must be thoroughly tested internally and integrated with the hardware system within which the computer operates.

Grand systems, composed of complex arrangements of hardware, software, and people performing procedures based on observations and circumstances (sometimes referred to by the troubling term peopleware), involve very complex system-testing demands because it is so difficult to intellectualize all of the possibilities. We are commonly left with few alternatives but to select some small finite number of alternatives from the infinite possibilities and extrapolate across these few test cases. Given that the hardware, software, and procedural elements are each examined carefully throughout their range of variation, there is a reasonable expectation that system testing, so constituted, will be effective in proving compliance with system requirements. Whether or not we will succeed in uncovering all the possible fault modes undiscovered analytically through failure modes and effects, criticality analysis is another matter. In a complex system, we will be hard-pressed to imagine all of the strings of activity that can occur in actual operation, especially where the user is intent on getting every possible scintilla of capability out of the system and is not averse to risk-taking to satisfy some borderline operational opportunity.

24.3.4.2 System test categories

The DoD recognizes several categories of system test that can also be applied to a broader range of customers. The category boundaries are based on who does the work and the conditions in effect rather than on military parameters. In all of these cases, the understanding is that the components and end items that comprise the system have been qualified as a prerequisite to system test. The DoD is actually blurring the lines between these categories by encouraging integrated test and evaluation. The participants have to realize, of course, that the two principal kinds of testing, DT&E (developmental test and evaluation) and OT&E (operational test and evaluation), have significant differences in the interests of the primary parties. In the former case, the contractor is interested in proving that the contractual requirements were satisfied, while in the latter case, the user is vitally interested in whether the system will satisfy their operations needs.

24.3.4.2.1 Developmental test and evaluation The DT&E is accomplished by the contractor in accordance with a test plan that focuses on engineering considerations. It should be driven by the content of the system specification, and the testing accomplished should produce convincing evidence that the system to be delivered does satisfy the requirements in this specification.

24.3.4.2.2 Operational test and evaluation The OT&E is accomplished by the customer in accordance with mission needs that dictated the procurement of the system adjusted over the development life of the system for its evolving characteristics.

24.3.4.2.3 Interim operational test and evaluation If the DT&E reveals a need to make significant changes to the system and these changes will take some time to develop and implement, the residual assets from the DT&E and any other end items produced to satisfy the customer's initial operating capability (IOC) requirement will not be in the correct configuration to use as the basis for the OT&E. This may lead the user to alter their original plan for the OT&E to focus on those parts that are not affected by the changes early on and postpone test elements that are affected until the changes can be implemented. The resultant testing may be called interim operational testing and evaluation.

24.3.4.2.4 Follow-on operational test and evaluation As the final resources begin to become available to the user subsequent to the DT&E changes discussed above, the user may choose to transition to a final set of test planning data that corresponds to the availability of fully operational assets leading to the identification of the test activity in the context of a follow-on activity. Another way to get into this kind of test and evaluation program is that during the OT&E, user testing uncovers problems that did not surface during the DT&E because of the different perspectives of the people and organizations responsible for the DT&E and the OT&E. DT&E plans are structured based on an engineering perspective, whereas the OT&E is structured based on a user's perspective. The user will set up test sequences focused on the missions that are known to be needed, and these differences may expose the system to modes not evaluated during the DT&E. These tests may uncover system problems that have to be fixed and follow-on testing accomplished after the fixes are installed.

The cost of making the changes driven by OT&E testing may or may not be properly assigned to the contractor. In order to allocate the cost fairly, it must be determined whether the testing attempted is appropriate for a system as described in the system specification. It is possible that the user has extended the mission desires beyond the mission capabilities they procured. This problem can easily materialize where a procurement office acquires the system and an operational unit operates it. The people in the latter commonly have very little interest in the contractual realities within which the procurement agent must function. They are strictly focused on their mission, and, if the system will not satisfy some portion of it, the system should be changed. It may even be the case that the system in question provides better capabilities than required, but it still falls short of the capabilities that the user has extrapolated from the capabilities observed during the DT&E that suggest other opportunities not yet satisfied nor covered in the contract. Users march to a different drummer than procurement people, and a contractor must remain alert to ensure that they are not steamrollered into making changes that are actually out of scope for the current contract.

24.3.4.3 Test results applications

The results of system testing are first useful in verifying that the system complies with system requirements. In addition, the results of system testing are useful in confirming the validity of models used to create the system. Throughout the development process while the system was an intellectual entity defined in a growing list of engineering drawings, models and simulations were crafted as part of the intellectual system description. These models and simulations were built based on some assumptions as well as rational assessments based on science and engineering. We should have done our best to validate these models and simulations as early in their life as possible, but the availability of the complete system is the first time that some of the features of these objects could be fully validated.

If system testing confirms that the models and simulations predict performance that is actually observed in the operation of the system, then we have a very valuable resource in our hands. The reader may conclude that since the system is already built, it is of little consequence whether or not the models agree with its operation so long as we are satisfied with system operation. Why do we not simply get rid of all the other representations of the system? We are finished.

The reality is that this system will have a long life, and in this long life, it will experience phenomena that we may not understand immediately. If we have available to us a simulation that has been validated through system testing, then we can manipulate the simulation to help us understand the cause for the problem observed in a very controllable situation. Without these resources, we may be forced to conduct dangerous flight tests to observe performance aberrations. Also, the simulation will permit us to examine a wide range of situations in a much shorter period of time than is possible through actual operation of the system.

So, we should make a conscious effort to correlate the results of system testing with any models and simulations we may have to use in future. In some cases, these will not agree, and the cause could be either the simulation is incorrect or we could have interpreted the results of system testing incorrectly. In any cases where there is not a good agreement, we should look for a cause on both sides of the divide.

24.3.5 *Other forms of system testing*

24.3.5.1 *Quality and reliability monitoring*

On a program involving the production of many articles over a long period of time, the customer may require that articles be selected from time to time to verify continuing production quality and reliability. This is especially valuable where the product cannot easily be operated to develop a history. A good example of this is a cruise missile. It would be possible to configure selected vehicles off the line for flight test purposes involving adding command control and parachute recovery systems, but while these changes would permit the very costly vehicle to survive the test, they would invalidate the utility of the test. In order to fully test the missile, we would have to cause it to impact a representative target and explode (conventional warhead assumed here, of course).

Military procurement of missiles for air-to-air, air-to-ground, ground-to-ground, and shipboard variations commonly involves purchase of blocks of missiles. Since the military cannot determine how many wars of what duration they will be required to fight, these missiles end up in storage bunkers and warehouses. The service may require that missiles be pulled from stock at random and actually flown to destruction from time to time. These actions, where the missiles work, encourage confidence in the store of hardware. These events also offer user units a great opportunity to conduct live fire training, so these two goals can be connected to great benefit.

24.3.5.2 *System compatibility test*

Teledyne Ryan Aeronautical developed a complete unmanned reconnaissance system for the U.S. Navy in the late 1960s. This was a shipboard rocket-assist-launched aircraft using an adapted anti-submarine rocket (ASROC) booster installed on the vehicle located in a near-zero length launcher angled at 15° to horizontal. Ryan had been supplying the Strategic Air Command with this equipment for air launch from DC-130 aircraft for several years at that point, but had never completely outfitted a new user. Past operations grew into their then current capability in the presence of a complete Ryan field team or field engineering representation with adjustments over time lost in the corporate knowledge. The program manager, A.A. Petrofski, and the project engineer, Joseph Doering, correctly sensed that there was a risk that the development team might not be capable of dealing with these problems from a purely intellectual perspective, so they required the performance of a system compatibility test. The author was asked to design and manage this test because he had experience in the field with the system and all of its support equipment.

The series of tests uncovered many small problems with vehicle and support equipment design but did not disclose several very serious ones that were only exposed in the subsequent flight test. The principal value of the test was to confirm the soundness of the maintenance process as a prerequisite to entering the flight test. Most of the problems that survived until the flight test entailed operational aspects that could not be examined during the compatibility test.

Some things that were discovered are as follows:

1. The system was to be operated on an aircraft carrier at sea continuously for weeks at a time. When the vehicle returned to the ship after its mission, it was recovered through parachute descent to water impact where it was picked up by the helicopters otherwise used for air-sea rescue on the carrier and placed on deck for postflight work that included freshwater decontamination of the aircraft and engine. An engine decontamination tank was provided to flush the external and internal parts of the engine with freshwater to remove the corrosive saltwater. When an attempt was made to place the engine in the tank, it was found that it would not fit. The problem was traced back to a conversation between the lead designer and the new engineer given the responsibility of adapting a previous tank design to this engine. The original tank design had been made for a J69T29 engine common to the Ryan target drone. The lead designer told the new engineer to go down to the factory floor where there were a number of the correct engines awaiting installation in vehicles. The new engineer, not knowing that there were several models in production with four different engines, happened upon a group of larger Continental engines used in a different model that did not have the same external interface and dimensions and used this engine as the model for his design. Therefore, the adapter the new engineer created did not fit the J69T41A used in this model. It had to be modified.
2. A number of other incompatibilities were uncovered, but the author has forgotten the details.
3. The test also acted as a technical data validation and numerous problems were discovered in the technical data and corrected.

Some things that were not discovered are as follows:

The Continental engine representative said that the J69T41A engine would not accelerate to launch thrust after start without an engine inlet scoop. The launcher design team located a retractable scoop on the front of the ground launcher such that it could be retracted out of the way subsequent to engine start and not be struck by the rocket motor moving past during launch. During the compatibility test (which included a launch simulation but not an actual launch), the scoop worked flawlessly while the vehicle sat in its launcher with the engine running at launch rpm (revolutions per minute). Several weeks later, the system went to the flight test operated out of Point Mugu, California. On the first launch, the rocket motor blast tore the engine inlet scoop completely off the launcher and threw it some distance. It was quickly concluded that the inlet scoop was not required, and the system worked flawlessly for many operational launches from the USS Ranger in that configuration.

A special circuit was designed for this vehicle such that if it lost the control carrier for a certain period of time, it would restore to a go-home heading so as to regain a solid control carrier at a shorter range. The only time this circuit was ever effective during operational use, it caused the loss of the vehicle rather than the salvage of it. The remote pilot during that mission was experiencing very poor carrier conditions and kept trying to issue commands to turn the vehicle back toward his control aircraft (an E2 modified with a drone control capability). What was happening was that every time the carrier got into the vehicle momentarily, it reset the lost carrier go-home heading timer so that it never was triggered. Poor control carrier conditions caused the vehicle to only go into a turn during the very brief moments when the control carrier got through. Since remote turns were not locking, these turns were almost nonexistent and the vehicle continued to fly further from the controlling aircraft. The controlling aircraft could not follow the vehicle to reduce the distance because the vehicle was flying over North Vietnam. Finally, the vehicle fueled out over Laos and crashed.

In preparation for launch, the jet engine was first started and brought up to launch rpm while other prelaunch checks were conducted. There was only a very brief time before too much fuel would be burned resulting in an unsafe launch, so the launch procedure was stressful for the launch operator. Upon launch command with the jet engine running at launch rpm, the ASROC rocket motor was fired, shearing a holdback. The jet engine thrust vector was supposed to be balanced with the rocket engine thrust vector in the pitch axis so that the vehicle departed the launcher at the 15° angle at which it rested in the launcher. On the first launch at Point Mugu, California, the vehicle departed the launcher and roared straight up through an overcast, and descended shortly thereafter behind the launcher in a flat spin, crashing fairly harmlessly between the launch control building and the launcher. System analysis had incorrectly estimated the pitch effect of jet engine impingement on the rocket unit and elevator effectiveness at slow speed, so the balance condition was beyond the control authority of the flight control system. It was unfortunate that a few weeks before, a Ryan target drone had been ground-launched from this facility and the rocket assisted take off (RATO) unit, with an undetected cracked grain, had stutter-fired, blowing up the drone just off the launchpad on its way to falling through the roof of a hanger on the flight line some distance behind the launchpad. But more problematic was the fact that the base commander was in the control building watching the launch of the new vehicle, and his wife and child were parked in his new Cadillac convertible in the parking lot between the building and where the drone impacted. These factors probably made it a little more difficult to regain range safety approval for the next launch.

The rocket igniter circuit was supposed to have been completely isolated from the rest of the electrical system. It tested okay during the compatibility and electromagnetic interference (EMI) tests. On the last launch of the flight test program from the USS Bennington, the igniter was set off when the launch control operator switched on the fuel boost pump in preparation for the jet engine start. The rocket fired without the jet engine operating, and the vector sum drove the vehicle into the ocean just in front of the launcher in one of the shortest flight times ever recorded. Subsequent EMI testing revealed a sneak circuit, which had not showed up in the previous launches and tests, that had to be corrected. The problem was traced to a damaged shield that had penetrated the insulation making contact with the wire leading to the igniter.

24.4 *Item acceptance verification*

24.4.1 *The final stage of verification*

The qualification process is intended to prove that the design synthesized from the item's development requirements satisfies these requirements. The development requirements, captured in a development, Part I, or performance specification, were defined prior to the design effort so as to be design independent, permitting the person or team responsible for synthesis to consider alternative competing design solutions and make a selection of the best overall solution through a trade study or other organized alternative evaluation process. This series of verification actions is only accomplished once to prove that the design is sound, and the evidence supporting this conclusion is audited on a DoD program in what is called a functional configuration audit (FCA). The tests accomplished during this work commonly stress the item beyond its capabilities to expose the extent of capabilities and extent of margins available in normal operation of the item and to validate the models used to create the designs. These stresses reduce the life of the unit and may even make them immediately unusable as a result of a successful qualification series. Therefore, these same inspections cannot be applied to every item produced for delivery, since the delivered items would not have their planned service life at the time of delivery.

What form of inspection might then be appropriate for determining whether or not the product is acceptable for delivery? This represents the second stage of verification, to prove that the product is a good manufacturing replication of the design that was proven sound through qualification. The acceptance process is accomplished on every article produced, on every product lot, or through some form of sampling of the production output. The process leading to the first manufactured item is subject to audit on DoD programs in what is called a physical configuration audit (PCA). Some customers may refer to this as a first article inspection process. The word "inspection" is used in this chapter to denote any of the four methods already discussed: test, analysis, examination, and demonstration.

24.4.2 The beginning of acceptance verification

The acceptance process should have its beginning in the content of the Part II, product, or detail specifications. This kind of specification is created based on a particular design solution and gives requirements for the physical product. This may include specific voltage measurements at particular test points under conditions defined in terms of a set of inputs, for example. Measurements will have to be made during a test and compared with requirements. As an alternative to two-part specifications, you may choose to use one-part specifications. In this case, the specification may have to have a dual set of values or tolerances, one for qualification and the other for acceptance. This difference is very important for the reason noted above. Subsequent to testing, the item must have a full lifetime before it. In either specification case, the beginning of the acceptance process should, like qualification, be in the specification content.

Unfortunately, this is not the case in many organizations. In many organizations, the acceptance test procedures are prepared independently of the specifications based on the experience of test and evaluation and quality assurance department personnel and their observations of the design concept. Subsequently, if a Part II specification is required, the test procedures are used as the basis for it. This is exactly backward. It is analogous to accomplishing the design of a product and then developing the specification. The Part I specification is the basis for the design of the product and the qualification process, while the Part II specification is the basis for the design of the acceptance process. If you first design the acceptance test process, then it makes little sense to create the requirements for the process subsequently. So, a fundamental element of acceptance management should be to insist on acceptance verification requirements as a prerequisite to acceptance planning activity.

Figure 24.7 illustrates the overall qualification and acceptance processes. In this chapter, we will focus on the blocks with the heavy border

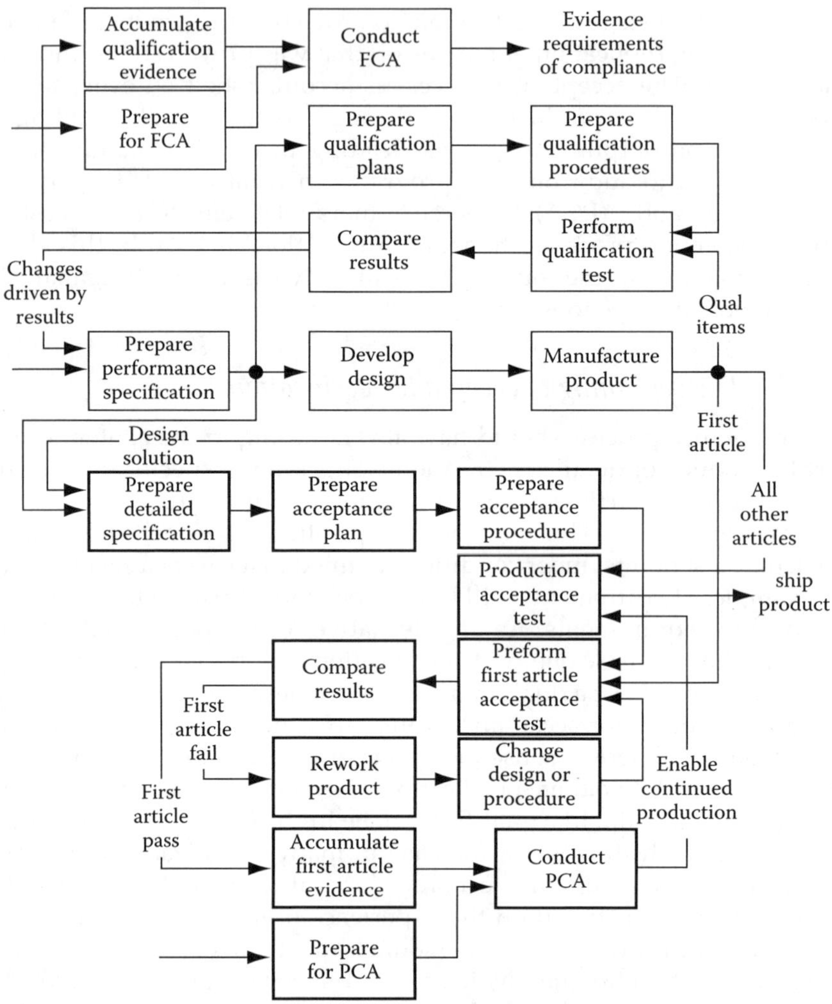

Figure 24.7 The path to acceptance. (From Grady, J., *System Validation and Verification*, CRC Press, Boca Raton, FL, pp. 213–275, 1997. With permission.)

beginning with the development of the detailed (Part II or product) specification. This process includes an audit of the acceptance evidence called, in DoD parlance, a physical configuration audit (PCA), but the process continues as long as the product remains in production.

24.4.3 The basis of acceptance

When dealing with a large customer, like the DoD or NASA, the acceptance process should be driven by what provides the customer with convincing evidence that each article or lot of the product is what they want, what they paid for. This evidence pool should also be minimized based on end-to-end parameters, system-level parameters, and parameters that provide a lot of insight into the condition of the product. So, of all the things that can be inspected on a product, we have to be selective and not base our evaluation on the possible, rather on the minimum set of inspections that comprehensively disclose the quality and condition of the product.

Some people feel that the content of a detail (Part II or product) specification should be determined prior to the design in keeping with the notion of "requirements before design." This, frankly, is impossible. As shown in Figure 24.7, we develop the performance (Part I or development) specification as a prerequisite to the design. Its content should be design independent. The detail specification must contain design-specific information because it is focused on the actual observations of the product features, like voltage measurements, physical dimensions, motions, and so forth. Therefore, the content of detail specifications must be created subsequent to the development of at least the preliminary design with a possible update after the detailed design has been approved.

24.4.4 Acceptance documentation

The suite of documentation needed as a basis for a sound item acceptance inspection includes the following:

1. Detail specification — Part II (detail) specification content could be integrated into the same specification used for development. These specifications should be prepared during the detailed design phase.
2. Quality inspection plan — Assurance should fashion an inspection plan based on the evolving design concept that identifies specific areas that should be inspected and the most appropriate times and conditions for these inspections in the evolving manufacturing plan. The goals of each inspection should be clearly defined.

3. Acceptance test plan — Assuming that the product has some functional capabilities rather than simply a physical existence, a test of some kind will be appropriate using instrumentation of some kind to convert reactions to a form suitable for human viewing. Each of the tests to be performed should be identified and defined in terms of goals, resources, budget, and schedule needs.
4. Acceptance procedures — Each inspection identified in the plan should have developed a procedure telling how to accomplish the work. This information may be integrated into a combined plan and procedure.
5. Acceptance reports — The acceptance reports may be as simple as sheets for recording values observed in tests and quality examinations or entail computer-recorded reading that can be reviewed on the screen of a computer terminal. The results should be collected for each article. The customer may require delivery of this data or require that the contractor maintain this data on file for a minimum period of time.

The program schedule should identify these documents and when they should be released. Responsible managers should then track the development of these documents and apply management skill as necessary to encourage on-time delivery of quality documents. Each document should pass through some form of review and approval.

24.4.5 Management of the work

All of the inspections work should appear on program planning and schedules for each article. The management should have available to them the cost/schedule control system and the manufacturing-tracking data indicating progress for each article or lot through the manufacturing process. As impediments are exposed, alternative corrective actions should be evaluated and a preferred approach implemented to encourage continued progress on planned work.

The aggregate of the work that must be managed in the acceptance process falls into two major components. First, the products are examined at various stages of the production process within the producing company and at all of its suppliers. In these examinations, quality assurance persons determine whether or not required manufacturing steps were performed and performed correctly to good result. If everything is in order in the traditional paper-oriented process, the QA person signs their name on or stamps a document to indicate acceptability. The other major inspection is in the form of acceptance testing where functional capabilities are subjected to input stimulus combinations and corresponding outputs compared with a standard. Commonly, checklists are used to record the results. The test technician records values observed, and a QA inspector initials or signs the list to indicate compliance with written procedures to good results.

The management must encourage the development of clear manufacturing and quality planning data as well as a sound achievable manufacturing schedule that defines the work that must be done and when it must be done including the quality inspections and the acceptance tests. These activities have to be integrated into the manufacturing process and coordinated between the several parties. In some organizations, manufacturing looks upon the inspection process as an impediment to their progress, and this wrong-headed attitude can be encouraged where an attempt is made to butter the inspection process onto a previously crafted manufacturing plan that did not account for the time and cost of the inspection process.

The reader should know what is coming next at this point. Yes, we need a team effort. Here is another situation where several specialized persons must cooperate to achieve a goal that none of them can achieve individually. The overall manufacturing and inspection process (word used in the broadest sense to include test) should be constructed by a team of people involving manufacturing, quality, material, tooling, and engineering. Ideally, all of these people would have participated in the concurrent definition of mutually consistent product and process requirements followed by a concurrent development of mutually consistent product and process designs. All of the planning data would have been assembled during the latter. Then, the management process is a matter of implementing the plans in accordance with planned budget and schedule constraints and adjusting for changing conditions and problems.

The teams effective in the production phase ought to be restructured from the development period, as illustrated in Figure 24.8. Figure 24.8a shows the suggested structure in development with several IPPTs on each development program along with a PIT and a business integration team (BIT).

An enterprise integration team reporting to the executive (by whatever name) acts to optimize at the enterprise level providing the best balance of enterprise resources to the programs based on their contractual obligations, available enterprise resources, and profit opportunities. This integration work is denoted by the heavy concurrent development bond joining the several activities. The PIT does integration work across the several programs through the PITs, and the PIT on each program carries on the integration work within the program. During development, the PIT acts as the system agent across the teams optimizing at the program product and process level. As the product maturity reaches a production-ready state, the IPPTs can collapse into the PIT with continued BIT (contracts, legal, etc.) program support. Production is managed through manufacturing facility managers under the direction of an enterprise's functional manufacturing manager with the EIT optimizing for balanced support to

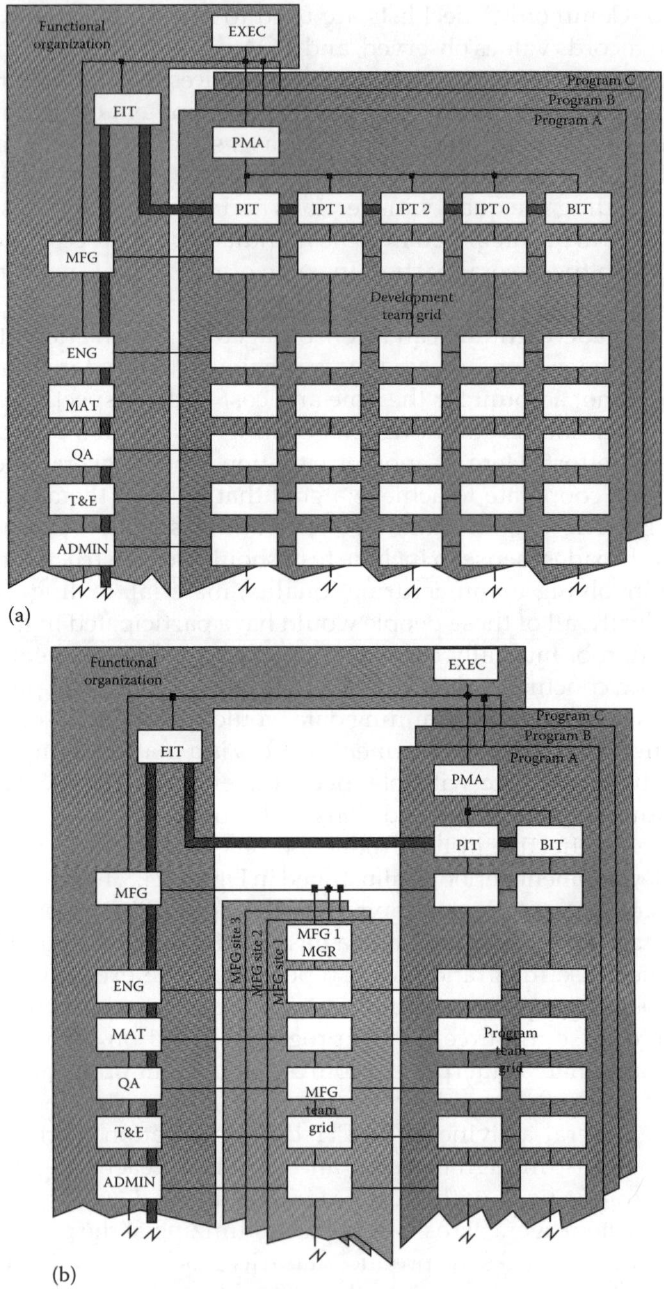

Figure 24.8 Development to production organizational transform: (a) development phase and (b) production phase. (From Grady, J., *System Validation and Verification*, CRC Press, Boca Raton, FL, pp. 213–275, 1997. With permission.)

satisfy program commitments. The PIT remains available for engineering in support of production changes and ECPs. PIT specialists support facility teams as required by these teams.

In this arrangement, it was a program responsibility to develop suitable plans and procedures to accomplish acceptance examinations and tests in the manufacturing environment. The production responsibility for particular items was assigned to particular manufacturing facilities and vendors that are all treated as vendors by the program. These facilities must carry out the planned acceptance work as covered in the approved plans and procedures.

24.4.6 FRACAS

24.4.6.1 Acceptance ethics

A key element of the production process is a means to acquire information on failures to satisfy acceptance criteria. When items fail to satisfy inspection requirements, the item should not be passed through to the customer without some kind of corrective action. A few companies have gotten themselves in great difficulty with the DoD in the past by treating failed items as having passed inspection requirements. In some cases, the tests called for were not even run in the interest of achieving schedule requirements. In other cases, failures were not reported and items shipped. Clearly, it is necessary to engage in deceit to put it mildly in order to perform in this fashion.

The management must make it clear to the workforce that the integrity of acceptance data and resultant decisions is important to the company as well as to the people who are doing the work. A company that falsifies records not only damages its customers and its own reputation in time, it also damages those who work there and are required to perform in this fashion. An ethical person simply will not work in this environment. So, either a person's ethics are attacked or the person replaced with someone who does not have a problem with this arrangement. Clearly, the company is a looser in either case. The moral fiber of the organization has been damaged.

The two causes for this kind of behavior are schedule and budget difficulty. Management persons throughout the chain dealing with acceptance should be alert for those doing the work coming to believe that delivery on time is more important than the company's good name. Unfortunately, this problem generally arises from management encouragement (intentional or otherwise) rather than grassroots creativity.

24.4.6.2 FRACAS implementation

The organization needs an effective failure reporting and corrective action system (FRACAS) to provide the management with accurate information on failures to satisfy acceptance requirements. Figure 24.9

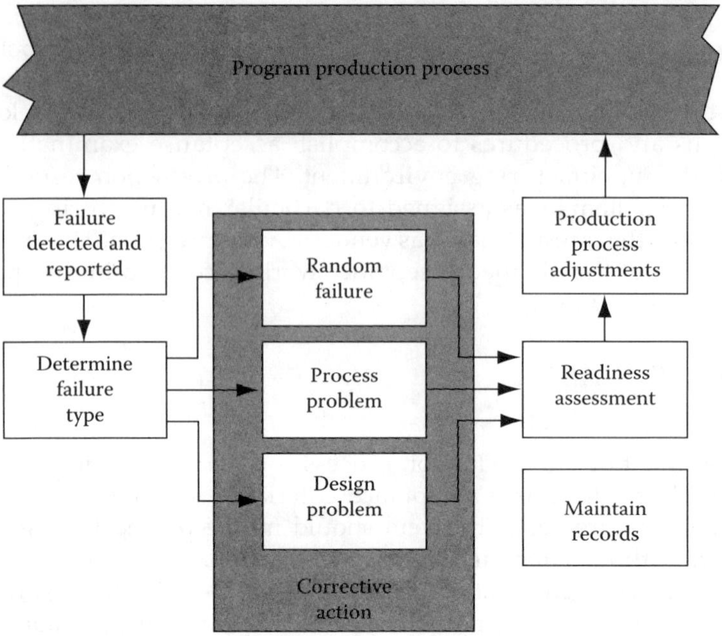

Figure 24.9 FRACAS flow. (From Grady, J., *System Validation and Verification*, CRC Press, Boca Raton, FL, pp. 213–275, 1997. With permission.)

illustrates a representative system for discussion. Failures can occur, of course, anywhere in the production stream, and the acceptance examination process should detect first that a failure has occurred and second what item or items have been affected by this failure. So, the first step in the FRACAS process is failure detection whereupon it must be reported in some formal way.

A failure can be caused by a number of different problems, and we should first determine the root cause before settling upon a corrective course of action. We should no more decide a corrective action without understanding the problem here than we should seek to develop a design solution to a problem that we do not yet understand. A fishbone, cause and effect, or an Ishikawa diagram is helpful in exhaustively searching for alternative causes. A Pareto diagram may be useful in prioritizing the corrective actions planned. Figure 24.10 illustrates these diagrammatic treatments, and the reader is encouraged to consult a book on process reengineering, TQM, or continuous process improvement for details.

Briefly, the Ishikawa diagram offers a disciplined approach for identifying the effect (failure) one wishes to investigate and possible causes for that problem, while the Pareto diagram shows us that a relatively few probable causes generally drive the negative effect we wish to avoid. In

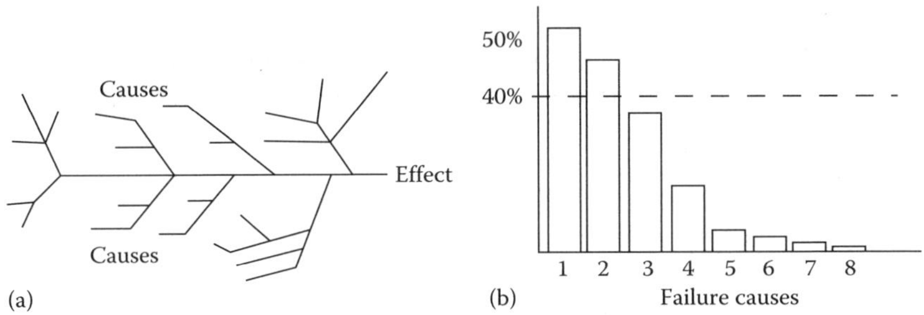

Figure 24.10 Supporting techniques: (a) Ishikawa (fishbone) diagram and (b) Pareto diagram. (From Grady, J., *System Validation and Verification*, CRC Press, Boca Raton, FL, pp. 213–275, 1997. With permission.)

that we seldom have all of the resources we wish were available, we are often forced to prioritize our actions to those areas that will produce the biggest positive effect.

Figure 24.9 suggests that there must be only a single cause of a failure, and this is not always the case. It may be possible to correct for a given failure in a number of ways. It may be possible to implement a design change, a manufacturing process change, or a procedural change to correct for an observed failure, or some combination of these. Our goal should be to select the best corrective action in the aggregate, generally in terms of life cycle cost effects; but on some programs and in some phases, near-term cost may be more important at the time.

24.4.7 Physical configuration audit

The PCA provides a way to reach a conclusion about the quality of the manufacturing process as a prerequisite to committing to high-rate production using this process. If there are problems in this process that have not yet been exposed, it is better to expose them earlier than later. Since the PCA is performed on the first production article, one cannot get any closer to the beginning of production. It should be pointed out that a sound development process up to the point of entering production of the first article will have a positive effect on reducing the risk that problems will occur in the production process. Another view of this translation was provided by W. J. Willoughby in his transition from development to production templates documented in NAVSO P-6071 and other places.

The PCA attracts all the evidence collected about the production of the first article into one place where it can be audited by a team of customers and contractors against the requirements. The result desired by the contractor is that the evidence will be convincing that the product satisfies the detail requirements, conforms to the engineering drawings and lists, and was manufactured in accordance with the approved planning data.

24.4.7.1 PCA planning and preparation

As in the FCA, the contractor should develop an agenda subject to customer approval, and prepare and organize the materials to be used at the audit. In that the audit covers a physical reality rather than the intellectual entity audited at the FCA, you will have to deal with things as well as paper in the audit. An audit space close to the production area will encourage the most efficient use of time during the audit. The first article should be made available in this space for direct visible examination. All of the acceptance report data should be available in paper form or via computer terminals depending on the format of this data. The exact arrangements that the contractor makes should be based on the mode of audit selected by the customer.

24.4.7.2 PCA implementation

The customer may require any one of three different modes for audit presentation. The customer preference should be known to the contractor at the same time the agenda is crafted because the different modes encourage very different audit environments. The contractor should also develop and reach an agreement with the customer on the criteria for successful completion of the audit that includes a clear exit criteria.

24.4.7.2.1 The physically oriented audit One of the three modes involves making the product available for customer examination and comparison with the engineering drawings. A U.S. Air Force customer may populate the audit with several staff noncommissioned officers with the right technical skills for the product line. They disassemble the product and compare the parts with the engineering drawings, confirming that the product satisfies its engineering. They may also conduct or witness functional tests on the product to ensure that it satisfies test requirements.

24.4.7.2.2 The paper-dominated audit The acceptance process generates a considerable amount of evidence in paper or computer media. If the customer elects to apply this mode of audit, it attracts totally different kind of people to the audit. They will be people with good research and analytical skills rather than good practical skills. They will review all of the evidence in one of several patterns to assure themselves that the product accurately reflects its controlling documentation. The auditors may selectively evaluate the evidence looking for inconsistencies. If they find one, they will search deeper; if more are uncovered, they may audit the whole set of evidence. If, on the other hand, no problems are found in the first layer, the audit terminates with a good report. Alternatively, the audit may focus on only one aspect of the evidence, such as qual test reports using the progressive technique just discussed. A third approach is to simply plan to audit all of the evidence in some particular pattern.

24.4.7.2.3 The combined-mode audit A customer could bring in both populations discussed above and perform both kinds of audit. This audit could be coordinated between the two groups to apply a progressive approach, or the plan could be to audit all data.

24.4.7.3 Post-PCA activity

The results of the PCA should be documented in minutes containing identification of any remaining problems and a timetable for resolving them. The precise mechanism for final closure should be included. These actions may be fairly complex entailing changes in specifications, changes in design, changes in acceptance test plans and procedures, and changes in manufacturing and quality planning. It will be necessary to track all these changes while they are in work and coordinate the package for integration into subsequent production articles. Some of these changes may not be possible in time to influence near-term production unless production is halted for an extended period. The effects of this on the continuity of the program team are generally so negative that the decision will be made to continue production with one or more deviations accepted by the customer for a brief period of time until the changes can be fully integrated into the design and planning data. Those units produced under a deviation may have to be modified in the field subsequent to the incorporation of the complete fix in the production process. If so, modification kits will have to be produced, proofed, and installed in the field to bring these units up to the final configuration. If the changes are sufficiently extensive, it may even be necessary to run a delta PCA on the first article through the upgraded production line.

24.4.8 Software acceptance

Software follows essentially the same acceptance sequence described earlier. The fundamental difference is that software does not have a physical existence except to the extent that it resides on a medium that will be delivered. The software will be delivered installed or on tape, disk, or other recording media, and it is this media that must be accepted. Once the computer software has been proven to accomplish the planned functionality, the acceptance test may be as simple as a checksum or as complex as a special test involving operation of the software in its planned environment with actual interfaces and human operators at work in preplanned situations similar to the anticipated operating environment.

24.4.9 Long-term acceptance implementation

Once the PCA has been completed and accepted, it remains to apply the proven acceptance process to all product articles produced. This will be done in a context of change over the life of the program production process. Therefore, configuration management practices are very important in the management of not only the design embodied in engineering drawings but also the manufacturing, quality, and acceptance test plans and procedures. As changes are implemented, all of this documentation must be brought up to the new configuration together and coordinated. On programs with a long production run, this process may be an extremely complex undertaking, because at any one time there may be more than one configuration in the production process.

chapter twenty-five

Risk management

25.1 Introduction

It is very difficult to comprehensively identify all of the ways that bad things can happen, unless you have a plan of action. Then you can contemplate things that could prevent you from achieving particular goals within the context of your plan. So, as the program plan goes together, a risk identification process should be going on in parallel. As risks are identified, the mitigation strategy for these risks should be woven into the plan, leaving a residue of risks about which nothing more can be done at the time and which have to remain on the books to be dealt with in the future under certain known or unknown conditions. One could rename system engineering risk management and not be too far from the truth. So many things that system engineers do on programs have the effect of reducing risk including the preparation of specification.

25.2 Risk identification and measurement

A sound risk management program identifies risk sources in the context of cost, schedule, and product performance. Some managers experienced in programs that pushed technology in one or more ways would add technology risk. There may be other useful categories in particular situations either at the top level or as subcategories to cost, schedule, and performance. Identifying risk categories is an example of the decomposition of a large problem into a series of smaller problem spaces within which we can concentrate more intensely to identify specific risks.

The program must have a policy of encouraging all program personnel to come forward with concerns from their perspective, and all persons in program leadership positions have to avoid instant criticism of those who do come forward no matter how silly the concern turns out to be upon reflection. All concerns that come forward should be evaluated by the PIT and, if warranted, added to a list of active risks to be considered and possibly mitigated. Mitigation of risks means to reduce the probability of the risk having a negative effect on the program. Figure 25.1 illustrates a risk list example.

Figure 25.1 identifies all of the current program risks by number and name, defines the type of risk using the trio of types noted above, gives numerical figures of merit for each risk, and identifies the responsible

Risk identification			Risk measure			Responsiblity	
ID	Name	Risk type	P(O)	Effect	Figure	Team	Engineer
1	Item A12 cost	Cost	2	2		1	Bill Adams
2	System schedule	Schedule	1	2		0	Cathy Spiker
3	Guidance accuracy	Performance	3	2		2	Alice Frome

Figure 25.1 Risk list.

team and engineer. The numbers in the risk measure column come from the category (CAT) columns in Tables 25.1 and 25.2 and the risk figures in the intersections of Figure 25.2. The engineer responsible for mitigating the risk in each case must maintain a risk abatement status chart providing the current status and the action plan as well as accomplish work identified in this plan.

For any one risk candidate on our list, we should identify the probability of it occurring (P(O) in Figure 25.1) during the program. One criteria that might be used to catalog risk probability on a program is shown in Table 25.2.

As we analyze potential risk, we can use this criteria to help us determine how seriously we should consider this potential problem. We have to categorize because we have limited resources. We will not be able to

Table 25.1 Risk Probability of Occurrence Criteria

CAT	Title	P(O)	Description
5	Nearly certain	0.95–1.00	Will occur at least once during program
4	Probable	0.75–0.95	Will probably occur once during program
3	Possible	0.50–0.75	May occur during program
2	Unlikely	0.25–0.49	Will probably not occur during program
1	Nearly impossible	0.00–0.24	Will not occur during program

Table 25.2 Risk Effects Criteria

CAT	Title	Description
5	Catastrophic	Program in jeopardy of termination
4	Serious	Serious damage to program
3	Moderate	Problems that cause program focus difficulties
2	Minor	Problems that can easily be overcome
1	Null	No problem

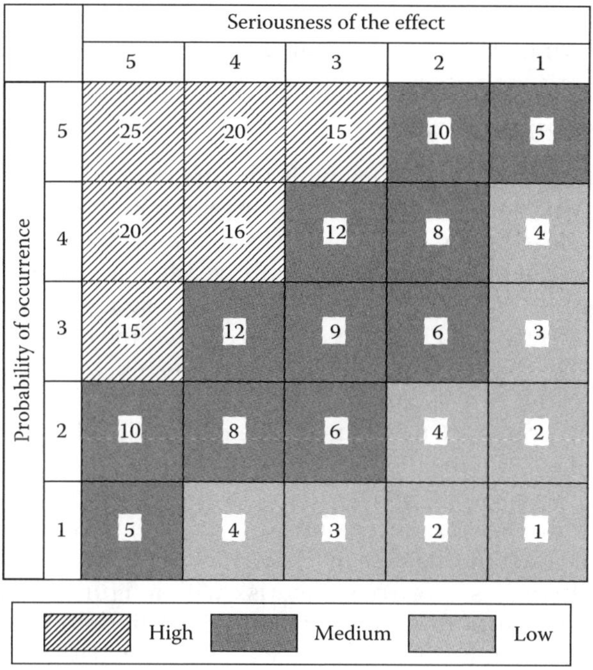

***Figure* 25.2** Risk categories.

attack every potential problem with full intensity, so we need to know which ones have the biggest potential of defeating our best efforts. These are the ones we have to attack with vigor. Others we may be able to set aside and allow to mature to some extent before we act. This is a case of attacking the tall poles, as some would say.

The other piece of the risk-cataloging information needed besides that identified in Table 25.2 is the amount of damage a particular risk can cause if it does occur. Some risks may have a high probability of occurrence but cause only a small nuisance when they do occur. Should we spend a lot of money and time combating these problems or should we focus on other potential problems that could do us more damage? Clearly, the risks of interest are those with a chance of occurring that will cause serious problems for us. These we might term high-risk items. Let use return to a two-dimensional categorization structure after we have completed the damage potential factor. Table 25.2 shows a risk effects criteria example.

In the literature, one can find many other categorical systems for risk identification, and the reader is encouraged to sample the literature in search of the one most appropriate for their situation. But, you should have one. Safety engineers are generally very knowledgeable about risk measurement, because this is their stock and trade and they use this same two-dimensional method for identifying safety hazards that are risks that may occur during operation and maintenance of a system.

25.3 Risk prioritization

Figure 25.2 illustrates a way to formally identify risk seriousness as a means of choosing which ones to attack with scarce program funds. The category 5-5 risks are clearly serious or high risks that must be attacked and mitigated. Risks in the 1-1 category are equally clearly not a significant problem for the program. We might independently choose to categorize these in between differently, but Figure 25.2 offers one regime for doing so.

The numbers appearing in the intersections reflect our degree of concern based on the two factors indicated. These were obtained simply by multiplying the category numbers on the margin in a linear fashion. The function used to combine these category numbers to form the intersection figures could have a slope greater than 1 or apply a nonlinear relationship resulting in a more pronounced spread of apparent seriousness between low and high risks. In this case, we have identified any risk with a figure of merit less than 5 as a low risk, any risk with a figure of merit between 5 and 12 as a medium risk, and any risk with a figure higher than 12 as a high risk. We then need a set of rules covering what to do about risks based on their figure of merit.

High risks should be actively mitigated in an attempt to drive the probability of occurrence and/or effects categories down such that the risk figure does not indicate a high risk (figure less than 13). Medium risks perhaps should be dealt with applying some form of conditional mitigation action identified in the risk management plan (or risk section of the system engineering management or program plan). Low risks should be monitored to see if their categories are adversely affected by future actions and conditions.

What we have done is to provide a way to prioritize all program risks such that we can focus our energy on those that have the chance of causing us the most damage. Is this a purely objective structure? Of course not. There is an opportunity for subjectivity in the assignment of a particular risk to a category in both dimensions. But, we have provided the people who will make these judgments with guidance that will help to remove some of the subjectivity. The errors that are made will result in unnecessary work being performed on some risks that have little chance of causing problems and a failure to mitigate some serious risks.

25.4 Risk mitigation action and status reporting

Each person assigned a high-risk mitigation action needs to create and maintain a risk mitigation action plan telling what is planned in the way of mitigation actions, the current status of these actions, and a schedule showing how these actions will run in time. This should be in the form of a single regularly updated chart (with possible backup data) that can be briefed on short notice.

The responsible engineer must pursue action plan activities and goals energetically and the management must monitor performance to these plans. It helps to establish the right granularity in this regard. If a program has five IPPTs reporting to the program manager, some of the risks might be identified as system risks while others as team-level risks. The former should be reported at system-level meetings while the latter are covered at team meetings and team actions countermanded by exception at system meetings if the program manager concludes the team solution was not program optimum. This granularity could involve two or more layers of teams on a very large program, and each team should be the system agent for all teams reporting to its manager.

25.5 Organizing the risk search

It is a mistake to induct a risk guru into a program because it causes everyone else to relax and fall into the trap of thinking that risk is not one of their problems. You must communicate the message that risk is an inseparable part of good management and cause everyone to feel like they are part of the problem and its solution. Therefore, while you need someone to accept the responsibility of leading the risk effort, you do not want them taking on cult status.

Everyone should be encouraged to bring forward their risk concerns without fear of criticism for their lack of skill or creativity in dealing with the problems they express. The best way for a manager to kill a risk program is to lash out at people who identify risks as being disloyal, silly, uninformed, or having some other form of bad character. Program managers should recognize that risk identification is not a natural act where the person exposing it is potentially exposing his or her inability to handle some situation they are charged with. So, the program manager must constantly be reassuring the staff that it is important that the staff understand all the ways things can go wrong and, once identified, they will work together to minimize the adverse consequences. Us together against the world, rather than you darn fool for worrying about that.

A risk program can act as a conduit for hostility, of course. If manager A does not like manager B, he or she may be very effective in identifying risks in the areas manager B is responsible. The right response for manager B is, "Thanks, I need all the help I can get!" If manager A is not focusing similar attention to his or her own area of responsibility, risks will occur, placing manager A in the wrong color spotlight. This problem should be self-correcting and not require a lot of attention on the part of the program manager. The problem of failure to voice risk concerns is a much more valid concern.

But, the program manager is also a human being, and program managers have been known to be their own worst enemies regarding risk. During the proposal effort, the proposal manager (program manager to be) may ask for serious risk identification work, but when they are exposed, quash the identification of them or encourage the downgrading of high-risk items to medium or low risk based on implausible logic. Certainly, if a program has high risks, it should be mitigated to the extent possible in the proposal effort, but any residual should be clearly identified and the program cost, schedule, and product performance factors adjusted accordingly. But, it is understandable why program managers allow themselves to be trapped in a proposal effort with high technical risks, limited funding, and serious schedule problems. They want the business desperately and want to lead their company's efforts in a real program, and they believe that exposing the realities will cause the customer to downgrade their proposal marks accordingly.

Is there an answer to this quandary? If so, it probably has to come from the top management. An enterprise or individual is most often best served by the truth in the long run. Our proposals should reflect the truth as we see it. Granted, we should be allowed some literary license in addressing our wonderful capabilities and the serious problems with our competitors. The principal agent that unintentionally encouraged enterprises to act otherwise in the past, the U.S. Department of Defense (DoD), has opened a new page in its procurement. The DoD has a much more realistic view of their programs today than in the past, in general. It tracks contractor performance and collects this data for future use in making selections better. It recognizes that it cannot expect contractors to solve their problems well where the problem difficulty does not match the available resources. It is more effective in spotting incompatible claims in proposals relative to risk.

The most effective risk management programs the author has been familiar with were a fundamental part of each program manager staff meeting. The risk action leader simply prepared and briefed the overall risk program status materials, and the team managers had to brief their status on all of their high-risk items. By making this the first 5 or 10 min of each program meeting, the program manager communicates very effectively that risk is important to the program and should be to the team leaders.

25.6 A program risk metric

Figure 25.2 was constructed with a view toward providing a program risk metric. It was motivated by the hazard risk matrix of MIL-STD-882, Safety

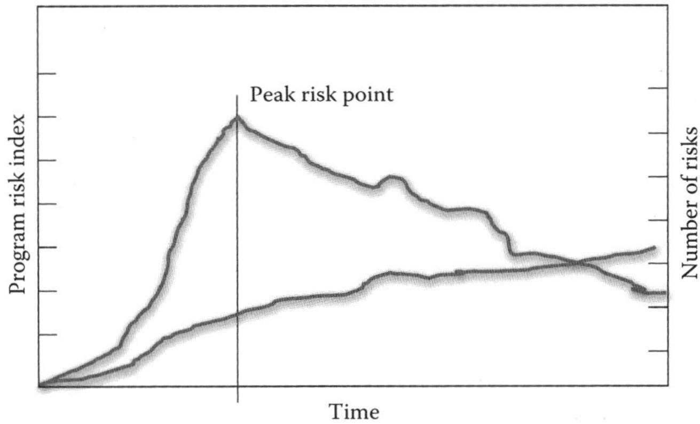

Figure 25.3 Program risk index.

Program, but this matrix assigns lower index numbers for higher risks, which makes it difficult to use the result as a metric. Figure 25.2 reverses the sense of the safety hazard matrix values such that higher risks have higher numbers. The best case on a program would be few risks, each with a low index, and the worst case would be many risks, each with a high index. Figure 25.3 shows what a risk metric tracking chart might look like partway through a program. In the beginning of the program, there may be relatively few risks identified, but these will commonly have high indexes in that no mitigation work has been accomplished yet. As the program proceeds, the number of risks continues to grow, but they are being mitigated such that their individual indexes are declining. It is reasonable to expect that there will come a point in the program that the aggregate risk index, formed by summing the individual risk indexes, will peak and start to decline, while the accumulated number of risks may continue to climb, reflecting the program having gained essential control over program risks.

There should be no serious attempt to reduce the number of risks identified; they must flow from the observations of people on the program based on the concerns they recognize. The energy must go into reducing the index of risks that are identified.

chapter twenty-six

Work performance management*

26.1 The complicated world of design

This book talks all around a central idea that is very difficult to talk about because few of us fully understand what is going on relative to it. This central idea is the consummation of a design. The dictionary tells us that to design is to plan or fashion the form and structure of an object. Engineering is said to be the practical application of science, mathematics, and available technology in the pursuit of a solution to a problem that can be solved by some form of machine. The question is just how exactly does this process occur within the mind of the design engineer. The author likes to believe that the design engineer comes to a clear understanding of the nature of the problem and then applies his or her command of the related knowledge and techniques of an engineering discipline to gain insight into ways that the problem could be solved. Ideally, the engineer would conceive two or more alternative possible solutions and seek to determine which of these is the better.

Over the several years that the author has taught courses in system synthesis, he has greatly enjoyed one part of the work required of students. They are asked to interview a design engineer and report back to the class how they think they accomplish this work. Often, the answers are that they do not know. Others have said that ideas come into the mind after intense thought about the problems that are explored further, leading to expressions of design features that might work. This is a creative process that involves what appears to be a chaotic disconnect in the system engineer's orderly world of deductive and inductive reasoning.

System engineers are vitally interested in establishing linear stings of traceability information, but it is very difficult to expose a clear case of traceability between the requirements and the design features. Eventually, as the tool companies consume one another into core businesses that cover both CAD and requirements work, it may come to pass that there will be a way to link the requirements with design features, but this will not change the existence of a discontinuity between the requirements and the creative process through which the design engineer conceives the features of the design solution.

* The material in this chapter is from Grady, J., *System Integreation*, CRC Press, Boca Raton, FL, Chapter 8, 1994. Used with permission.

Given that this is an acceptable description of what is going on in the mind of the individual design engineer, we have to recognize that today the world in which engineers live in is so much more complicated than suggested by this simple description. The nature of the problem is commonly very complex to the degree that no one person can individually fully conceive the nature of the problem or a possible solution to it. Therefore, we have resorted to teams of people who collectively have the knowledge coverage needed to understand the problem space. But, because we have spread out the understanding of the problem space across several domains that different people have to deal with, it is necessary to patch together these thought processes in an integrated fashion. Whole design engineers are formed on programs through human communication that knits together the aggregate knowledge base of all of those cooperating on a particular task. Often this process of integration and optimization will proceed better if a system engineer, a generalist, is involved.

26.2 Controlling the well-planned program

26.2.1 Program execution controls

We will first focus on program integration when things are going according to plan. We arrive at a condition like this through good planning, faithful execution of the plan, and the good fortune to have developed precisely the conditions that we had planned for within an environment of imperfect knowledge of the future or the management agility to adjust when they do not. Throughout the period when things are going according to the plan, we need to have our sensors turned up to detect indicators of potential problems or risks, for risk changes over time as a function of what we do, and what happens in the world around us. During this initial discussion, we will assume that these sensors are reporting nothing but a Go condition.

Our program plan along with the integrated master plan (IMP) and the program-level schedule are the two principal program execution–controlling documents at the top level. The author believes that the work breakdown structure (WBS) and the statement of work (SOW) are subsets of the IMP, properly structured so that the IMP includes them. The combination of these documents tells what has to be done, when it must be done, and by whom to what end, providing a structure within which the program cost and schedule may be managed, collected, and reported to the customer. The contract data requirements list (CDRL) defines data products that must be delivered to the customer, and the schedule in the contract tells what other deliverables are required and when.

26.2.2 *Alas, good planning is not everything*

These documents do tell what should be happening, but they do not, of course, tell what is actually happening during execution. So, in addition to good planning data, program management personnel need good program status tracking data. Good management skills and judgment are also needed to take advantage of accurate status information to steer the program along a sound course. Program status data must be compared with planning data and appropriate direction determined and given to encourage continuation on the plan. Most companies have cost/schedule control systems (C/SCS) or earned value systems (EVS) that relate to the WBS. These systems are designed to provide the management with financial and schedule data that tells how the program is doing relative to the plan. Many of them at the time this book was published, provided exactly that, data, in reams of computer printouts. There is a lot of work for computer programmers to do in refining the presentation of C/SCS or EVS information for managers and workers in an easy-to-understand way that will lead to effective decision making. There are some good examples of a sound direction for this work in the form of the technical performance measurement (TPM) approach and functional management metrics. Some C/SCSC systems are still operating in the batch-processing era while the world has moved to online access.

Program management must also have access to the third leg of program management information dealing with product technical performance progress. One very effective tool for managing product performance risk is the TPM process. In the TPM process, the customer and the company management agree on a relatively small number of quantifiable product performance parameters that will be controlled under the TPM program. These should be key requirements that collectively span the development effort, ones that will be adversely affected when trouble is encountered. The parameter values are tracked in time on TPM charts that are annotated to note critical events related to changes in values. When a reported parameter value exceeds an allowable deviation from a planned value, it is clearly visible on a graphical chart. Specific actions are taken to drive the value back within limits. The value continues to be monitored to see if the corrective action has the desired effect.

Weight, maximum speed, guidance accuracy, predicted turnaround time, throughput, and memory margin are examples of possible TPM parameters from several different fields. In picking these parameters, it helps to know what kind of product requirements your company has had trouble in satisfying, in your history. Some company people may be concerned that if the customer has access to this data, it might make the company look bad to use these parameters for formal performance measurement. Some company program managers would prefer to have a contract set of parameters that are not so effective at spotting company problems supplemented by an additional list of parameters that are evaluated only internally that may recognize traditional problems. The customer that knows its supplier will not stand for this, of course.

The TPM technique is a closed-loop process and this is what program managers need for cost and schedule control as well. In these systems, an unacceptable deviation is visually obvious from the management information, specific action is taken to correct the indicator of unsatisfactory program performance, and the indicator then provides, with some time constant, feedback of the effect of the action taken. TPM and the application of this same approach to C/SCS are examples of what are called metrics. They all work in this same way. Key numerical parameters are charted in time and used as a basis for management feedback to the process. The process reacts and the results are displayed by the charted parameter. We will return to TPM later in this chapter.

26.2.3 Implementing the IMP/IMS

In Part II, we encouraged the use of the U.S. Air Force program planning initiative called the integrated management system using an IMP and an integrated master schedule (IMS) or a variation developed by JOG System Engineering as the principal management tools. In so doing, we need the information inputs provided by the systems discussed above focused on the events identified in the IMP. The IMP provides not only an organized set of events to which we have hooked the tasks defined in the SOW, but for each event a list of accomplishments that we should expect to see materialize and specific criteria by which we may objectively judge whether or not these accomplishments can be claimed. The IMS fixes these events and accomplishments in sequence and time.

The IMP/IMS should have been organized such that it is perfectly clear which product team is responsible for the work covered. There should be an overall IMP sheet, covering each team, that may have to be expanded in detailed team-planning data, but from which program management personnel may determine what should be happening on the isolated team and relative to work on other teams.

26.2.4 Controlling the advancing wave

One of the most confusing and distracting things about the development of a complex system is that at any one time all of the pieces of the development work are not at the same stage of maturity. They cannot possibly be if we are serious about the notion of the top-down development, since a sequence is implied. As we develop the product entity structure for our system, it expands into the lower reaches, as noted in Figure 26.1, originating with the simple allocation of the customer need to the block called system. Subsequent allocations of identified functionality result in

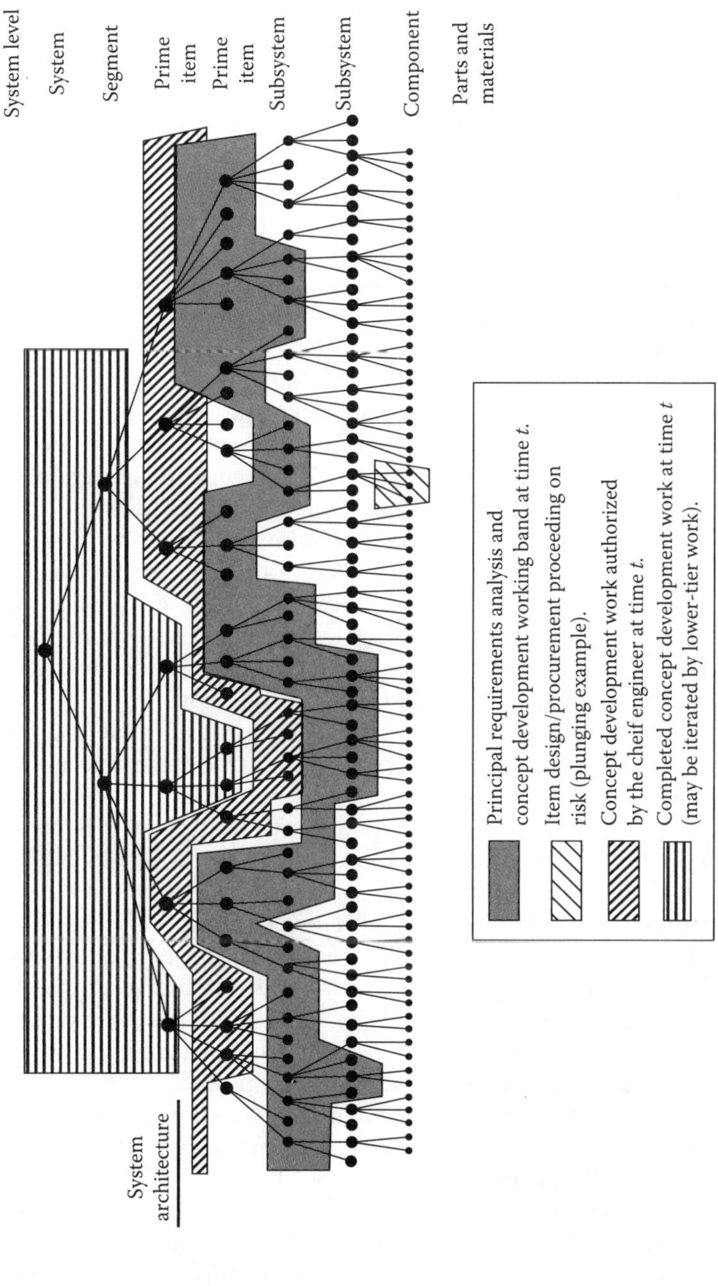

Figure 26.1 The advancing wave. (From Grady, J., *System Integration*, CRC Press, Boca Raton, FL, pp. 139–147, 1994. With permission.)

the identification of lower-tier elements and requirements corresponding to these items. During the decomposition process, there is an advancing wave of development downward through the product entity structure to identify items and requirements for these items.

When this process hits bottom, the wave rebounds upward and we move to the design and integration process. Hitting bottom is characterized by the following: (1) Requirements are identified for everything in the product entity structure and (2) everything on the lower fringe will surrender to a detailed design by an identified specialized design team or associate contractor or it will be purchased at this level. These two strokes or wave motions must be embedded in our IMP/IMS, but we also need an effective means to track how well we are doing during execution. This means should also encourage a structured approach defined by requirements approved before design.

Figure 26.2 illustrates one version of a development control matrix designed for this purpose. The matrix tracks the development of each selected product entity structure item in three primary development modes: (1) requirements documentation, (2) concept development (preliminary design), and (3) design development (detailed design). In each case, due dates are identified from the IMS or lower-tier supporting schedules. The dates work is actually completed should be included, but were dropped here for lack of space. Each development mode portion of the matrix also calls for an engineering review board (ERB) number entry, where the item was approved to progress to the next mode, and a status column, where you could use numbers, letters, or colors.

There are many other things that a particular development program might choose to place on such a matrix. But matrix design is inevitably a compromise. You need to select things that help you to focus on status visually. You want to be able to tell quickly what the status is on every level and branch of the product structure shown in Figure 26.2 down to the lowest level at which you choose to manage. Naturally, the PIT might have a system-level matrix and each IPPT leader could have a matrix for their item, thus reducing the vertical dimension of any particular matrix.

Another technique would be to eliminate the status columns on the matrix (freeing up space for other text data) and place the status information on a graphic of the product entity structure block diagram by coloring the blocks in accordance with an agreed-upon code.

Whichever status-coding scheme we select, we should always see the requirements for an item go green before detailed design work is allowed to move to an in-work status. The author would break up the requirements status tracking reporting into two steps: (1) a requirements list that captures the functional requirements for the item (what does it have to do and how well does it have to do it) and (2) the item specification. Concept development should be allowed to proceed based on an approved requirements list.

Architecture identification		Principal			Requirements documentation						Concept development			Remarks
Arch id	Item name	Contractor	Dept	Engineer	Type	Number	Rev	Due date	ERB NBR	STAT	Due date	ERB NBR	STAT	
A14	Cosgrove upper stage	GSTS	120-1	Jones	CI	72-1934	—	02-25-90	051	G	03-10-90	063	G	
A141	Equipment module	GSTS	120-1	Burns	DRD	72-2430	—	03-10-90	062	G	04-2090	068	G	
A1411	Em structure	GSTS	231-3	Adams	DRD	72-1122	A	04-05-90	066	G	05-10-90	080	O	
A1412	Avionics subsystem	GSTS	342-0	Perkins	DRD	72-3323	—	04-10-90	067	O	05-15-90	081	R	Platform mounting scheme
A14121	Guidance & control subsystem	GFG	350-1	Smith	DRD	72-2579	B	05-20-90	083	R	06-25-90		O	
A141211	On-board computer	TELEBOND	780-2	Brown	PS	72-2582	—	06-10-90	091	O	07-10-90		O	Chip GIDEP alert
A141212	Interial navigation set	REYNOLDS	780-2	Windham	PS	72-2585	A	06-15-90		O	08-10-90		O	
A141213	Star tricker	GFE	100-0	Fletcher	—	S-1267-3	A	02-10-90		G	02-10-90		O	
A14122	Electrical subsystem	GSTS	420-0	Blackmer	DRD	72-9333	—	05-10-90	082	G	06-10-90	093	G	
A141221	Battery	RPS	420-1	GOMEZ	PS	72-9354	C	06-10-90	090	R	06-15-90	095	O	Strike scheduled
A141222	Wire harness	GSTS	420-2	CHIN		72-9365	—				06-20-90			
A141223	Power control unit	ESI	420-3	Johnson		72-9202	B	06-10-90	089	G	07-25-90		O	
	Structure					72-1211	—	05-20-90	084	G	06-10-90		O	

Figure 26.2 Development control table. (From Grady, J., *System Integration*, CRC Press, Boca Raton, FL, pp. 139–147, 1994. With permission.)

Detailed design development should not be allowed to proceed without an approved specification. We could have a lively debate about whether preliminary design should be allowed to proceed in the absence of an approved specification. The author believes that it is during preliminary design that many requirements are finally validated, suggesting that there could be some requirements volatility during preliminary design. Certainly, the specification should be formally reviewed and released before the beginning of detailed design, but the author has no firm suggestion about the timing relative to preliminary design. If your tendency is to approve the specification before preliminary design, you simply have to be open to the likelihood of maintaining it actively during preliminary design.

26.2.5 Summing up

In summary, this worn-out maxim of good management comes to mind: "Plan the work and work the plan." To which the author would add, "… while keeping one eye on risk." Risks realized can cause a discontinuity in program execution relative to the plan. Therefore, the program must have some way of foreseeing discontinuity triggers and working to reduce the probability of them occurring and/or the degree of difficulty should they occur, as Chapter twenty-one discusses. Finally, we will discuss how to get back on track after having fallen off the plan.

26.3 Discontinuity management

Perfection is seldom attained. If a program was perfectly planned, executed with good fidelity to this good plan, and everything unfolded during plan implementation as conceived by the planners, then everything would go according to plan. Has there ever been such a program experience? Probably not. We have to plan programs with imperfect knowledge of the future, and this reality practically guarantees that everything will not go precisely according to plan. Even if your plans were proceeding swimmingly, your competitors may remain actively tuned to your progress and work to counter your best efforts with new products and advertising. Therefore, we must have a way to detect that we have fallen, or are in danger of falling, off the plan and have the machinery for getting back on. We must have a capability for replanning any remaining work whenever serious problems arise. Perfection can be approached through good planning, as discussed in Parts II and III, and good management leading to good execution can go a long way toward encouraging success, but it is just very hard to know everything that will happen during program execution at the time you are making your plan. We must plan a program with imperfect knowledge of the future.

26.3.1 Discontinuity defined

A program discontinuity is an interruption in planned activities. It is a condition that precludes continued efficient execution of work in accordance with the predetermined plan. There are three principal factors involved in determining whether conditions of discontinuity will develop during program execution:

1. Planning quality
2. Degree of correlation between the assumed conditions that will be in place during execution while planning, and the reality during execution
3. Fidelity of plan execution

Figure 26.3 is a three-dimensional Venn diagram illustrating the possibilities from these three factors. In the simple case of binary possibilities for each variable, there are two cubed, or eight, outcomes. The preferred outcome during execution is, of course, that we had planned well, executed the good plan faithfully, and the conditions in effect during execution were those for which we had planned. All of the other combinations will yield less desirable results. One possible exception is that a failure to execute a bad plan faithfully amidst radically different circumstances

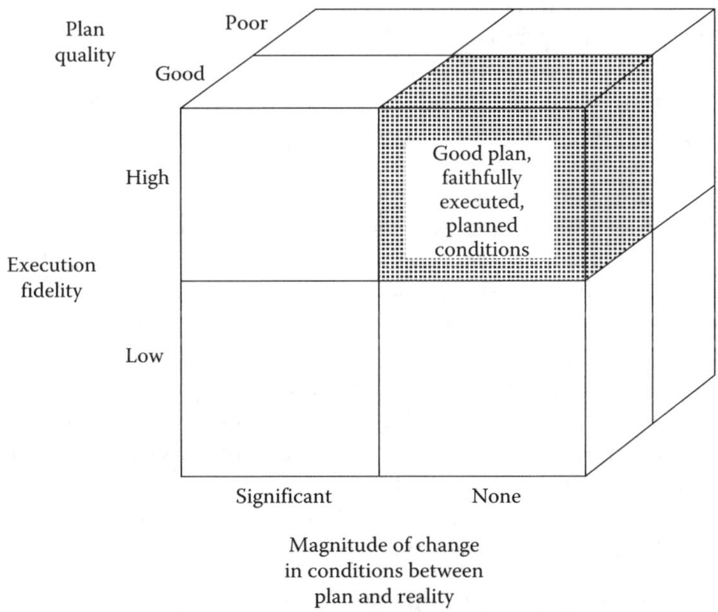

Figure 26.3 Discontinuity cause cube. (From Grady, J., *System Integration*, CRC Press, Boca Raton, FL, pp. 139–147, 1994. With permission.)

than planned could conceivably turn out very well. This is a condition of almost total discontinuity, however, involving instinctive management without a plan that requires a lot of good luck and unusually fine intuitive management skills. The reader will have to look elsewhere for guidance for this possibility. If the management team is this good, think what they could do with a good plan.

We generally think of discontinuities as being bad, but it is possible that conditions can change for the better and we may wish to change our plan to gain full benefit from these changed conditions. For example, our plan could involve using a very expensive technology because another is thought to be too far out in development. During our program execution, this other technology might become available in practical terms, making it possible for our customer to save hundreds of thousands of dollars. Certainly, we would want to embrace this new technology no matter what our current plan calls for.

26.3.2 Discontinuity detection

A program discontinuity can arise from several sources suggested in Figure 26.4, and we need a detector tuned to each of these sources. The three principal program planning focuses are technical or product performance, cost, and schedule. Most of everything that can go wrong in terms of plan quality, execution faithfulness, and changed conditions can be expressed in these terms. Some people, including the author, would add a technology list to this for many programs even though technology unavailability can be expressed in terms of cost, schedule, and performance. A new development program that is pushing the technology will benefit from this added trigger. A program that can rely on existing technology need not include it. This gives us four triggers for which we must arrange detectors. For particular programs, or just because of historical precedent, a particular company may choose to add other specific triggers to this list.

The detectors we need are really tuned to what commonly are called risks. Risks are potential program discontinuity triggers that should be avoided, mitigated, or corrected for. Wherever possible, we should avoid them by detecting their potential appearance in the future and taking action in the present to ensure that the conditions needed to bring them into existence never occur. When a risk has been recognized, but is not yet disrupting our plans, we need to be actively mitigating it to prevent it from becoming an open discontinuity. A risk that is fulfilled is a program discontinuity that must be removed through replanning.

26.3.2.1 Cost and schedule triggers

The Department of Defense (DoD) requires its contractors to use a certified C/SCS or an EVS for tracking and reporting program cost and schedule

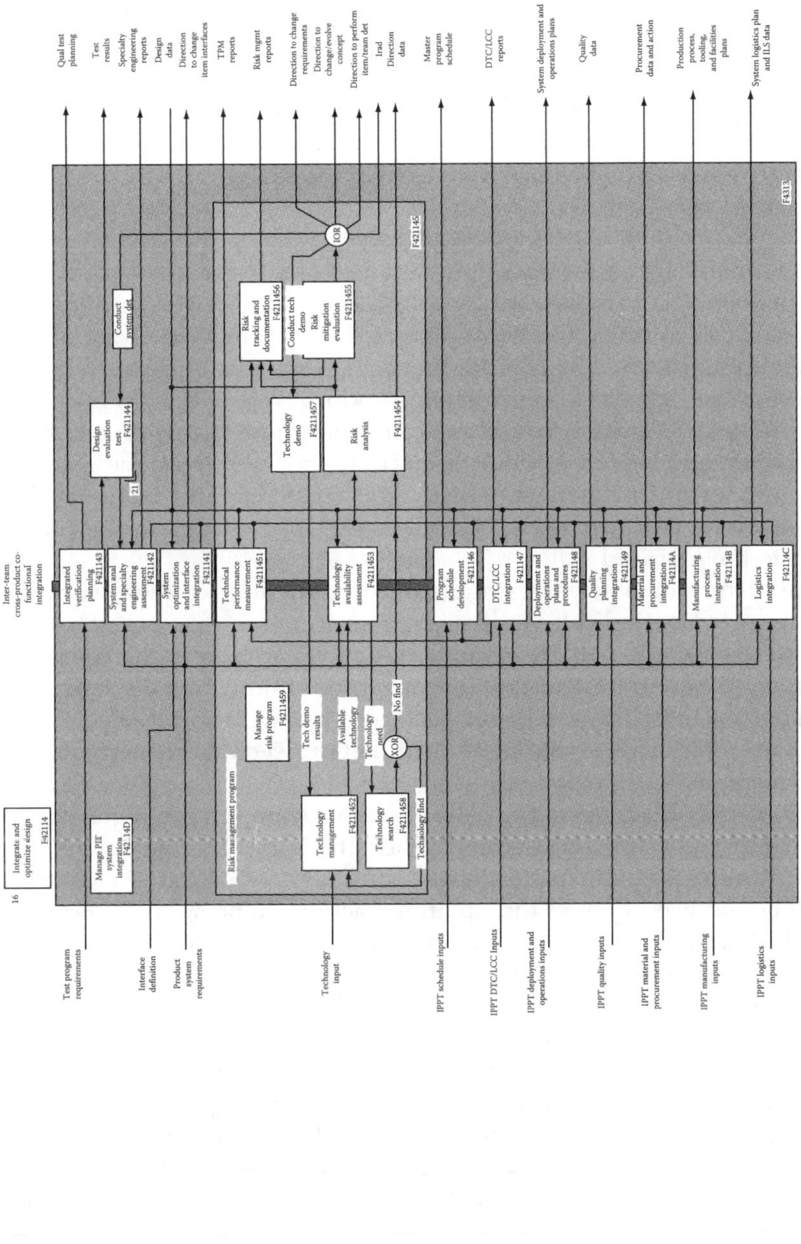

Figure 26.4 System integration process. (From Grady, J., *System Integration*, CRC Press, Boca Raton, FL, pp. 139–147, 1994. With permission.)

information. The information these systems produce is adequate to detect possible or realized discontinuities from these two triggers.

Figure 26.4 shows one function of the PIT to be Program Cost/Schedule Control. This responsibility could, of course, be assigned to a program business team, alternatively. This activity must constantly monitor the performance of all of the IPPTs and the PIT against planned expenditure and achievements using the EVS. Deviations beyond predetermined boundary conditions trigger a potential risk that has to be studied by someone responsible for program risk analysis. The corrective action generally applied is that the offending party must complete a variance report that explains why the variance occurred and what that person is doing to cause future reports to be in compliance with the plan.

26.3.2.2 Product performance trigger

DoD customers also often encourage the use of a technique called TPM to detect problems in satisfying the key technical product requirements. TPM involves customer and contractor agreement on a small list of key quantified parameters, the condition of which signals the general health or illness of the complete program and product system. Weight is a common parameter chosen where this is an important characteristic, such as in aircraft and missile systems. The contractor will track the value of these parameters in time against the required value and make predictions, based on planned work, how the value will change in the future relative to the required value.

The history of past values and future predictions of these parameters is reported periodically to the customer in graphical form. If a TPM parameter is outside predetermined boundary conditions, the customer will expect the contractor to explain the reasons and what they are going to do about getting the parameter back on track with its planned value. As shown in Figure 26.4, the PIT-TPM function feeds its concerns to the person responsible for risk analysis.

Figure 26.5 illustrates a typical TPM chart showing the requirement and current value historical traces. When the current value or the predicted future value exceeds the agreed boundary conditions, the contractor will be required to brief their planned actions to bring the parameter on track. Boundary conditions may be defined to permit the contractor some slack within which they may manage the program without undue customer micromanagement.

26.3.2.3 Technology trigger

There is no comparable highly organized mechanism in common use for technology-triggered discontinuities. It is not difficult, however, to invent one here and now. First we need to know what technologies our design will have to appeal to. Each development team or principal engineer should be

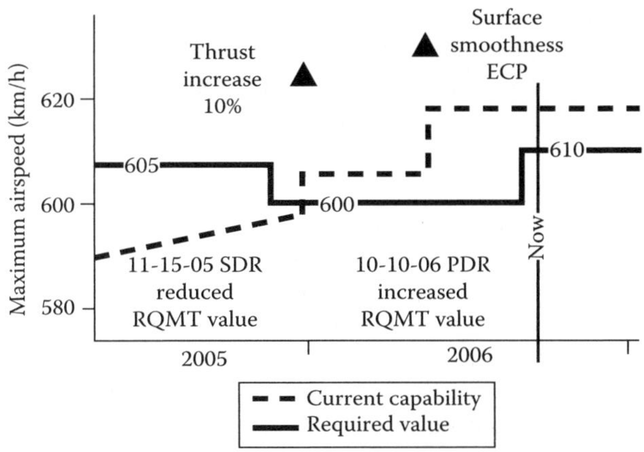

Figure 26.5 Typical TPM parameter chart. (From Grady, J., *System Integration*, CRC Press, Boca Raton, FL, pp. 139–147, 1994. With permission.)

called upon to survey the needed technology for the planned design concept. Concurrently, the manufacturing, quality, logistics, and operations people should be doing the same thing for their own fields based on the current product and process design. We then must compare the content of these lists against a list of technologies available to us. These technologies may be available to us because these are freely available for anyone's use, we know how to gain access to them on the open market, or we hold them as proprietary properties.

There is another class of technologies that drive program managers to insanity that are said to be just over the horizon. If we are depending on one or more of these coming in during our program, we are probably setting ourselves up for a grand discontinuity by plan. All immature technologies should be recognized in our risk management concerns and accompanied by risk mitigation plans corresponding to the case where they do not come online at the needed time.

As the first gear in our mechanism for a technology trigger, we should require design groups (manufacturing and logistics as well as product) to identify the technologies they intend to appeal to for each design concept during preliminary design. The system engineering community (in the form of the PIT) should review this mix against available technologies, as indicated in the Technology Availability Assessment block of Figure 26.4. The program must, first, have a way of knowing what technologies are available that might have a program relationship, and this may require a specific technology search or simply the application of an edited list from a comparable prior program.

In all cases where there is a match between required and available technologies, there is no triggering condition present. Where it can be demonstrated that a technology needed by an IPPT does not now exist on the program list or there is a likely probability that it will not be available in a practical sense at the time it will be needed on the program, there exists a technology discontinuity trigger event and corresponding risk. The appropriate response is to conduct a technology search followed either by adding a found technology to the available list or signaling a technology risk to the risk analyst.

The outcome of the risk analysis process may be a direction to the designer to change the design to avoid the unavailable technology or a technology demonstration to develop and report the new technology needed for the existing design concept. The later path should carry with it a periodic monitoring of the status of this new technology as the time approaches when it must be mature. If there is a risk that it may not be available in a timely way, it may be necessary to run parallel development paths to mitigate the risk of the technology not coming in on time. In this case, it may be possible to convince the customer that the alternate technology will be adequate for initial capability, even though it does not meet the system requirements, and that it can be replaced at a particular cut-in point in production. The customer may agree to a deviation from specification requirements for a particular number of articles that will be corrected when the needed technology becomes available.

The U.S. DoD has shown a great deal of maturity in the life cycle model being applied when this was written, calling for the employment of mature technology feeding into each cycle of the spiral development life cycle process. The author can recall programs where as many as three critical technologies had to be developed concurrently inside the program that depended on magic happening on schedule.

26.3.3 Risk assessment and abatement

When any of these triggers are activated for a specific risk, they should stimulate the PIT and/or program management to a particular course of action. First, the risk should be studied to determine the nature and scope of exposure. What specifically will happen if the risk becomes a reality? Then, we must try to ascertain the probability of the risk materializing. A risk with a combination of very serious consequences and a high probability of occurrence should be dealt with as a serious matter. A risk with minor consequences and low probability of occurrence may be set aside. In between these extremes, we need a policy for selecting risks for mitigation. The problem is that we have limited resources and cannot afford to squander them on every possible problem. We wish to focus on real problems that we have to deal with now. Refer to Chapter twenty-five for risk management details.

chapter twenty-seven

Cost and schedule management

27.1 The fundamentals

There are three primary program management factors most often cited. They are cost, schedule, and product performance. This chapter deals with the first two. The rationale for the importance of these two factors is very obvious, of course. A customer for a development program will generally be very interested in taking delivery as soon as possible because the product is important to them for particular reasons. It may result in better performance of the service they provide, whether that service be to destroy the enemy more completely or to communicate over a larger area with a broader band signal. Earlier possession of the new system rather than later will lead to some advantage over competitors, whether they be military or commercial in nature. It would be most uncommon if a system developer were to have a customer that told them that they had plenty of money and they would like to take delivery just whenever the system was ready. Our world is simply not built this way.

The developer that can build and deliver systems that function as desired more quickly and at lower cost will succeed relative to those that require more time and money to do so every time. Therefore, it is an extremely important aspect of management to understand how the money available for development should be spent over time and to be able to track the actual expenditure of funds over time. Most organizations use some form of cost-estimating system to determine an appropriate cost for a new system and some form of scheduling or networking construct to allocate planned work to available time, and apply these techniques to the program planning work during the process of bidding a program. This work must produce a cost and schedule plan that has two fundamental characteristics: the cost and schedule plan must be achievable and it must be competitive. If it is not competitive, the bidder may not win the contract. If it is not achievable, then the developer may fail in the implementation.

There is a natural desire to want the bid to be as high as possible to encourage success and profit, but in a competitive situation there is the countervailing force to want to reduce these numbers to encourage a win. During the planning work, there is a desire to identify all of the necessary work, but there are two fault modes that those planning the work may enter. First, essential work may not be identified, meaning that the development effort will not include cost and schedule components critical to success. Second, those building the plan may include unnecessary work either in the interest of a functional department trying to make a home for as many of its members as possible or through error or misunderstanding of the customer's needs. Therefore, there is a need for integration and optimization during the planning process to encourage that the results hit the target of complete coverage with no excess content.

An enterprise should apply some form of cost–schedule control or earned value system (EVS) to track program performance relative to the planned time and money expenditures linked to planned achievements. It generally is not good enough to just know how much money has been spent relative to the clock and calendar, but to know what progress has been made relative to the plan in doing so. Therefore, cost and schedule figures have to be linked to specific concrete achievements. This is the basis for the discussion on generic planning in Part four of this book. A common process for all programs has the benefits described therein, and each task in this common process should have identified as part of its description what is achieved in accomplishing the task as well as how to determine the extent to which the task has been accomplished on a program. These factors are what the U.S. Air Force called significant accomplishments and accomplishment criteria, respectively, in their integrated management system. In the JOGSE enterprise and program planning method, these are simply part of the task description.

27.2 Building the cost and schedule plan

The cost and schedule plan for a program must be based on the product and its characteristics; hence, during the proposal effort it is essential that the proposal team develop an insight into the product characteristics. This insight is also required, of course, in order to explain to the customer how the contractor intends to satisfy the customer's needs. There will be a limit on the amount of time available to develop this insight and a limit on the amount of money the proposal leader will be permitted to spend in developing it. As a result, it is possible that our estimates of cost and schedule will include errors caused by our ignorance and a need to guess to some extent. In general, these errors will commonly cancel out in terms of excesses and deficiencies, but it is wise to include a risk factor of a magnitude proportional to the risk we feel appropriate.

27.2.1 Cost planning

The cost component of planning work must focus on two major components: material and labor. The former is driven by the cost of the materials and finished goods we have to purchase to actually build the product, and this cost generally comes into play as we move into production, but there are material costs prior to that. The latter is a function of the number of people and how many work hours it will take for them to accomplish each task we include in the plan.

27.2.1.1 Material cost

In order to compile a representative material cost figure, it is necessary that engineering be able to tell estimators what the system will consist of and to supply some understandable basis for estimating related costs. During proposal work, it is necessary to define a planned product entity structure down to a level sufficient for cost planning for the next program phase. On the very earliest phase, the system and the immediately subordinate entities may be sufficient in that no production will occur on the phase being planned. Material cost estimating in this case is intended to provide the customer with an honest estimate of how much out-year cost they should expect if they select your proposal.

Material cost estimating is critical only when the phase being planned will result in some form of production. You may only be building the qualification-testing assets or be called upon to enter a high-rate production of 500 articles, but in any case it is necessary to identify all of the entities that must be produced and the intended sources of the material needed to do so in each case.

The author has done the engineering work related to this activity by establishing a product entity structure and pairing up principal engineers on the proposal team with each item at some level of indenture. These principal engineers are called upon to identify the material cost basis of a typical article and the numbers of articles needed, coordinated with the program office.

To the extent that there are prior programs that required similar product entities, the principal engineers can refer to these earlier designs as a basis for cost estimating with notes citing the differences that will influence the cost on the new program to a higher or lower per unit cost. On a largely unprecedented program, it may be necessary to offer estimates of the number of printed circuit boards for avionics boxes, the number of bulkheads in an aircraft structure, or some other factors for which the estimation has cost relationships established.

27.2.1.2 Labor cost

In Parts four and five of this book, methods were described for identifying tasks to be performed on programs. Each of these tasks will require some number of people over some period of time to perform the related work. If we have been using a common process for some time, then we will have moved beyond raw estimates of cost to actual figures derived from prior programs that can be used as the basis for the new program. If we stack up the instantaneous labor cost for all planned work and the cost of labor for all of the people needed was the same, then the total instantaneous cost is rate times the number of people and the total cost is the integration of the resultant curve as a function of start and stop times. The reality is that different skills commonly have different labor rates, so we have to either use some experience-derived average cost figure or deal with these different rates.

27.2.2 Schedule planning

The parameter of interest in scheduling is, of course, time. During the proposal effort, we must associate start times and time durations with each program task. This data is needed in order to calculate the labor cost and to determine how long the tasks will require for accomplishment. Schedulers must link tasks into sequences that make sense using one of the three planning schemes covered in Chapter nineteen and sum implementation times for higher-tier tasks. Actually, this process can be accomplished from the top down or the bottom up. In top down, one decides on the amount of time available and allocates this total to subtasks. In the bottom-up approach, you simply add up the serial time being careful to recognize the series–parallel relationships between the tasks.

Through the combination of schedule planning and labor cost planning, you have for each task in the program plan knowledge of (1) start time, (2) duration, (3) cost in terms of dollars and/or man-hours, and (4) how many people are needed plus the distribution of these personnel in terms of the functional department source and the specific periods of time when these people are needed (throughout or only during specific periods).

27.3 Managing the cost and schedule plan

Once the program receives a go-ahead, the full complement of required personnel must be brought on board in accordance with a plan that recognizes when these people are needed to accomplish the tasks. The author would recommend that the program system team (the program integration team, or PIT, in his terms) be made responsible for coordinating the program-staffing problem between the selected team leaders and the functional department managers. The team leaders come to an understanding of their needs ideally during their work on the proposal and should have been instrumental in helping to determine these needs as members of this team.

The computer system supporting the management of cost and schedule is commonly called an EVS today and may have to be compliant with the ANSI/EIA standard 748. In years past, the DoD commonly referred to these systems as cost/schedule control systems (C/SCS), and contracts required that contractors have a C/SCS that complied with C/SCS criteria.

The EVS should be implemented immediately with program start. All tasks should come under the management responsibility of some team leader (a PIT or an IPPT). Each team leader must ensure that all active work orders are opened on time and that the responsible task leaders and implementers are aware of their responsibilities. As the team leaders cause work to be initiated and continued, those accomplishing the work will complete their time cards on a daily or weekly basis and the computer system will accumulate cost and schedule data for presentation to the management through an EVS.

The EVS depends on defined planned and actual start and stop dates and planned and actual budget data in terms of man-hours declared by those accomplishing work, and claiming of milestones by those responsible. When a milestone is claimed, no further charges relative to this milestone should be entered or allowed. In between task start and stop, the man-hours accumulate against the task. A particular task may have only one or several milestones assigned. A milestone corresponds to the achievement of some specific planned work product status. Ideally, all planned tasks would have associated with them a clear identification of what it means to accomplish the task and an unambiguous way to determine completion that can be used on programs for these purposes. These matters can be defined generically in enterprise planning and specifically identified within the program context as the generic data is imported to form the IMP.

EVSs employ three primary terms named and defined below. The acronyms for these terms are followed by the names of these terms popular in the DoD, followed by an alternative name in each case in parentheses.

1.	BCWS	Budgeted cost of work scheduled (planned value)	The planned cost of the total of work scheduled to be performed on a given task by the milestone date
2.	BCWP	Budgeted cost of work performed (earned value)	The planned (not actual) cost to complete the work that has been done
3.	ACWP	Actual cost of work performed (actual cost)	The cost incurred to accomplish the work that has been done to date

The two most important metrics derived from these three parameters are schedule variance (SV) = BCWP – BCWS and cost variance (CV) = BCWP – ACWP.

Note that the SV is actually stated in terms of cost. A negative value of this metric tells you that there has been more cost budgeted in the

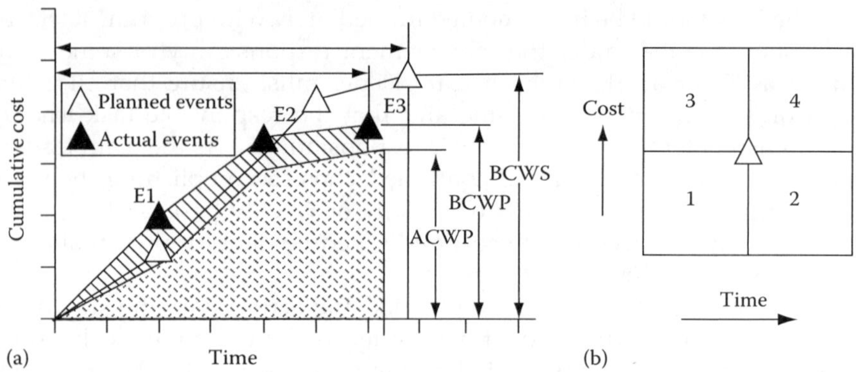

Figure 27.1 Cost and schedule relationships: (a) graphically presented and (b) milestone focused.

schedule to achieve a milestone than budgeted toward that milestone. Thus, you are behind schedule. The CV is easier to relate to reality in that you are overspent if the actual cost is greater than the budgeted cost and underspent in the opposite case.

There are some parts of the program that the SV does not offer clear guidance on in that it is determined only from the cost figures. Granted, the cost figures are linked to achieving particular milestones but dates do not enter into the resultant metrics. In these cases, one can use planned and actual dates.

Figure 27.1a shows the cumulative cost mounting over time measured by BCWS, BCWP, and ACWP, reflecting that as a program continues to consume budget, these totals will of course change. The program manager can observe resultant variances and appeal to those responsible for negative cost and/or SVs to explain to them and describe how they will correct over time for these variances. Figure 27.1b observes the possible conditions at a particular milestone. If the achievement of the milestone does not occur precisely on schedule and cost, the ideal condition is condition 1, where you are underrunning in cost and time in that the milestone has been claimed early and after having consumed less than the budgeted cost. The next best condition is arguably 2, where you are under budget but overrun in time. Some might claim that it is better to be in condition 3, where you are overrun in cost but under in time. The preference between these two conditions is probably situational. Clearly, condition 4 is unfavorable, having spent more money and time than available.

chapter twenty-eight

Configuration and data management

28.1 Configuration management during development

In the beginning of the development of a new system, there is a void of descriptive information and there exist no physical entities that comprise the system. As the user evolves a description of its need, documentation begins to pile up. Once the acquisition agent makes a decision to acquire a system to satisfy a need expressed by a user, a process begins that leads to the selection of a contractor that is believed to possess the capability to satisfy the user's need by successfully implementing a contract imposed by the acquisition agent. One of the conditions for a successful contract implementation should be that the contractor delivers a product, every part of which is in a known configuration, as defined in a baseline set of documents including specifications, engineering drawings, and lists.

Throughout the development effort, there must be an effective application of sound configuration control work to clearly and unambiguously define the product entities in a baseline set of documents, and as the physical product starts to become available, it must be assured that the physical deliverables comply perfectly with the baseline definition. The discipline called configuration management is charged with this task.

There is a pair of ideas at work in opposition during early development work. The engineers do not want any constraints on their creative work to develop a synthesis to the content of the specifications. If they were permitted to work in this kind of environment, they and everyone else on the program would suffer from the failure of product definition. Configuration management people wish to lock down the definition of the product absolutely as early as possible. If we fund this effort too early on a program, we may have great difficulty complying with the early phase schedule. One way to balance these conflicting views of the early development work is to provide for a series of baselines that progressively introduce rigor into the design process, leaving it fairly wide open in the beginning and tightening down control later in the program to make it purposefully difficult to change the result of our investment in a decision-making process that has resulted in the design defined in an expanded baseline definition.

The progressive approach suggests that we will gradually increase the configuration management budget over the program run starting with a very low budget. Some program managers believe that this number should begin at zero budget and start ramping up at preliminary design review (PDR). There is, however, a downside in beginning at a zero budget, which we will discuss shortly.

28.1.1 Information and baselines

Programs produce more than just physical products or lines of code on a delivery media. In order to produce these products, a program must develop and maintain many other work products in the form of information about the primary products that will be used directly by the customers. There will, therefore, exist on a program a large number of entities that must agree relative to some standard. This standard is often referred to as a baseline. The baseline will change over time as more details are created and introduced into it. The Department of Defense (DoD) has for many years identified three formal baselines of interest in its contracts with suppliers. These are named and defined as expressed in Table 28.1.

Table 28.1 also identifies when these milestones should be initiated on a program using the traditional DoD major review names. The functional baseline defines what the system must do, and this is captured in the approved or authenticated system specification and should be declared after a successful system requirements review (SRR) or a combination of the SRR and the system design review (SDR). At this point, the contractor and the customer will have reached a mutual understanding about the content of the system specification. The allocated baseline is defined in the content of the full set of performance specifications, including

Table 28.1 Formal Baselines

Baseline	Initiate	Description
Functional	SRR/SDR	Consists of the system specification
Allocated	PDR	Consists of the system and item performance specifications
Product	FCA/PCA	Consists of the system, item performance, and item detail specifications, as well as engineering drawings, software code lists, supporting analytical basis for the design, manufacturing and quality planning and verification plans, procedures, and reports

the system as well as item performance specifications. This should be declared as a result of the PDR. The contractor will possess at this point a clear understanding of the validated requirements appropriate for the detailed design of the system that will also be the basis for the subsequent item qualification work. The final baseline is referred to as the product baseline, and it consists of all the documentation both functionally and physically defining the product, including the addition of the detailed specifications and engineering drawings to the package.

A baseline in general is a collection of specific documents, each in a particular version and formally released with the masters under configuration control. Each document must be reviewed and approved as part of its release process, and all changes to it must pass through the same process. The documents composing the baseline will typically each have a unique document number and a version or revision identification.

Program and team managers are responsible for reviewing and approving the documents and changes that are allowed to flow into the baseline and, so, should employ some form of meeting where the document is reviewed requiring a presentation of the document by those responsible plus critical review of the document by appropriate disciplines. In this forum, the responsible manager must reach a decision about the absolute correctness of the content of the document and the correctness relative to the other documentation within the baseline.

28.1.2 Baseline information relationships

The content of the baseline should be organized in accordance with the product entity structure, with team leaders held accountable for their portions of the baseline. This includes the program integration team (PIT) relative to system-level documentation and an interest in integration across the team boundaries. But, there is another axis of this information, and it is the functional department perspective. The members of the PIT from the several functional departments should exercise an overarching interest in all the data in the baseline that is created and maintained by persons from their functional department on the several program teams.

The baseline is a subset of the total program data set illustrated in the three-dimensional Venn diagram in Figure 28.1. Ideally, both the baseline data set and the non-baseline or supporting set would all be protected in an overarching program information storage capability controlled by a data management person on the PIT with clear configuration management control of document access to the baseline data set that has been formally released. Two other subsets exist for both baseline and supporting sets, and these are data that is currently in work and that which is planned but has not yet been started. These two data sets should be under the control

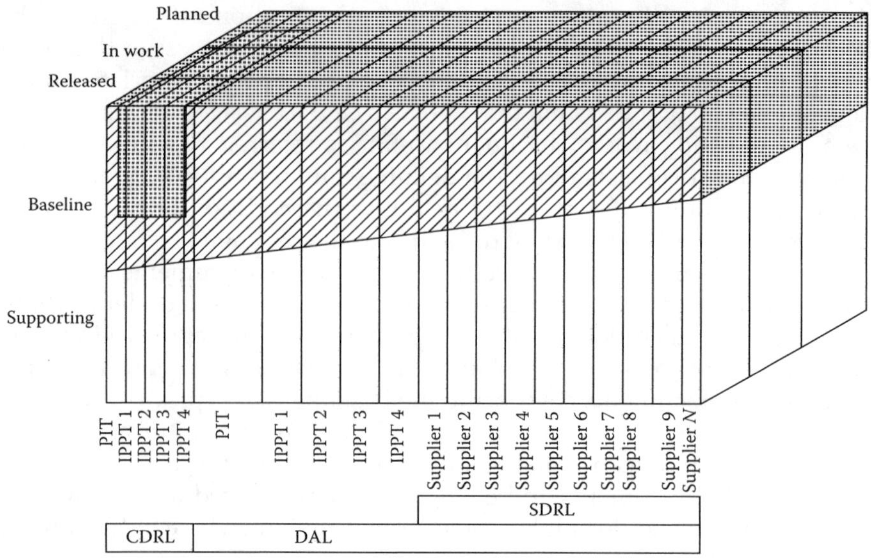

Figure 28.1 Data sets.

of the responsible party until items are formally released, at which time they will enter the released baseline or supporting category as a CDRL (contract data requirements list) or a DAL (data accession list) item.

The principal interest of data management has traditionally been with the data that crosses contractual boundaries with either the CDRL items delivered to the customer by the prime contractor or the supplier data requirements list (SDRL) items delivered from suppliers to the prime. But a customer often also recognizes one of the CDRL items as a DAL that includes all the program data not on the CDRL in its own right including all the SDRL items. Normally, the customer must pay a fee to receive items from the DAL. Since one can say that all program data may cross program contractual boundaries, then data management should accept the job of a total program librarian with special rights granted to configuration management relative to the baseline subset.

We might ask if it is true that there is no data produced by the contractor that should not be included in the DAL. One can imagine a program manager wishing to suppress a true but very negative report about the design, but is that the act of an ethical manager? If the argument in ethics fails to convince, we should all learn something from the trials of past presidents of the United States who have tried to suppress embarrassing data only to find that there are always people who would expose their faults. Better in these situations is to expose the information at a time of their own choosing and as favorably as is possible, even if it is career ending in nature.

A very special subset of the baseline CDRL collection of documents does not play the role of explaining and defining the product, but is actually part of the delivered product. This includes all of the technical data telling how to operate and maintain the system. Rather than being a CDRL, this data could conceivably be delivered as a contract line item similar to the physical product itself.

28.1.3 Data source

Each data item produced on a program should appear as a work product resulting from a budgeted task in the integrated master plan. Each of these items should have an intended purpose and one or more persons or teams that are considered customers for the item. These items may be involved in the earned-value-claiming criteria for the source task.

28.1.4 Formal review and release process

One of the author's employers, General Dynamics, employed on its programs an engineering review board (ERB) concept that was the formal engineering decision-making forum chaired by the program manager or by his or her chief engineer, who would today be the PIT manager or the senior engineer on the PIT. The ERB can be held on the same day, or days of the week when the decision rate is high, or called when an issue arises. This forum is suggested as the reviewing agent for the initial movement of baseline documents or their changes into a released status. The chairperson could be the program manager, the chief engineer (however named), the PIT manager, or a particular IPPT manager as a function of the significance of the content of the document or the scope of its application. A particular board meeting would have people invited to support the decision-making process, one or more presenters, and people who might be expected to critically review the ideas expressed by the presenters.

28.2 Physical product configuration control

Up through PDR, there will be few, if any, physical products created by a program. The focus of configuration management work can be on the data that describes and defines the physical product to be. Beyond PDR, the product design data will mushroom and force the program to tighten up its attitudes toward changes. Beyond critical design review, the physical product will start flowing from engineering laboratories in the form of qualification verification test articles. It is intended that these items represent the final production articles, and it is important that the program be able to show the customer at an item FCA how the article is tested compared with the final production articles. It may come to pass that the article tested had a part number of 453E3439-3, and by the time the FCA is held, the production articles that were going to be manufactured on the production line were in a 453E3439-5 configuration. With good configuration control, the contractor may be able to convince the customer that the changes do not warrant reverification of the design or any special testing beyond what was accomplished. If the customer senses a weakness in configuration management, they may require extensive retesting.

28.2.1 Physical product representations

In addition to the data produced in the patterns exposed in Figure 28.2, the development program will also produce a physical product and product representations suggested in Figure 28.2 that should be configuration controlled. These representations include simulations, physical models, development evaluation testing articles, manufacturing mock-ups, special test equipment, and manufacturing tooling. There may be some small overlap between this special equipment and deliverable physical product entities, but it will often be a null. One rare case would be realized if it was necessary to deliver depot support equipment also used in the factory.

One of the problems in covering the early engineering modeling and simulation work products under configuration management is the common practice of funding configuration management work only after PDR. Ideally, every engineering work product would have assigned an engineering drawing number and at least be configuration controlled by the party responsible for it to the extent that this person could unequivocally tell you what product baseline their versions relate to.

One of the most dangerous configuration management faults that can happen to a program is where the product representations are not rigorously controlled and are allowed to depart unknowingly from the intended baseline, causing the program to accept erroneous data from their use as accurate, thus corrupting the baseline. After two or three of

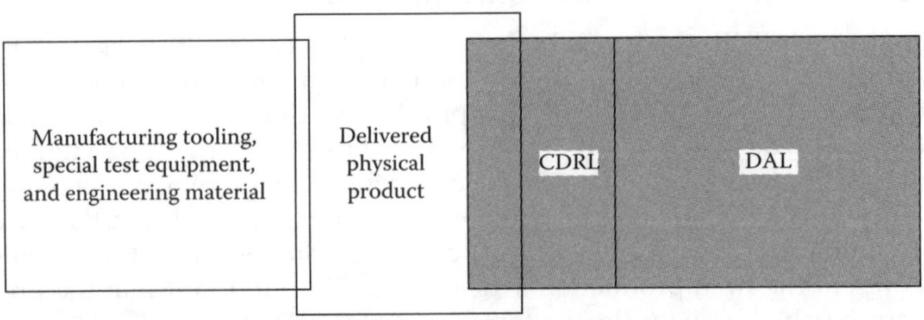

Figure 28.2 Aggregate program-related property.

these events, it will be nearly impossible to retrace your steps to figure out how the difference between actual test results and predicted results came about. A program where test results match the specification content and previously predicted outcomes marks a program with good engineering integrity, a result worthy of a celebration of no small proportions.

28.2.2 *Procurement, production equipment, and software*

The equipment and software purchased from tool companies needed to support production and test of the deliverable products should be configuration controlled by the suppliers of these products and respected by the enterprise using these products in its processes. The program may have to develop some tooling for use with the purchased equipment that will have to be configuration managed by the program. Other tooling created by another program may be generically used on several programs within the enterprise, and this can be a serious concern because another program may be interested in a tool design change that is not compatible with your program. In these cases, the enterprise will have to exercise configuration control over the tooling across two or more programs.

Numerically controlled tools will require the development of computer-aided manufacturing (CAM) data driven by computer-aided design (CAD) data, both of which must be configuration controlled individually and mutually.

The data supporting the procurement process that includes procurement specifications should be configuration controlled in the specification library, and the contractual data (SOW, WBS, contract, and SDRL) should be maintained through contracts working in the context of the supporting parent team.

28.2.3 *The deliverable product*

The deliverable product will commonly be identified by name and drawing number as contract line items, telling how many of each item to deliver and the schedule for deliveries. The listing of these items must be kept in sync with changes in the formal baseline description, and it should be a condition of delivery that an article does in fact satisfy its baseline description. This is no easy feat, because the end items will consist of possibly hundreds or even thousands of parts, every one of which must conform to a specification and a product baseline description. Something as small as a rivet in the wrong configuration used a thousand times could conceivably cause a structural failure in operation as could the flight software running in the onboard computer. Every manufacturing action is intended to be controlled by manufacturing planning subjected to configuration control. A manufacturing action incorrectly applied that escapes detection in inspection as well as a failure to implement properly a quality assurance step could similarly result in a failure. Clearly, there is a tremendous collection of data and material that must be configuration controlled and caused to remain in synchronization with a baseline definition.

Once the physical product including software on a delivery media starts rolling off the production line and into the control of the customer, the configuration control process may take on an added dimension. It is nearly a certainty that the system will require modifications during its production run. In the development of an engineering change proposal and resultant contract changes, it will be necessary to determine the production cut-in point for each change relative to the items changed. It will also have to be determined to what extent the product shipped prior to the cut-in point will be subsequently changed to reflect the updated configuration. Will we subsequently support two different product blocks with possible operational and maintenance technical data and training coverage or will we update all prior equipment and software to the new configuration such that the updated technical data and training apply to the whole system?

If the contractor retains through the contract some continuing responsibility for knowledge of the configuration of the product delivered, it will require some kind of reporting input from the customer. The rationale for this is in properly accounting for the modification of fielded product entities. It may develop that the onboard computer and its software requires a modification to preclude an operational anomaly observed during operational use in a combat action. It may be that three of these aircraft have been lost in training and mission operation in the 3 years since this aircraft was fielded. These three aircraft and the onboard computers in them will not have to be subjected to modification, clearly. Knowledge of these facts will help in understanding the cost of the modification, allow the design groups to properly account for the next assembly coverage on drawings, and aid delivery of the correct number of modification kits to the field.

28.3 Product flexibility

Some systems are developed to satisfy multiple needs. Many of the parts of the Army Future Combat System vehicles are intended to be used in a family of vehicles. Caterpillar determined many years ago that they might be able to reduce cost if they could modularize their yellow boxes permitting the vetronics systems for bulldozers, scrapers, and skip loaders to be assembled, to form a family of common hardware and software entities. There are often several aircraft that employ a particular jet engine and space transport stages employing the same rocket engine. Configuration control of these products can be very difficult for the suppliers because all of the applications for their product will find valid reasons for changes that may not be compatible with the other applications.

28.4 The baseline again

At the center of all these matters is the formal baseline defined in the form of a stack of documents in a specific version all formally released with the masters under configuration control. There exist companies where the importance of the baseline is not fully appreciated. If the product is at all complicated, subject to changes, or used in different combinations, buyers should be very wary of doing business with such companies.

chapter twenty-nine

Tools base

29.1 Program focus tool relationships

The author expects the four system development focuses that he has used in this book and the other books listed in Table 1.1 to continue to be useful far into the future, but the computer will move from being a servant of each individual engineer to becoming a member of the teams. Some enterprises may find themselves trapped in a dead-end paradigm if they do not recognize this fundamental change that the successful enterprises will grasp.

29.1.1 System requirements focus

The evolution of requirements analysis support tools has been very interesting because it has gone full circle, as shown in Figure 29.1. We all applied some form of functional analysis in the beginning of the computer software development. Subsequently, the software engineers evolved several modeling approaches found useful, including modern structured analysis, several early versions of object-oriented analysis (OOA), unified modeling language (UML), and Department of Defense Architecture Framework (DoDAF), while the system engineers remained with traditional structured analysis (TSA), employing one of several forms of functional flow diagramming combined with several other artifacts. At the time this chapter was being written (October 2007), we were poised to return to a universal modeling approach consisting of the UML combined with the system modeling language (SysML) and perhaps a few artifacts from the TSA.

Within the working lifetime of many current system engineers, they will be using a universal modeling approach to gain insight into the product entity structure of a system to be created and the appropriate requirements for the system and its entities. This modeling work will be captured in computer models and be used as the basis for solution space modeling that results in the development of lines of code and designs for hardware entities. These models will also be integrated with simulation capabilities such that the simulations draw upon the requirements captured in the problem space models in a more sophisticated fashion than is possible in the excellent tools RDD-100 by Ascent Logic and CORE by Vitech.

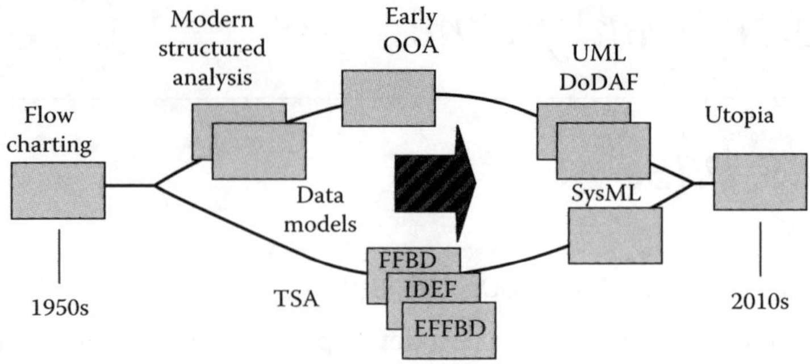

Figure 29.1 A short history of modeling methods.

29.1.2 System synthesis focus

There are very few designs that are not captured in the form of lines of code. The software, of course, takes this form in that it is entirely an intellectual product. But so does hardware. The development of application-specific integrated circuit (ASIC) obviously follows this pattern where the engineers craft a design on a workstation by introducing icons on the screen, while the computer application adds lines of code to the evolving design. This code passes through a series of adjustments and emerges in a format to drive the machine that actually manufactures the ASIC. Hardware entities that will be machined from a block of metal also exist in a computer-aided design (CAD) computer as lines of code that resulted from the engineer manipulating the CAD workstation graphically in a fashion similar to how an engineer would have used a drafting board in years past. The code thus formed can be applied to the computer-aided manufacturing software (CAM) to control the manufacturing tooling to actually convert that block of metal into the final manufactured part.

29.1.3 System verification focus

The verification process described in Chapter twenty-four will in the future still require some testing of physical product entities, especially at the system level, but most of the management support activity will be completely accomplished within computer software. Much of the analytical verification work will also be completed within the computers. One company for which the author did some consulting work several years ago was vitally interested in moving from what they called iron verification using a series of prototypes to what they called simulation verification in the interest of cost and time to market reduction.

29.1.4 Management focus

Not only is there an emerging cohesion between modeling tools and between modeling and manufacturing capabilities, but the business models are progressing such that they can be more tightly linked to the product system development support systems. A significant component in the advancement of this trend is the adoption of a common process within an enterprise. The author can recall as a functional manager in industry that he would have to spend a day or two once every few months to reach a conclusion about his department's current manpower level relative to the available budget in the near term. The business systems were program focused to serve the needs of program managers and were difficult to extract the information functional managers needed. Also, each program had its own process definition that was sufficiently different to add difficulty to this process.

29.2 The use of computers

As suggested, we will become even more dependent upon computers to accomplish system development in future. The author believes that our future reliance on computers can be cataloged into three phases that will unfold in time, with each depending on success in the prior phase. This sequence will involve manual, semiautomatic, and finally automatic operation of the computers in our networks. In all cases, humans will continue to ply their trade developing design solutions to problems characterized by requirements, but there will come to be a later stage of computer capability supporting the humans that we simply do not possess at present.

29.2.1 The precomputer past

In the rapidly fading past, engineers had the luxury of substantial independence in the development of product designs for products isolated on one or a very few technologies. They accomplished their work with computing equipment that included pencil and paper, slide rules, mechanical calculators, and electromechanical analog computers. Knowledge resided on programs in the minds of the engineers assigned to the program and paper documents created by them that reported interim or final results from the work they performed. The information was not easy to acquire in that it was located on islands that engineers had to visit to understand the content.

The human conversation was much more important then than it is today (not to depreciate its importance today), because so much of the exchange was communicated in that fashion. Clearly, the humans represented the primary nodes in the network of knowledge available to the program. Meetings were critically important because it was very difficult for individuals to share their knowledge. People had to come together and verbally discuss problems. Some of the most important conversations involving system engineers occurred with design engineers over the drafting board. Design engineers were very protective of the engineering drawings that they were working on and did not appreciate anyone unrolling them on their drafting table and looking at them in their absence. There was no easy way to gain access to in-work analysis work products either, except to engage the analyst in conversation. In some companies like the divisions of General Dynamics, engineers were encouraged to maintain a journal in which they wrote the results of their work, meetings they attended, and other matters of interest relative to their work. These journals were thought to be personal property even though the company generally provided the materials and the content was directly related to the work the company was accomplishing under contract. A supervisor might require that engineers in their department show them the content of their journal at annual or semiannual employee evaluation time.

29.2.2 Islands in the stream

As this book is being written in 2008, many enterprises dealing in system development were running programs using cross-functional teams, each populated by specialized engineers provided by a functional department structure within a matrix-managed company. Each specialized discipline has a home room provided by their functional manager responsible for providing all programs with good practices, people who know what they are doing relative to the functional department charter, and a set of good tools all coordinated with similar resources from other functional departments within the context of a common process definition respected by all programs. In the author's view, the enterprises so characterized are the fortunate ones. Many others are far less well organized. A few are experimenting with a new paradigm referred to by some as model-driven development, which we will discuss shortly.

In the current reality, the structural design engineers apply CAD tools to fashion a two- or three-dimensional model of the product structure with the many views needed, stored in files coordinated with an engineering drawing structure. Electrical design engineers apply a similar approach, possibly using a different CAD application, as do all other hardware design engineers. In reality, what is going on in this process is that the CAD applications are converting the workstation operator actions on the keyboard and the mouse into lines of code that collectively represent the design, and from which can be printed on screen or paper the equivalent of an engineering drawing prepared with a pencil and a T square on a velum on a drafting board not too many years earlier. The software engineers develop designs using solution space models driven by the lessons learned from their problem-space-modeling work.

The responsibility for each engineering drawing captured in CAD is assigned to a specific principal engineer who has read and write privileges, and all other persons have read-only privileges. But notice the fundamental difference from the precomputer days permitting easy access by anyone to the ongoing design work. Within organizations that are properly equipped, anyone holding a meeting should be able to project the current design drawings, released or not, onto the screen in a meeting for a discussion among specialists.

In addition to the work of these primary design definers, there are many other engineers evaluating the evolving design from many specialized perspectives. The reliability engineers build a mathematical model of the design, associating a failure rate with each entity in the product design. The maintainability engineers build a math model of the mean time to repair also coordinated with the product entity structure. Dozens of other specialists build specialized views of the evolving product design in all cases linked to the product entity structure. Ideally, the work of these other engineers would be similarly accessible, but this is uncommon. The work that is commonly accessible is that which is retained in formal libraries, like the CAD software. There is no reason why the mass properties, reliability, maintainability, system safety, and other models should also not be accessible in the read-only mode.

Most of these specialized product views are retained in computer applications, including CAD, databases, word processors, graphics packages, spreadsheets, or specialized computer applications including models and simulations. In general, these views are isolated from each other in the form of islands of information. They exist and evolve in a stream of engineering, administrative, and management work is a less-than-perfect cooperative relationship between those populating these islands with information.

While more accessible to a wider audience than in the precomputer period, the information in these islands is isolated one from another and the process of changing the information that can be shown to be in error requires some kind of human conversation leading to a conclusion that a change must be made and possibly coordinated with other information.

The capability thus described can be referred to as a manual mode. Each collection of specialists owns their own tools and data and are commonly responsible for maintaining the data in a configuration coordinated with the current product baseline, at least up through the preliminary design review (PDR). The computers in the manual case are simply containers for the data that results from human mental effort. Each specialist applies his or her skill and knowledge to the problem space and develops a part of the final solution that resides for all to view on a networked computer. Others, including system engineers, view this information and they may find inconsistencies between two or more of these information sources. These inconsistencies or errors are brought to the attention of decision makers (team leaders and the program manager) with recommendations for changes that will resolve the inconsistencies. A decision is reached and the information either changed or those who incorrectly reported the inconsistency caused to recognize their error. In either case, the consistency of the aggregate collection of program information is restored.

In this manual mode, the information stored in networked computers is but an extension of the content of the minds of the specialists working on the program, with the added benefit that everyone on the program can easily discover the content of the stored data on the networked computers. It is the responsibility of the humans to seek out and detect inconsistencies in the information, and the corrective action is initiated by management persons exposed to the problem. In the precomputer era, it was not so easy for this information to be shared or for inconsistencies to be discovered.

As a program moves beyond the PDR, at least some of the information should be placed under a more comprehensive configuration control. Some examples of this data are engineering drawings, computer software, and specifications. The configuration management should establish libraries for these critically important product-descriptive data items. In these formal libraries, rules have to be respected that requires a formal review and the approval of a data item in order to gain entry into the library, the configuration of the data in the library must be protected, and changes to an item in the library must go through the same formal review and approval process required for the original item to gain entry.

29.2.3 A new paradigm

Previously, we discussed what was called a manual mode that represents the best one can expect of existing system development infrastructures. But, what of the future? Should we expect the evolution of a more dynamic and effective development engine? Well, of course we should; but how might we characterize this future? The author has postulated a trio of development infrastructures following the pattern of three kinds of rifles: manual, semiautomatic, and automatic. When using a manually operated rifle, the human must do all the work. In a semiautomatic rifle, gas pressure is often used to operate the mechanism to eject the spent cartridge and reload a new cartridge. The operator must take action to fire each shot. In the automatic rifle, it fires as long as the operator holds the trigger depressed, and rounds remain in the rifle. Translated to the development process, these analogies are effective in characterizing unique differences in ways that what is being called model-driven development can be employed.

29.2.3.1 Semiautomatic operation

In a semiautomatic operation of the development process, islands of data exist as in the manual mode but these islands are joined by computer software that accomplishes logical and mathematical operations on the data on the islands identical to the manual model process, where these actions are accomplished by engineers. As a result, this computing capability may detect that some data is in error or inconsistent, causing a flag to rise possibly in the form of an e-mail message to the responsible engineer indicating the detected problem. The responsible engineer and perhaps his or her supervisor must take action to either correct the error or justify the existing content. An infrastructure may also include the computer capability to calculate new information that is added to the islands after a review and approval. This process of checking the consistency and accuracy of information entered by program workers progresses alerting the management of concerns.

In the manual mode, we are dependent on system engineers knowing the relationships between the islands of data and taking the time to evaluate these relationships looking for inconsistencies. In the semiautomatic mode, the integration and optimization work can be accomplished by computers that have been programmed to perform the cross-disciplinary work across the island gaps. However, sensed errors and inconsistencies cannot be repaired through an automatic process. Given that a problem is detected, the semiautomatic response must be reviewed by the management and the proposed changes approved.

The only difference between manual and semiautomatic operation is that in the former, we depend on humans to sense problems in the accumulated information, and in the latter, computer software intervenes between the islands of information looking for inconsistencies and freeing up the humans to focus on things they are better than machines in accomplishing, involving creativity. The semiautomatic mode also permits machine computation of some information based on that created by the humans in populating the islands of data. Obviously, a lot of work is required to determine what information should be developed by humans and what should be developed by machines, and this may be different for different enterprises.

The semiautomatic mode will evolve in enterprises that master the manual mode and start to clearly understand the information relationships between the islands. Its introduction will begin the reality of the networked computer as a team member. Currently, the computer is only the individual engineer's servant.

29.2.3.2 Fully automatic operation

As you have already surmised, full operation does something more than simply calculate data between islands and sense inconsistencies. An enterprise employing fully automatic operation will have to have passed through the semiautomatic phase, through which they will have gained confidence in their machinery reaching correct conclusions to the point that they will allow changes to be accomplished without formal approval by managers. The computers will have become trusted employees to the same degree that the human engineers have proven themselves.

This possibility will appear a little scary to many managers who read this and with good reason. As a program manager offered to the class in a course the author taught at the United Defense in Hartford, Connecticut, model-driven development will not move forward until contracts, legal, and management people have determined how contracts can be written to cover acquisition accomplished in that fashion.

29.3 Where are you?

Today, an enterprise should take stock of its current capability and develop a road map for their future growth. The first goal should be a sound manual capability that the enterprise can use to gain a better understanding of the relationships between the data residing on the several islands of information. This involves an inventory of the computer capabilities that are being employed and an ongoing study of the relationships between the data retained in these applications.

One way to gain an insight into current capability that opens a door for movement into future capabilities is to form an interim common database (ICDB) that is described in *System Synthesis: Product and Process Design* (Grady, in press). This construct entails two components using equipment and software most enterprises already possess. Figure 29.2 illustrates these two components. The work is accomplished on workstations by humans that can be thought of as forming a ring. While these machines are networked, they function more as islands, each containing unique information developed by one or more engineers working in a particular specialized discipline. The development information grid (DIG), residing in some available server capacity, is organized into columns by the product entity structure (and, therefore, IPPTs) and into rows by specialty disciplines, as illustrated in Figure 29.2. The vertical columns are called development data packages (DDPs) under the responsibility of the IPPT leaders. The horizontal rows correspond to the several specialty disciplines assigned to the teams, such as reliability, mass properties, manufacturing, quality, and so forth. This construct actually mirrors the enterprise structure. The columns relate to the program structure formed

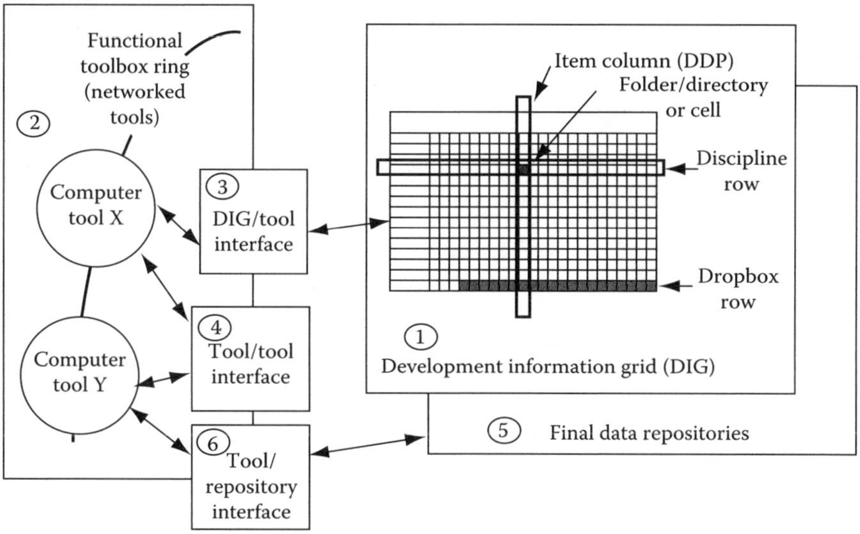

Figure 29.2 Interim common database components. (From Grady, J., *System Integration*, CRC Press, Boca Raton, FL, p. 55, 1994. With permission.)

into teams, and the rows relate to the functional department structure providing resources to programs.

This structure can be helpful up through the PDR, but, after that, much if not all of the data retained should start migrating to formal libraries, each under effective configuration management influence. However, the enterprise employing this construct will find it useful to help them understand the relationships between the data on the islands on programs. The author believes that the DIG may eventually be replaced by direct read-only access to the data on the workstations and an expanding understanding of ways that data on one island is related to that on others. In the case of some of these relationships, the enterprise will find ways to develop a code that can check for consistency between these sets. Further along in time as confidence mounts, they may find it acceptable for the computers to create some of the data.

part seven

It just keeps getting better

chapter thirty

System engineering assessment and improvement*

30.1 Is a static identity adequate?

Given that we are successful in once completing our process description, can we then rest and apply it for all time thereafter? Some companies have taken this route in the past. It is not uncommon when a company is flushed with contracts and has the indirect budget to spare for them to engage in procedure writing. The business cycle eventually returns to lean times and the practices fall on hard times along with the firm. The procedures are not maintained. Another reason for a departure between written and practiced methods is the replacement of the manager who stimulated the written practices by one who has a different value system. It may take another managerial generation to return to the maintenance of these documents. In one company the author visited, their system engineering practice was not bad but it was identified by the company that had owned the division some time previously twice removed, a signal that some other things may have changed as well. One would find very few enterprises with a current set of practices.

The enterprise identity described in *System Engineering Planning and Enterprise Identity* (Grady, 1995), and assumed in this book as discussed in Chapter one, is not intended to be a fixed entity but a continuously improving one. Yes, we will have to maintain our process definition for it to have any lasting value. Change is an essential part of the life of a person or a corporation, and we have to accept it and adapt to it. This is the reason for mounting and sustaining a continuous process improvement feature in our process. We have to be careful, however, to avoid a revolutionary change in a continuous stream. There has to be enough sound content to begin with that we can incrementally change our practices so as to preserve the vast amount of the current content at any one change so as to preserve the learning benefits from process repetition. Now this may be very hard to do. We wish to have a documented process but be constantly changing that process description.

* The material in this chapter is from Grady, J., *System Engineering Deployment*, CRC Press, Boca Raton, FL, Chapter 17, 2000. Used with permission.

30.2 Procedures media

In what form shall we publish and retain our process definition? A computer word processor leaps to mind immediately since the computer is perfectly adapted to continuous improvement. It is very easy to edit these documents stored in electronic files. Alternatively, some computer toolmakers, like Ascent Logic, at the time this book was being written, were becoming very interested in applying their tools to enterprise modeling as well as the product system engineering process. It is possible to enter all your procedures into a database from which practices can be printed. Some of these systems even allow you to simulate the operation of your enterprise in accordance with the way you have described it in an effort to find better ways to define your process.

The past standard approach to this would involve printing approved procedures and distributing them to manual holders. There is a lot of unnecessary labor involved in this process, including reproduction of paper, distribution of paper, manual update, and periodic audits of manuals to ensure that they are current.

You are encouraged to use an online networked medium for publication, distribution, and access for your procedures. In this approach, your approved procedures are uploaded from the workstation where they are maintained to a network server library of procedures. When anyone needs to gain access to one of these procedures, they only have to call it up on their workstation. Everyone has access to the very same configuration of the current procedures.

The combination of computerized procedures, a central electronic library, and universal read-only access through a network will enable an economical approach to maintaining your procedures to take advantage of incremental improvements. If one needs a copy of a complete practice or a few pages, they can simply print them out, use them as needed, and throw the paper away. In order to signal the reader that a paper version may be out of date, some companies place a warning on the cover or in a header that this version may not be current and to check the online library for the current version.

30.3 Lessons learned and continuous improvement

30.3.1 Continuous improvement process

Figure 30.1 illustrates a suggested process for incrementally improving our process continuously. Lessons learned from programs are studied (function F21) for ways to improve the current process description maintained in function F26. In function F23, we study external process descriptions and compare them with our internal procedures. This may lead to

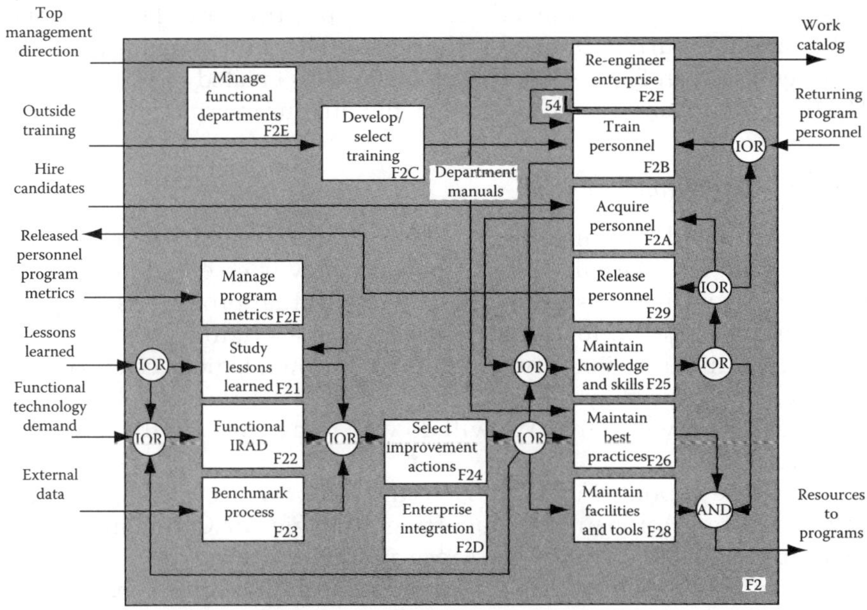

Figure 30.1 Overall continuous improvement process. (From Grady, J., *System Engineering Deployment*, CRC Press, Boca Raton, FL, pp. 251–267, 2000. With permission.)

insights into improvements or knowledge about our process maturity relative to our competitors.

The results from these two steps are opportunities for improvement actions. In function F24, we determine which of several alternatives improvement approaches we should implement. This may involve personnel training, tools improvements, or procedural improvements. The selected action or combination of actions is accomplished in functions F25, F26, and/or F28. All of this work should be the responsibility of the functional portion of the matrix providing an infrastructure within which programs can apply a common process possibly adjusted for program peculiarities.

30.3.2 The servo loop analogy

Control theory, as applied by engineers in the development of systems or networks to control processes and machines, offers us a mechanism of vital importance in the continuous improvement of our processes. We should seek out closed-loop processes with degenerative feedback that tend to follow up to a standard. In these kinds of control loops, an error signal is created through comparison of the commanded value and the actual value achieved. The result is an error signal that should drive the actual value closer toward the commanded value. Our practices are the command value, and our current practice is the actual. The error signal is the delta between these. In Figure 30.1, this error signal is created in F21, F22, and F23 by comparison of some measurement with our internal process.

30.3.3 Benchmarking

One must avoid defining their current capability and progress completely from an internal view. An outside view is essential just as it is when we as individuals produce some intellectual product as a report, it is wise to subject it to a peer review. We, in a singular or team form, have great difficulty identifying faults in our own logic because we are completely captured by these flaws. It is true that criticism is very inexpensive and very effective. We will do well to pay for it in the form of a little humility.

As a case in point, in the late 1960s the author was a field engineer for Teledyne Ryan Aeronautical supporting Strategic Air Command use of their unmanned photo reconnaissance aircraft operating out of Bien Hoa airbase in South Vietnam. He convinced himself that some rubber plugs that the Air Force troops would often leave out of holes in the wingtips might permit the North Vietnamese air defense system to more easily knock the birds down due to acoustic noise thus generated. These holes were for mounting flares on target aircraft using the same wingtips such that when a fighter aircraft targeted the vehicle, the heat-seeking missile would take off a wingtip rather than going up the engine exhaust completely terminating the target. These flares were never used on a reconnaissance bird, so stoppers were placed in the holes and would not always be replaced when lost. The author wrote to the plant asking for a survivability evaluation and was told in a subsequent letter that it was unlikely that the bad guys would hear the sound over the sound of the engine.

Benchmarking is a term used to obtain a comparison of your capability with some outside standard and thus determine how well one's internal process stacks up with this standard. This may entail some kind of reciprocal visit between two companies that agree to mutually benchmark each other or the comparison of one's process with some generally accepted external standard. Also, commonly, the enterprise will establish a goal for a particular parameter, such as a maximum number of avoidable drawing errors significantly less than currently registered, and establish an action plan to move to that condition. The number in this case may have been suggested by the mutual benchmarking exercise, published standards, surveys of industry performance, knowledge of competition capabilities, or any number of other sources.

The enterprise then implements the action plan and tracks the parameter over time, shooting for the goal on a particular date. Some enterprises applying this technique basing their performance goals on competition have been disappointed in the past because when they arrived at their goal they found that their competition had moved to an even more favorable value. The term bench trending was coined to cover this case. Instead of establishing the goal at a competition current capability value, one tries to determine where the competition might be at the time the improvement effort is scheduled to complete and uses that value as the goal.

30.4 Performance measurement systems

In order to take advantage of the servo loop approach, we must have a means of measuring our performance relative to our process specification. A common term for these measurements is metrics. There are several kinds. We should have metrics identified for our generic processes that tell how well we are performing them on programs. It should be possible to make comparisons within a program over time to track performance with these metrics and across program boundaries to see differences in performance that may lead to insights into how best to improve the practices being reported upon. They have to be numerical in nature in order for these benefits to be realized.

Two measurement systems are commonly applied on programs to identify how well the program is responding to its contractual requirements. These systems are cost/schedule control systems (C/SCS) and technical performance measurement (TPM). The former catalogs all program budgets in accordance with the work breakdown structure, the integrated master plan (IMP), and the integrated master schedule (IMS) discussed in Part four of this book. The latter tracks the values assigned to key requirements in specifications.

An enterprise also needs one more system providing program feedback to the functional organizations. This system should be driven by the content of the written practices. For each task at some level of indenture, one or more metrics should be defined that can be tracked on programs with relative ease.

30.4.1 Cost/schedule control systems

As work is accomplished on a program, the personnel fill out time cards noting what work order they are working on and this work is entered in a computer tracking program cost and schedule performance. All program tasks are also identified with the completion criteria for each milestone or event, and when this criteria is satisfied the corresponding milestone is claimed. These systems will track performance in terms of man-hours and schedule time consumed relative to milestone achievement. If you claim a milestone before the allocated budget and schedule are exhausted, you are in a favorable underrun condition with respect to both. The worst case is that you have overrun both cost and schedule allocations when you achieve the related milestone.

Figure 30.2 illustrates a typical cumulative track of cost and schedule on a program where the milestone is claimed after having consumed more schedule and cost than planned. If a task being tracked is in an overrun condition beyond the gray threshold area on the figure in either budget or schedule (threshold selected by program management), the C/SCS will issue a variance report that requires the person responsible for the task to report why the overrun condition exists and requires submission of an action plan that tells what will be done to correct the overrun condition. These variance reports should be reviewed by the management and either accepted or further inquiries made to determine the veracity of the report, its assignment of cause, and the chances of success in achieving the action plan goals. Where these are accurate and valid showing a cost and/or schedule problem that cannot be closed within the current resources, the management may consider turning loose some budget margin or schedule float to boost the task resources to a realistic level, otherwise simply require corrective actions within existing resource constraints. Where the claims are invalid or fraudulent, the program may have to consider replacing the person responsible for the task.

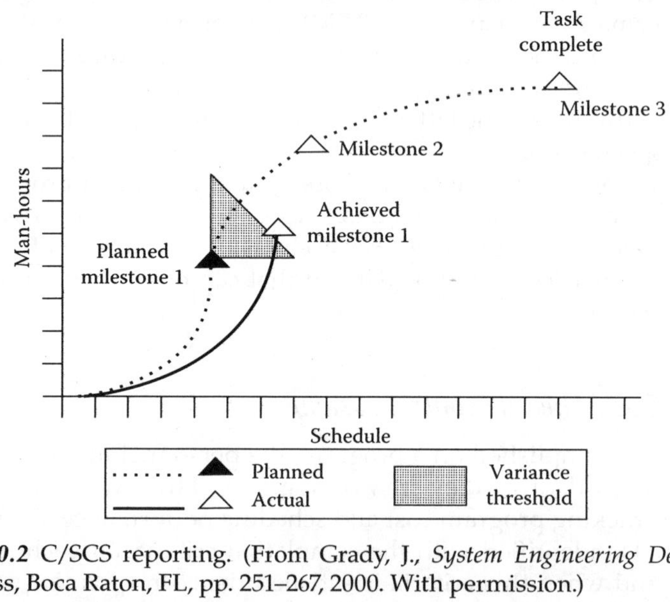

Figure 30.2 C/SCS reporting. (From Grady, J., *System Engineering Deployment*, CRC Press, Boca Raton, FL, pp. 251–267, 2000. With permission.)

30.4.2 Technical performance measurement

TPM complements C/SCS on programs by providing a set of metrics keyed to risky system and item requirements in program specifications. We first select a small number of requirements that satisfy a criteria defining our concerns. An example of such a criteria might be as follows:

1. The requirement is in the system or an end item specification.
2. There is some risk that the requirement cannot be satisfied within the current program constraints.
3. The enterprise has not been fully successful in satisfying this requirement on past programs.
4. The requirement is closely linked to program success or failure or has a great deal of leverage in determining program success.

On a small program, we may select as few as 10 system requirements for this purpose. On a large program with multiple teams at two or more levels, we might identify as many as 100 parameters with each team responsible for several.

For each TPM parameter, we should assign a responsible engineer who must create and maintain two charts: (1) a TPM-tracking chart and (2) an action plan chart. The first one should show a graph with the requirement value range and trace the history of the requirement value and the achieved value in time identified on the horizontal axis, as illustrated in Figure 30.3. Notes offer keys to significant events in the change of the requirement and demonstrated values. The second chart, only necessary where the demonstrated value has not achieved the required value, should offer a specific plan for closing the gap between demonstrated and required values as well as a timetable for achieving closure.

30.4.3 Generic process metrics

Each generic task selected for program application should have one or more metrics identified for it and a simple means whereby the data can be collected without excessive burden for the program. These metrics should be collected on each program and made available to the EIT for formal capture and reporting.

30.5 Formal postmortem

It is common practice for a program to hold some kind of review in association with major milestones or program events. In these reviews, the program compares performance against program goals for that program

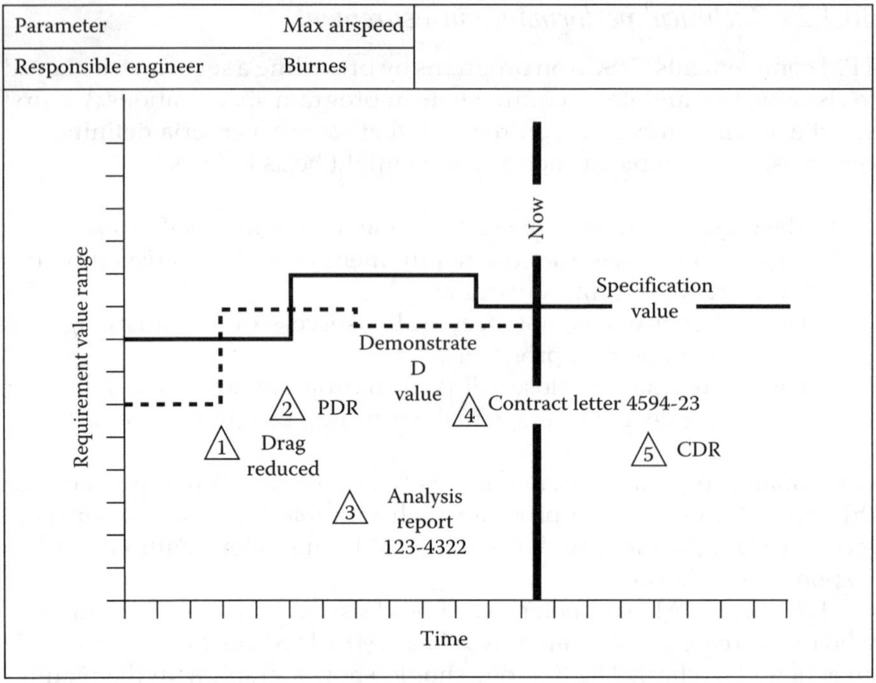

Figure 30.3 Typical TPM chart. (From Grady, J., *System Engineering Deployment*, CRC Press, Boca Raton, FL, pp. 251–267, 2000. With permission.)

task, event, or phase; reaches conclusions about that performance; and makes decisions about subsequent planned activity. This same approach should be applied to program performance relative to the quality of the process applied.

The U.S. Army has developed a technique for a formal evaluation of their performance subsequent to a training or real military action, called an after action review (AAR), that might help us do well to understand and apply to programs as a means to gain insight into our performance on programs, leading up to task milestone, program event, and program phase terminals. If we have planned these activities in accordance with the content of this book, then each task at all levels on indenture and for all milestones and events has associated with it a clear definition of the work to be done and a clear criteria for the conditions under which the work can be said to be complete. It remains for us to capture our actual performance statistics, and this can be done through our C/SCS and other metrics. Given that we have the plan and the actuals, we have the detailed information needed for a thoughtful and rapid evaluation of our performance.

The Army's AAR focuses on three important factors: (1) What was the plan? (2) What did we do? and (3) What was the outcome? The answers to these questions can form a useful narrative coordinated with the metrics collected for the activity under review. This review must be held close to the conclusion of the related work in order to access the memory of the participants about related activities. Also, some or all of the people involved may be departing the program upon completion of the activity.

The AAR, by whatever name (perhaps after the task or activity review, if the word "action" is too strident for your taste), would have to be planned as a part of an activity so that there is some time and budget to accomplish it. One way to do this is to simply make it a part of every formal review, the final element in a review. For example, at the end of the preliminary design review, the participants should answer the questions noted above candidly and report the results. This could be done with the customer in attendance or subsequent to their departure, as one chooses. If the customer has been a real team member during the work performance, they should be included in the review. This may be more than a little painful in that the review will expose things that went wrong and, in some cases, because of flawed judgment of those participating.

This is one of the reasons that some persons in a position of leadership would not want to encourage this kind of review of work. Mistakes will be made; many important mistakes are made by persons in a position of responsibility, and some people cannot stand the ego impact of the exposure of these mistakes. This is something that an organization and its members must overcome if it is serious about improving its process and performance. Interestingly, the Army allows criticism of superiors by junior officers and enlisted personnel in these meetings, and it required a lot of adjustment in the Army to tolerate this, as you can imagine. Somehow the Army's discipline system survived this change, and so can yours. Army personnel apparently came to understand that they needed the truth about how well they had done, and this truth would include exposure of problems caused by its own people as well as the enemy. The basis of change should be reality, not fantasy.

There are other expressions of this approach in the management literature as well. There is a technique where the participants quickly list the positive and negative things that happened in a meeting and discuss them. The AAR simply extends this notion to the preceding activity about which the meeting or review was called.

In addition to finding out the truth about our performance and having the truth as a basis for the corrective actions we will take to improve subsequent performance, we are likely to also find that the policy of brutally honestly reviewing our performance at the end of a task will lead to more thoughtful performance of the work during the task.

Programs would do well to enlist the help of functional department managers to participate in these AARs as well as the program reviews they should follow. This permits the functional managers to remain current on how their people, tools, and processes are doing on programs and provides program work to unburden a probable overhead limitation. The AAR could be a quality inspection function to check how well the process specification was implemented on the program tasks covered by the review.

The results of this review along with metric captured on the related tasks provide the EIT with the information they need to assess the quality of performance of the program and determine what if any process changes should be made based on the resulting lessons learned.

30.6 Two-dimensional process audit

We have gone through a lot of trouble to insist on a common enterprise process. Our rule is that all programs must implement this standard process. How do we ensure that this actually happens? First, proposals should be screened to ensure that they reflect this rule. Second, we must audit programs in accordance with some predetermined sampling rule to verify that they are applying the standard process. A two-dimensional audit process is suggested. The first dimension verifies that programs are properly implementing the generic process. The second dimension determines how well our practices compare with an accepted standard, thus benchmarking process maturity.

Figure 30.4 shows how this process can operate. We select specific practices to audit from our generic documentation, such as a particular SEM/SEMP process, that maps to particular program process steps. An auditor studies how the program process is performed using a question list prepared for this purpose. Several alternatives are possible from this audit. The auditor may conclude that the generic process is good and it is being implemented well, resulting in no need for corrective action. Detected practice problems should result in practice changes. Unfaithful implementation on the program should result in changes in implementation. Figure 30.5 shows a process diagram that accomplishes the task implementation faithfulness assessment.

It is possible that bad practices are being faithfully implemented and we need another process to detect this event. This same practice discussed above should also be compared with an accepted standard, resulting in a conclusion either that our practice compares favorably or unfavorably with the standard. If unfavorable, it should lead to an improvement in our practice.

Our audit process should collect numerical data that can be used to track performance over time. Figure 30.6 suggests a means of converting audit results into metrics that can be tracked over time. The answers to implementation assessment questions are converted into a score between

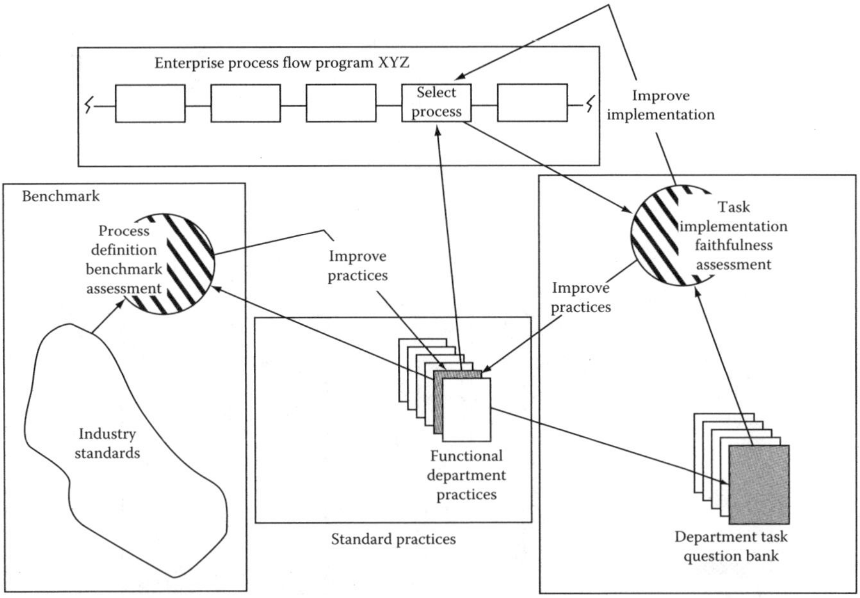

Figure 30.4 Dual-track assessment. (From Grady, J., *System Engineering Deployment*, CRC Press, Boca Raton, FL, pp. 251–267, 2000. With permission.)

0 and 100, just like an exam grade is computed. The process-benchmarking grade must be more subjectively determined in the same range, though it would be possible to define ranges for particular classes of comparisons. Figure 30.6 illustrates several extreme but possible results during audits. We would prefer to receive scores of 100,100 on all audits. Where we fail to achieve a dual high score, a need for changes in our practices, implementation, or both is suggested.

The implementation quality assessment checklists are relatively simple to construct. You begin with the standard practice that covers the activity in question and form a list of questions, much like building a quiz that tests class members' knowledge of the most important points covered in a lecture. The next question is, "Who creates these questions and when?" When you begin this process, you will have no audit questions at all. You can either form a team to generate the questions for selected activities as an audit process implementation step, delaying the beginning of audits until finished with the questions, or proceed with the audit building the questions as you go. The latter course is encouraged because it begins the process that can then be improved incrementally as you go.

Let us say that you choose to audit one process per month per program per major functional department. Each functional department must select the persons to conduct the audits and assign responsibility and schedule for these audits. The responsible person for a particular audit must first

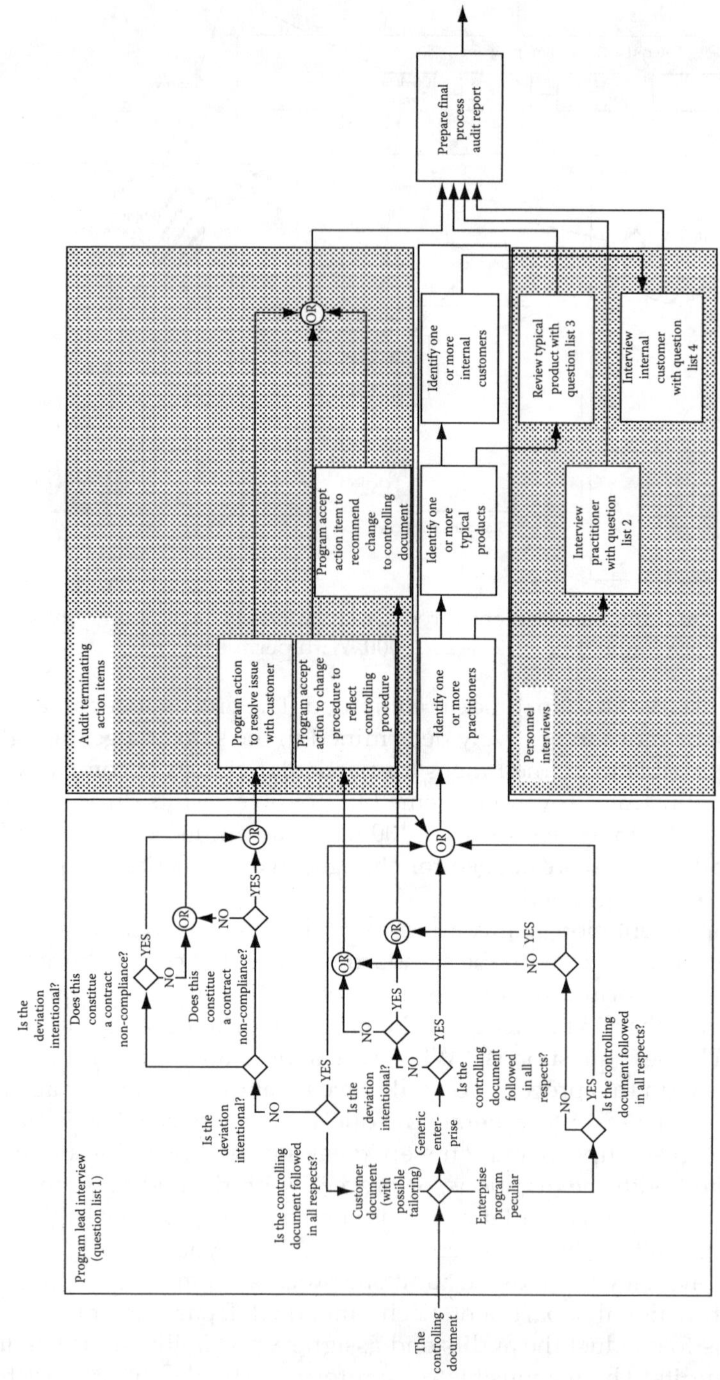

Figure 30.5 Detailed assessment decision network. (From Grady, J., *System Engineering Deployment*, CRC Press, Boca Raton, FL, pp. 251–267, 2000. With permission.)

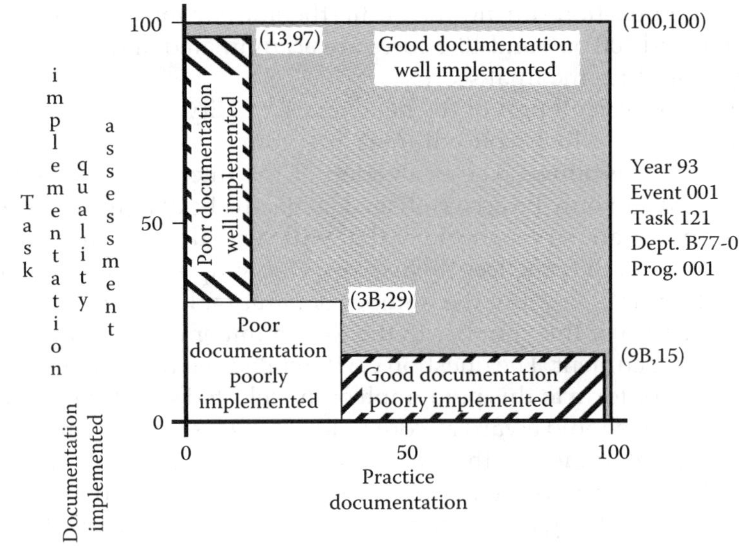

Figure 30.6 Two-dimensional assessment scoring system. (From Grady, J., *System Engineering Deployment*, CRC Press, Boca Raton, FL, pp. 251–267, 2000. With permission.)

determine if a set of questions exists for the audit. Initially, a set of questions will not exist. If there are none, the auditor must first make the list as discussed above and gain the functional management's approval. This list becomes the baseline for subsequent audits with adjustments over time.

The suggested list of questions has four parts, as noted in Figure 30.5. The first part is defined generically in Figure 30.5 and is answered by the program lead person for that function. If the problems encountered involve contract noncompliance, then the audit should be curtailed and that compliance issue resolved. If no contract compliance issues are detected, the auditor identifies a practitioner on the program for that activity and applies question list 2 focused on task performance in accordance with the controlling procedure. The practitioner identifies a typical product generated on the program in accordance with the practice, and the auditor evaluates the product against question list 3. Finally, the practitioner identifies an internal or external customer for the activity, and the auditor applies question list 4 containing of perhaps only three questions: "Does the product conform to your requirements?", "Are you satisfied with product quality and accuracy?", and "Was it made available to you in a timely way?" Other questions could be added.

Scoring for these questions can be easily done. The questions may all be weighted the same or differently depending on their content and management attitudes of relative importance. The total score is scaled for 100.

For any one question during the audit, the auditor must define a grade between 0 and 100 or 0 and 10. Scores are accumulated and influenced by scaling to produce the final score.

The most difficult part of the benchmark quality audit is to select the standard against which you will measure your capability for a particular functional discipline. The evaluation of the internal practice against this standard can only be accomplished subjectively, barring an extensive analysis and creation of a checklist that will very likely be more difficult to maintain than the practices themselves. One simple way to place this in numerical terms is to count the number of paragraphs where you comply and form a ratio of this number to the total number of paragraphs in the benchmark document. The question can be stated relatively simply, "Does our practice conform to this paragraph of the selected standard?" If it does in all cases, you should award a score of 100. If it does not, you must assign some numerical grade less than 100. You could do this based on a simple paragraph count as noted above or adjust the ratio as a function of the difficulty in closing the gap through changes in the in-house practice. This factor would not only involve the written practice but possible associated social issues in our organization. The score should be accompanied by an explanation.

Simple computer database systems can be built to capture the results of these audits, provide administrative support, and perform the arithmetic to generate the final scores. The author knew of no readily available products on the market to support such audits at the time this book was written.

Given that we have accumulated information about our performance through this audit process, we must have the machinery in place to do something about the results derived from it. This machinery must include methods to change our practices where needed and enforce work performance on programs in accordance with our standards where deviations are not in our best interest.

30.7 External maturity models

Several organizations have developed models of excellence against which one can compare the features and capabilities of their own enterprise and derive an indication of how your organization is performing relative to a well-recognized standard. Some of these models focus tightly on system engineering, while others are software oriented or broadly tuned to the whole enterprise. In all cases, questions have to be answered, and the answers or products made available in response to questions are evaluated by someone to determine organizational qualifications for a particular rating.

30.7.1 Malcolm Baldrige National Quality Award

Malcolm Baldrige was the Secretary of Commerce during the Reagan presidency. On August 20, 1987, the 100th Congress passed the Malcolm Baldrige National Quality Improvement Act of 1987, establishing the machinery to identify and reward American firms characterized by quality. This action was motivated by the results of congressional studies, indicating that the leadership of the United States in product and process quality had been challenged strongly by foreign competition with the potential for very serious consequences in our economic viability in the long run.

The act established a board of examiners to evaluate the performance of candidates for award and a board of overseers to define the standards for award, to review the evidence of company qualifications for award, and make recommendations for award each year to the director of the National Bureau of Standards within the Commerce Department. The criteria for selection in the original act was as follows:

1. The company must apply for the award in writing.
2. The company must permit a rigorous evaluation of the way in which its business and other operations have contributed to improvements in the quality of goods and services.
3. The company must meet the requirements and specifications identified by the board of overseers and the director of the National Bureau of Standards.

Figure 30.7 shows the evaluation process applied by the examiners. The categories, eight of which are indicated in the figure, can be changed by the board of overseers as can the details supporting each category. The examiners are given a free hand to evaluate materials supplied by the firm and to visit the firm for more detailed evaluation of these materials if desired. There is no cost for the evaluation that must be paid to the government, but the company attempting to win this award most often ends up spending a tremendous amount of money to attain it through improvements it has to make to be a contender.

The management literature contains papers on the sinking spell winners sometimes have right after winning the award, linking it to the same effect that makes it difficult for super bowl winners to repeat the following year. The intensity, concentration, and focus that one has to sustain have a draining effect requiring a period of psychological rest. If you have not been on a super bowl winning team or won the Malcolm Baldrige award, you can perhaps identify with this feeling because of your experiences on a recent proposal effort.

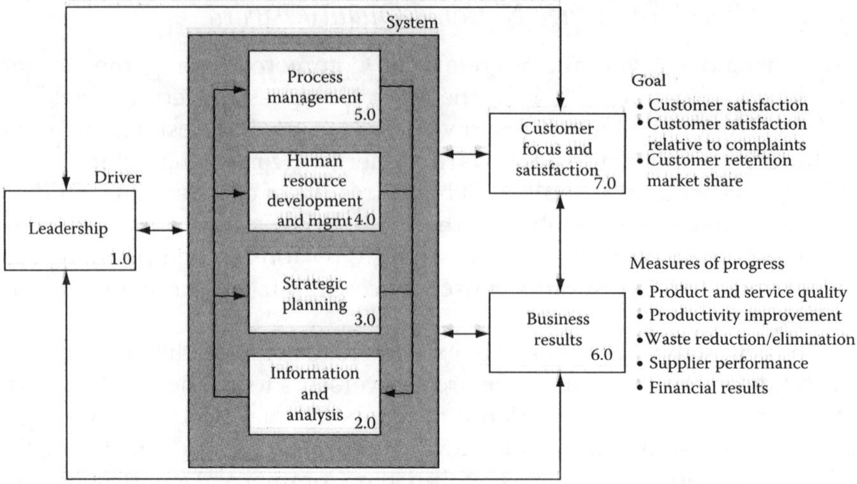

Figure 30.7 Malcolm Baldrige Award evaluation pattern. (From Grady, J., *System Engineering Deployment*, CRC Press, Boca Raton, FL, pp. 251–267, 2000. With permission.)

30.7.2 ISO 9001 certification

The International Standards Organization has crafted a standard for the development of products numbered ISO 9001, titled *Quality systems—Model for quality assurance in design/development, production, installation and servicing*, within the quality standards series numbered 9000 that has been applied by many organizations. Some have gone to the expense of becoming certified in its application, while others have claimed conformance through internal audit and adjustment of their process. There is nothing in this standard to which any reader of this book would take exception. All of the content makes good sense. The authors of the standard have distilled the basis for excellent enterprise performance down to its fundamentals. In reading a standard like this, and it is only six and a half pages long, one cannot help wonder why companies have such a difficult time implementing these simple and sound practices and why any company would want to do anything else.

Any company can purchase a copy of the ISO quality standards and audit their own performance against these standards, changing practices, culture, and implementation to comply. Alternatively, it is possible to hire a certification firm to accomplish the audit, report the results, and help you attain certification status. One of the fundamental steps in the ISO certification process is that one has a written process and that you actually follow it. This is quite simple, but very few companies comply with this requirement, especially the part about following their own internal practices.

While helping one company with some system engineering work where the company had become ISO 9000 certified not too long before, the author asked to see their system engineering manual to make sure he was not creating anything in conflict with their written practices. It turned out that the company did not have one. Their reasoning was that since they did not have a system engineering organization, it would not be proper for them to have a manual for an organization that did not exist. The ISO certification firm apparently agreed with them. The author agrees that it is not necessary to have a functional organization to perform system engineering work well on programs in that system engineering is more of a process than a professional discipline or department name, but it would be of help to write down what it is and how one does it whether it is required by a standards body or not.

30.7.3 Carnegie Mellon software capability maturity model

Carnegie Mellon University was approached by the DoD several years ago to establish a Software Engineering Institute (SEI) to develop a means of determining the maturity of an enterprise software development capability. The DoD had become very concerned about the difficulties they were having in maintaining software for fielded systems and anticipated these problems getting worse as their software dependency grew. The SEI developed an organized process for satisfying DoD concerns involving a means of awarding a numerical score in the integer range 1 through 5 inclusive, 5 being the best. This is called a capability maturity model (CMM).

This process has worked out well with a few detractors. In general, it has been well received and most defense contractors have worked very hard to move their software development capability at least to level 3 as encouraged by the DoD. Table 30.1 defines the five levels in this system. The certification process involves answering a series of questions and producing evidence of compliance. The questions used in evaluating maturity are based on a structure of maturity levels (1 through 5), supporting key process areas (KPAs) and key practices that specify key indicators, as illustrated in Figure 30.8 from SEI CMM literature.

In 1997, one could count the number of firms in the world rated 5 under the SEI CMM software model on one hand. Of course, every firm developing software had not been evaluated by SEI, but the percentage of those evaluated level 5 was very low. Despite the simplicity of the requirements and their great good sense, it is very difficult to achieve something close to perfection in one's process and its implementation. This is true in software, in system engineering, and in industry in general, in all fields. In more recent years, the number of organizations rated 5 has increased based on improved standards and the good offices of SEI in correlating good results with a high rating.

Table 30.1 SEI Software Capability Maturity Model

Level	Name	Keywords	Description
Level 5	Optimizing	Continuous improvement	The organization has quantitative feedback systems in place to identify weaknesses and strengthen them proactively. Project teams analyze defects to determine their causes; software processes are evaluated and updated to prevent types of defects from recurring.
Level 4	Managed	Predictable	Detailed software process and quality metrics establish the quantitative evaluation foundation. Meaningful variations in process performance can be distinguished from random noise, and trends in process and product qualities can be predicted.
Level 3	Defined	Standard and consistent	Processes for management and engineering are documented, standardized, and integrated into a standard software process for the organization. All projects use an approved, tailored version of the organization's standard software process for developing software.
Level 2	Repeatable	Intuitive	Basic project management processes are established to track cost, schedule, and functionality. Planning and managing new products is based on experience with similar projects.
Level 1	Initial	Ad hoc and chaotic	Few processes are defined, and success depends more on individual heroic efforts than on following a process and using synergistic team effort.

Source: Grady, J., *System Engineering Deployment*, CRC Press, Boca Raton, FL, pp. 251–267, 2000. With permission.

30.7.4 Carnegie Mellon system engineering CMM

Subsequent to the development of the software CMM, Carnegie Mellon extended the model to system engineering using similar techniques.

30.7.5 INCOSE system engineering capability maturity model

The International Council On Systems Engineering (INCOSE) created a CMM as well. Its structure, illustrated in Figure 30.9, is composed of four

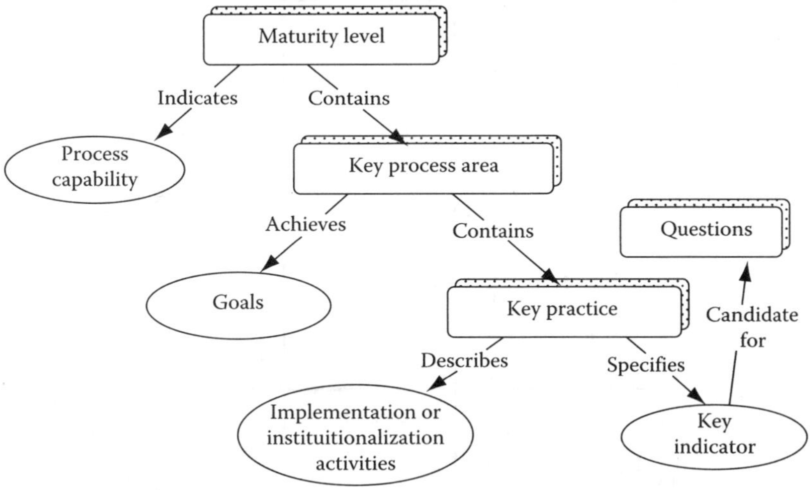

Figure 30.8 SEI software CMM. (From Grady, J., *System Engineering Deployment*, CRC Press, Boca Raton, FL, pp. 251–267, 2000. With permission.)

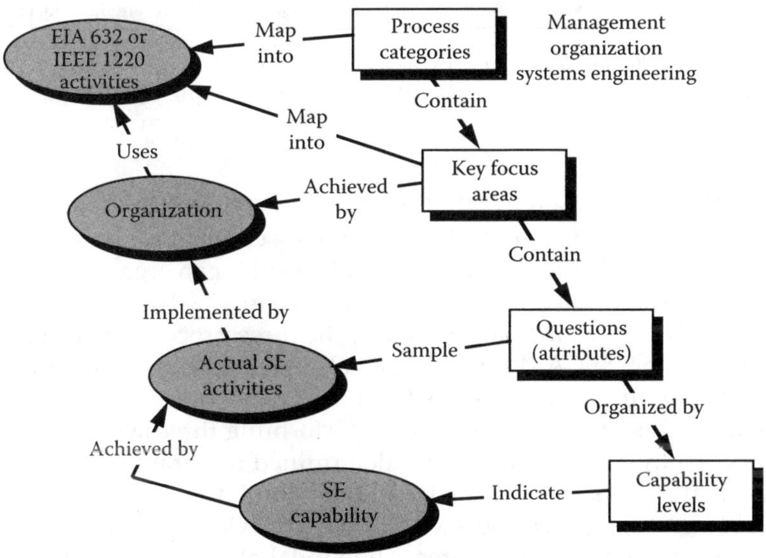

Figure 30.9 INCOSE SECAM interim model. (From Grady, J., *System Engineering Deployment*, CRC Press, Boca Raton, FL, pp. 251–267, 2000. With permission.)

model structures depicted by rectangles linked to four corresponding real-world system engineering elements depicted as ovals, joined by directed line segments, indicating relationships between these entities. The whole system engineering process is partitioned into the three process categories indicated in Figure 30.9, and each of these consists of several KPAs.

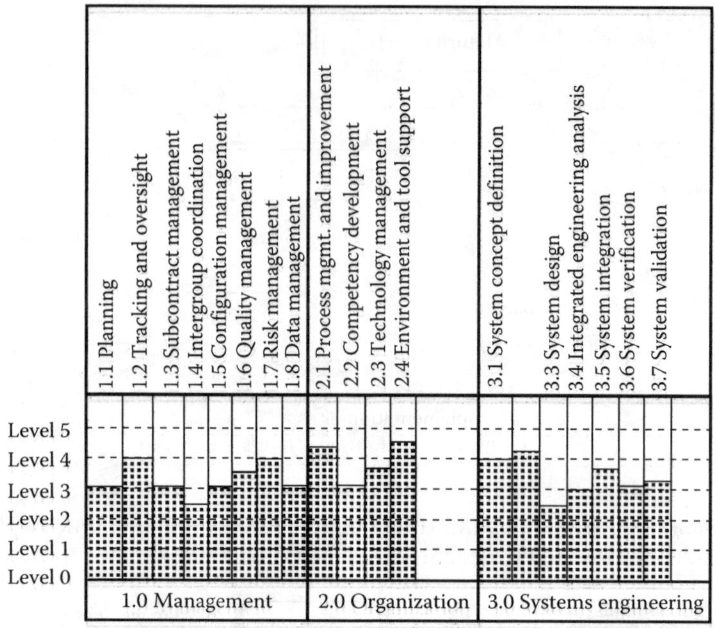

Figure 30.10 INCOSE SECAM scoring profile example. (From Grady, J., *System Engineering Deployment*, CRC Press, Boca Raton, FL, pp. 251–267, 2000. With permission.)

Each KPA has associated with it a series of questions, organized by six assessment levels, that are the basis for the assessment process.

Figure 30.10 illustrates a sample set of scores for an organization based on the assessment of their process in accordance with the questions in each of the KPAs subordinate to each of the three process categories. Note that the INCOSE SECAM carries six levels rather than the five used in SEI CMM. One of the author's clients for training determined through a self-audit using this model that they were performing three system engineering tasks badly. In each case, it was determined that training was needed to elevate the skills and knowledge of their employees.

Table 30.2 identifies the six INCOSE SECAM capability levels and corresponding process attributes. The model also includes non-process attributes not included in Table 30.2. Level 5 is the highest or best capability.

30.7.6 The EIA 731 model

The Electronic Industries Alliance (EIA) has published its own system engineering CMM document. Unfortunately, in the author's view, 731 is

Table 30.2 INCOSE SECAM Capability Levels

Level number	Capability term	Process attributes
5	Optimizing	Program process effectiveness goals are established upon business goals Continuous process improvement of program processes Continuous process improvement standards
4	Measured	Metrics derived from process data Quantitative understanding of program processes Ability to predict performance Program-process-induced defects identified Program processes improved
3	Defined	Processes are defined by organizational standards Standards are tailored and used Tailoring is reviewed and approved Program processes data is collected Customer feedback is obtained
2	Managed	Policies define need for activities Activities are planned, tracked, and verified Work products are reviewed for adequacy Corrective actions are taken Work products are controlled
1	Performed	Activities done informally Non-rigorous plans and tracking Dependency on heroes Work products are in evidence General recognition of need for activity
0	Initial	General failure to perform activities No easily identifiable work products No proof something was accomplished

Source: Grady, J., *System Engineering Deployment*, CRC Press, Boca Raton, FL, pp. 251–267, 2000. With permission.

not in any special way coordinated with EIA 632 that offers 33 requirements for a sound system engineering process. It would be a wonderful parallel for 731 to provide a verification process for a system engineering process created in response to the requirements of 632, but these standards were not built with this perspective. Table 30.3 lists the 731 criteria under three major headings referred to as TC 1, 2, and 3. Under each of these major headings, one finds specific functional areas (FA) listed.

Table 30.3 EIA 731 System Engineering CMM Criteria

EIA 731 Element	Title	Description
TC 1.0	System engineering technical category	
FA 1.1	Define stakeholders and system-level requirements	A validated set of system-level requirements that reflect stakeholder needs and expectations is defined
FA 1.2	Define technical problem	Develop and evolve the functional architecture, requirements, management, establishment of lower-level requirements baselines, and generation of requirement specifications
FA 1.3	Define solutions	A process of transforming system-level and lower-level requirements into documentation of components and interfaces, such that their implementation and integration would satisfy the requirements
FA 1.4	Assess and select	A process of making good decisions by (1) identifying issues that require application of decision theory techniques in order to accomplish timely technical decision making, (2) selecting an appropriate decision-making technique for technical issues, (3) involving the right mix of technical disciplines, and (4) following the decision-making process during the system life cycle
FA 1.5	Integrate system	Ensuring that the system solution elements function as a whole
FA 1.6	Verify system	Ensure system stakeholders that the incremental and final work products conform to the requirements
FA 1.7	Validate system	Apply an end-to-end approach to ensure that the completed, integrated system will operate as needed in the environment for which it is intended
TC 2.0	Systems engineering management category	
FA 2.1	Plan and organize	Identification of needs and constraints at the program level
FA 2.2	Monitor and control	Provides visibility into program progress and risks, and detection of variances needing corrective action

FA 2.3	Integrate disciplines	Select and blend the interdisciplinary people necessary for system development such that they jointly and effectively work together
FA 2.4	Coordinate with suppliers	Address the needs of the organization and programs to effectively select and manage those portions of product work that are conducted by outside suppliers
FA 2.5	Manage risk	Identify, assess, monitor, and mitigate risks in the performance, cost, and schedule activities of a program throughout the life cycle
FA 2.6	Manage data	Administrative control of program data, both deliverable and non-deliverable
FA 2.7	Manage configuration	Identification of configurations, change control, status accounting, and auditing of the product and product elements that define the product
FA 2.8	Ensure quality	Address the quality of the system and the quality of the processes being used to create the system
TC 3.0	Systems engineering environment	
FA 3.1	Define and improve the systems engineering process	Involves those activities needed to establish, maintain, and improve processes required to accomplish systems engineering
FA 3.2	Manage competency	Establish and nurture a learning environment that results in the needed evolution of staff knowledge and skills over time, and establish and maintain a capability to deliver the required knowledge to programs via training or other sources when there is a shortfall in available resident staff knowledge
FA 3.3	Manage technology	Identify technologies applied to current products and processes; monitor the progression of currently used technologies through their life cycle; identify select, evaluate, and invest in new technologies and incorporate the appropriate technologies into the organization's products and processes to achieve competitive advantage
FA 3.4	Manage engineering	Provide the environment needed to develop the product and perform the support environment process

Source: Grady, J., *System Engineering Deployment*, CRC Press, Boca Raton, FL, pp. 251–267, 2000. With permission.

30.7.7 Integrated capability maturity model

As the reader can observe from an incomplete sampling of maturity modeling standards in this chapter, management in many enterprises became concerned that it might be costly beyond their ability to satisfy potential customer demands for their score under these several criteria. In response to this concern, Carnegie Mellon teamed with other organizations to develop a capability maturity model integrated (CMMI) that is intended to provide an enterprise-wide way of assessing maturity. It does not yet stretch across the whole enterprise, but it does deal with management, software, system engineering, and concurrent teams. An enterprise with all of these working well would be a powerhouse by anyone's measure.

chapter thirty-one

Motivation of the systems approach*

31.1 The logic is not everywhere accepted

Throughout the prior chapters of this book and in so many other books on system engineering written by many authors, the logic of the systems approach to the solution of complex problems rings true to the faithful, but there are many people who do not accept the notion that the systems approach well applied will result in reduced cost, better product performance, and on-time delivery relative to the results obtained by an ad hoc approach. The available statistics quoted in many books on program management, system engineering, and system acquisition seem to support the utility of careful planning, definition of the problem as a prerequisite to solution, active integration across cross-organizational interfaces, an energetic search for potential problems (risks) followed by active reduction in their potential for occurrence and degree of damage to the program, and clear proof of requirements compliance as ways to develop better systems.

Clearly, spending money well in the early program phases to mitigate risks and prevent them from appearing later in the program has the effect of solving problems in their least expensive form when the investment is minimized. Solving problems after the engineering drawings, test plans and procedures, manufacturing and quality planning, customer training materials, and technical data are complete, simply costs more than solving problems when a simple conversation in the hallway can preclude the problem from ever materializing. For example, a 15 min conversation between four engineers costing the company $50/hour results in a total of (4)(1/4)($50) = $50. This same problem could cost $100,000 or more to correct in the late program phases and have a significantly adverse impact on the delivery schedule. The logic of these ideas still fails to penetrate the defense systems of some of the people who work on these programs.

If it is true as claimed by psychologists that humans make an error in each 1000 operations and that the work involved in solving a very complex problem involves millions of operations, then we should expect that

* The material in this chapter is from Grady, J., *System Engineering Deployment*, CRC Press, Boca Raton, FL, Chapter 18, 2000. Used with permission.

we will make thousands of errors in solving the problem. Some of these errors may easily be overcome or never even come to the surface as a problem, but others discovered late may force us into very costly corrective actions. We would therefore be wise to have an active mechanism to detect these errors as early as possible and avoid others from happening. This is what the systems approach is intended to do by providing a structure and encouraging risk identification and mitigation.

It is very unfortunate that many of the people who harbor a hostile feeling for the systems approach are in positions to cause deployment of the systems approach to be stillborn or otherwise make it a success or a failure in that they are program managers. One of the key reasons companies are not successful in the application of the systems approach is a failure of program managers to demand that it be applied and be applied well. One other responsible position within an organization where we will often find resistance and sometimes open hostility to the systems approach is, of all places, the vice president or director of the engineering department, a position commonly held by a person who rose through the ranks of functional management in design or analysis departments of engineering. A third string of resistance is often found down through the directors and managers of the design and analysis domain engineering departments in the functional organization. It is seldom the case that any active resistance comes from the working level.

31.2 The causes of poor system engineering performance on programs

System engineering is poorly done on programs for reasons that can be grouped into supply, demand, and deployment causes. Either, the enterprise has a questionable system development process, there is no demand for an effective process (insufficient budget provided to do it well, for example), or the machinery is not in place to transfer a capability for doing the work well into a program that wants to apply it well.

31.2.1 Supply problems

In order to perform the system development work well, the people in the organization must have a set of skills and knowledge that is fairly well defined and agreed upon in the many books and standards on system engineering. Unlike the past, there are many books on this process now available. There is a growing list of universities that provide instructions in system engineering either in the form of an elective course that can be taken in an engineering degree program or a complete system engineering certificate. Three campuses of the University of California (San Diego,

Irvine, and Riverside) have certificate programs through their extension. The University of Southern California (USC), the George Mason University, Universities of Washington, and the University of Arizona have been teaching the systems approach for years. There are many more across the country that offer the student some insight into this process. There are many consultants available in this field stimulated by the premature departure of many seasoned employees before their time from industry through downsizing. This is a vanishing resource, but it will be available through the waning years of the final decade of the twentieth century and perhaps a little longer.

The point is that there is a way for an organization to improve the skills and knowledge of its personnel as well as individual professional system engineers to independently improve their understanding of the process even if their company is not so inclined. Nonetheless, there is a great need for system engineering education in industry, because the process is not self-renewing in most companies. There probably has not been a written process description nor a consciously provided way for the people in the organization to acquire the knowledge needed to excel in implementing it. Seasoned system engineers depart the company for retirement with no way to transfer their acquired knowledge to those beginning their careers. While there are some universities that provide some form of educational opportunities in system engineering, it is nothing like the availability of mechanical, electrical, or chemical engineering educational opportunities. Also, these fields have been practiced well in industry in fairly standardized ways long enough that the transference of knowledge within the company in these fields is relatively easy.

However, the supply of effective system engineering capability is not in abundance in industry despite the increasing availability of the resources necessary to make it happen. This condition is probably improving since 1997, but has a long way to go to become a renewable resource within most companies.

The good news is that the engineers at the working level in the system engineering field seldom offer any resistance to learning and applying the systems approach to the best of their ability. The supply-side problem is caused more by poor management in functional system engineering departments or the absence of such departments charged with maintaining the company's system engineering knowledge base than by working-level staff working purposely to thwart the development of an effective systems capability. There are also a refreshing number of companies that are working to prepare system engineering practices manuals and find ways to effectively deploy these practices, and the number is increasing, but many others have not discovered the simple secrets booming out of ISO 9001 and system engineering capability assessment models.

31.2.2 Demand problems

The author believes that the single-biggest problem in system engineering is a failure of programs to demand excellence in this field. This failure can be traced to the desk of the program manager who, either through ignorance of the benefits to be derived or in spite of them, continues to insist on the application of the ad hoc approach. Believers in the systems approach just have to shake their head in disbelief that anyone could be so stupid as not to understand that the systems approach provides programs with obvious benefits that result in programs that are easier to manage while satisfying all of the customer's requirements on cost and on schedule. But, there are reasons why program managers refuse to accept this logic.

Many people who seek out program management and make it to that position are creative people and good ad hoc thinkers. Balancing on the edge of the cliff can be such a great challenge for these people that they may secretly not want to replace this opportunity with good planning that prevents these situations from maturing. Many of these people believe that high risk is a normal part of management rather than risk management (and mitigation) being a component of good management. The author knows of no way to retrain these people for service in later phases of programs. They may be very useful in earlier program phases except that they will have already buried many mistakes in the program by the time they hand off to a more stable individual to take the program through manufacturing and engineering development. The characteristics that drive these people to become program managers are the same ones that lead to their failures to apply sound system engineering practices.

The only way that the author would encourage this kind of person be placed in a program manager role is with a system thinker as his or her deputy responsible for the day-to-day leadership role. One risky alternative that could be tried is to convince the risk-loving program manager of the amount of perverse enjoyment they might gain from applying the systems approach the first time. There may still be some people in companies who lead programs from a position of ignorance about the benefits of the systems approach. The recommendation is that these people be identified and an attempt made to educate them. If this attempt is successful, and they have other qualities to encourage their continued employment as a program manager, then the problem is solved!

There is another class of program managers who know about claims for the improved management environment on programs applying the systems approach who have not yet experienced it. They continue to encourage the ad hoc approach because that is where their experience is. Despite the knowledge that their current methods will continue to result in problems, they do not have the self-confidence to apply the systems approach on their first program. These people need some hand-holding from the company system engineering manager to build their confidence in the process. One of the things that encourages this attitude is that once money is spent, it cannot be restored. Thus, program managers know that if they withhold money early in the program, it will be available to spend later, and if they spend it early doing the system work when it can be effective in reducing later risks, it will not be available later if the early work does not have the desired effect in later program phases.

One of the great truths of system engineering is that one must accept the notion of delayed gratification. Most often, the benefit derived from spending program money for system engineering work is not immediate; rather, it is delayed by weeks, months, or years. Every parent understands how difficult it is to convince a teenage son or daughter of the wisdom of foregoing a social event in favor of doing the assigned homework based on delayed gratification. Many people in earlier generations were forced to respect this discipline because of the depression and World War II interrupting their lives. But, with the United States at the pinnacle of world power and the good times continuing to roll as the fourth-place digit changes on our calendar, it is hard for people to relate to delaying gratification by choice. This leads to an attitude of encouraging only those things that lead to immediate benefit, and system engineering simply does not work this way.

31.2.3 Deployment problems

This book will, hopefully, provide a resource for those who find themselves in an organization with an acceptable process defined in the functional organizations and qualified people to accomplish the work on programs that demand good performance in this field, but they do not seem to be able to deploy an effective capability into these programs. The earlier chapters of this book are laid out to support the use of the book in a prescriptive way, building each layer of the deployment capability until complete. The reader is encouraged to apply it in this fashion. But, the enterprise must have an effective functional system engineering capability, programs that demand it, and people ready to do the work. Generally, this is enough to encourage effective deployment, but not always. It takes some successful experience to be able to do it well, and, if this has not happened yet, it may be difficult to do it well. As a minimum, you need to use these early programs as a source of critical feedback for ways to improve the process and your organization's ability to deploy it.

Many companies that have not had a proud and effective history of system engineering have established a system engineering council, by this or other name, charged with bringing the company to a point where it is practicing the process well. This group writes a system engineering standard, monitors and studies program performance in this field, selects training needed to bring up staff knowledge, and researches computer tools useful in performing the work. In companies with an existing system engineering organization, it can act as this agent, of course. The author's preferred approach in this matter is the establishment of an enterprise integration team (EIT) to act as this catalyst, as discussed in earlier chapters.

31.3 Changing manager attitudes

Those in positions of responsibility for system engineering in companies should make an effort to find out the attitudes presently among managers in program and functional roles within their company as a basis for mapping out a bottom-up strategy for improving the company's systems capability. Your company will not improve in this area without the active support of these people. These people will not necessarily have a positive outlook just because you think it makes sense for that to be the case. While the logic of the case for the systems approach will carry the day with some people, many others will be guided by their own experiences with the systems approach. Since it has seldom been applied well, that experience may not be good. This is the fundamental problem in improving the systems approach in companies. You need to build on past successes, but if you have not had any successes, how do you get this process moving in the first place? This is the typical bootstrap problem.

Persons interested in deploying an effective systems approach on programs in a company can make the mistake of looking for the same motivation among all those in opposition, and this is not the case. So, we must apply a broadband approach with different messages for different people. Two specific targets of our interest should be program managers and functional domain managers.

31.3.1 Functional domain managers

It sometimes happens in organizations that the functional domain managers (structural design, electrical engineering, etc.) come to feel a real dislike for system engineers and their organization. There may be many reasons for this, but let us trace one of them. The author has talked to many managers about this since leaving captive employment, and found a good deal of similarity in their responses. A common response is one of dislike for system engineers who failed to do a good job of requirements analysis (often referred to as requirements flowdown by the managers) leading to poor design solutions discovered after the fact in qualification testing, manufacturing, or subsequent customer use of the product. This situation is made all the more memorable for the domain manager when the system engineer is then rewarded for coordinating the solution to the big problem that the domain manager believes he or she caused in the first place.

While the author has talked to many domain managers subsequent to leaving captive employment, the fact is that while employed as a system engineer, the author seldom went out of his way to find out what domain managers thought about system engineers or the way the process was applied in companies within which he worked. This was an opportunity lost that you should not repeat. You need to collect information on current attitudes both as a way of finding out how difficult the conversion process is going to be and to provide an opportunity for you to spread a positive message to be discussed below.

A survey involving forms to fill out in isolation is not recommended. You need to meet one to one with the domain managers to suggest the depth of your concern, which cannot be faked easily, so do not begin this until you have managed to elevate your own motives to what is honestly best for the enterprise. Prior to these discussions, you do need to map out the questions that will provide the information you need and to prepare the supportive message you should deliver in the process of inquiring about the manager's attitudes and experiences.

The person making these contacts should be from as high up in the functional system engineering tree as possible, such as the director, the top staff, or department managers. Of course, if the organization reflects the author's belief in lean functional organizations, there may not be very many people who fit the bill. No matter, even if the only person available to do this work is the director of system engineering, it should be recognized that there are a few things more important to him or her than finding out how the current capability is perceived as a basis for improvements in perception or reality, as appropriate.

Ideally, the person making the survey has a good personal and professional relationship with the domain managers. If this is not the case, regardless of what else is done, improving interdepartmental management relationships should be given some effort. In this regard, the system engineering manager will find that all of his or her peers are vitally interested in what they do and will most often be delighted to speak about their job and the importance of it indefinitely. So, one approach to building a good relationship is to listen to these explanations. Another way to kick off an improvement is to ask these managers to participate in a no-cost lunchtime series of lectures for members of the system engineering department (and anyone else wishing to attend) on the different engineering departments and what they do. The author did this at the General Dynamics Space Systems Division, and it worked so well that engineering chiefs would sometimes attend from other departments, and in one case an argument erupted between the speaker and another manager in attendance who thought his department did that particular work.

Many functional design department group managers have concluded that the system engineering process progressively damages their creative designers because of the brain-deadening appeal to rigor and discipline. These are characteristics of a misapplied system engineering process. The reality is that a program needs a variable environment with respect to creativity and discipline, which can be thought of as on opposite extremes of a continuum. In early program phases, one needs to appeal to the creativity of the staff to come up with good alternative approaches. In later program phases, as the investment in past decisions becomes great, the program needs to transfer into rock-hard discipline, making it purposefully hard to change the design. It is very hard to master the skills and techniques to be equally effective as a system engineer in both of these situations. Many system engineers came to this field through early or late program experiences over many years and have concluded that there is only a single balance point of worth between creativity and order. When such a person is first assigned to a program on the opposite end of the development spectrum, they will generally run into trouble because their model does not fit the need. Either they are being too open to change in later phases or too closed to alternatives in the early phases.

The reader may feel that this is not a significant problem, but as evidence one should consider what several communist governments did to their people and country over the period 1920 through 1990. The practice of communism in the twentieth century is not a bad model of system engineering run amuck. Engineers who have experienced this environment while working as designers become very hostile to the process as managers.

The best way to cause design and analysis managers so afflicted to reconsider their attitudes toward system engineering is for the system engineering manager to convey to them the system engineering policy and belief that the system development process is based on providing the creative genius of design engineers as much solution space as possible and, otherwise, facilitating the communication of specialized knowledge between engineers to support that genius in the development and selection of preferred design concepts. This is difficult for some system engineers to do because they have incorrectly come to believe that all programs revolve around them. It is the creative genius of the design engineer from whom the design flows. Take that away and everyone else becomes frozen in their tracks. Take away the system engineer and life goes on, not efficiently perhaps, but it goes on.

31.3.2 *Program managers*

We have already discussed the special problems of lack of demand for effective system engineering traceable to the program manager earlier.

If this is the only weakness in the chain of acceptance of the systems approach, it may be helpful simply to draft company policy signed by the chief executive that the systems approach will be applied on all proposals and programs in accordance with a released written standard. It is much harder for a program manager to ignore this kind of direction than when the policy has not been written down.

Another approach that can be helpful as well is to work with the estimating department, which may not have good skills and experience in developing system engineering estimates. It often happens that the estimate turned in by the system engineering organization does not survive a proposal team scrub because of lack of understanding of the utility of the services offered. It is much easier to convince the proposal manager of this where the estimating department is supportive than when they do not have the information they need to advise the manager on the real benefits of delayed gratification.

If the company has a department of program managers, from which all program managers come, the person who manages this department needs to be approached first to find out whether they have a supporting, detracting, or flexible attitude about system engineering and to make an effort to change this attitude in the latter two cases. If this person is a supporter, it may be possible to establish policies encouraging the demand for sound system engineering practices on programs and a staff that knows how to deliver it.

31.3.3 Managers within programs

In addition to the overall program managers, we will also find on programs several managers with system responsibilities dependent upon how the program is organized and run. The program may be organized into functional departments (functional department subsets) responsible for technical development and include WBS managers to coordinate the product axes of the program in terms of cost and schedule. Alternatively, and more ideally, the program may employ cross-functional teams where the team managers have technical, cost, and schedule responsibilities. All of these managers within programs should have a positive attitude toward the systems approach, but this is often not the case.

One manager at McDonnell Douglas referred to those who were not supportive of the process as misdirectors. This is not to point a critical finger at this company that was doing a lot to improve its practice of system engineering prior to purchase by Boeing, rather to recognize a reality also poignantly made clear in an article by U.S. Air Force Captain Tom Schorsch (1996). This article added a null and three negative maturity identifiers to the positive 1 through 5 identified in the Carnegie Mellon

SEI software maturity model. Captain Schorsch was offering this model extension with tongue in cheek to a certain extent, but those familiar with system development programs in industry can relate to the levels he identified. Table 31.1 modifies both the software CMM level descriptions (1 through 5) as well as Captain Schorsch's additions (0 through –3) to translate them from a software-specific to a system development generic meaning. Some of these words might be shocking, but this may be what is needed to cause some persons to reconsider their attitude toward effective development practices.

31.4 Customer demand for improvement

Failure to apply practices in the best interest of the organization has a lot in common with an individual's failure to change their habits to eliminate cigarette smoking, excessive eating, or excessive intake of alcohol. You know logically that these things have an adverse impact on your well-being and it is very simple to stop the bad practice. It is not easy to do so, however. It is often the case with alcohol and illegal drugs that the person must hit bottom before they can break out of the destructive habit pattern. They must be near death's door. So it is with some companies and their managers with regard to the poor practice of system engineering. One will hold onto the ad hoc approach almost to the company's grave and sometimes all the way to the end of the company. Perhaps we should form an organization for recovering ad hoc managers completely with a 12-step process of renewal.

In aerospace industry, it is often the government customer that provides the incentive for a company to cease and desist from its self-destructive behavior. Many companies have been told to either improve their system engineering capability or have a large contract withdrawn. Following the replacement of a few clueless misdirectors, hiring of some qualified system engineers, training of the staff, establishment of a system engineering department to sustain improvement in this field, and preparation of written practices, these kinds of companies have some chance of survival.

In the past, it has been fairly easy for a company to hide the fact of their system engineering incompetence for some time in that the fruits of their failures ripen late in a program at a time when the customer's need for the product may be sufficiently desperate even with serious performance, cost, and schedule problems to encourage them to take whatever they can get. It is relatively easy to describe an effective systems capability in a proposal because there is a lot written about it, but it is very difficult to deploy an effective capability even for a contractor with past experience and success. The Department of Defense (DoD) is rapidly becoming very skilled

Table 31.1 System CIMM

Level	Name	Keywords	Description
Level 5	Optimizing	Continuous improvement	The organization has quantitative feedback systems in place to identify weaknesses and strengthen them proactively. Project teams analyze defects to determine their causes; processes are evaluated and updated to prevent types of defects from recurring.
Level 4	Managed	Predictable	Detailed development processes and product quality metrics establish the quantitative evaluation foundation. Meaningful variations in process performance can be distinguished from random noise, and trends in process and product qualities can be predicted.
Level 3	Defined	Standard and consistent	Processes for management and engineering are documented, standardized, and integrated into a standard development process for the organization. All projects use an approved, tailored version of the organization's standard process for developing systems.
Level 2	Repeatable	Intuitive	Basic project management processes are established to track cost, schedule, and functionality. Planning and managing new products is based on experience with similar projects.

(*continued*)

Table 31.1 (continued) System CIMM

Level	Name	Keywords	Description
Level 1	Initial	Ad hoc and chaotic	Few processes are defined, and success depends more on individual heroic efforts than on following a process and using synergistic team effort.
Level 0	Negligent	Indifference	Failure to allow successful development process to succeed. All problems are perceived to be technical problems. Managerial and quality assurance activities are deemed to be overhead and superfluous to the task of system development. Reliance on silver pellets.
Level –1	Obstructive	Counterproductive	Counterproductive processes are imposed. Processes are rigidly defined, and adherence to the form is stressed. Ritualistic ceremonies abound. Collective management precludes assigning responsibility.
Level –2	Contemptuous	Arrogance	Disregard for good development engineering institutionalized. Complete schism between system development activities and system development process improvement activities. Complete lack of a training program.
Level –3	Undermining	Sabotage	Total neglect of own charter; conscious discrediting of peer organizations' system development process improvement efforts. Rewarding failure and poor performance.

Source: Grady, J., *System Engineering Deployment*, CRC Press, Boca Raton, FL, pp. 278, 2000. With permission.

at keeping track of past performance and sensing early signs of failure. They often require the application of the planning machinery necessary to clearly manage programs and identify the true status. Cost/schedule control systems linked to sound planning data defining clear task goals and exit criteria accomplished by a product-oriented team structure combined with team-oriented product technical performance measurement (TPM) parameters encourage accurate assessment of the current situation and early identification of program risks. On large contracts, the DoD will also insist on having members of their program office in the contractor's teams, making it very difficult to disguise any attempt to mislead.

31.5 Moving the top person's attitude

All of the above techniques to improve the practice of system engineering are effective if the top executive in the enterprise is not hostile to the idea. In the face of high-level hostility or indifference, people in the lower tier will only be locally successful at best and then temporarily. The organization takes its lead from the top and will generally support the leader's attitudes, especially where they are shared. Those who try to make improvements from the bottom often place their career in jeopardy, for they are making waves amidst a sea of tranquillity dominated by incompetence. Small victories are generally overcome by great disappointments and a general background of rejection.

The best strategy for winning over potential detractors in an organization is to have a firm president or division general manager who is a solid supporter of the systems approach. People in this position will always have many things about which to feel passionate, but, if system engineering is one of them, the enterprise is blessed. If this person also understands how to translate their passion into action on the part of the people reporting to them through the application of a simple list of priorities repeated to the point of redundancy, then the organization is doubly blessed. If this is not the case, anyone who would want to make improvements in the practice of system engineering should put the first priority on improving the top manager's attitude toward the systems approach. Failing this, one might next check the job opportunities at a company already managed by a believer.

But, let us say that the reader is intent on improving the systems approach at his firm, wants to stay there, and it is currently managed by a person who does not see the wisdom of the structured approach to the development of the product. How might this person enlighten the reader without adversely affecting their career? The first question is where is this person in the current chain of management? If this person reports to the top executive, then there is an opportunity for the direct approach. If not, then the person must work their way up through the chain starting with the person they report to either on a program, in their functional organization, or both. At each stage of the organizational climb, the resistance may be different, so one should not be so overwhelmed by an easy victory at the manager level as to fail to consider changes in the approach at the director level. The initiative has to be well thought through before it is taken to the lowest level in the chain in the interest of minimizing the overall time span. Ideally, the management hierarchy will remain stable while the initiative is being worked up the chain. A change at the director level after the previous director has been won over may cause a reset in the plan, forcing a recycle through the new director. In a volatile organization, the improvement initiative may fall by the wayside just from exhaustion of those leading the effort.

People who make it to the top levels of industry commonly think in terms of cost benefits, so your proposal for improvement should show clear measurable benefits, probably in terms of cost savings. This is hard to quantify where this data has not been collected in the past and especially hard to allocate potential savings from the practice of an effective systems approach. The place to look for savings that could have been realized on past programs is in the context of the cost of correcting for major errors on previous programs. Try to find ways that the normal systems approach would have prevented the mistake and consequent costs. For example, you may find that there were 5000 engineering drawing changes subsequent to critical design review where 95% of all of the originally planned drawings should have been released signaling the near completion of the engineering design effort. This level of changes will have been driven by a faulty understanding of the requirements. It may be possible to trace at least some of these errors to omissions or mistakes in particular specifications and show that an effective structured analysis process would have prevented the mistake. Other design changes may be traceable to great problems that the program only became aware of late in the program, and it will likely be possible to show how an effective risk management process would have encouraged the identification of the cause of the problem early in the program leading to a much lower cost of correction or mitigation.

A little humor can help so as not to overwhelm the top manager with negatives. The brief story about the software engineer who says that he does not have the time to prepare a specification because he has so much debugging to do is one for which it is hard to avoid the point. The Dilbert cartoon strip has produced some classic statements, like the one where pointy hair is telling Wally that he does not want to spend any money on defining the requirements because they are so rushed. However, pointy hair says he wants Wally to start designing so that people will not think they are doing nothing. The next view shows Wally, feet on the desk, reading the paper, daydreaming, "Of all of my projects, I like the Doomed ones best."

It is not enough to parade to the top executive, or any of those en route to this post, with a demand for improvements. It is going to be necessary to offer a complete solution. This complete solution should have at least the following parts.

1. Formation or continuation of a functional system engineering organization responsible for nurturing the capability, building its effectiveness, developing its process and documentation, acquiring needed tools, and training the people who will do the work. In a large enterprise with multiple programs, a matrix organizational structure is inescapable in order to provide the focus for both continuous process improvement and product-oriented leadership. Some organizations with no system engineering organization history have established a system engineering council by whatever name as a transition agent.
2. Building the deployable capability to do system engineering work on programs using the resources to be provided by the functional system engineering organization.
3. The outline for a standard process in the form of a flow diagram, a brief description of each block, and a map between the functional organization and each block of the process diagram at some level of indenture such that it is clear how the department responsibilities form up into charters in the context of the process definition. Where more than one functional department maps to a particular process task, it should be made clear which of these has the principal responsibility and, therefore, must show leadership in process definition, tool selection, and personnel training.
4. The identification of one or two smaller near-term projects where the new process can be applied to prove the process.
5. A list of risks in implementing the proposed changes that the management should be aware of and how the likelihood of these risks coming to pass can be reduced along with the severity of their consequences.

It should not be necessary to have the complete solution for all of these matters, but the plans should be laid out for accomplishing them along with any cost estimates and an estimated payback time when improvements in productivity and reduction in program risk will compensate for the cost of the changes proposed. To the extent that your company has collected data on past program costs categorically, you may be able to craft some believable numbers. Otherwise, you may have to appeal to estimates that appeal only to faith in the generic savings generally accepted, though this will probably be a hard sell to the management.

31.6 The workers

The enterprise that would do this work well needs people who understand the process and how to apply it on programs. If the staff does not have the right skills at the beginning, a way to improve staff capabilities must be found. One route is through changing the people in the staff. Those without the necessary prerequisites should be moved to other company functions or separated. Those with the right background, but not yet educated, need to be trained. If there are insufficient numbers of qualified people to perform system engineering work on real and anticipated programs, then a hiring or retraining process must be undertaken. If we must acquire new people, then we should have in mind the characteristics of the people we want to attract and select to do system engineering work. The list of characteristics below is a start of such a list:

1. Determination—Recognizing that it sometimes is very hard to accomplish system engineering work because of the numbers of people involved and their different views, system engineers must have the determination to see something through to successful conclusion.
2. Discipline—System engineers must work toward a condition where others follow a grand plan in good order, so they should not think, act, or speak in a way to detract from this essential aspect of system development.
3. Creativity—It may be more than we should ask of one person to be both creative and of good discipline in that the two are often at cross-purposes, but system development work is complex and system engineers should be alert to ways to change the product or the process to enhance movement toward program goals faster at less cost without compromising product performance, if possible.
4. Communicative—System work crosses boundary conditions of the product and among the people working on the development. Therefore, those who would be system engineers must be good listeners, good speakers, and capable of writing clearly for understanding.
5. Logical thinking—Given one or more related facts, one should be able to apply the rules of logic, whether or not they can be listed from memory, to achieve a valid conclusion based on the facts.
6. Technical competence—No one can know everything, but a system engineer must have a founding engineering knowledge in some field. Good abilities in language (generally, English in the United States), mathematics, and physics encourage the ability to communicate with

other engineers who have skills and knowledge in a specialized field where the system engineer is weak with full comprehension of the ongoing work and its consequences.

7. Functional thinking—No one person can master every technology, so the broad system engineer must be able to think about product entities in terms of input–process–output without necessarily being able to master the details of the internal functionality in a particular technology.
8. Broad Interest—A system engineer cannot focus on the narrow interests of any one engineering domain, rather across several. You do not want engineers who want to design circuit boards doing system engineering work at the system level.
9. Inquisitive—The system engineer must be eager for knowledge.
10. Tenacious—One should not easily release an inquiry before it is complete. Problems have to be solved and issues closed despite distractions and purposeful attempts to deflect.
11. Mental agility—People who would do this work should have the ability to change their mind based on the facts and adjust their approach quickly where warranted.

The vice president for system engineering for the Northrop Grumman segment in El Segundo, California, once told the author that he was looking for the deoxyribonucleic acid (DNA) of a system engineer, and it was this conversation that suggested the list above.

All of these kinds of lists appear to be a resume for a god. We should not hope to find people who match our prescription precisely. But, this list can be helpful in rejecting people who fail to match any of these characteristics. While an engineering manager in industry, the author interviewed many prospective employees and sometimes made mistakes in failing to hire qualified people and in hiring unqualified people. Sometimes, however, he made the right choices in excluding candidates. Two of these experiences remain vividly in mind. In one case, the recent college graduate said three times during the interview that he could not wait to get to the drafting board and start designing printed circuit boards. Clearly, this young man had not come to understand system engineering in school nor did he satisfy the broad interest criterion. In another case, the author could not tell the candidate where the work would be performed because it was a secret at the time that Teledyne Ryan Aeronautical unmanned reconnaissance aircraft were being flown into China and North Vietnam from Bein Hoa, South Vietnam. But, the author thought he should at least tell the candidate that some employees had developed stomach and other problems from apparent stress caused by hostile action around the base of operations (mortar and rocket attacks). The candidate replied that there was no reason to worry on that score because he had already had ulcers apparently immunizing him to recurrence. The logic was not one of this candidate's strong suits. As a minimum, we should at least prevent the wrong people from getting into the system engineering staff.

31.7 Need for motivation

The most we can expect from the general population at work in the system engineering field is that they will apply themselves to learning the relevant techniques and practices and do their work on programs in accordance with this teaching as well as they possibly can. One might ask, "Yes, but what is the motivation for so doing?" The author believes this is called accepting a paycheck. Those on the payroll in a system engineering function should, like any employee, do their work professionally. Those who cannot or will not do so should be first approached in an attempt to correct performance problems, and, if this is not successful, these people should be separated from the system function as well as any other function.

Beyond this simple binary motivation, it is possible to expand the appeal to those not influenced by simple compensation. Pay increases and promotion should be allocated to those who do or can lead this work well relative to those who cannot or will not. In order for this to happen, there has to be someone with enterprise responsibility for system engineering, a functional department manager who can maintain continuity and individual records of performance. This is the same person who must also be responsible for building the enterprise system engineering competence in cooperation with the EIT through process definition, staffing decisions, tool selection, education, and practices development and maintenance.

Almost everyone working in industry is interested in a career path through which they feel that there is some hope of advancement in a direction appealing to the employee. A sound career path is not always identified in companies for people doing system engineering work. Three pathways should be identified: project leadership, functional management, and technical excellence. The use of cross-functional teams provides many opportunities for team leadership, and people with system knowledge are good selections for these roles. Those who excel in team leadership should have an open road to program management. Functional management opportunities are generally fewer in number than program opportunities because there is only one functional entity but many programs. But, ideally, these leadership roles should be open to promotion from within. Every organization includes people who are not interested in leading the work of others but are motivated to continuously improve their skill and knowledge in their chosen field. There should be opportunities for the advancement of these people as well, but these people need to understand that this pathway will seldom be as financially rewarding as leadership roles.

Many system engineers are mission-oriented people motivated inwardly to do the best they can without external encouragement, but these people are probably not in the majority. Everyone needs and probably deserves recognition for their accomplishments, preferably in public. If we accept this idea, then we should also accept that those who do not do well need and deserve critical feedback, preferably in private.

All of these means of motivating people to do work well are extensions of good leadership that we all deserve. There is no greater motivating force for most of us than the knowledge that we are working for someone who knows what they are doing and is interested in causing us to do our very best in working toward known goals. Persons who offer good leadership are involved in the education, encouragement, and guidance as well as direction of those in their charge. Some feel that teams should be self-directed, but the author has not succumbed to this attitude. Self-directed teams will never be as efficient as teams with good leadership by someone skilled in giving it.

chapter thirty-two

Closing

If it is true that a structured approach in the development of product systems will yield better results than an ad hoc approach, why should we also not choose to apply these same techniques on our process system? In so doing, we will evolve a clear definition of our process, recognize a valid need for a functional organizational structure and how best to task that structure, and acquire an efficient program transform from generic program planning data to program-specific planning data.

This book has shown how to identify all of the elements of an enterprise planning infrastructure and connect these elements into strings that can be selectively pulled into program-peculiar planning, where time and manpower intensity factors are added.

In that we have encouraged the use of cross-functional teams on programs that are aligned with the product entity structure and insisted that all program work and planning data be linked to this same structure, we have simplified the planning data as much as possible. By cataloging all of this work within the context of an understandable pattern, we have found a way to align and communicate subsets of the whole to those responsible for accomplishing it.

The content of this book is thought to be consistent with any and all system engineering assessment models, ISO 9001, and process standards. We have encouraged that an enterprise prepare its own process specification or standard mapped to one or more selected standards known to be favored by their customer base. It is imperative that having once prepared a process standard that it be followed, so management energy must be expended to gain acceptance and willing compliance with the content of this process standard.

It is insufficient to create a one-time condition of excellence and expect that the condition will survive in a changing world, without maintenance. An enterprise needs a means of assessing how well it is doing on programs that results ideally in numerical data that can be compared over time. Based on information and lessons learned from program experiences, we should determine changes that will be most effective in improving our program performance and implement these changes.

An outsider often finds in an organization that does not have a history of excellent systems capability, an attitude of near despair openly commented upon. Comments are expressed in gallows humor like, "I'll give you odds we'll screw this one up too." This is a very self-destructive behavior in any organization whether it be a boy scout troop, a football team, a military unit, a family, or a business enterprise. Where this kind of behavior is permitted to continue by management, it eats away at the foundations of any hope for rebirth. Workers who think these thoughts have most often been badly led, and it may be impossible for these same people to change worker attitudes restoring or creating for the first time an attitude of exuberance about their work.

Yes, a lot of the work described in this book is not always fun, or even rewarding, but the development of solutions to complex problems should be a joy, an adventure, and contain the makings of a fine, professionally fulfilling career. When management finds the right structure for work organization, manages this work well, and is at least halfway decent to the people doing the work, they perform a great service not only to those who own the business and its customers, but to the working troops as well. The differences between the employee who works for a paycheck despite bad work experiences and the employee who cannot wait to get to work for the next chapter in a long and fascinating book, flow far afield from the workplace gates.

The consequences of failure in performing these things must rest at the feet of the manager, whether engineering chief, CEO, or someone in between. The direction to change and improve must come from the manager as well, for change stimulated from the bottom cannot be sustained. It burns out even the individual zealot because there is so much resistance to change. The only way to sustain improvement from below is a conspiracy that has the discipline to rotate its leader as burnout nears, reflecting the logic of flocks of birds that fly in formation. It takes more energy to fly the lead position than it does to be a wingman. Where this process must go on covertly, however, it challenges the normal patterns of leadership and management and cannot help but create an unhealthy working environment by undermining management no matter the best motives.

While each person in the management stream must share the blame for a poor development process and should do what they can within their management space, the enterprise is a system and isolated positive actions will seldom cause this system to move monotonically toward the mountain top. If a system is performing badly, one can defend and improve the system by finding out and correcting all of the systemic causes or reengineer the whole. This book focuses on the latter in order to communicate the whole pattern of behavior, but there is nothing that prevents the practitioner from applying structured analysis to the organization as you would to a major modification of your product. In this approach, you are mindful of the existing reality and seek the least-cost, most effective way to change system behavior to achieve new functionality.

In any case, the management must take the lead in recognizing the failures of an organization and either act smartly to begin to correct them, have the good grace to step out of the way, or empower a conspiracy. If you can see beyond the next management buzzword to a long term, never-ending improvement process following a sound systems approach to process improvement, you are probably well suited to lead the charge and good luck to you.

Bibliography

Booch, G., 1993. *Object-Oriented Analysis and Design with Application*; New York: Addison-Wesley.

Coad, P. and Nicola, J., 1993. *Object Oriented Programming*, vol. 3; Upper Saddle River, NJ: Prentice Hall.

Coad, P. and Yourdon, E., 1990. *Object Oriented Analysis*, vol. 1; Upper Saddle River, NJ: Prentice Hall.

Coad, P. and Yourdon, E., 1991. *Object Oriented Design*, vol. 2; Upper Saddle River, NJ: Prentice Hall.

Defense System Management College, *System Management*.

DeMarco, T., 1979. *Structured Analysis and System Specification*; Upper Saddle River, NJ: Prentice Hall.

Davidow, W.H. and Malone, M.S., 1992. *The Virtual Corporation*; New York: Harper Collins.

Eisner, H., 2002. *Essentials of Project and System Engineering Management*; New York: Wiley.

Grady, J., 1994. *System Integration*; Boca Raton, FL: CRC Press.

Grady, J., 1995. *System Engineering Planning and Enterprise Identity* (SEPEI); Boca Raton, FL: CRC Press.

Grady, J., 1997. *System Validation and Verification*; Boca Raton, FL: CRC Press.

Grady, J., 2000. *System Engineering Deployment*; Boca Raton, FL: CRC Press.

Grady, J., 2005. *IEEE Std 1220™-2005 IEEE Standard for Application and Management of the Systems Engineering Process*.

Grady, J., 2005. *System Requirements Analysis*; Amsterdam, the Netherlands: Elsevier Academic Press.

Grady, J., 2007. *System Verification*; Burlington, MA: Elsevier Academic Press.

Grady, J., in press. *System Synthesis: Product and Process Design*; Boca Raton, FL: Taylor & Francis Group CRC Press.

Grady, J., in press. *System Management: Planning, Enterprise Identity, and Deployment*; Boca Raton, FL: Taylor & Francis Group CRC Press.

Hatley, D. and Pirbhai, I., 1988. *Strategies for Real-Time System Specification*; New York: Dorset House Publishing Company, Inc.

Juran, J.M. and Godfrey, A.B., 1999. *Juran's Quality Handbook*; New York: McGraw Hill.

Kerzner, H., 2003. *Project Management*; Hoboken, NJ: Wiley.

Machol, R., 1965. *System Engineering Handbook*; New York: McGraw Hill.

McFadden, F. and Hoffer, J., 2000. *Modern Database Management*; New York: Addison-Wesley.

Rumbaugh, J. et al., 1990. *Object Oriented Modeling and Design*; Upper Saddle River, NJ: Prentice Hall.

Schorsch, T., 1996. The capability immaturity model (CIMM), *Crosstalk—The Journal of Defense Software Engineering*, 9(11), November, 1996, Software Technology Support Center at Hill Air Force Base (reprinted with permission from the August 1996 issue of Nexus).

U.S. Army Field Manual, 1979. *System Engineering*; FM 770–78, Washington, DC.

Yourdon, E., 1989. *Modern Structured Analysis*; Upper Saddle River, NJ: Yourdon Press.

Index